Rubber Technology

Rubber Technology

Compounding and Testing for Performance

Edited by

John S. Dick

owner:
Gopal Krishnan

With Contributions from

R.A. Annicelli, C. Baddorf, K. Baranwal, C.J. Cable, G. Day, R.D. Demarco, J.S. Dick,
T.M. Dobel, L.R. Evans, R.W. Fuest, M. Gozdiff, W. Hacker, J.R. Halladay, F. Ignatz-Hoover,
G.E. Jones, W.H. Klingensmith, J. Laird, S. Laube, J.M. Long, P. Manley, S. Monthey, L. Outzs,
L. Palys, A. Peterson, C. Rader, R. School, D.R. Schultz, K.K. Shen, C. Smith, R.D. Stevens,
C. Stone, B. Stuck, A.L. Tisler, B.H. To, D.S. Tracey, R. Vara, A. Veith, W. Waddell, M.-J. Wang,
L. Weaver, W. Whittington, M.E. Wood

HANSER

Hanser Publishers, Munich

Hanser Gardner Publications, Inc., Cincinnati

The Editor:
John S. Dick, Alpha Technologies, Akron, OH 44314-1392, USA

Distributed in the USA and in Canada by
Hanser Gardner Publications, Inc.
6915 Valley Avenue, Cincinnati, Ohio 45244-3029, USA
Fax: (513) 527-8950
Phone: (513) 527-8977 or 1-800-950-8977
Internet: http://www.hansergardner.com

Distributed in all other countries by
Carl Hanser Verlag
Postfach 86 04 20, 81631 München, Germany
Fax: +49 (89) 98 12 64

The use of general descriptive names, trademarks, etc., in this publication, even if the former are not especially identified, is not to be taken as a sign that such names, as understood by the Trade Marks and Merchandise Marks Act, may accordingly be used freely by anyone.

While the advice and information in this book are believed to be true and accurate at the date of going to press, neither the authors nor the editors nor the publisher can accept any legal responsibility for any errors or omissions that may be made. The publisher makes no warranty, express or implied, with respect to the material contained herein.

Library of Congress Cataloging-in-Publication Data
Rubber technology : compounding and testing for performance / edited by John S. Dick
   p. cm.
Includes bibliographical references and index.
ISBN 1-56990-278-X (hardcover)
1. Rubber chemistry. 2. Rubber–Testing. 3. Rubber industry and trade. I. Dick, John S.
TS1890 R855 2001
678$'$.2–dc21   2001024948

Die Deutsche Bibliothek - CIP-Einheitsaufnahme
Rubber technology : compounding and testing for performance / ed. by John S.
Dick. With contributions from R. A. Annicelli .... - Munich : Hanser;
Cincinnati : Hanser Gardner, 2001
   ISBN 3-446-19186-0

© Carl Hanser Verlag, Munich 2001
Production Management in the UK by Martha Kürzl, Stafford
Typeset in the UK by Alden Bookset, Westonzoyland, Somerset
Printed and bound in Germany by Fa. Kösel, Kempten

Preface

This book is intended to be a practical guide for cost effective formulating and testing of rubber compounds in order to help achieve optimal processing and product performance. Attention is also given to tradeoffs in properties. The book demonstrates some of these tradeoffs in various chapters.

In this book we covered rubber compounding as a series of interdependent "systems" such as the elastomer system, the filler-oil system, the cure system, the antidegradant system, etc. We endeavored to use more of a "holistic" approach to show how changes in these inter-related systems can affect specific compound properties.

Rubber compounding is a very diverse subject which draws its body of knowledge from many different areas of science and technology. To address this diversity in the most effective way, we have 42 authors who are experts in many specific areas of rubber compounding, all contributing to the usefulness of this book. Emphasis was placed on the practical; however, theoretical explanations were used mainly when it was necessary to explain an important principle or concept.

This book consists of 23 chapters and over 500 pages of text. In order to make this book an effective reference, a very extensive index of key words and rubber technology terms is provided.

It is my sincere hope that this book will benefit both new and experienced rubber compounders and technologists in their efforts to improve the art and science of rubber technology.

John S. Dick

Contents

Contributors

Ralph A. Annicelli
ralph_annicelli@uniroyalchemical.com
CK Witco, Uniroyal Chemical Co., TSSC Building, Naugatuck, CT 06770, USA

Charles Baddorf
charles.r.baddorf@usa.dupont.com
DuPont Dow Elastomers, Stow, OH 44224, USA

Krishna Baranwal
krisb@ardl.com
Akron Rubber Development Laboratories, Akron, OH 44305, USA

Clark J. Cable
cable@zclp.teamzeon.com
Zeon Chemicals L.P., Louisville, KY 40211, USA

Gary Day
day@hankook-atc.com
Hankook Tire, Uniontown, OH 44685, USA

Robert D. Demarco
demarco@zclp.teamzeon.com
Zeon Chemicals LP., Louisville, KY 40211, USA

John S. Dick
john_dick@alpha-technologies.com
Alpha Technologies, Akron, OH 44314-1392, USA

Theresa M. Dobel
theresa.m.dobel@usa.dupont.com
DuPont Dow Elastomers, Stow, OH 44224, USA

Larry R. Evans
hglre.@huber.com
J.M. Huber Corp., Chemicals Division, Havre de Grace, MD 21078, USA

Ronald W. Fuest
ronald_fuest@ckwitco.com
Uniroyal Chemical, Middlebury, CT 06749, USA

Michael Gozdiff
Zeon Chemicals L.P., Louisville, KY 40211, USA

William Hacker
ppoult@ameripol.com
Ameripol Synpol Corp., Port Neches, TX 77651, USA

James R. Halladay
james_halladay@lord.com
Lord Corporation, Mechanical Products Division, Erie, PA 16514-0038, USA

Fred Ignatz-Hoover
frederick.ignatz-hoover@flexsys.com
Flexsys America L.P., Akron, OH 44333-0444, USA

G.E. Jones
Exxon Chemical Company, Baytown Polymers Center, Baytown, TX 77520-5200, USA

William H. Klingensmith
wklinge610@worldnet.att.net
Akron Consulting Co., Akron, OH 44319, USA

Janet Laird
DuPont Dow Elastomers, Stow, OH 44224-1094, USA

Steve Laube
stephen_laube@cabot-corp.com
Cabot Corp., Industrial Rubber Blacks, Alpharetta, GA 30005, USA

John M. Long
longjm@dsmcopo.com
DSM Copolymer, Stow, OH 44224, USA

Paul Manley
manley@teamzeon.com
Zeon Chemicals, L.P., Louisville, KY 40211-2126, USA

Steve Monthey
steven_monthey@cabot-corp.com
Cabot Corp., Industrial Rubber Blacks, Alpharetta, GA 30005, USA

Leonard Outzs
leonard.l.outzs@usa.dupont.com
DuPont Dow Elastomers, Stow, OH 44224-1094, USA

Leonard Palys
lpalys@ato.com
Elf Atochem North America, Inc., Building 12, King of Prussia, PA 19406-0936, USA

Alex Peterson
apeterson@inspec-chem.com
INDSPEC Chemical Corp., Pitttsburgh, PA 15238, USA

Charles Rader
crader2000@yahoo.com
Advanced Elastomer Systems, Akron, OH 44311, USA

Rudy School
rschool@rtvanderbilt.com
R.T. Vanderbilt Co., Inc., Fairlawn, OH 44333, USA

David R. Schultz
schultzd@harwickstandard.com
Harwick Standard Distribution Corp., Akron, OH 44305, USA

Dr. Kelvin K. Shen
kelvin.shen@borax.com
U.S. Borax, Inc., Valencia, CA 91355-1847, USA

Charles Smith
smithc@teamzeon.com
Zeon chemicals, L.P., Louisville, KY 40233-7620, USA

Ronald D. Stevens
ronald.d.stevens@usa.dupont.com
DuPont Dow Elastomers, Stow, OH 44224, USA

Christopher Stone
tsc@btinternet.com
Schill + Seilacher ''Struktol'' Aktiengesellschaft, Hamburg, Germany

Bonnie Stuck
bstuck@sovereignchemical.com
Sovereign Chemical Co., Akron, OH 44320, USA

Andrew L. Tisler
Exxon Chemical Company, Baytown Polymers Center, Baytown, TX 77520-5200, USA

Byron H. To
byron.h.to@flexsys.com
Flexsys America LP, Akron, OH 44333-0444, USA

D.S. Tracey
donald.s.tracy.exxon.com
Exxon Chemical Company, Baytown Polymers Center, Baytown, TX 77520-5200, USA

Rajan Vara
rajan.g.vara@usa.dupont.com
Dupont Dow Elastomers, Stow, OH 44224-1094, USA

Alan Veith
arveith@msn.com
Technical Development Assoc., Akron, OH 44321, USA

Walter Waddell
walter.h.waddell@exxon.sprint.com
Exxon Chemical Co., Baytown Polymers Center, Baytown, TX 77522-5200, USA

Meng-Jiao Wang
meng-jiao_wang@cabot-corp.com
Cabot Corp., Billerica, MA 01821-7001, USA

Laura Weaver
laura.b.weaver@usa.dupont.com
DuPont Dow Elastomers, Stow, OH 44224, USA

Wesley Whittington
wesacacia@aol.com
C.P. Hall Company, Bedford Park, IL 60499-0910, USA

Michael E. Wood
woodm@zclp.teamzeon.com
Zeon Chemicals L.P., Louisville, KY 40211-2126, USA

1 Rubber Compounding: Introduction, Definitions, and Available Resources

John S. Dick

1.1 Introduction

Rubber compounding is the art and science of selecting various compounding ingredients and their quantities to mix and produce a useful rubber formulation that is processable, meets or exceeds the customer's final product requirements, and can be competitively priced.

The "three Ps" of rubber compounding are price, processing, and properties [1,2]. In our globally competitive business environment, the cost of compounding ingredients is very important. Some compounding ingredients might impart superior performance to the cured compound and product; however, the *price* might be so high that it is not competitive. Chapter 4, "Rubber Compound Economics," discusses this subject in more detail.

Processing is also a very important consideration in the development of a compound. A compound may impart fantastic performance properties to a product; however, if you cannot process it effectively in your plant, you may have serious quality problems, lost production, and increased scrap, and much of your time at work will be tied up solving the latest factory crisis. Chapter 2 discusses rubber compound processing characteristics and testing. Other chapters in this book discuss how the selection of various compounding ingredients affect compound processing behavior.

The physical *properties* of the cured compound are very important considerations because they determine how the resulting product performs. These properties can relate to the product's fitness for use, reliability, service life, dynamic properties, etc. If these properties are not satisfactory, the compounder may be faced with customer dissatisfaction, a high percentage of returned goods, and complaints because of lower sales. Chapter 3 reviews cured compound performance characteristics and testing. Other chapters give suggestions for achieving improved properties. As is noted, many times, improvements in one set of cured compound properties can lead to losses in other compound properties.

For example, increasing the loading of a reinforcing grade of carbon black leads to a higher modulus compound, but also increases the hysteresis and heat build-up from deformational cycling. Therefore, good compounding is finding the best compromise, i.e., the best balance of properties. This is where the "art" of compounding comes into play.

While the three Ps are very important in rubber compounding, a fourth consideration is also becoming important. This consideration involves the health, safety, and environmental aspects of the compounding ingredients, the compounds produced from them, and the rubber end product. For example, some rubber compounds may give off volatile chemicals such as certain nitrosamines that may pose a health hazard in the workplace. New governmental concerns extend to numerous common compounding ingredients. Some examples of compounding ingredients which government agencies are reviewing and/or regulating include protein in natural rubber; residual monomers in BR, SBR, NBR, and CR; alpha quartz in clays and whiting; antimony trioxide; phthalate ester plasticizers; tricresyl phosphate; various halogenated flame retardants; accelerators that give off nitrosoamines during cure; aromatic oils; "labeled" naphthenic oils; ingredients that contain lead or cadmium impurities; certain water soluble phenylene diamine antiozonants; resorcinol; and even zinc oxide and carbon black [3]. Government rules and regulations are constantly changing. This area is totally outside the scope of this book. The reader should contact the appropriate governmental agencies, raw material suppliers, and literature references regarding health, safety, and environmental concerns for rubber compounding ingredients.

1.2   The Recipe

The heart of rubber compounding is the formulation, usually referred to in the industry as a recipe. An example rubber recipe for a radial passenger tire tread compound is shown in Table 1.1. This recipe is based on two raw rubbers, a carbon black reinforcing filler, a plasticizing oil, a antidegradant package (based on 6PPD, TMQ, and wax), and a cure system (stearic acid, zinc oxide, sulfur, and TBBS). Rubber compounds are commonly based on a blend of two or more different raw rubbers to achieve the best balance of properties. Also, the "parts" shown for each of these ingredients are, by accepted convention, expressed in "parts per hundred rubber" or "phr." The total of the "parts" for one, two, three, or more different rubbers in a given recipe is always defined as 100. All other non-rubber ingredients are "ratioed" against the 100 parts of rubber hydrocarbon. This way, rubber compounders can compare easily the levels of curatives, loadings, etc., between different compounds based on the same relative proportion of rubber hydrocarbon and do not have to recalculate percents for every component after adjusting levels of only one or two components.

If the rubber being used is oil- or carbon black-extended (a pre-made oil or carbon black masterbatch), then the part level for the extended rubber is assigned a value sufficiently above 100 to adjust the total calculated rubber hydrocarbon content to equal 100.

Table 1.1   Sample Rubber Recipe (Radial Passenger Tire Tread)

Ingredients	Parts per hundred rubber (phr)	Compound function
SBR 1502, synthetic rubber	80.0	Rubber
SMR 20 natural rubber	20.0	Rubber
N299 carbon black	60.0	Reinforcing agent
Naphthenic oil	10.0	Processing oil
6PPD (phenylene diamine antiozonant)	3.0	Antidegradant
TMQ (antioxidant)	1.0	Antidegradant
Wax blend	2.0	Antidegradant
Stearic acid	1.0	Activator
Zinc oxide	4.0	Activator
TBBS	1.2	Accelerator
Sulfur	2.5	Vulcanizing agent

1.3   Classification of Rubber Compounding Ingredients

ASTM D5899, published in 1996, gives 18 different functional classifications for rubber compounding ingredients. This standard lists the most commonly used chemical classes of compounding ingredients under each of the functional classifications. These functional classifications are discussed below.

- Accelerators are organic chemical additives used to accelerate the cure and reduce vulcanization time. Some examples include sulfenamides, thiazoles, thiurams, dithiocarbamates, and guanidines. Chapters 15 and 16 discuss the use of these ingredients.
- Adhesion Promoters are additives included in the recipes to improve rubber adhesion, usually to brass-coated, steel tire cord. These ingredients include methylene donors, resorcinol donors, and cobalt salts. Chapter 20 reviews these compounding components.
- Antidegradant components include antioxidants and antiozonants, including protective waxes. These agents retard the deterioration of the cured rubber compound from exposure to oxygen, ozone, heat, light, and mechanical flexing. These include chemical classes such as p-phenylene diamines, substituted phenols, and quinolines. Chapter 19 discusses these additives.
- Antistatic Agents are chemical compounds used to reduce electrostatic buildup in a rubber product.
- Blowing Agents are used in cellular rubber formulations. These chemicals decompose at the cure temperature to liberate gas in the compound which forms the cellular structure. Classes of these agents include certain azodicarbonamides, carbonates, and sulfonylhydrazides. These agents are discussed in Chapter 21.
- Colorants are either inorganic colorants, such as iron oxide or titanium dioxide, or organic colorants. These additives are generally used only in non-black rubber compounds (formulations which do not contain carbon black).
- Fillers, Extenders, and Reinforcing Agents: the most often used material in this group is carbon black. Other materials are clays, ground coal, flocs, silicas, silicates, and reinforcing resins. Carbon blacks are discussed in Chapter 12 and non-black fillers are reviewed in Chapter 13.
- Flame Retardant additives include halogen donors, certain metallic oxides, and hydrates. Some of these are discussed in Chapter 22.
- Fungicides are very useful in some outdoor applications.
- Odorants are used to impart a unique odor to a rubber compound. Sometimes these odorants are used to help factory workers identify a specific compound.
- Processing materials are used as softeners to reduce compound viscosity and/or to improve the compound's processing behavior. Examples include petroleum oils, various ester plasticizers, and various soaps. Chapter 12 discusses petroleum process oils while Chapter 14 reviews synthetic ester plasticizers and processing aids.
- Promotors and Coupling Agents for Fillers and Reinforcing Agents are commonly used to improve filler and reinforcing agent compatibility and/or dispersion in the rubber hydrocarbon medium. These agents can be particular organosilanes, amines, and titanates.

- Rubbers are the most important group of compounding ingredients. Without rubber, there can be no rubber compound. There are more than 24 different types of rubber. These are discussed in Chapters 6, 7, 8, and 9.
- Retarders and Inhibitors impart longer scorch safety to the compound, which usually allows further processing of a mixed stock. Chapter 15 discusses these additives.
- Tackifiers impart or improve the building tack (rubber-to-rubber stickiness) of the compound before cure. Examples include phenolic resins, hydrocarbon resins, and rosins. Chapter 18 reviews these materials.
- Thermoplastic Elastomers are not normally compounding ingredients *per se*, but represent an alternative to conventional rubber compounding. Thermoplastic elastomers can be processed similarly to plastics at elevated temperatures, but behave like cured rubber at room temperature. Chapter 10 provides more information on this subject.
- Vulcanizing Agents and Activators: vulcanizing agents are directly responsible for the formation of crosslinks during the rubber curing process. Examples of these agents are sulfur, organic sulfur donors, and certain organic peroxides. Activators are chemical additives which activate the accelerator in a cure and improve its efficiency. Stearic acid and zinc oxide together are the most widely used activators. Some vulcanizing agents and activators may be the same in this broad group because certain agents, such as zinc oxide, can be an activator for the cure of most general purpose rubbers, but functions as the vulcanizing agent in specific halogenated elastomers. Chapters 15 through 17 discuss the use of these chemicals.

1.4   Standard Abbreviations for Compounding Ingredients

When compounders prepare a recipe, many times they avoid writing out the chemical name and instead use accepted abbreviations. This is a kind of "shorthand," which is commonly used. For example, it is much easier to write "77PD" rather than "*N,N'*-bis-(1,4-dimethylpentyl)-*p*-phenylenediamine." ASTM D3853 lists more than 133 standard abbreviations while ISO 6472 gives more than 72 abbreviations for rubber chemicals commonly used in compounding. Also, ASTM D1418 gives more than 50 standard abbreviations for natural and synthetic rubbers while ISO 1629 shows 48 abbreviations [4].

1.5   The Diversity of Rubber Recipes

There are tens of thousands of uniquely different recipes used in the rubber industry today. This diversity exists because there are so many choices to meet a set of target properties. For example, different rubbers and rubber blends, different cure systems, and different filler/oil combinations can be selected. Because it can cost thousands of dollars to develop a rubber compound that has good performance properties and is competitive, rubber companies tend to keep their recipes proprietary. This secrecy adds to the variety of formulations.

Table 1.2   Summary of Composition of 14 Tire Compounds from the Literature

Tyre component	Total phr	Rubber base	phr Carbon black	Type carbon black	phr Oil	Other comments
Apex	197	60 NR 40 IR	65	N326	12	Reinforcing resin 12 phr
Bead insulation	230	SBR	100	N660	20	Two oils 10 phr aromatic 10 phr rosin oil
Steel tire cord skim	178	NR	45	N326	0	15 phr silica 3.8 RF resin 3.2 HMMM Co Naphthenate
Black sidewall	160	50 NR 15 IR 35 BR	40	N330 N550	4	Phenolic and hydrocarbon tackifiers at 6 phr
Carcass	174	50 NR 30 SBR 20 BR	50	N660	18	Paraffinic oil
Innerliner	233	65 CIIR 35 NR	70	N660	12	50 phr Whiting
Truck undertread	172	80 NR 20 BR	50	N550	12	Naphthenic oil
White sidewall	174	30 NR 10 EPDM 60 CIIR	0	–	0	25 phr titanium dioxide, 40 phr clay
Truck tread	169	60 NR 40 BR	50	N330	10	Aromatic oil
OTR tread	181	NR	40	N220	3	20 phr Silica
Performance tread	226	SBR	68	N330	37	
Passenger tread	218	70 SBR 30 BR	70	N339	14	Naphthenic oil
ASTM standard tread	197	25 BR 75 SBR	64	N351	22	
Inner tube	201	IIR	70	N660	25	Paraffinic oil

Table 1.2 gives general descriptions for various tire compounds which are published in the literature [5–18]. Also, Table 1.3 describes in addition, typical industrial product formulations found in the literature [19].

1.6   Compatibility of Compounding Ingredients

The diverse groups of compounding ingredients used in rubber formulations generally are not completely solubilized in the base rubber or rubber blend. Moreover, even blends of similar rubbers, while compatible, still are not soluble in one another [20]. In fact, blends of rubber usually form continuous and discontinuous phases with microscopic domains. Many of the compounding ingredients dispersed in a rubber formulation during mixing

Table 1.3   General Industrial Product Compounds

Compound ID	Total phr	Rubber phr	Filler phr	Oil phr	Comments
Heat resistant conveyor belt cover	165	CIIR	N330 50	5	MBTS/TMTD cure
V-belt friction oil resistant	163	CR	N550 30	20	Zinc oxide and magnesium oxide No accelerator
Hose tube compound	280	NBR	Talc 40 Silica 90 N990 5	20	CBS/DTDM/TMTD Marching modulus
Oil field compound	216	NBR	N231 35 N774 50 Silica 15	0	MBTS/TMTD/Sulfur Cure
Red sheet packing	640	SBR	Clay 240 Whiting 250 Iron oxide 10	25	CBS/DOTG/sulfur cure
SBR gum compound	113	SBR	0	0	CBTS/sulfur cure
Hydraulic hose cover	266	CR	N762 55	28	Cured with DPG/TMTM/ ZnO/sulfur
Black EPDM roofing	311	EPDM	N660 85 Clay 100	95	MBT/TMTM/sulfur cure
Peroxide cured SBR compound	182	SBR	N330 52	10	High molecular weight peroxide at 4 phr
Eraser	612	NR	MgO 30 Lithopone 100 $BaSO_4$ 80 Whiting 150 TiO_2 10 Factice 100	30	Reversion from overcure
Leather innersole footwear	229	SBR	Clay 115 TiO_2 10	0	
Translucent soling	208	SBR	Silica 65	12	
Highly extended low cost EPDM compound	819	EPDM	N650 250 N774 100 Whiting 200	165	MBTS/TMTD/ZnDBC/ sulfur cure
Rubber bands	115	NR	TiO_2 1.5 Whiting 2	1	MBTS/TMTD/DTDM/ sulfur cure
Printing rubber, 45 duro	191	NR	Barytes 40 Clay 15	12	CBTS/Sulfur/and ZnDBC cure
High damping, low creep damper	127	NR 75 NBR 25	N762 20	0	MBS/TMTD/sulfur cure
Idler roll compound	189	NBR	N330 60	6	
Vibration isolator compound	146	NR	N330 10 $CaCO_3$ 20	5	CBS/sulfur cure
Fluoroelastomer reference compound	124	FKM	N550 10 N326 5	0	
Pharmoceutical closure	207	CIIR	Whiting 100	0	
Hot water bottles	200	NR	Whiting 90	0	
Hose tube	175	IIR	N550 35 N770 25	0	MBT/TMTD/TeDEC cure
Radiator hose	513	EPDM	N650 130 N762 95 Whiting 40	130	TMTD/DTDM/ZnDBC/ ZnDMC/sulfur

have different solubility parameters than the base rubber [21,22]. This is why, after a compound is cooled from mixing, some of these compounding ingredients may start to separate as surface exudation. These exudates are commonly referred to as *bloom* in the rubber industry. Bloom can cause processing and product appearance problems. This is why, for example, a compounder may use as many as five different accelerators to cure an EPDM compound. If only two accelerators had been selected, a higher concentration might be required for each accelerator. This required concentration might exceed the accelerator's "bloom point," the critical concentration above which exudation occurs. Common examples of bloom are seen with sulfur, antioxidants, antiozonants, accelerators, plasticizers, oils, and zinc stearate. Therefore, compounders should select combinations of ingredients to avoid blooming problems.

1.7   Rubber Compounding Ingredients' Specifications

As is well known by many experienced compounders, it is possible to mix a given recipe with the same type of internal mixer under identical mixing conditions at different plants, but find that these different batches have different processing and physical properties. This probably occurs because the raw materials (compounding ingredients) are not identical. For example, two generically "identical" grades of nitrile rubber with the same tradename, same Mooney viscosity, and the same bound acrylonitrile content can impart quite different properties to a compound [23]. Differences in sources of rubber that are nominally the same grade can be a leading cause of variation. Also, sources of carbon black of the same N number (grade number) may impart different compound properties. Even a raw material such as zinc oxide may impart different properties, depending on its source of production. Zinc oxides can differ in surface area, particle shape, and chemical impurities. This, too, can affect compound properties [24].

To assure the uniformity of these raw material sources, it is best to establish specifications. Table 1.4 shows 19 ASTM or ISO specifications and classifications which can be used to control the quality of the compounding ingredients. There are also 131 ASTM and ISO raw material tests for accelerators, antidegradants, carbon blacks, ground coal, calcium carbonate, clay, plasticizers, natural and synthetic rubbers, silica, stearic acid, sulfur, protective waxes, and zinc oxide. These chemical and physical test methods are too numerous to list here; however they can be found in ACS Paper No. 97, presented at the Fall, 1994, meeting of the ACS Rubber Division [25]. In many cases, these ASTM and ISO standards for classifications, specifications, and test methods can be used to establish specific company material specifications.

1.8   Raw Material Source Books

The following are source references which may help in searching and selecting raw

Table 1.4   ASTM and ISO Material Specifications/Classifications

Class of raw material	IISRP, ASTM, or ISO classifications/specifications	Description
Natural rubber	ASTM D2227	Specifications for natural rubber
	ISO 2000	Specifications for natural rubber
	ISO 1434	Amount of coating on NR bales
Synthetic rubber	IISRP	IISRP synthetic rubber manual
Carbon black	ASTM D1765	Classification system for carbon black
	ISO 1867	Specifications for sieve residue in carbon black
	ISO 1868	Specifications for heat loss in carbon black
Process oils	ASTM D2226	Classification for petroleum oils used in rubber compounding
Zinc oxide	ASTM D4295	Classification for zinc oxide
Stearic acid	ASTM D4817	Classification of stearic acid
Sulfur	ASTM D4528	Classification for sulfur
Accelerators	ASTM D4818	Classification of accelerators
Antiozonants	ASTM D4678	Description of p-phenylene diamine AO's
Protective waxes	ASTM D4924	Classification of petroleum waxes for use in rubber compounding
Silica	ISO 5794/3	Specifications for silica
Clays	ISO 5795/3	Specifications for clay
Calcium carbonate	ISO 5796	Classification of calcium carbonate
Ground coal	ASTM D5377	Classification of ground coal
Titanium dioxide	ASTM D4677	TiO_2 classification

materials or compounding ingredients for consideration in a new rubber compound. Also included are tradename directories for the chemical identification of these materials.

- *Blue Book* – published by Rubber World, Lippincott and Peto, Inc., 1867 West Market St., Akron, OH 44313. Phone: (330) 864-2122. This annual publication groups compounding ingredients by functionality and gives chemical descriptions. It has a very useful tradename index and is available in hard copy or CD ROM.
- *Rubber Red Book* – published by Lippincott and Peto, Inc., 1867 West Market St., Akron, OH 44313. Phone: (330) 864-2122. This is a comprehensive directory of manufacturers and suppliers in the rubber industry.
- *Rubbicana* (now called Rubber Directory and Buyer's Guide) – published by *Rubber and Plastics News*, 1725 Merriman Road, Suite 300, Akron, OH 44313-5251. For new subscriptions call (800) 678-9595. This is a rubber directory and buyer's guide.
- *RAPRA New Trade Names in the Rubber* and *Plastics Industries* and *Rubber Compounding Ingredients Source Book* – available from Publications Sales Office, RAPRA Technology Limited, Shawbury, Shrewsbury, Shropshire, SY4 4NR.

1.9   Key Source References for Formulations

The rubber industry is very secretive regarding commercial recipes because it can cost a great deal to develop an effective recipe that yields optimum physical properties, processes well, and is cost effective. These proprietary recipes become assets to the company and in many cases, can provide a competitive advantage in the marketplace. Therefore, finding good, realistic recipes published in the public domain is difficult. However, some of these "public domain" recipes can be useful starting points for further compound development.

Some of the best sources of model recipes are raw material suppliers. Many of them have formularies of recipes based on their rubber compounding ingredients. Some rubber companies also try to reconstruct a rubber recipe from a competitor's product through chemical analysis. There are, however, limits to how much detail this analysis can provide. Companies in general must be careful not to violate any active patents. Rubber recipes can also be developed through a custom mixer. Many times custom mixers have libraries containing thousands of commercial recipes.

For the 15 year period from 1977 to 1992, 69 categories of commercial rubber recipes are found in the literature. Table 1.5 lists these categories as well as the number of recipes in each category. Table 1.6 shows the literature sources for the 840 specific commercial recipes in descending order of number of recipes provided. This table indicates that *NR Technology* is one of the best sources of published rubber recipes; however, the obvious disadvantage of this source is that these recipes are all based on natural rubber. The other

Table 1.5   69 Categories of Commercial Rubber Recipes Found in Literature 1977–1992

Adhesives 12	Hose, automobile 10	Tank Tracks 2
Apparel, Diving Suit 1	Hose, Cover 18	Tape 3
Balls 4	Hose, Friction 2	Tire, Bead filler 2
Bearings for Earthquake,	Hose, General 28	Tire, Bead insulation 1
Bridge Bearings 5	Hose, Tube 19	Tire, Breaker 22
Belts, Coat Stock 1	Latex Compounds 25	Tire, Black Sidewall 10
Belts, Cover 22	Linings 14	Tire, Carcass 15
Belts, Cushion Compound 1	Medical Applications 6	Tire, Curing Bladders 5
Belts, Friction 11	Engine Mounts, Bushings,	Tire, Innerliner 8
Belts, General 7	Isolators 7	Tire, Innertube 10
Belts, V 15	Molded Goods, General 30	Tire, Motorcycle, Bicycle 8
Bumpers/Fenders 6	Oil Field Compounds 4	Tire, Repair gum 1
Cable, General 19	Polymerization, Emulsion 5	Tire, retread 17
Cable Jackets 20	Printing Rubber 2	Tire, solid 7
Coating 1	Rocket Insulation 3	Tire, General 26
Coating, Textiles 12	Roller Compounds 42	Tire, OTR tread 3
Defouling Compound (clean	Roofing 7	Tire, passenger tread 18
molds) 2	Rubber Bands 4	Tire, treads, truck 18
Ebonite and hard rubber 13	Sealants 15	Tire, undertread, (tread base) 1
Eraser, Pencil and/or ink 2	Seals & Gaskets 53	Tire, valve stem 2
Fabric Dips 22	Sheeting 15	Tire, White Sidewall 12
Foam, Cellular Rubber 29	Special Compounds 33	Weather Strip 3
Footwear, General 15	Spread Fabrics 3	Windshield wipers 1
Footwear, Soling 33	Springs 2	

Table 1.6   Literature Sources for Commercial Recipes (1977–1992)

	Number of recipes found
NR Technology	113
Vanderbilt Rubber Handbook	111
Rubber World	107
Rubber Chemistry and Technology	92
Rubber India	86
Synaprene Formulary	81
Rubber and Plastics News	71
Elastomerics (no longer published)	54
Desk-Top Data Bank	32
Monsanto Compounder's Pocket Book	28
ICI Rubber Technologists' Pocket Book	20
Akron Rubber Group Technical Symposium	12
European Rubber Journal	11
Silica Pigment Formulary	10
Ashland Carbon Blackboard	<10
Rubber Industry	<10
Phillips Petroleum	<10
ASTM Vol. 9.01 and 9.02	<10
Rubber News (India)	<10

sources give recipes based on natural as well as synthetic rubbers, which is much more realistic.

Many times, recipes found in the literature are only model formulations which may have never been actually used in commerce. They may be "ideal" formulations which may not work in a specific product application without an extensive amount of further compound development and "fine tuning". The ultimate final compound that is actually produced may not even resemble some of these recipes. Nevertheless, these "model" recipes from the literature sources given in Table 1.6 may be good beginnings.

1.10   Technical Organizations

The single best source of information regarding rubber technical and trade organizations is compiled by Joan C. Long, Librarian, Rubber Division, ACS, and published in the *Rubber Red Book*. This listing gives information on 113 organizations associated with rubber.

To learn more about the rubber industry, you may want to consider joining or contacting the following technical organizations:

- Rubber Division, American Chemical Society, University of Akron, P.O. Box 499, Akron, OH, 44309-0499. Phone: (330) 972-7814. Also, the Rubber Division has 26 local Rubber Groups through out the United States, Canada, Mexico, and Colombia.
- American Society for Testing and Materials (ASTM), Subcommittee D11 (Rubber), D24 (Carbon Black), and F9 (Tires), 100 Barr Harbor Dr., West Conshohocken, PA

19428-2959. Phone: (610) 832-9585. This is the organization which develops many of the standard test methods used in the United States.

- International Society of Automotive Engineers (SAE), 400 Commonwealth Dr., Warrendale, PA 15096-0001. Phone: (724) 776-4841.
- Rubber Roller Group (see listing in latest Red Book for current contact).
- The Tire Society, P.O. Box 1502, Akron, OH 44309-1502. Phone: (330) 253-TIRE.
- Polyurethane Manufacturers Association (PMA), 800 Roosevelt Rd., Bldg. C, Ste. 20, Glen Elyn, IL 60137-5833 Phone: (630) 858-2670. PMA conducts technical workshops.
- Rubber Manufacturers Association, 1400 K St. NW, Washington, DC 20005. Phone: (202) 682-4800. Membership consists of companies from the tire, hose, belt, and other rubber product industries, as well as suppliers.
- International Institute of Synthetic Rubber Producers, Inc. (IISRP), 2077 South Gessner Rd., Ste. 133, Houston, TX 77063-1123. Phone: (713) 783-7511. Membership consists of synthetic rubber producers. The IISRP publishes the *Synthetic Rubber Manual* which classifies and lists synthetic rubber products.
- Rubber Recycling Topical Group, Rubber Division, ACS, Michael Rouse, Chairman.
- Malaysian Rubber Research and Development Board, Tun Abdul Razak Laboratory, Brickendonbury, Hertford SG13 8NL, UK. Phone: +44(0)1992 584966.
- RAPRA Technology Limited, Shawbury, Shrewsbury, Shropshire SY4 4NR, UK. Phone: +44(0)1939 250383.
- Laboratoire de Recherches et de Controle du Caoutchouc et des Plastiques, (LRCCP), 60, rue Auber, 94408 Vitry sur Seine, Cedex, France. Phone: +33 1 49 60 57 70.
- Deutsche Kautschuk-Gesellschaft e.V., Postfach 90 03 60, D-60443, Frankfurt am Main. Germany. Phone: +49(069) 79 36-153.
- Malaysian Rubber Board, 3rd Mile Jalan Ampang, P.O. Box 10150, 50908 Kuala Lumpur, Malaysia. Phone: +60 3 456 7033.

1.11   Key Technical Journals and Trade Magazines

The best single source for these publications is the listing compiled by Joan C. Long, Librarian, Rubber Division, ACS, and published in the *Rubber Red Book*. This listing gives 95 publications related to rubber. There are many very useful rubber trade journals and magazines. Some of the ones to consider for rubber compounding information are given below:

- *Rubber Chemistry and Technology* – Rubber Division, ACS, University of Akron, P.O. Box 499, Akron, OH 44309-0499. Phone: (330) 972-7814. This journal is published five times per year. Many of its contributions are from papers presented at the spring and fall meetings of the Rubber Division. Traditionally, the third issue each year is the ''Rubber Reviews'' issue.
- *Rubber World* – Lippincott & Peto, Inc., 1867 West Market Street, Akron, OH 44313. Phone: (330) 864-2122. This is a monthly magazine and contains many useful articles related to rubber compounding.

- *Rubber and Plastics News* – Crain Communications, 1725 Merriman Rd., Suite 300, Akron, OH 44313-5251. Phone: (330) 836-9180. This publication usually has a technical section commonly related to rubber compounding. Also, R&PN publishes *ITEC Select*, which contains certain papers presented at the International Tire Exposition and Conference (ITEC).

- *Journal of Elastomers and Plastics* – Technomic Publishing Co. Inc., 851 New Holland Ave., Box 3535, Lancaster, PA. This journal publishes papers relating to rubber compounding.

- *European Rubber Journal* – Crain Communications Ltd., A division of Crain Communications Inc., New Garden House, 78 Hatton Garden, London, EC1N 8JQ, UK. Phone: +44 (0)171 457 1400. This is a widely read trade magazine in Europe.

- *Rubber Technology International* – UK & International Press, a division of AutoIntermediates Ltd., Talisman House, 120 South Street, Dorking, Surrey RH4 2EU, UK. Phone: +44(0)1306 743744. This annually contains compounding ingredient and compound performance enhancement articles focused mainly on non-tire applications.

- *Tire Technology International* – UK & International Press, a division of AutoIntermediates Ltd., Talisman House, 120 South Street, Dorking, Surrey RH4 2EU, UK. Phone: +44(0)1306 743744. The newer "sister" publication to *Rubber Technology International*. It, too, is published annually. It discusses rubber compounding ingredients and compounding technology associated with tires.

- *Plastics and Rubber Weekly* – EMAP Maciaren Ltd., 19 Scarbrook Rd., Croydon, Surrey CR9 1QH, UK. Phone: +44(0)181 956 3017

- *Progress in Rubber and Plastics Technology* – Rapra Technology Ltd., Shawbury, Shrewsbury, Shropshire SY4 4NR, UK. Phone: +44(0)1939 250383

- *RAPRA Abstracts* – Rapra Technology LTD., Shawbury, Shrewsbury, Shropshire SY4 4NR. UK. Phone: +44(0)1939 250383. RAPRA stands for "Rubber and Plastics Research Association of Great Britain".

- *Polymer Testing* – Elsevier Science, P.O. Box 211, 1001 AE, Amsterdam, The Netherlands. Also in the United States contact Elsevier Science Inc., P.O. Box 945, New York, NY, 10010. Phone (212) 633-3730.

- *China Synthetic Rubber Industry*/Hecheng Xiangjiao Gongye. Lanzhou Chemical Industry Corporation of SINOPEC, the Synthetic Rubber Technology Development Center of SINOPEC. Distributed by: China International Book Trading Corp., P.O. Box 399, Beijing. Phone: +86 (0931)7555368

- *GK-Kautschuk Gummi Kunststoffe*; International Technical Journal for Polymer Materials. Huethig GmBH, Postfach102869, 69018 Heidelberg, Germany. Phone: +49(062 21)4 89-242. This is one of the larger rubber and polymer publications in Europe. This German Journal publishes Abstracts and some papers in English.

- *Gummi, Fasern, Kunststoffe* – Dr. Heinz Gupta Verlag, GbR, Postfach 10 41 25, 40852 Ratingen, Germany. Phone: +49 (02102) 93 45-0.

- *Rubber India* – All India Rubber Industries Association, 3/8, Navjivan Society, Lamington Rd., Mumbai 400 008, India. Phone +91 308 50 32/30621.

- *Materie Plastiche ed Elastomeri* – O.VE.S.T. srl, Via Simone d'Orsenigo, 22-20135 Milano, Italy. Phone: +39 02/54.69.174.

1.12   Regularly Scheduled Technical Conferences

One of the best ways for a rubber compounder to stay up to date and well informed regarding new technical innovations and compounding materials is to attend annual or semiannual technical conferences which are usually scheduled, including:

- Rubber Division, ACS Meetings are held every spring and fall in the U.S., Canada, or Mexico. Usually between 50 and 150 technical papers are presented. Copies of these papers are available at a nominal price. Usually the fall meetings also have an Exposition of the latest chemicals, polymers, instruments, and equipment. In years ending with odd numbers, the Expo is "full scale" while in even-numbered years, it is a "mini" Expo. Contact the Akron Rubber Division office at (330) 972-7815.
- International Rubber Conference (IRC) is usually held at least once a year. Scheduled meetings are announced in *European Rubber Journal* (ERJ), in the "Diary" section. Sometimes an exposition is also held with this conference.
- International Tire Exposition and Conference (ITEC) is usually held in Akron, OH, in the fall of every even numbered year. The primary focus of the meeting is tire technology. These meetings are sponsored by *Rubber and Plastics News*. Presented papers can be individually purchased at the meeting. Some selected papers are published by R&PN in *ITEC Select*. Contact R&PN for more details.
- "K" (Internationale Messe Kunststoff + Kautschuk), or the "International Plastics & Rubber Trade Fair," is held every three years in Düsseldorf, Germany. While this exposition is an important show for rubber technology, it has a much larger focus on plastics. It is perhaps the largest exposition of its kind in the world. Usually a scheduled meeting is announced in *European Rubber Journal* (ERJ) in the "Diary" section.
- Brazilian Congress is a rubber technology conference meeting every two years in Brazil. The most common meeting location is São Paulo. Papers may be presented in Portuguese, Spanish, or English. The paper presentations normally have simultaneous language translations.

1.12.1   Regularly Scheduled Courses

There are several continuing education or "enrichment" courses on rubber technology and compounding. Some of these courses are listed below.

- Akron Polymer Training Center, University of Akron, offers several continuing education courses on rubber technology. Their phone number is (330) 972-8625. Courses offered are given below:
  - Chemistry and Physics of Elastomers
  - Using Thermoplastic Elastomers
  - Tire Mechanics
  - Mixing and Processing of Rubber
  - Rubber Technician Training
  - Testing of Rubber Processability and Dynamic Properties
  - Polymer Compounding Materials

- Rubber Division, ACS Courses are commonly offered at the regularly scheduled Rubber Division meetings including:
  – Rubber Technology Workshop
  – Compounding, Processing and Testing of Elastomers
  – Rubber Technology – Basic, Intermediate, and Advanced (offered via correspondence)

- University of Wisconsin, Milwaukee, Courses, at the Center for Continuing Engineering Education, offers the following courses on rubber:
  – Compounding of Rubber
  – Dynamic Properties of Rubber
  – Product Design and Application of Silicone Elastomers
  – Molding of Rubber and Design of Rubber Molds
  – Advanced Rubber Molding Technology
  – Elastomeric Products: Materials, Testing and Design
  – Design of Rubber Products
  – Finite Element Analysis of Elastomeric Components
  – Rubber Extrusion Technology
  – Rubber Materials and Process Selection
  – Thermoplastic Elastomers
  – The Importance of Flow in Rubber Processing

1.13   Web Sites Available

The Internet and the World Wide Web have become important sources of information. A few sites are listed below. Additional sites can be found as ''links'' from these web pages.

- *European Rubber Journal* (*ERJ*), http://www.crain.co.uk/ERJ/ – This web site has an index of back issues.
- *Rubber & Plastics News*, http://www.rubbernews.com/ – By entering your account number which is on the mailing label to your subscription, you can search the back issues of the ''Technical Notebook'' section of *R&PN*. Also, you can have online access to Rubbicana and to ''hot links'' to various polymer-related universities, rubber product manufacturers, suppliers, and trade and technical associations.
- *Rubber World*, http://www.rubberworld.com/ – This web site is free. It has a supplier index, a bulletin board (with general information concerning the rubber industry), a review of the news of the week, ''hot links'' to many companies and organizations, and a ''Tech Forum.'' You can place a specific technical question needing an answer on this ''Tech Forum.'' These questions and your address can be read by any visitor to this site. If that visitor feels he or she has an answer for you, then answers to questions can be posted by them.
- *Rubber Division*, *ACS*, http://www.rubber.org – This site is very useful for obtaining information about future meetings, and searching and ordering previously presented papers from the last two or three Rubber Division meetings. Also, this site has links to other sites in rubber science and to other local rubber groups.

- *American Society for Testing and Materials* (*ASTM*) http://www.astm.org/ – This is a useful site to find out about Committee D11, on Rubber; D24, on Carbon Black; and F9 on Tires. Also, it gives information on the dates and locations for future meetings and on the availability of different Industry Reference Materials (IRMs).

References

1. J. R. Beatty, M. L. Studebaker, The Rubber Compound and Its Composition, Part 1, *Rubber World*, August, 1975.
2. C. M. Blow, *Rubber Technology and Manufacture*, Newnes-Butterworth, London, 1971, p. 308.
3. Bertil Haggstrom, "How to Cope With New Environmental Rules in European Union, International Rubber Conference, Oct. 6–10, 1998, Kuala Lumpur, Malaysia.
4. J. S. Dick, "ASTM In the Globalization of Rubber Standards and Specifications", ACS Rubber Div. Paper No. 97, Fall, 1994.
5. L. A. Walder, W. F. Helt, "High-Temperature Curing of Radial Passenger Tires", *Rubber Chem. Tech.*, Vol. 59, No. 2. (1982), Apex, p. 298
6. R.O. Babbit, *Vanderbilt Rubber Handbook*, 1978, Bead Insulation, p. 657.
7. A. Peterson, M. I. Dietrick, "Resorcinol Bonding Systems for Steel Cord Adhesion", *Rubber World*, Aug, 1984, Steel Tire Cord Skim Coat, p. 24.
8. L. A. Walker, W. F. Helt, "High-Temperature Curing of Radial Passenger Tires," *Rubber Chem. Tech.*, Vol. 59, No. 2, Black sidewall, p. 298.
9. M. L. Deviney, L. E. Whittington, B. E. Whittington, B. G. Corman, "Migration of Oil in Elastomers," *Rubber World*, Jan. 1972, Carcass, p. 35.
10. W. F. Fischer, D. G. Young, "Contributions of Innerliners and Sidewalls to Tubeless Radial Ply Tyre Performance", *Rubber Industry*, Aug. 1975, Innerliner, p. 141.
11. L. A. Walker, "Increasing Truck Tire Endurance Through Compounding", *1983–1984 Technical Symposiums*, Akron Rubber Group, Inc. Undertread, p. 25.
12. A. L. Barbour, "New Clays Improve Compound Properties", *Rubber and Plastics News*, Aug. 24, 1987, White sidewall, p. 48.
13. L. A. Walder, "Increasing Truck Tire Endurance Through Compounding, *1983–1984 Technical Symposiums*, Akron Rubber Group, Inc., Truck Tread, p. 27.
14. S. Wolff, "Optimization of Silane–Silica OTR Compounds. Part 1: Variations of Mixing Temperature and Time During the Modification of Silica with Bis-(3-triethoxisilylpropyl)-tetrasulfide", *Rubber Chem. Tech.*, Vol. 55, No. 4, OTR Tread, p. 969.
15. A. F. Werner, "SBR Polymer Help Users in Solving Rubber Compounding Problems", *Rubber and Plastics News*, Aug. 16, 1982, Performance Treads, p. 21.
16. ASTM E1136-88, Standard Specification for a Radial Standard Reference Test Tire, *Annual Book of ASTM Standards*, Section 04.03, ASTM Standard Tread.
17. *Compounders' Pocket Book*, Monsanto Co., 1981, Innertube, p. 199.
18. P. C. Vegvari, W. M. Hess, "Measurement of Carbon Black Dispersion in Rubber by Surface Analysis", *Rubber Chem. Tech.*, Vol. 51, No. 4 (1978), Passenger Tread, p. 819.
19. J. S. Dick, H. Pawlowski, Rubber Div. ACS Paper 98, Fall, 1993.
20. E. T. McDonel, K. C. Baranwal, and J. C. Andries, "Elastomer Blends in Tires", Chapter 19, *Polymer Blends*, Vol. 2, Academic Press, Inc., 1978, p. 276.
21. Hildebrand, *J. American Chemical Society*, Vol. 38, p. 1452, 1916.
22. C. M. Hansen, "The Three Dimensional Solubility Parameter", *J. Paint Technology*, Vol. 39, No. 505, February, 1967, p. 104.

23. Wayne Cousins, John Dick, "Effective Processability Measurements of Acrylonitrile Butadiene Rubber Using Rubber Process Analyzer Tests and Mooney Stress Relaxation", ACS Rubber Div. Meeting in Cleveland, Paper No. 90, Fall, 1997.
24. Op cit, Jacques, p. 308.
25. J. S. Dick, "ASTM in the Globalization of Rubber Standards and Specifications", ACS Rubber Div. Paper No. 97, Presented at the Pittsburgh Meeting of the ACS Rubber Div., Oct. 10 (Table 5).

2   Compound Processing Characteristics and Testing

John S. Dick

2.1   Introduction

The processing of a rubber formulation is a very important aspect of rubber compounding. Also, it is important to understand the rubber properties needed for good processability and the tests performed to measure these properties.

2.2   Manufacturing Process

Uncured rubber, whether natural or synthetic, behaves as a viscoelastic fluid during mixing. Under processing conditions, various rubber chemicals, fillers, and other additives can be added and mixed into the rubber to form an uncured "rubber compound." These compounding ingredients are generally added to the rubber through one of two basic types of mixers, the two roll mill or the internal mixer.

2.2.1   Two Roll Mill

The two roll mill in Fig. 2.1 consists of two horizontal, parallel, heavy metal rolls which can be jacketed with steam and water to control temperature. These rolls turn towards each other with a pre-set, adjustable gap or nip to allow the rubber to pass through to achieve high-shear mixing. The back roll usually turns at a faster surface speed than the front roll; this difference increases the shear forces. The difference in roll speeds is called the *friction ratio*. The rubber generally forms a "band" around the front roll. Mill mixing is the oldest method of rubber mixing, dating back to the very beginning of the rubber industry; however, it is a relatively slow method and its batch size is limited. Internal mixers overcome these problems.

2.2.2   Internal Mixers

Internal mixers were first developed by Fernley H. Banbury in 1916. Today, these internal mixers are commonly used because they are much more productive than two roll mills. Internal mixers consist of two rotors or blades turning toward each other in an enclosed

Figure 2.1   Two roll mill.

metal cavity. These rotors can be either tangential or intermeshing in arrangement (Fig. 2.2). The cavity is open to a loading chute through which rubber, fillers, and various chemicals are placed. Upon completion of the mix cycle, the mixed rubber stock is discharged through a door in the bottom of the mixer.

Mixing time is determined by the shape and size of the rotors, the rotor speed, and horsepower of the motor turning them. The rotors generally turn at a high friction ratio. Some internal mixers can handle batches in excess of 1000 pounds (greater than 455 kg), and in some cases, can completely mix a compound in less than two minutes. Of course, with so much energy being absorbed by the rubber stock, the batch temperature can rise well above 120 °C (250 °F) before it is dumped and cooled. The temperature rise that results from viscous heating of the rubber compound often means the compounds must pass through the internal mixer more than once to disperse fillers and other compounding

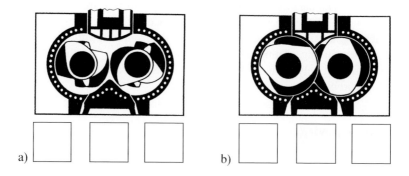

Figure 2.2   Tangential vs. intermeshing rotors for internal mixers.

ingredients. Batches are usually dumped from an internal mixer onto a mill where they may be further worked while being cooled. Sometimes additional compounding ingredients, such as curatives, are added at this point.

2.2.3   Further Downstream Processing

After the rubber stock is mixed, it may be remilled and fed into either a calender or extruder. A three or four roll calender (Fig. 2.3) is generally used when uncured rubber is applied to a textile fabric or steel cord as a coating. Calenders are also used to form sheets of rubber. Extruders (Fig. 2.4) are used when the uncured rubber stock is to be shaped into a tire tread, a belt cover, or a hose tube, for example. A screw rotating in a cylinder transports the rubber compound from the feed port to the die of the extruder. During this process, additional heat and work histories usually affect the viscoelastic behavior of the compound and how it extrudes through the die. Extruders with lower screw length to screw diameter (L/D) ratios are considered ''hot feed'' extruders while ones with high ratios are considered ''cold feed'' extruders.

After calendering or extrusion, the uncured rubber parts may be brought together in a building step. For example, uncured tire components are put together by hand or automatically on a tire building machine where the calendered carcass plies are placed over the innerliner; then, the belt and sidewalls over the carcass, etc. In a similar manner, conveyor belts are constructed on large building tables on which each calendered ply is

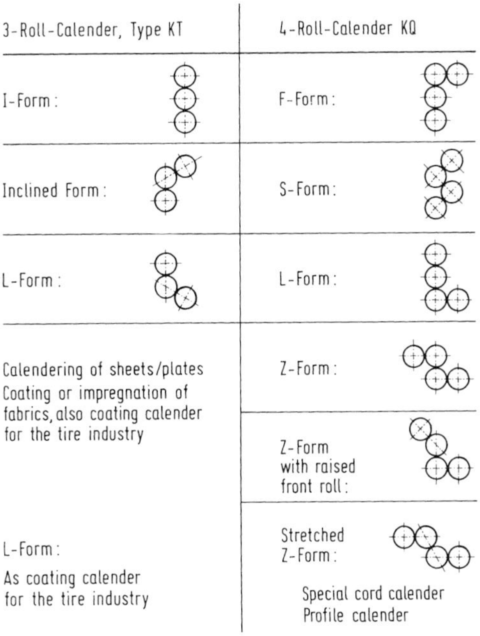

Figure 2.3   Three and four roll calenders.

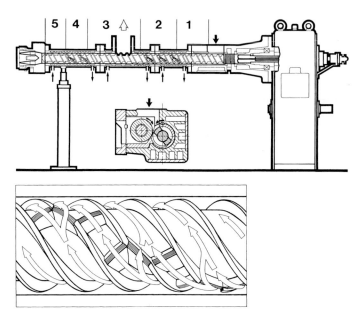

Figure 2.4   Rubber mixing extruder.

manually placed over another, followed by the extruded cover stock. Some hose may be hand built as well. Other rubber products may require that the mixed compound be preshaped and molded.

2.2.4   Curing Process

After the shaping and/or construction of the uncured (''green'') rubber product, it goes through the cure process. With tires, the uncured constructed assembly is placed in a special tire press, such as a Bag-O-Matic press, where it is shaped in a tire mold and cured at elevated temperatures under pressure. Other products, such as hose and rollers, are sometimes cured in autoclaves with steam pressure. Many automotive rubber parts are also cured through compression molding, transfer molding, or injection molding. Rubber cure temperatures can generally range from as low as 100 to more than 200 °C. The higher cure temperatures mean shorter cure cycles. Many rubber products are therefore cured at higher temperature ranges for greater productivity, provided that reversion or thermal degradation is not a problem at these higher temperatures. Most rubber products are cured under pressure as well to avoid gas formation and porosity.

2.2.5   Factory Problems

If rubber compound processability characteristics are not controlled for the different stages of the manufacturing process, various factory quality problems can result. Poor control can

Table 2.1   Examples of Factory Problems

Mixing
  Long black incorporation times (BIT)
  Lower filler dispersion
  Higher compound viscosity
  Crumbly batches
  Lumpy stocks
  Slow mixing
	Stock bin storage
	  Scorch
	  Green strength
	  Sticky slabs

Milling
  Back rolling
  Bagging
  Poor mill release
  Thickness control
	Compression/transfer/injection molding
	  Poor appearance
	  Shrinkage
	  Cured hardness
	  Porosity
	  Mold release
	  Mold fouling

Extrusion
  Rough surface (not smooth)
  Poor dimensional stability
  Uncontrolled extrudate shrinkage
  High die swell
  Low extrusion rate
  Pin holes
  Scorch
	Autoclave cures
	  Appearance
	  Mandrel release
	  Shrinkage

	Second step tire building
	  Poor green strength
	  Blow outs

Calendering
  Scorch
  Bare spots, holes
  Poor dimensional stability, width, thickness
  Heat blisters
  Trapped air
  Calender release (from rolls)
  Rubber fabric penetration
  Liner release
	Tire molding
	  Underfills
	  Poor appearance
	  Uniformity
	  Porosity
	  Mold fouling
	  Vent problems

result in higher scrap rates, higher internal and external failure costs, and lost plant productivity, all of which can command a great deal of the rubber compounder's attention as daily plant problems. Some examples [1] of these problems are given in Table 2.1.

Lord Kelvin once said that "applying numbers to a process makes for the beginnings of a science." If rubber processability characteristics can be measured effectively, then the rubber compounder can find the cause of many of the factory problems given in Table 2.1. Various test methods help determine the cause of many problems and aid in the design of new compounds and/or better processing conditions to reduce factory problems.

2.3   Processability Characteristics and Measurements

The following are rubber compound processability characteristics that determine how well a given compound processes:

1. Viscosity
2. Shear Thinning
3. Elasticity (or V/E Ratio)
4. Time to Scorch
5. Cure Rate
6. Ultimate State of Cure
7. Reversion Resistance
8. Green Strength
9. Tack
10. Stickiness
11. Dispersion
12. Storage Stability
13. Mis-Compounding (Compound Variation)
14. Cellular Rubber Blow Reaction

2.3.1   Viscosity

Viscosity is the resistance of a fluid, such as rubber, to flow under stress. Mathematically, viscosity (η) is shear stress divided by shear rate as shown below in Eq. (2.1).

$$\eta = \frac{\text{shear stress}}{\text{shear rate}} \qquad (2.1)$$

Viscosity is very dependent on temperature; at higher temperatures materials are less viscous. The viscosity of rubber can be measured by four methods:

1. rotational viscometers
2. capillary rheometers
3. oscillating rheometers
4. compression plastimeters

2.3.1.1   Rotational Viscometers

By far, the most commonly used rotational viscometer in the rubber industry is the Mooney viscometer. Melvin Mooney of U.S. Rubber Co. developed this instrument in the 1930s [2]. Since then, it has become one of the most widely used test methods in the industry. It is used for testing both raw rubber and mixed stocks.

This method is described in detail in ASTM D1646 or International Standard ISO 289, Part 1. Two precut rubber test pieces with a combined volume of 25 cm^3 are placed into a two-part compression cavity mold. With the dies closed, a sealed, pressurized cavity is formed, in which a special rotor is imbedded in the rubber. This rotor and the dies are grooved to help prevent the rubber from slipping at the rotor or die interface while the rotor is turning.

Usually there is a pre-heat time after the dies are closed to allow the rubber to approach the set temperature of the instrument. Then the test specification calls for the rotor to turn at two revolutions per minute (2 rpm) for a specified time period. The instrument records

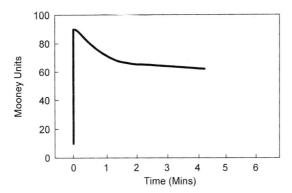

Figure 2.5   Mooney viscosity curve of raw rubber.

viscosity in Mooney Units (MU), which are arbitrary units based on torque. Generally, the measured viscosity of the rubber under test decreases with running time because of the thixotropic effects of the rubber tested. However, depending on the type of rubber and the test temperature, the *rate* of decrease in the measured Mooney viscosity with time should slow down greatly (Fig. 2.5). Usually, the Mooney viscosity test is performed with a one minute preheat and a run time of either four or eight minutes. The final Mooney viscosity value is reported as the lowest value recorded in the last 30 seconds of the test.

As an example, a Mooney viscosity value may be reported as 55 ML (1 + 4). This is in accordance with the convention recommended by the ASTM standard cited above. The term ''ML'' indicates a large standard rotor was used. ''55'' represents the measured Mooney viscosity value reported for the specified conditions of the test. The ''1'' represents the preheat time before the rotor starts to turn. The ''4'' indicates the running time of actual rotation of the rotor before the final Mooney viscosity measurement is made.

Mooney viscosity crudely relates to the average molecular weight of a raw rubber and to the state-of-mix or quality-of-mix for a masterbatch or final uncured rubber stock. If the Mooney viscometer with a large rotor is measuring values well above 80 MU, it can be somewhat insensitive to subtle differences among raw polymers or mixed stocks. When the rubber specimen is too tough, slippage and tearing may occur. One method to avoid this is to simply run the test at a higher temperature. For example, many EPDM polymers do not test well at 100 °C. Instead, they are tested at 125 °C. However, a higher temperature can be a problem for rubber compounds containing curatives that are ''scorchy.'' Also, if a rubber has a high Mooney viscosity from the large rotor (ML), then repeating the test with a small Mooney rotor (MS) should be considered. One basic problem with the Mooney viscometer is that it measures viscosity at a low shear rate of only $1\,s^{-1}$, which is far lower than many rubber manufacturing processes.

2.3.1.2   Capillary Rheometer

The capillary rheometer measures the viscosity of mixed rubber stocks at relatively high shear rates. ASTM D5099 describes this method for rubber testing. There is no ISO

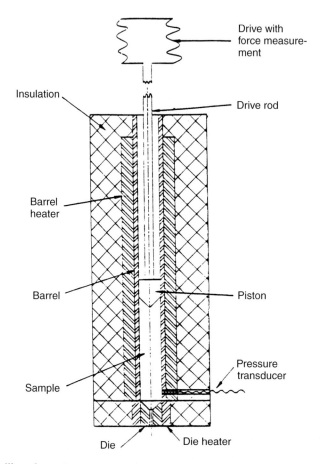

Figure 2.6   Capillary rheometer.

International Standard for the capillary rheometer's use with rubber. The ASTM method consists of placing (or packing) cut pieces of a rubber sample in a heated barrel. Then, a special piston pushes the rubber out of the barrel through an orifice of a special die (with a given capillary length to orifice diameter ratio or L/D) to form an extrudate. The apparent shear rate is determined from the speed of the piston traveling in the barrel (the ram speed). The apparent shear stress is determined from the resulting barrel pressure measured by a transducer (Fig 2.6). From Eq. (2.1), the apparent viscosity η_{app} can be calculated. This apparent viscosity can be converted to "true" viscosity by applying the Rabinowitsch correction to obtain the "true" shear rate and the Bagley correction to obtain the "true" shear stress [3,4].

The main advantage of the capillary rheometer for measuring rubber viscosity is the wide shear rate range it can apply to the rubber specimen. Many capillary rheometers can measure viscosity shear rates at over $1000\,s^{-1}$. The disadvantages of these rheometers are that they are difficult to operate, require more time to run a single test, and require extensive time to clean the barrel and set up for the next test.

Figure 2.7   RPA 2000® rubber process analyzer.

2.3.1.3   Oscillating Rheometers

An instrument called the Rubber Process Analyzer (RPA) by Alpha Technologies [5] measures dynamic viscosity through the application of a sinusoidal strain to an uncured rubber specimen molded in a sealed, pressurized cavity. ASTM D6204 describes this unique method for measuring processability. Figure 2.7 shows the instrument used for this method. Figure 2.8 shows a sinusoidal strain being applied to the rubber test specimen. The complex torque response (S^*) is observed to be out-of-phase with the applied strain because of the viscoelastic nature of the rubber being tested. The phase angle δ quantifies this out-of-phase response.

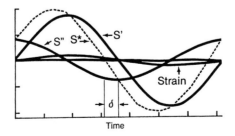

Figure 2.8   Applied sinusoidal strain and resulting stress response.

From the complex torque S^* response and the phase angle δ, the elastic torque S' (in-phase with the applied strain) and the viscous torque S'' (90° out-of-phase with the applied strain) can be derived (Fig. 2.8). The elastic response S' is a function of the amplitude of the applied strain while the viscous torque is a function of the rate of change of the applied strain.

The storage shear modulus (G') is calculated as follows:

$$G' = k \cdot S'/\text{strain} \tag{2.2}$$

The loss shear modulus (G'') is calculated as follows:

$$G'' = k \cdot S''/\text{strain} \tag{2.3}$$

where k is a constant that takes into account the unique geometry of the die cavity.

The complex shear modulus (G^*) is equal to the square root of the sum of G' squared and G'' squared.

$$G^* = [(G')^2 + (G'')^2]^{1/2} \tag{2.4}$$

The complex dynamic viscosity η^* is calculated as follows:

$$\eta^* = G^*/\omega \tag{2.5}$$

where ω is the frequency of the sinusoidal strain in radians/second.

The real dynamic viscosity η' is also calculated in the following manner:

$$\eta' = G''/\omega \tag{2.6}$$

The complex dynamic viscosity η^* from the Rubber Process Analyzer is analogous to the apparent viscosity (η_{app}) from the capillary rheometer, while the real dynamic viscosity η' is analogous to the "corrected" viscosity from the capillary rheometer.

Cox and Merz in the early 1950s published the following empirical relationship between capillary rheometer viscosity η_{app} measured under conditions of steady shear rate and dynamic complex viscosity η^*, which is measured through sinusoidal deformation (and constantly changing shear rates) applied by a dynamic mechanical rheological tester [6]:

$$\eta_{app}(\dot{\gamma}) = \eta^*(\omega)|_{\omega=\dot{\gamma}} \tag{2.7}$$

- η_{app} is the apparent (uncorrected) capillary rheometer viscosity at a steady shear rate (in s^{-1})
- η^* is complex dynamic viscosity measured at an oscillatory frequency of ω (in radians per second).

This empirical relationship sometimes works, but not every time [7,8]. Sometimes the success of this empirical relationship is based on the nature and concentration of the reinforcing fillers used.

The important advantages of the Rubber Process Analyzer in measuring rubber viscosity over the other methods discussed are its versatility in measuring viscosity at both low and high shear rates, ease of use, and excellent repeatability.

2.3.1.4   Compression Plastimeters

While viscosity is defined as the resistance to plastic deformation, the term *plasticity* refers to the "ease of deformation" for a rubber specimen. In a way "plasticity" and "viscosity" define the same property, but have the opposite meaning. A plastimeter measures the plasticity of an uncured rubber specimen. Plastimeters are very simple, crude methods for measuring the flow of a rubber sample. The main problem with most plastimeters is that they operate at extremely low shear rate ranges of only 0.0025 to $1\,s^{-1}$, much lower than even the Mooney viscometer [9]. To reinforce this point, the Rubber Process Analyzer can be correlated to these methods but *only* when very low frequencies (very low shear rates) are applied to the rubber specimen [10]. The principle of plastimeter procedures is basically to measure the deformation of a cylindrically cut, uncured rubber specimen after it has been subjected to a constant compressive force between two parallel plates for a specified time period at a specified test temperature.

The initial part of ASTM D 3194 (the standard describing the Plasticity Retention Index for natural rubber) and International Standard ISO 2007 describe the use of the Wallace rapid plastimeter method. The Williams plastometer parallel plate method is described by ASTM D926 and ISO 7323.

2.3.2   Shear Thinning

Shear thinning is the characteristic of non-Newtonian fluids (such as rubber compounds) to decrease in measured viscosity with an increase in applied shear rate. Not only is it important to measure the compound viscosity, it is also important to know how rapidly viscosity decreases with increasing shear rate. All rubber compounds are non-Newtonian in their flow characteristics and their viscosity usually decreases according to the power law model. If the log of viscosity is plotted against the log of shear rate, a straight line usually results. Rubber compounds with different reinforcing filler systems have different log-log slopes. These different slopes can be quite important because compounds are commonly processed at different shear rates. As shown in Fig. 2.9, the ordinal relationship of Compound 1 vs. Compound 2, for example, can change between a low shear rate to high shear rate. So, Compound 2 might have higher viscosity than Compound 1 in an

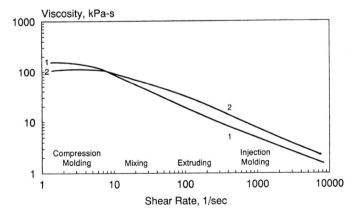

Figure 2.9   Comparison of two compounds with different shear thinning profiles from capillary rheometer viscosity measurements with increasing applied shear rates.

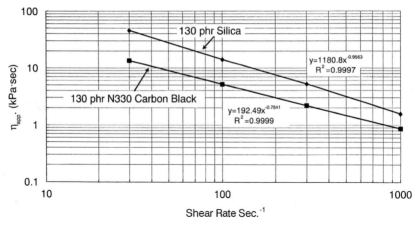

Figure 2.10   Log-log plot of MPT capillary rheometer.

injection molding operation, but Compound 2 might have lower viscosity than Compound 1 in the mold after injection.

Two effective methods of measuring shear thinning behavior are the capillary rheometer and oscillating rheometer.

2.3.2.1   Shear Thinning by Capillary Rheometer

The capillary rheometer can apply a wide range of shear rates. As discussed earlier, according to ASTM D5099, the piston can be programmed to travel in the barrel at a series of faster and faster speeds which result in higher and higher shear rates. Figure 2.10 shows the log-log plot for a capillary rheometer. The applied shear rate for a capillary rheometer can be more than $1000\,s^{-1}$.

2.3.2.2   Shear Thinning by Oscillating Rheometer

As discussed earlier, the Rubber Process Analyzer (which is defined in ASTM D6204) increases shear rate by increasing the frequency of the sinusoidal strain. Figure 2.11 shows a log-log plot of the complex dynamic viscosity vs. shear rate (frequency in radians/second) for the same set of compounds shown earlier in the figure for capillary rheometer measurements. In fact, Fig. 2.12 shows how the power law slopes from these measurements for capillary rheometer vs. Rubber Process Analyzer agree quite well [11].

Shear rate with the Rubber Process Analyzer can also be increased by using a higher applied strain. Sometimes applying higher strains to achieve higher shear rates may have some advantages over higher frequencies because these higher strains may more effectively destroy carbon black aggregate-aggregate networks which form in the rubber compound while in storage. Figure 2.13 shows the very good correlation achieved between a capillary rheometer shear stress values measured at $100\,s^{-1}$ and Rubber Process Analyzer G' values at very high applied strain [12].

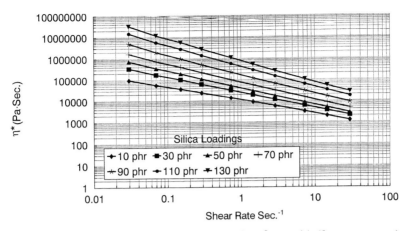

Figure 2.11   Log-log plot of RPA dynamic complex viscosity η^* vs. rad./s (frequency sweep).

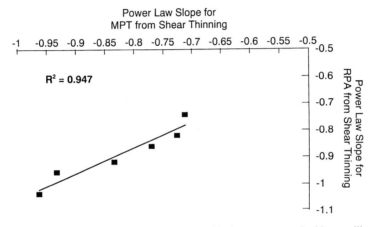

Figure 2.12   Comparison of power law slopes for shear thinning as measured with a capillary rheometer vs. an oscillating rheometer (RPA).

The Rubber Process Analyzer is very simple to operate and versatile in applying different shear rates by either varying the frequency or varying the strain. It is a unique instrument because it can apply very high strains to an uncured rubber specimen in a sealed pressurized cavity and achieve very repeatable results. Also, the Rubber Process Analyzer has an advantage over the capillary rheometer in that it can more effectively test *raw* rubber samples for shear thinning behavior and achieve better repeatability.

2.3.3   Elasticity

Elasticity is that material property that conforms to Hooke's Law. A purely elastic material, such as some metals at low strains, conforms perfectly to Eq. (2.8) below [13]:

$$\sigma = E\gamma \qquad (2.8)$$

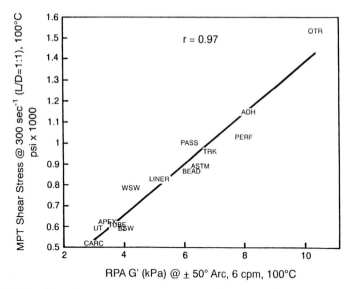

Figure 2.13   MPT capillary rheometer shear stress at $100\,s^{-1}$ vs. RPA S' at 90° arc strain.

where:

- σ = stress, or force per unit area
- γ = strain (displacement) as measured from change in length
- E = the static modulus of elasticity

With a perfectly elastic material, the rate of applied deformation has no effect on the measured stress response. Of course, rubber is not perfectly elastic even in the cured state. Rubber is viscoelastic, possessing both viscous and elastic qualities in both the uncured and cured states. However, the ratio of the viscous quality to the elastic quality (the V/E ratio) decreases greatly when a rubber compound is cured.

Uncured rubber possesses an elastic quality mainly as a result of chain entanglements. Uncured rubber with a high elastic quality has what is commonly called *nerve*. Rubber with high "nerve" (high elasticity) resists processing. It is not uncommon for two rubbers to have the same viscosity, but one is "nervier" (has more elasticity) than the other. This elastic quality difference affects how well the rubber processes. This higher elastic quality affects how well a rubber mixes, how well it incorporates fillers, the length of a mix cycle, and the viscoelastic properties imparted to the final batch. For example, a mixed stock with a higher elastic quality may have greater die swell, poorer dimensional stability during a downstream extrusion process, or mold differently from a stock with higher nerve.

The elasticity of rubber can be measured by five test methods:

1. Mooney stress relaxation
2. oscillating rheometer
3. capillary rheometer die swell

4. compression plastimeter elastic recovery
5. direct shrinkage measurements.

2.3.3.1   Mooney Stress Relaxation

In 1996, a new Part B for measuring stress relaxation was added to ASTM D1646, the
Mooney Viscosity Standard. This new method is widely used in the rubber industry
because it can measure Mooney viscosity as well as stress relaxation decay rates. Stress
relaxation tests are run on the same specimen immediately after completing viscosity
measurements. For example, an ML $1 + 4$ test (large rotor, 1 minute preheat, 4 minutes
run) can be performed on a raw rubber sample followed by a two minute stress relaxation
at the end of the test. The total test time is seven minutes. Mooney stress relaxation can be
performed automatically after the "final" Mooney viscosity measurements by very
rapidly stopping the rotation of the rotor and measuring the power law decay of the
Mooney viscosity output with time. Mathematically, this power law decay is described in
Eq. (2.9).

$$M = kt^{-\alpha} \tag{2.9}$$

where

- M is the torque value in Mooney units
- k is the torque value at 1 second
- t is the time in seconds
- α is the rate of relaxation (the slope of the relaxation function)

In a log-log plot, this expression takes the following form:

$$\log M = -\alpha \log t + \log k \tag{2.10}$$

The slope α is commonly used as a measure of the stress relaxation for raw rubber and
rubber compounds. Figure 2.14 illustrates how this slope α can be used to quickly compare
the elasticity of two EPDM polymers. These two EPDMs have the same Mooney viscosity

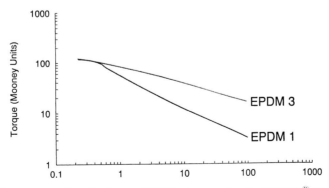

Figure 2.14   Mooney stress relaxation for two EPDM polymers using the MV2000$^{\text{®}}$ Mooney viscometer.

but different elasticity. A steeper slope (a faster rate of decay) indicates that the polymer has a higher V/E ratio and a lower elastic quality than the polymer shown with a flatter slope. Therefore, these two polymers are likely to process differently [14].

2.3.3.2   Elasticity by Oscillating Rheometer

ASTM D6204, which describes the Rubber Process Analyzer (RPA), can provide a procedure to measure the elasticity of rubber. As described in Section 2.3.1.3, the storage modulus G' and loss modulus G'' are measured by this technique from the testing of raw rubber as well as uncured mixed stocks. G' is a direct measure of elasticity. A higher G' response means a higher elastic quality at a defined temperature, frequency and strain. Dividing G'' by G' calculates a parameter known as $\tan \delta$ as shown in Eq. (2.11).

$$\tan \delta = G''/G' \qquad (2.11)$$

Tan δ is the tangent of the phase angle discussed in Section 2.3.1.3 (see also Fig. 3.5). A higher uncured $\tan \delta$ for a raw elastomer or mixed stock implies a higher V/E ratio, which is also analogous to a steeper slope from a stress relaxation test. The $\tan \delta$, as a processability parameter, has been found to be at least twice as sensitive to real differences in rubber processability than the α slope from the Mooney Stress Relaxation test discussed earlier [15]. Also, an RPA $\tan \delta$ as a parameter is more reliable and can be measured at a wider range of shear rates than the α slope. Therefore, $\tan \delta$ and G' are both very effective processability measurements. An example of comparing viscoelastic properties with the RPA is shown in Fig. 2.15. Here, two different sources of SBR 1006 have almost exactly the same Mooney viscosity but have quite different $\tan \delta$ values. Other studies have shown similar differences [16,17].

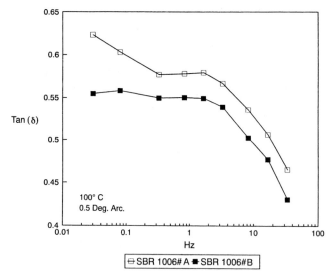

Figure 2.15   Uncured $\tan \delta$ response from the RPA for two SBR 1006 polymers with the same Mooney viscosity.

2.3.3.3   Capillary Rheometer Die Swell

As discussed earlier, ASTM D5099 describes the use of the capillary rheometer for testing the viscosity of rubber compounds at different shear rates. When the rubber compound is extruded through a standard die orifice, it swells from its elastic memory. A rubber compound with high elasticity usually displays greater die swell. ASTM Test Method D5099 does not describe a method for measuring this die swell. However, a capillary rheometer, such as the Monsanto Processability Tester, can measure die swell directly using a special optical detector. This method has a disadvantage in that its statistical test sensitivity and repeatability are not nearly as good as the RPA method described earlier. Also, the capillary rheometer is not normally effective for testing raw rubber.

2.3.3.4   Compression Plastimeter Elastic Recovery

As discussed earlier, the principle of these plastimeter procedures is to measure the deformation of a cylindrically cut, uncured rubber specimen after it is subjected to a constant compression force between two parallel plates for a specified time period at a specified test temperature. This relates to plasticity, or the viscous component, as discussed earlier. However, these plastimeter methods usually have an additional procedure for elastic recovery, which is measured dimensionally in a given time period after the deformational load has been removed. This elastic recovery is supposed to relate to the elasticity of the rubber specimen. Some problems with these procedures in measuring elasticity are that they are performed at low strain, low shear rate, and at a temperature that is usually much lower than the factory processing temperature. Because of these conditions, the data may not be relevant to plant processes. Variations in sample preparation also greatly affect test results. Recovery data can be quite variable.

2.3.3.5   Direct Shrinkage Measurements

Another way of measuring nerve is to directly measure percent shrinkage. An example of this is presented in ASTM D1917, which is the standard method for measuring shrinkage of raw and compounded hot-polymerized SBR. Rubber strips from a mill are prepared and placed on a specified type of low-friction surface. The strips are exposed to elevated temperatures for specified lengths of time, and finally cooled, dimensionally measured, and their percent shrinkage calculated. Even though this is an old method, no precision statement has been published. This method is not widely used and some believe it is not very repeatable.

2.3.4   Time to Scorch

The time to scorch is the time required at a specified temperature (or heat history) for a rubber compound to form incipient crosslinks. When a scorch point is reached after a compound is exposed to a given heat history from factory processing, the compound cannot be processed further by milling, extruding, calendering, etc. Therefore, scorch measurement is very important in determining whether a given rubber compound can be processed in a particular operation. The scorch time can be measured by:

1. Rotational viscometer
2. Oscillating rheometer
3. Capillary rheometer

2.3.4.1   Scorch by Rotational Viscometer

The Mooney Viscometer has been used to measure scorch since the 1930s. It was the first instrument method used to measure scorch safety of a mixed stock. ASTM D1646 and International Standard ISO 289, Part 2 describe how Mooney scorch time is reported. Usually, the Mooney viscometer is set at a temperature higher than 100 °C, such as 121 °C or 135 °C, for example. From a practical viewpoint, some rubber technologists consider the best scorch test temperature is selected when the compound routinely reaches scorch within 10 to 20 minutes. However, others feel the temperature for the Mooney scorch test should be close to the normal temperature of the factory process in question.

Mooney scorch is generally reported for a large rotor as the time required for the viscosity to rise five Mooney units above the minimum viscosity (referred to as t_5). However, when the small rotor is used, the scorch time is reported as the time required for viscosity to rise three Mooney units above the minimum (referred to as t_3). It should be noted that the Mooney rotor is unheated, which means that Mooney Scorch values are not true isothermal measurements.

2.3.4.2   Scorch by Oscillating Rheometer

The oscillating disk rheometer (ODR) was introduced in 1963 and is considered a great improvement over the Mooney viscometer because the ODR measures not only scorch, but also cure rate and state of cure [18]. The ODR method is described in ASTM D2084 and in International Standard ISO 3417. Unlike the Mooney viscometer, which continuously rotates its rotor, the ODR oscillates its biconical disk (rotor) sinusoidally. This gives the ODR an advantage in that it can measure the complete cure curve.

ODR scorch time is usually defined as the time until one torque unit rise above the minimum is achieved (t_S1) when 1° arc strain is applied, or time until two torque units rise above the minimum is achieved (t_S2) when 3° or 5° arc strain is applied. Also, through the computer software used today with ODRs, it is relatively easy to calculate time to 10% state-of-cure (t_C10). This cure time t_C10 also relates to scorch. The parameter t_C10 may have an advantage over t_S1 because it is not based arbitrarily on a one torque unit rise (which can be either dNm or in-lb torque units). Because the ODR measures the entire cure curve, it is usually performed at a higher temperature than the Mooney scorch test discussed earlier. As a result, the ODR may not be as sensitive to scorch differences as the Mooney Scorch test [19].

Even though thousands of ODRs have been used worldwide, the ODR itself has a design flaw which involves the use of the ODR disk itself. This is why the rubber industry is shifting away from the ODR to rotorless curemeter designs (discussed next). The problems associated with the disk rotor are as follows [20]:

1. The ODR disk itself functions as a "heat sink," preventing the rubber specimen from reaching the set temperature quickly. Because the rotor is not heated, this means that the ODR scorch values are not a true isothermal measurement.
2. The ODR torque signals must be measured through the shaft of the rotor. This design results in poor signal-to-noise response resulting from the friction associated with the rotor.
3. This same friction associated with the ODR rotor prevents true dynamic properties from being measured during cure.
4. After an ODR test, the cured specimen must be physically pried off the rotor. Because of the rotor shaft, it is not easy to use a barrier film. This also means that the ODR dies are easily fouled and require constant cleaning.
5. The ODR is very difficult to automate because the cured specimen must be removed from the rotor.

These problems were greatly reduced with the introduction of new rotorless curemeters such as the Moving Die Rheometer (MDR) [21]. ASTM D5289 and International Standard ISO 6502 describe the use of rotorless curemeters. In an instrument such as the MDR, the lower die oscillates sinusoidally, applying a strain to the rubber specimen which is contained within a sealed, pressurized cavity. The upper die is attached to a reaction torque transducer which measures the torque response. This design greatly improves the test sensitivity (signal-to-noise) measurements so that real changes in a rubber compound can be detected faster. Also, because there is no rotor, the temperature recovery of the test specimen is less than 30 seconds, compared to approximately 4 to 5 minutes for the older ODR design. Because curing a given rubber specimen in the MDR is closer to a true isothermal cure, the $t_S 1$ scorch time values from the MDR are significantly shorter than the $t_S 1$ scorch time values from an ODR for the same compound at the same test conditions. At higher cure temperatures, such as 190 °C, scorch information from the MDR can be obtained in about one-half the time as from the ODR [22]. Also, barrier film can be used with rotorless curemeters which greatly reduces clean-up time.

With the MDR, two curves are produced during cure, as shown in Fig. 2.16. The elastic torque S' is the traditional cure curve which is more commonly used as an indication of cure state (the ODR produces only an S' cure curve). The viscous torque S'' curve is a second curve generated simultaneously. Sometimes, the S'' peak can be used as an alternate method for measuring scorch. Studies have shown that the percent drop from this S'' peak as the cure progresses gives information regarding filler loadings as well as the ultimate crosslink density [23].

A more advanced rotorless curemeter is the Rubber Process Analyzer (RPA). Improved sensitivity to scorch is achieved by increasing the applied shear rate. When the applied frequency or strain is increased, the resulting shear rate is increased. When a rubber compound is tested at higher shear rates, the sensitivity to subtle scorch differences is increased (which is also true in factory processes which apply higher shear rates to a rubber compound). Studies have shown that higher frequencies or strains applied to a rubber compound provide earlier warning of a scorch condition [24].

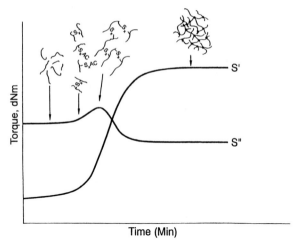

Figure 2.16   Elastic torque S' and viscous torque S'' cure curves.

Another way to improve test sensitivity to scorch is to use the Variable Temperature Analysis feature of the RPA. This feature allows the user to program the RPA to linearly ''ramp'' the test temperature to achieve an increase in a very controlled way from, for example, 100 to 180 °C in 8.0 minutes. In this way, the RPA first measures processing properties at 100 °C, which is a temperature typically low enough to prevent any vulcanization. Then the temperature is ''ramped'' up to measure scorch and cure properties. This ''ramping of temperature'' results in better test sensitivity to scorch differences than measuring scorch under traditional isothermal conditions [25]. In one test, both viscoelastic processing behavior and scorch are measured with excellent precision and test sensitivity.

2.3.4.3   Scorch by Capillary Rheometer

As discussed earlier, tests performed at higher shear rates are more sensitive to the beginnings of scorch than tests at lower shear rates. A capillary rheometer used at a sufficiently high enough temperature (close to the temperature of the factory process), at a shear rate of 300 to 1000 s^{-1}, and a capillary die with a length to diameter (L/D) ratio in excess of 10:1, has high test sensitivity to scorch. Studies have shown that capillary rheometer scorch values relate better to computer simulation of injection molding than do the lower shear rate ODR or Mooney scorch values [26,27]. One problem with the capillary rheometer is that it is more complex and time-consuming than other tests, which limits its use mainly to research and development.

2.3.5   Cure Rate

Cure rate is the speed at which a rubber compound increases in modulus (crosslink density) at a specified cure temperature or heat history. *Cure time* refers to the amount of time required to reach specified states of cure at a specified cure temperature or heat history. An example of cure time is the time required for a given compound to reach 50% or 90% of the ultimate state of cure at a given temperature. Of course, determining what is the instrumental optimum cure time for a small curemeter specimen is not the same as determining what is the optimum cure time for a high mass, thick, rubber article cured in the factory. This is because usually an instrumental cure is closer to an isothermal cure (being cured at the same temperature), whereas the center portion of a thick rubber article sees a variable temperature heat history.

The rubber technologist must take into account other factors such as the temperature recovery time of the curemeter being used, the thickness of the rubber article cured in the factory, the compound thermal conductivity, the variable heat history of the center of the article, the compound cure kinetics, the compound overcure stability, etc., to determine the optimum cure temperature and time for a rubber article in the factory. Sometimes an empirical approach with test trials at different cure times and temperatures may be a practical way to determine the optimum cure temperature and time for a given rubber article. If the cure residence time is too short, then poor cured physical properties result, especially toward the center of a thick article, because it sees less heat history. On the other hand, if the cure residence time is too long for a thick article, then there is deterioration in the cured physical properties, especially at the surface of the article where it receives too much heat history. Also, a long cure time results in poor productivity. Thick articles made from a composite of different rubber compounds represent a "balancing act" by the R & D compounder in adjusting the cure properties for the different component compounds that receive different heat histories.

There are two methods which can be used for measuring cure profile properties such as cure rate and cure times:

1. rotational viscometers
2. oscillating rheometers

2.3.5.1   Cure Rate by Rotational Viscometer

As discussed earlier, Mooney Scorch with the large rotor is measured by the time required for the viscosity to rise 5 Mooney units above the minimum (referred to as t_5), and when the small rotor is used, the scorch time is reported as the time required for viscosity to rise 3 Mooney units above the minimum (referred to as t_3). ASTM D1646 describes what is called the "ASTM cure index." These indices are expressed in Eqs. (2.12) and (2.13).

$$\text{For small rotor} \quad \Delta t_S = t_{18} - t_3 \tag{2.12}$$

$$\text{For large rotor} \quad \Delta t_L = t_{35} - t_5 \tag{2.13}$$

A lower cure index means the cure is faster. However, the Mooney scorch test is not very effective at giving cure information above t_{18} (for small rotor) or t_{35} (for large rotor) because the rotor tears and slips at the rubber interface as the rotor rotates. The Mooney viscometer is not effective at providing a complete cure curve because it uses a rotating rotor, not an oscillating one. Also, the Mooney Viscometer uses large sample weights of approximately 25 g and an unheated rotor which functions as a ''heat sink.'' Therefore, the temperature recovery of the Mooney viscometer (the time required for the temperature of the rubber specimen to reach the set temperature of the dies) is relatively long compared to an oscillating rheometer such as the MDR. This temperature recovery is more critical at higher cure temperatures.

2.3.5.2   Cure Times and Cure Rate by Oscillating Rheometer

The ODR method as described in ASTM D2084 or ISO 3417 has been used since the 1960s to measure cure times and cure rate. The $t_C x$ is simply the time to reach a given percent x state of cure (for example $t_C 50$ is the time required to reach 50% of the state of cure). This is illustrated in Fig. 2.17. So, mathematically, $t_C x$ is calculated as follows:

$$t_C x \text{ is the time to produce torque } (T) = (x/100) \cdot (M_H - M_L) + M_L \qquad (2.14)$$

where:

- M_L = minimum torque
- M_H = maximum torque
- x = percent state-of-cure
- $t_C x$ = time to a given percent (x) state-of-cure

As discussed earlier, the unheated rotor used with the ODR is functioning as a heat sink when it is embedded in the rubber specimen. This can cause the temperature recovery time for the rubber specimen (the time required for the temperature of the specimen to reach the set temperature of the dies after the dies are closed) to be 4 to 5 minutes. Therefore the cure times from the ODR are generally inaccurate in that they do not truly reflect the cure times for a true isothermal cure.

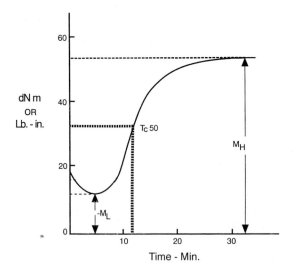

Figure 2.17   Illustration of the calculation of $t_C 50$ (time to 50% state-of-cure).

With computer software used today, the *maximum cure rate* as well as the average cure rate or the ASTM/ISO *Cure Rate Index* can easily be calculated. The maximum cure rate is the slope of the tangent line at the inflection point on the cure curve while the Cure Rate Index is calculated as follows:

$$\text{Cure Rate Index} = 100/(\text{Cure Time} - \text{Scorch Time}) \qquad (2.15)$$

The user of this index can select preferred cure and scorch times, such as $t_C 90$ and $t_S 1$.

Rotorless curemeters, such as the MDR, use essentially the same methods for calculating the cure times and cure rates as specified in ASTM D5289 and ISO 6502. However, the cure times from the rotorless curemeters are significantly shorter than those measured by the ODR for the same compound under the same set of test conditions. As mentioned before, this is because rotorless curemeters, such as the MDR, have a cure which is much closer to a true isothermal cure. The temperature recovery for the MDR [28] is only 30 seconds and the temperature drop when loading a sample is much less than for an ODR. Therefore, cure kinetic studies based on the German Standard DIN 53529 can be performed much better with the MDR [29]. If the ODR were used, there would be a high degree of error because of the ODR's much longer temperature recovery.

While the MDR is curing a compound at near-isothermal conditions (the same temperature over time), thick rubber article cures are not isothermal. The center of a thick section of a rubber article cured in a mold is exposed to a variable heat history. A thermocouple in the center of a rubber article can measure the rise in temperature as the rubber article is cured. This time-temperature profile can be used by the RPA software to give the same heat history to the compound as measured by the thermocouple. Because the RPA has low mass dies and very efficient heaters, the temperature of the upper and lower dies can exactly match the required time-temperature profile. The RPA can also match a drop in temperature after demolding through the use of a forced air cooling system. This Variable Temperature Cure technique is an empirical way to estimate state-of-cure for the center of a thick section. It should be understood that the cool down portion of the test can introduce an error, i.e., the S' can increase from both crosslinking and the drop in temperature.

2.3.6   Ultimate State of Cure

Ultimate state of cure refers to the "ultimate crosslink density." Strictly speaking, the "best" state of cure can be quite different for one rubber property, such as tear resistance, compared to another rubber property, such as rebound. When discussing processability, however, the ultimate state of cure is usually measured as the maximum elastic torque (M_H). This is not a perfect method of measuring the ultimate state of cure, but it is very practical. One problem with using maximum torque to measure ultimate state-of-cure occurs when the compound displays a "marching modulus." This phenomenon occurs when the S' torque never reaches a plateau during cure. In this case, an arbitrary cure time must be set to measure the maximum S' torque response. This selected cure time should be located where the rate of S' increase has slowed significantly.

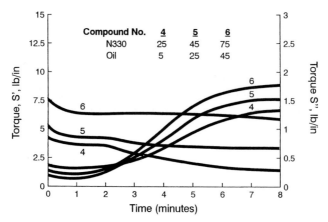

Figure 2.18 S' and S'' cure curves for selected vibration isolator compounds.

There are two tests used in production to measure the ultimate state of cure:

1. ring testing
2. oscillating rheometer

2.3.6.1   Ring Testing

Ring testing can be used in a production setting provided the ring specimens can be cured fast enough. This method is discussed in Chapter 3 as ASTM D412 and ISO 37 for tensile testing.

2.3.6.2   Oscillating Rheometer

The most common and practical test to measure the "ultimate state of cure" is by using the maximum S' torque (M_H) from the oscillating rheometer. The ODR (ASTM D2084 and ISO 3417) is commonly used to measure M_H. The MDR can also measure M_H (Ref. ASTM D5289); however it can measure the viscous torque S'' @ M_H, as well. This is illustrated in Fig. 2.18. As can be seen, the S'' response sometimes changes more from carbon black and oil variations than the S' response. In addition, the RPA can give not only an S' and S'' cure curve, but it can also quickly drop the die cavity temperature and measure after-cure dynamic properties at a designated lower temperature. These properties often relate better to product performance. An example of an RPA after-cure temperature sweep with the tan δ responses, is given in Fig. 2.19. In this example, the differences in the cured tan δ values are much greater at the lower temperature.

2.3.7   Reversion Resistance

Reversion resistance is the resistance of a rubber compound to deterioration in vulcanizate properties usually as a result of extended curing times. This property is particularly

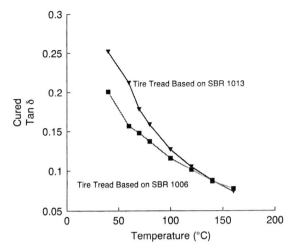

Figure 2.19   RPA after cure temperature sweep for two different tire tread stocks.

important when a rubber part or compound experiences too much heat history during curing. For example, the outside surface of a thick article may be exposed to an excessive heat history, causing reversion to occur. Rubber compounds based on natural rubber commonly revert at higher cure temperatures.

Oscillating rheometers are the most practical method for measuring reversion. The ODR, MDR, and RPA can measure the drop in the S' response after M_H. When the S' elastic torque peaks and then drops because of reversion, the S'' torque response and the tan δ both rise with reversion. There is some evidence that these dynamic properties are more sensitive to reversion [30]. The RPA can increase the test sensitivity to reversion even more by performing an *in situ* post cure aging test at an elevated temperature such as 190 °C and then measuring the percent change in the cured tan δ at a lower temperature such as 60 °C [31].

2.3.8   Green Strength

Green strength is the tensile strength and/or tensile modulus of an uncured rubber compound. This important processing property relates to the compound's performance in extrusions, calendering, and conveyor belt or tire building, particularly the second-stage building machine for radial tires. If tires are constructed with rubber compounds with poor green strength, they may fail to hold air during normal expansion in the second stage of tire building prior to cure. High molecular weight, strain crystallizing elastomers (such as natural rubber) tend to exhibit good green strength.

The only standardized test for measuring green strength is International Standard ISO 9026, which specifies the preparation of dumbbells for tensile testing. This method calls for testing five dumbbell specimens and reporting the medians and ranges found.

Research done in 1996 showed that high strain testing with the RPA could correlate to green strength for a series of natural rubber compounds [32].

2.3.9   Tackiness

Tackiness refers to the ability of an uncured rubber compound to stick to itself or another compound with a short dwell time and a moderate amount of applied pressure [33]. This property is very important whenever rubber products, such as tires or conveyor belts, are built by laying one calendered or extruded rubber ply on top of another. The uncured product must hold together before it is placed in a mold or press for cure. Usually compounds based on natural rubber have good building tack. Compounds based on EPDM, on the other hand, usually have poor building tack. Many times, tackifiers are added to a compound to improve tack.

Rubber compounds consist of many ingredients with differing degrees of solubility. Some of these ingredients may separate from the compound under certain cooling conditions, exuding to the compound's surface to impart a surface *bloom*. Many times, this bloom destroys building tack. Some compounding ingredients which may bloom are sulfur, accelerators, antidegradants, petroleum oils, zinc stearate, and waxes.

There are no ASTM or ISO test standards for measuring the tack of rubber compounds. However, the most widely used tack testing instrument is the *Tel-Tak Tackmeter*, introduced by Monsanto in 1969 [34].

2.3.10   Stickiness

Stickiness refers to the nature of a rubber compound to stick to non-rubber surfaces such as metals or textiles. Too much stickiness to metal surfaces can result in poor mill and calender release and problems in other processing equipment. However, not enough stickiness can result in rubber compound slippage against metallic surfaces in extruders or internal mixer rotors. Adjusting temperature can sometimes control stickiness. Certain compounding ingredients such as external lubricants or release agents are sometimes used to control the stickiness level. These agents, however, should be used with caution because adhesion and other compounding properties might be affected.

There are no ASTM or ISO test methods developed to measure stickiness. However, the *Tel-Tak Tackmeter* developed by Monsanto in 1969 can be used to measure not only compound tackiness, but stickiness to a stainless steel surface as well [35].

2.3.11   Dispersion

Dispersion is a property which defines how well filler aggregates and particles are dispersed in a rubber compound from a mixing process. This property relates not only to percent carbon black dispersion, but also to non-black fillers such as clays, silica, titanium dioxide, calcium carbonate, etc. Rubber curatives, such as accelerators and sulfur, can also be poorly dispersed. Poor dispersion can sometimes be a particularly critical characteristic with curatives because they are commonly added late in the mixing cycle. Poor dispersion for a rubber batch can result in poor stock uniformity and highly variable cured physical properties such as ultimate tensile strength. It is well known that poor dispersion can decrease abrasion, tear, and fatigue resistance as well as hurt flexometer heat buildup and other dynamic properties. Test method ASTM D2663 lists three different methods for quantifying percent carbon black dispersion [36].

2.3.12   Stock Storage Stability

Stock storage stability is defined as the period of time a given mixed stock can be stored on the factory floor and remain usable. Stock can be affected by scorch time [37]. Usually, but not always, the longer a mixed stock is stored, the lower its scorch safety. With storage, rheological properties change as a result of an increase in interaction between the rubber and carbon black (called bound rubber). This can be seen by a rise in compound's Mooney viscosity or oscillating rheometer minimum torque M_L (see Section 2.3.1). LeBlanc and Staelraeve reported on the advantages of large strain RPA tests for improved sensitivity to storage maturation [38]. Not only mixed stocks, but raw elastomers, such as natural rubber, can manifest changes in their rheological properties with storage. For example, the storage of natural rubber can produce a rise in viscosity and storage hardening [39]. The RPA can also be used to measure the storage hardening properties of natural rubber [40].

2.3.13   Mis-Compounding

Mis-compounding occurs when an error is made in the labeling or weighing of compounding ingredients before mixing or when not all of the ingredients are properly loaded into the mixer. Establishing specifications for batch weight is one check for compounding ingredient weighing errors. Many processing tests are affected by such errors. The careful monitoring of dynamic property variations and pattern changes with the MDR or the RPA can help determine which compounding ingredient might have been weighed in error [41,42]. Also, the Compressed Volume Densimeter described in ASTM D297 can easily measure the density of unvulcanized rubber compounds. If, for example, the compound density did not change, but the dynamic properties from the MDR or RPA suggested a change in the fillers, this may indicate that the wrong type of carbon black was used.

2.3.14   Cellular Rubber Blow Reaction

This chemical reaction occurs as a result of the decomposition of one or more blowing agents which generate gas during the cure. The generation of this gas is necessary to produce a cellular rubber product (see Chapter 21). To achieve the proper cellular structure, the cure reaction and blow reaction must be in balance. Therefore, it is important to measure the blow reaction.

Currently there are no standards available for tests to monitor blow reactions. However, instruments, such as the MDR-P [43] from Alpha Technologies, measure both the cure reaction and the blow reaction simultaneously. This is important because if these two reactions are not balanced, then unacceptable cell size or structure may result.

Parameters somewhat analogous to what is used to describe the cure reaction are used for the blow reaction. For example, the minimum and maximum pressure as well as time to 50% of the maximum pressure are calculated automatically [44,45].

References

1. Class participation responses from "Testing For Rubber Processability and Dynamic Properties," organized by J. Dick, U. of Akron., 1996 and 1997.
2. Symposium on rubber testing, ASTM Special Technical Publication No. 74, 50th Annual Meeting of ASTM, Atlantic City, NJ, June 16–20, 1947, p. 36.
3. Robert C. Armstrong, *Dynamics of Polymeric Liquids*, John Wiley & Sons, NY, 1987, pp. 527–528.
4. John Sezna, *Monsanto MPT Applications Manual.*
5. RPA2000® Rubber Process Analyzer by Alpha Technologies, Akron, Ohio, USA.
6. W.P. Cox, E. H. Merz, *J. of Polym. Sci.*, 28, 619 (1958).
7. Y. P. Khanna, Dynamic Melt Rheology. 1: Re-examining Dynamic Viscosity in Relationship to the Steady Shear Flow Viscosity, *Polymer engineering and Science*, March 1991, Vol. 31, No. 6, p. 440.
8. John Dick, Henry Pawlowski, "Applications of the Rubber Process Analyzer in Characterizing the Effects of Silica on Uncured and Cured Compound Properties" , Presented at a Meeting of the Rubber Division ACS at Montreal, May 4, 1996, Paper No. 34 (Published in *ITEC'96 Select* by Rubber and Plastics News, Sept. 1997.).
9. Roger Brown, *Physical Testing of Rubber*, Chapman & Hall, London, 1996, p. 58.
10. J. S. Dick, Clair Harmon, Alek Vare, "Quality Assurance of Natural Rubber Using the Rubber Process Analyzer," May 9, 1997 Meeting of the Rubber Div. ACS at Anaheim, California, Paper No. 97.
11. John Dick, Henry Pawlowski, "Applications of the Rubber Process Analyzer in Characterizing the Effects of Silica on Uncured and Cured Compound Properties" , Presented at a Meeting of the Rubber Division ACS at Montreal, May 4, 1996, Paper No. 34 (published in *ITEC'96 Select* by *Rubber and Plastics News*, Sept. 1997).
12. John Dick, Henry Pawlowski, *Rubber World*, Jan. 1995.
13. C. M. Blow, *Rubber Technology and Manufacture*, Newnes-Butterworths, 1971, p. 54.
14. H. Burhin, W. Spreutels and J. A. Sezna, "MV 2000 Mooney Viscometer – Mooney Relaxation Measurements on Raw and Compounded Rubber Stocks," presented at the Detroit ACS Rubber Div. Meeting, Oct. 1989.
15. Wayne Cousins, John Dick, "Effective Processability Measurements of Acrylonitrile Butadiene Rubber Using Rubber Process Analyzer Tests and Mooney Stress Relaxation," paper presented at the Fall Meeting of the Rubber Div., ACS, Cleveland, OH, October 24, 1997 (Published in *Rubber World*, January, 1998).
16. J. Dick, H. Pawlowski, "Applications for the Rubber Process Analyzer," *Rubber and Plastics News*, Apr. 26 and May 10, 1993.
17. W. Cousins, J. Dick, "Effective Processability Measurements of Acrylonitrile Butadiene Rubber Using Rubber Process Analyzer Tests and Mooney Stress Relaxation," *Rubber World*, January, 1998.
18. G. E. Decker, R. W. Wise, D. Guerry, "An Oscillating Disk Rheometer for Measuring Dynamic Properties During Vulcanization," *Rubber Chem. and Tech.*, Vol. 36, 451 (1963).
19. J. S. Dick, H. A. Pawlowski, "Alternate Instrumental Methods of Measuring Scorch and Cure Characteristics," *Polymer Testing* 14 (1995) 45–84.
20. Roger Brown, *Handbook of Physical Polymer Testing*, Chapter 8 by John Dick and Martin Gale, 1998.
21. MDR 2000® Curemeter, Alpha Technologies, Akron, Ohio, USA.
22. J. S. Dick, "The Optimal Measurement and Use of Dynamic Properties from the Moving Die Rheometer for Rubber Compound Analysis," *Rubber World*, January, 1994.

23. J. S. Dick, H. A. Pawlowski, "Alternate Instrumental Methods of Measuring Scorch and Cure Characteristics," *Polymer Testing* 14 (1995) 45–84.
24. *Ibid.*
25. John Dick, Chris Sumpter, Brian Ward, "New Effective Methods for Measuring Processing and Dynamic Property Performance of Silicone Compounds," Rubber Div. ACS Meeting, Paper No. 10, Fall, 1998.
26. John Somner, "Physical Properties and Their Meaning," *Rubber World*, June, 1996.
27. J. Sezna, *Rubber Chem. and Tech.*, 57(4) (1984) 826.
28. MDR 2000® Moving Die Rheometer, Alpha Technologies, Akron, OH, USA.
29. DIN 53 529, "Testing of Rubber and Elastomers, Measurement of Vulcanization Characteristics (Curometry)."
30. P. J. DiMauro, J. DeRudder, J. P. Etienne, "New Rheometer and Mooney Technology," January, 1990.
31. J. Dick, H. Pawlowski, "Applications for the Rubber Process Analyzer," *Rubber and Plastics News*, Apr. 26 and May 10, 1993.
32. J. S. Dick, Clair Harmon, Alek Vare, "Quality Assurance of Natural Rubber Using the Rubber Process Analyzer," presented at the ACS Rubber Div. Meeting at Anaheim, CA, May, 1997.
33. C. K. Rhee, J. C. Andries, "Factors Which Influence Autohesion of Elastomers," *Rubber Chem. and Tech.*, 54, 101 (1981).
34. Roger Brown, *Physical Polymer Testing*, Chapter 8 by J. Dick and M. Gale, 1998, p. 216.
35. *Ibid.*
36. *Ibid.*
37. John Dick, Allan Worm, "Storage Stability of Fluoroelastomer Compound Based on a Bisphenol AF/Onium Cure System and its Potential Use as a Standard Reference Compound," ACS Rubber Div. Meeting in Cleveland, Oct. 6, 1997.
38. J. L. LeBlanc, A. Staelraeve, "Studying the Storage Maturation of Freshly Mixed Rubber Compounds and its Effects on Processing Properties," The Polymer Processing Society, April 1994.
39. Sekaran Nair, "Characterization of Natural Rubber for Greater Consistency," *Rubber World*, July, 1988, p. 27.
40. J. Dick, C. Harmon, A. Vare, "Quality Assurance of Natural Rubber Using the Rubber Process Analyzer," presented at the Anaheim CA Meeting of the Rubber Div. ACS, May 6–9, 1997.
41. J. S. Dick, "The Optimal Measurement and Use of Dynamic Properties from the Moving Die Rheometer for Rubber Compound Analysis," *Rubber World*, January, 1994.
42. J. S. Dick, H. A. Pawlowski, "Alternate Instrumental Methods of Measuring Scorch and Cure Characteristics," *Polymer Testing*, 14 (1995) 45–84.
43. MDR20000-P® From Alpha Technologies, Akron, Ohio.
44. J. S. Dick, R. A. Annicelli, "Compound Changes to Balance the Cure and Blow Reactions Using the MDR-P to Control Cellular Density and Structure," Paper 53 Presented at the Indianapolis Rubber Div. ACS, May, 1998. (Published in *R* and *PN*, Nov. 2, 1998.)
45. J. Sezna, H. Burhin, "The MDR 2000P, a New Rotorless Curemeter for Testing Cellular Rubber," ACS Rubber Div., Oct. 1994.

3   Vulcanizate Physical Properties, Performance Characteristics, and Testing

John S. Dick

3.1   Introduction

The physical properties of cured rubber compounds are major determinants of product performance. The following physical properties are discussed as follows in this chapter:

- Density (in Section 3.2)
- Hardness (in Section 3.3)
- Stress Strain Properties Under Tension (in Section 3.4)
- Stress Strain Properties Under Compression (in Section 3.5)
- Stress Strain Properties Under Shear (in Section 3.6)
- Dynamic Properties (in Section 3.7)
- Low Temperature Properties (in Section 3.8)
- Stress Relaxation, Creep, and Set (in Section 3.9)
- Gas Permeability (Transmission) (in Section 3.10)
- Adhesion (in Section 3.11)
- Tear Resistance (in Section 3.12)
- Degradation Properties (in Section 3.13)
  - Flex Fatigue Resistance (in Section 3.13.1)
  - Heat Resistance (in Section 3.13.2)
  - Ozone Resistance (in Section 3.13.3)
  - Weathering Resistance (in Section 3.13.4)
  - Resistance to Liquids (in Section 3.13.5)
  - Abrasion and Wear Resistance (in Section 3.13.6)

Not all of these cured physical properties are optimized at the same point on the cure curve (see Chapter 2). As shown in Fig. 3.1, the optimum properties for tear resistance, rebound, ultimate tensile strength, hysteresis, etc. do not occur at exactly the same time during vulcanization.

3.2   Density

Density is simply weight (or mass) divided by volume at a specified temperature. This property determines the mass (weight) of a given rubber compound required to fill a specific mold cavity. Compounds with higher densities require greater weights of the

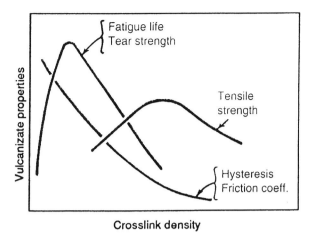

Figure 3.1   Optimum cure profiles for different vulcanizate properties (courtesy John Sommer, Elastech).

compounded stock to fill a given size mold cavity. Because raw materials are usually purchased by unit weight and molded rubber products are produced from a mold cavity with a fixed volume, knowing the compound density is very important in product cost calculations. Usually, increasing compound filler loading, such as carbon black, silica, or clay, results in a higher compound density. However, many times increased filler concentration still reduces the product's cost. Also, measuring compound density is an effective quality procedure to detect variations in the rubber compound composition resulting from changes in ingredient weighing and mixing, among other reasons.

The density of a vulcanized rubber compound specimen can be measured and calculated by Archimedes' Principle in which the specimen is weighed in water and weighed in air. ISO 2781 [1] gives detailed procedures for calculating the density of a cured specimen. Additionally, special cases are also addressed. For example, if cured tubing or cable insulation is tested with trapped air, this may lead to erroneous results. To overcome this problem, Method B of ISO 2781 calls for the specimen to be cut up into small pieces and tested in a density bottle, using an analytical balance.

3.3   Hardness

Hardness is a simple, inexpensive, and fast test used throughout the rubber industry. Hardness is measured from the cured rubber's resistance to deformation when a force is applied to a rigid indentor. This results in a measure of "a modulus" of a rubber compound under very limited deformation (strain). When the force is applied to the indentor with a dead-load, this method is called the International Hardness in IRHD units (International Rubber Hardness Degrees), which is described in ASTM D 1415 or ISO 48. This test normally uses a hemispherical indentor.

If the force is applied to the indentor by a spring, it is called the Durometer Hardness Method (usually a small pocket-size apparatus), described in ASTM D2240 and ISO 7619.

[Refs. on p. 67]

This method uses the Shore A scale, which is similar, but not identical, to the IRHD scale, and the Shore D scale, which is used for testing rubber vulcanizates with high hardness. Also, these methods refer to other hardness scales, as well. There is no completely accepted conversion between a Shore A and a Shore D scale, just a crude approximation. Also, the Shore type indentor has a different geometry from the IRHD indentor: truncated cone vs. hemispherical. Shore hardness is the more popular method because the hand held durometer is more portable and can be used in the laboratory or in the factory.

These hardness tests are somewhat crude and measure only under very limited deformations that may not relate to end product applications. Also, data from these tests can show much scatter. This variability and poor repeatability can be the result of variations in sample thickness, operator dwell time, how the instrument is set up and applied, sample edge effects (readings taken too close to the sample edge), or differences in sample geometry, to name a few. Therefore, these hardness tests should *not* be considered a reliable measure of a design or engineering property, but a quick and simple method of detecting gross differences in cured compound properties.

3.4   Tensile Stress–Strain

Tensile stress–strain is one of the most commonly performed tests in the rubber industry. These tests are performed on tensile testing instruments where a cured, dumbbell-shaped rubber specimen is pulled apart at a predetermined rate (usually 500 mm/min.) while measuring the resulting stress. Figure 3.2 shows a commonly used dumbbell shaped specimen. ASTM D412 and ISO 37 detail the standard procedures used to measure tensile stress–strain properties of a cured rubber compound.

Generally, (1) ultimate tensile strength, (2) ultimate elongation, and (3) tensile stress at different elongations are reported. Ultimate tensile strength is the maximum stress when the dumbbell specimen breaks during elongation. Ultimate elongation is the applied strain when the break occurs. The tensile stress is usually measured and reported at different predetermined strains (such as 100 and 300%) before the break occurs.

Figure 3.3 shows a stress–strain curve for a "typical" rubber compound. Unlike metals, this stress–strain curve shows no (or a very limited) linear portion. Therefore, it is usually not practical to calculate Young's modulus, which would be the slope of a straight line drawn tangent to the curve and passing through the origin. Instead, stress at selected elongations is usually reported. These stress values for different elongations are erroneously reported by some rubber technologists as 100% modulus, 300% modulus, etc. However, these measures are not actually modulus values.

Figure 3.2   Rubber dumbbell test specimen.

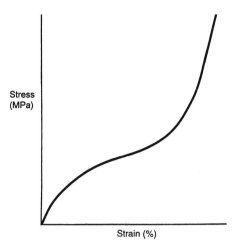

Figure 3.3   Typical stress strain curve for rubber.

Stress-strain properties, such as ultimate tensile strength can be easily affected by poor mixing and dispersion, the presence of contamination, under-curing, over-curing, and porosity, among other factors. Undispersed particles of different compounding ingredients, such as carbon black agglomerates, cause stress concentrations during the stretching of a rubber dumbbell, causing premature breaks at lower stresses. Impurities, such as dirt or paper fragments, can also cause the dumbbell to break at a lower stress. Likewise, volatile compounding ingredients can cause porosity to form during cure. These voids can also cause lower tensile strength [2]. Lastly, laboratory-mixed batches usually have higher tensile strength than factory-mixed batches because laboratory mixes often are better dispersed.

The tensile stress–strain testing discussed here involves non-prestretched specimens. However, if the dumbbell specimen is prestretched, for example, to more than 400% of its original length, and then tested in a normal manner, stress–strain would probably be significantly affected. This is particularly true of compounds containing high reinforcing filler loadings such as carbon black [3]. Prestretching causes ''stress softening,'' which results from breakdown of the carbon black agglomerates. Many times, if prestretched dumbbells are allowed to rest, their modulus (or tensile stress) increases. Because many rubber products are exposed to repetitive stress–strain cycling, this phenomenon can affect end use performance.

Many rubber products are not extensionally deformed more than 30%. So tensile stress–strain is usually not of great importance for product design, unless the product is a rubber band. On the other hand, tensile stress–strain testing of a given compound can be a valuable quality assurance tool to detect compounding mistakes in the factory and is very useful in compound development [4].

3.5   Stress–Strain Properties under Compression

Compression stress–strain testing often relates to actual product service conditions better than extension testing. Usually, test methods involve measuring the stress resulting from a

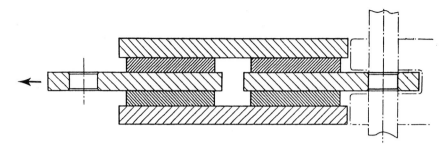

Figure 3.4   Quadruple lap shear test piece (ISO 1827).

compressive deformation applied to a standard, cured cylindrically-shaped rubber specimen between two plates. Compression test results depend on such factors as the shape of
the rubber specimen, preconditioning, rate of deformation, and the degree of bonding or
slippage of the specimen to the two metal surfaces. The more slippage experienced with
the test piece means less ''barrelling.'' The degree or lack of ''barrelling'' greatly affects
the test results [5,6]. ASTM D575 and ISO 7743 are both standard methods for measuring
stress–strain properties under compression, although they are quite different. The ASTM
method uses sandpaper to prevent slippage, while one part of the ISO method allows a
lubricant to be used, and another part requires the samples to be bonded to the parallel
metal plates. On course, these different conditions result in different results.

3.6   Stress Strain Properties under Shear

Measuring the stress–strain properties under shear can also be very relevant to some rubber
product applications. Generally, most rubber product applications do not exceed a strain of
75% [7]. The resulting stress–strain curve may be linear up to about 100% for ''soft''
compounds and up to 50% for ''hard'' rubber compounds [8]. ISO 1827 is a commonly
used test method for measuring the stress–strain properties of a rubber compound under
shear. Figure 3.4 shows the quadruple shear test piece which is separated, as noted by the
arrow.

3.7   Dynamic Properties

Rubber products are used in many dynamic applications such as tires, belts, isolators,
dampers, etc. The best way to measure and quantify the cured dynamic properties of a
rubber compound is to mechanically apply a sinusoidal strain to a cured rubber specimen
and measure the complex stress response and the resulting phase angle (δ), as was
illustrated in Fig. 2.8. As discussed earlier in Section 2.3.1.3, this phase angle δ and

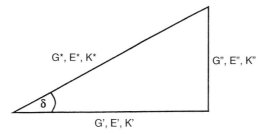

Figure 3.5   Vector diagram for dynamic properties.

complex modulus response (either G^* for shear or E^* for compression or extension) are used to calculate the storage modulus (G' or E') and the loss modulus (G'' or E''). Also, $\tan \delta$ is calculated by dividing G'' by G' or E'' by E'. In addition, the same principles can be applied to determine spring rates k^*, k', and k''. The vector analysis for these complex, elastic, and loss functions is shown in Fig. 3.5. Standards that define dynamic properties are ISO 2856, ASTM D2231 and the newer ASTM D5992.

The following are definitions for some common dynamic property terms used in the rubber industry.

- Storage Normal Modulus E' (Elastic Normal Modulus) is the component of normal stress exactly in phase with the normal sinusoidally applied strain, divided by the strain.
- Loss Normal Modulus E'' (Viscous Normal Modulus) is the component of normal stress exactly 90° out-of-phase with the normal sinusoidally applied strain, divided by the strain.
- Complex Normal Modulus E^* is the resultant normal stress divided by the resultant normal strain. By vector analysis, $(E^*)^2 = (E')^2 + (E'')^2$ (see Fig. 3.5).
- Storage Shear Modulus G' (Elastic Shear Modulus) is the component of shear stress exactly in-phase with the sinusoidally applied shear strain, divided by the strain.
- Loss Shear Modulus G'' (Viscous Shear Modulus) is the component of shear stress exactly 90° out-of-phase with the sinusoidally applied shear strain, divided by the strain.
- Complex Shear Modulus G^* is the resultant shear stress divided by the resultant shear strain. By vector analysis, $(G^*)^2 = (G')^2 + (G'')^2$ (see Fig. 3.5).
- Phase Angle δ (Loss Angle) is the angle by which dynamic force leads the dynamic sinusoidal deflection (see Fig. 2.8).
- Tan δ (Loss Factor) is the ratio of loss modulus to storage modulus. For normal stresses, $\tan \delta = E''/E'$; while for shear stresses, $\tan \delta = G''/G'$. Higher $\tan \delta$ values usually denote a more hysteretic compound at a given complex modulus.
- Hysteresis refers to a process occurring within the rubber in which there is a loss of mechanical energy as heat from an applied cyclical deformation of a rubber body.
- Hysteresis Loop refers to the closed curve formed from a plot of dynamic force vs. dynamic deflection in a complete cycle.
- Damping refers to a component of a complex dynamic force which is 90° out-of-phase with the strain.

When designating test conditions for dynamic testing of rubber, it is extremely important that the exact temperature, frequency, strain amplitude, type of strain, preconditioning, and strain history be specified. Other factors, such as test piece shape, can also affect test results. Typically, but not always, the elastic modulus for a vulcanizate decreases with a rise in temperature or a decrease in applied frequency. Rheologically, the effects of increasing the temperature while dynamically testing rubber are usually equivalent to decreasing the frequency and vice versa. This is the principle of the time-temperature superposition, which can be performed on the elastic modulus, loss modulus and loss factor, $\tan \delta$ using the Williams, Landel, and Ferry (WLF) equation [9].

The applied strain amplitude and the strain history on a test specimen are also very important, particularly for rubber vulcanizates containing fillers because of the filler–filler and filler–polymer interactions affected by applied strains [10]. Because of these filler effects, the elastic modulus generally decreases when the amplitude of the applied strain is increased, the well known Payne effect [11]. Also, prior strain history and preconditioning can have a great effect on the dynamic properties measured [12].

Futamura [13] and Gatti [14,15] have published extensively on the use of cured dynamic properties measured on tire tread compounds to predict tire rolling resistance, steering response, and dry, wet, and snow traction. These different tire properties are usually measured at different conditions of temperature, frequency and strain. Warley [16], Novotny [17], Gregory [18], and others have researched the advantages of using dynamic property measurements in predicting the performance of rubber automotive parts, such as bushings, mounts, bumpers, dampers, caps, isolators, and drive belts. Calculating percent *transmissibility* for a rubber part in forced vibration (the ratio of transmitted force to applied force) is reviewed by Warley and Novotny [16,17].

As is discussed in other chapters, specific formulating ingredients have profound effects on the dynamic properties of the vulcanizate. For example, base elastomers which have higher glass transition temperatures (T_g) may impart higher hysteresis to a compound [19]. Also the addition of a specific plasticizer to a compound can lower the compound's T_g and affect dynamic properties of the compound by reducing hysteresis. The type and concentration of a plasticizer are important factors in determining the rubber vulcanizate's dynamic properties [20]. The specific grade and loading of carbon black also have profound effects on the compound's dynamic performance [21]. The type of crosslinks and the crosslink density can affect dynamic properties of a rubber compound, as well [22].

The dynamic properties discussed here are usually measured by forced vibration methods because the strain amplitude is controlled. ISO 4664 specifically gives a method for determining dynamic properties by forced sinusoidal shear strain. Other methods for measuring the dynamic behavior of cured rubber compounds include *rebound resiliency* and *free vibration* methods. Generally, when a compound is formulated to have lower hysteresis, it also has a higher rebound. Specific rebound test methods are discussed in ASTM D1054 (Goodyear-Healey Rebound Pendulum method), ASTM D2632 (falling weight method), and ISO 4662 (which mentions apparatus designs based on the Lupke Pendulum, the Schob Pendulum, and the Zerbini Pendulum). The free vibration methods are given by ASTM D945 (Yerzley Oscillograph) and ISO 4663, which includes three different methods. While the rebound resiliency and free vibration methods are generally not recommended as sources of engineering data, they are usually simpler to perform than forced vibration methods.

3.8   Low Temperature Properties

As the temperature of a rubber vulcanizate is decreased, the material becomes stiffer and its modulus increases. If the temperature is decreased enough, there is no longer sufficient energy for molecular rotation of the vulcanized rubber. The material takes on the properties of a glass, and becomes very hard and brittle. Figure 3.6 shows a log plot of the vulcanizate modulus vs. temperature for a typical rubber compound. As can be seen, the modulus can increase over 1,000 times when the material is exposed to low temperatures. Rubber behaves as a glass at very low temperatures because of a lack of thermal kinetic energy. With an increase in temperature, the thermal energy increases the material's flexibility, as shown in Fig. 3.6, where the modulus drops off the glassy plateau. At these temperatures, the rubber takes on characteristics similar to leather. Here, there is enough thermal energy to allow chain rotation and flexibility, but not enough energy for complete chain mobility and the good resiliency commonly associated with vulcanized rubber. As the temperature increases further, the vulcanizate finally reaches what has been called the "rubbery plateau" where normal rubber behavior is seen. So unlike what is shown in Fig. 3.6, in an unvulcanized rubber, the modulus would drop off this rubbery plateau because, at higher temperatures, the thermal kinetic energy is sufficient to disentangle rubber chains that are not chemically crosslinked.

When the temperature of a vulcanized rubber compound is so low that no further molecular rotation can occur, then it has reached the compound's glass transition temperature (T_g). The T_g of a rubber compound is greatly dependent on the chemical structure of the base elastomer. Structural aspects such as polarity, bulkiness, and

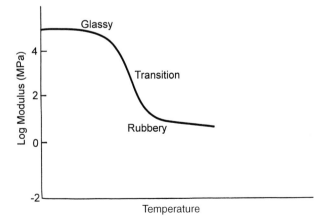

Figure 3.6   Vulcanized rubber modulus vs. temperature.

flexibility of side groups; flexibility of the backbone, symmetry; and steric hindrance can affect the T_g of the compound [23,24]. Also, the selection of other compounding ingredients, such as plasticizers [25], greatly influence the compound's T_g. Changing T_g and low temperature properties of a compound by varying components also affects cured dynamic properties. Differential Scanning Calorimetry (DSC) is a very common method for measuring the T_g of a rubber compound.

As discussed, when a rubber vulcanizate is exposed to very low temperatures, it can become stiffer from a lack of molecular kinetic energy. However, certain compounds based on elastomers such as natural rubber and polychloroprene may stiffen from crystallization. Crystallization, if it occurs, usually requires a longer exposure to the cold. Usually, an optimal temperature maximizes crystallization rates. These rates are different for natural rubber and polychloroprene [26]. Applying a strain to these crystallizable rubbers also increases the crystallization rate.

Because most physical properties of a vulcanized rubber compound change with a decrease in temperature, then theoretically many of these properties can be used to measure changes occurring at lower temperatures. For example, rebound and tensile strength are sometimes measured at sub-ambient temperatures. It is simply a matter of building the proper apparatus or environmental chamber. Also, some dynamic mechanical rheological testers are designed to measure the dynamic properties of the vulcanizate at temperatures well below room temperature. For example, the literature reports dynamic properties at $0\,°C$ for predicting tire wet traction performance imparted by a tread vulcanizate [27]. Other standard test methods are specifically designed to measure low temperature properties of cured rubber compounds.

3.8.1   Brittle Point

ASTM D2137 and ISO 812 are two similar, but not identical, methods for measuring the *brittle point* of a vulcanized rubber. As test temperature is increased from a very low value, the brittle point is the lowest temperature at which *none* of the cured rubber test pieces shows any cracks and breaks or develop any fissures or holes when hit with a striker at a specified velocity. Test conditions, such as the liquid or gaseous heat transfer medium, can affect reported values. This is a simple method for relating the temperature at which a cured rubber specimen becomes stiff enough to approach a glassy, brittle state.

3.8.2   Gehman Test

The *Gehman* Test also measures the relative stiffness of a cured rubber compound over a wide temperature range, as described in ISO 1432 and related to ASTM D1053. The Gehman Test determines at what temperatures specific "relative moduli" or "relative stiffness" occur. The "relative modulus" here is calculated from the torsional modulus at the specific subnormal temperature divided by the torsional modulus at 23 °C. Usually, the temperatures at which the rubber test specimen has stiffened to relative moduli of 2, 5, 10, and 100 are reported from this test. Once again, test conditions such as the liquid or gaseous heat transfer medium can affect reported values.

3.9   Stress Relaxation, Creep, and Set

- Stress Relaxation is the decease in the stress of rubber over time from a given applied constant strain.
- Creep is the change in the deformation of rubber over time from an applied constant force or load. It is also called strain relaxation.
- Set is the residual strain remaining on rubber after the removal of the force which produced the deformation. Usually there is a given recovery time allowed after the removal of the force before final measurements are made.

Figure 3.7 shows two viscoelastic models that illustrate stress relaxation and creep. The Maxwell model [28], with the dash pot and spring in series, illustrates the stress relaxation phenomenon from the decay in stress after the application of a step strain. This is very similar to stress relaxation with rubber in which the stress decreases (decays) at a fast rate initially, but at decreasing decay rates later. Therefore, it is common to plot this stress decay against time on a log scale. On the other hand, the Voigt model, with the dash pot and spring in parallel arrangement, illustrates the creep (or strain relaxation) phenomenon which occurs when a load or force is suddenly applied. This mechanical model also is very similar to the creep phenomenon which occurs with rubber. Here, the strain increases at a fast rate initially, but this rate of change decreases with time.

These rubber properties of stress relaxation, creep, and set are all time-dependent viscoelastic characteristics that are broadly related to each other. However, none of these properties can be used to predict any of the other properties. In other words, stress relaxation, creep, and set are all unique, independent properties.

For short term/low temperature tests, the cured rubber compound differences seen in each of these three properties are mainly the result of viscoelastic differences. On the other hand, long term/high temperature tests show not only viscoelastic differences, but also differences resulting from chemical changes among the compounds. These chemical changes include oxidative chain scission, oxidative crosslinking, and further vulcanization. A higher test temperature results in a faster stress relaxation rate or strain relaxation rate.

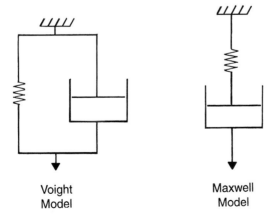

Voight
Model

Maxwell
Model

Figure 3.7   Two viscoelastic models, Voigt and Maxwell.

Sometimes, stress relaxation tests or creep tests are run at elevated temperatures to try to estimate what happens at lower temperatures over longer time periods. This high temperature testing to predict longer term, lower temperature behavior sometimes may not be valid because of different chemical reactions which may occur at the higher temperatures. Also, air diffusion rates of thermal oxidative aging may differ. In addition, if comparisons are made between laboratory test conditions and actual service conditions, there may be differences resulting from other significant causes of deterioration, such as ozone exposure and sunlight. So extreme caution should be used if Arrhenius plots are generated, as discussed in ISO 11346, to estimate long term service.

A stress relaxation test is very relevant for measuring sealing pressure of rubber seals and gaskets. Some rubber compounds impart a faster drop in sealing pressure than other compounds over time. Traditional compression set tests are very poor predictors of the sealing pressure imparted by a given rubber seal. Stress relaxation tests more directly relate to sealing pressure. Also, the environment to which a seal is exposed also has a major effect on its service life. For example, some fluids may swell a rubber and affect its apparent stress relaxation. Therefore, some stress relaxation tests are conducted on rubber seals in a fluid medium. Ring specimens are the best choice for stress relaxation tests in fluids because they provide a high surface-area-to-volume ratio enabling faster equilibrium swelling during the test.

The creep test is extremely useful when rubber is to be used in applications such as motor mounts, bridge bearings, and building bearings. These rubber parts are under a constant force and their service life can be estimated through creep studies.

The compression set test is the most common set test in the rubber industry today. This test itself does not relate directly to rubber product performance conditions as do the stress relaxation and creep tests. However, the compression set test is probably performed more often than stress relaxation or creep because compression set is relatively simple, making it ideal for quality control.

As is noted in later chapters, rubber compounding has a great effect on stress relaxation, creep, and set values. Some elastomers give better sealing pressure, creep resistance, and set resistance than other elastomers. The selection and concentration of filler and plasticizer, if any, have a profound effect on these properties. Also, the type of crosslinks and the crosslink density have a great effect. For example, carbon–carbon crosslinks from peroxide cures sometimes impart an advantage in sealing pressure for seal applications.

Theoretically, stress relaxation tests, creep tests, and set tests can be individually performed under compression, tension, or shear. However, most commonly, these tests are done under compression. Results from these tests are generally quite variable as indicated by interlaboratory crosschecks. It is very important that details be specified, such as pretest conditioning, sample geometry, tester geometry, the type and manner of lubricant application applied (if any) to the platens or plates, and temperature controls during the test and when samples are removed, among others.

A standard method for compressive stress relaxation was developed in 1956 as ASTM D1390; however this method was withdrawn in 1986 because it was rarely used. Because of a new interest in measuring stress relaxation, a special ASTM D11 Task Group developed a new method: ASTM D6147, now published. Also, a different method, ISO 3384, can measure compression stress relaxation of cylindrical test pieces and rings. The rings are suitable for making stress relaxation measurements in a liquid environment.

ISO 8013 is a standardized test developed to measure creep of vulcanized rubber in either compression or shear. There is currently no ASTM Standard for measuring rubber creep.

For compression set measurements, ASTM D395 is commonly used with Methods A and B. Method A defines compression set under constant force in air, while Method B defines compression set under constant deflection in air. Also, ASTM D1229 is commonly used when measuring compression set at low temperatures. ISO 815 covers testing compression set at ambient, elevated, and low temperatures. It is similar to the ASTM procedures which call for constant strain; however one should pay attention to the differences among these procedures, such as the temperature at which specimen recovery occurs; the sample size; and whether or not lubricants are applied to contact surfaces of the plates. Lastly, ISO 2285 is available for measuring tension set values at normal and elevated temperatures.

3.10   Permeability (Transmission)

Gas permeability or transmission tests measure how easily a given gas penetrates through a specified rubber sheet or membrane. Gas permeability is a function of both the solubility and diffusion rate of the gas in the rubber compound. It is important because, in some applications, rubber must function as a container of or barrier to certain gases. A given gas permeates through a rubber membrane in the direction of lower pressure. One important example of permeability resistance is in pneumatic tires. Tires commonly possess an innerliner compound based on a halobutyl rubber, which imparts higher air permeability resistance to a compound than other general-purpose elastomers.

ISO 1399 is the constant volume method for measuring gas permeability, while ISO 2782 is the constant pressure method. Also, the related ASTM D1434 test method (under the jurisdiction of the Plastics Committee) is sometimes used to measure air permeability of vulcanized rubber membrane specimens.

For measuring vapor permeability, ISO 2528 is commonly used for water vapor and ISO 6179 for volatile liquids. ASTM D814 is a related standard which is used for measuring volatile liquid permeability through barrier sheets of moderate thickness.

3.11   Cured Adhesion

Many rubber products are composites where rubber-to-metal, rubber-to-fabric, or rubber-to-cord adhesion is very important. Examples of these composites are seen in tires, belts, hose, isolators and dampers, and various other products fabricated from rubber-coated fabric. Usually, the best measures of adhesion for these examples result from testing actual products. But this type of testing is not always feasible. Therefore, several standardized tests have been developed, which are very useful in quality control as well as compound development.

ISO 813 is a 90° (angle) peel test of a cured rubber piece from a metal strip to which it is bonded. ASTM D429, Method B, is similar, but there are differences in specific dimensions and other factors.

ISO 814 is a tension adhesion test of a rubber disk specimen bonded between metal plates. This test is designed not to peel. ASTM D429, Method A, is similar, but again there are differences such as specific dimensions.

ISO 5600 is another adhesion tension test using two specified conical metal end pieces that are bonded to the rubber specimen. This method is related to ASTM D429, Method C. On separation, the stress is concentrated at the cone tips.

ISO 1827 is the shear stress–strain test for a vulcanizate that was discussed in Section 3.6. As mentioned, a special quadruple test element is constructed (Fig. 3.4) to measure shear modulus. By running this test to failure, rubber-to-metal adhesion in shear can also be measured.

ISO 36 is a peel test measure of the bond strength between rubber and fabric for ply separation peelings at approximately 180° (angle). ISO 6133 can be used to analyze the "trace" resulting from the output for ISO 36, which consists of various "peaks" and "valleys," somewhat similarly to the tear "trace" which is discussed in the next section. ASTM D413 is somewhat similar to ISO 36, but there are differences.

ISO 4647 describes the *H-Pull* Test for tire textile cord-to-rubber adhesion. Likewise, ISO 5603 is for steel tire cord-to-rubber adhesion. Related ASTM Standards are D2138 for textile cord adhesion and D2229 for steel tire cord adhesion.

ISO 6505 tests metal corrosion induced by a given rubber compound. Some rubber compounds cause metallic corrosion by just being close to or touching a metal surface. Corrosion destroys adhesion, which is why this test is important. The metal substrates commonly tested by this method include brass, copper, mild steel, and aluminum.

With any of the standard adhesion tests just discussed, sometimes the mode of failure (i.e., cohesive failure, adhesive failure, etc.) may be as important as the actual adhesion values themselves. These standard adhesion tests may *not* necessarily relate to or predict adhesive failure during a product's use, because many other factors affect the integrity of the bond line of a given rubber product. These factors include the product's dynamic service history (which generates rubber fatigue), corrosion, and high service temperatures. With the exception of the *Scott* Flexer (ASTM D430), which was specifically designed for measuring tire and belt ply separations, there are no other established standardized "dynamic" adhesion tests. Yet such "dynamic" adhesion tests would probably relate to rubber product performance better than any of the existing standardized adhesion methods just discussed. R. Brown has given a good review of some non-standard, "dynamic" adhesion tests [29].

3.12   Tear Resistance

High stress concentration on a rubber product applied at a cut or defect area during service can lead to the propagation of a tear or rupture. Different rubber vulcanizates show different resistance to tear. Tear characteristics for a compound can be related to the

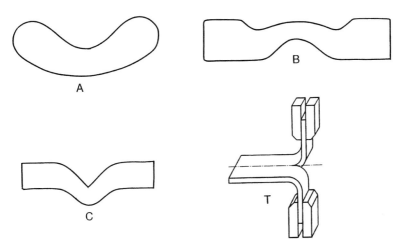

Figure 3.8   Shape of standard tear specimens, A, B, C, and T.

compound's crosslink density and state of cure, as well as filler type and loadings. The force to initiate a tear is quite different from the force required to propagate a tear. Various tear tests place a deliberate flaw in a rubber specimen to try to relate to the tear propagating force.

Figure 3.8 shows the shapes of some commonly used tear test pieces which are described in ASTM D624. Die B is a crescent-shaped test piece with large ends for better gripping in a tensile tester. This specimen is nicked with a razor blade to a specified depth to help initiate a tear. The Die C test piece has an angle to help initiate a tear and does not necessarily require a cut nick. The Die T Trouser tear specimen is separated by a shear force from the tension applied in opposite directions to each leg at right angles to the plane of the test piece. The tear strength (T_s) is reported as kilonewtons/meter of thickness from the formula:

$$T_s = F/d$$

where F = maximum force (in N), for dies B and C, and the median or mean for die T (Trouser), and d = test piece thickness in mm.

The tear values from these three tests are very dependent on the geometry of the test piece. They are *not* engineering properties. Generally, these tests do not yield the same tear values. Because of the complex nature of tear testing, *none* of these tear tests relate to rubber product performance directly. Generally, Die B gives higher values than Die C, with the Trouser test piece generating the lowest tear values. Tear test results are very dependent on consistency of sample preparation, including accuracy of sample thickness, depth of the nick cut, and the sharpness of the razor blade used to make the nick.

ASTM D624 also describes a Die A, which is a smaller crescent not used as frequently as the other tear test pieces just discussed. ISO 34 is similar to the ASTM method; however, there are differences. The ISO method allows the Die C test piece to either be un-nicked or nicked, for example. Also, ISO 816 describes the Delft test piece, a rectangular-shaped test piece with a slit in its center, not its edge.

In accordance with fracture mechanics, *tearing energy* theoretically is a basic material property that is truly independent of the cut geometry and geometry of the rubber specimen [30]. Most standard tear tests do not measure tearing energy. A modified Trouser tear comes closest to relating to true tearing energy; however, features such as the width of the cut and variations in cut geometry, among others, can greatly affect accuracy of results. The Trouser Test has the advantage of a relatively long tear path. However, knotty tears can be encountered with some compounds, which can add to variability. ISO 6133 can be used to determine the median force for the Trouser Test from its tear trace.

3.13   Degradation Properties

3.13.1   Flex Fatigue Resistance

Rubber flex fatigue can be defined as a deterioration in any physical property resulting from extended exposure to a cyclical deformation, such as occurring in bending, shearing, compression, or extension.

There are two types of fatigue tests which measure quite different properties. The first type is represented by various *flex cracking* tests. These tests generally repeatedly bend and/or extend a cured rubber specimen and measure the initiation and/or growth of cracks. The heat from this type of deformation is dissipated and the specimen temperature stays close to ambient. The second type of flex fatigue test is basically a *heat build-up test* where the heat from flexing cured specimens does not dissipate. Usually, the temperature rise and permanent set values are measured under severe conditions of cyclical compression for a cured specimen of a predetermined geometry.

Various standardized flex cracking tests are commonly used in the rubber industry to measure this important characteristic. Usually, these standardized methods require that the specimen pass through zero strain during flexing. A common property of crystallizing rubber is that it is less likely to crack unless the cyclical deformations pass through zero strain. Also, there are two types of flex cracking characteristics, i.e., *crack initiation* and *cut growth*. For the former property, the cured rubber specimen is not cut, but flexed and initial cracks and their rates of propagation observed. However, for the latter property, the cured specimen is carefully cut in a standard manner and the growth of the cut from flexing is measured. Some natural rubber compounds show poor crack initiation resistance, but display very good cut growth resistance, because natural rubber crystallizes on straining and forms crystallites at the tip of the crack. On the other hand, many SBR-based compounds are just the opposite, with good crack initiation resistance but poor cut growth resistance.

The standardized flex cracking tests are very specific to the specimen geometry and may not correlate to the rubber product's performance. Results from these standardized methods are not engineering properties, nor are they fundamental material properties. Instead, they are used for quality control and one-on-one compound comparisons.

Standardized flex cracking tests are inherently variable because, in some cases, it is hard to objectively measure the extent of cracking. Also, other variables can greatly affect test results. For example, flex cracking is very sensitive to the presence of ozone for specimens based on unsaturated rubber. Many standards severely restrict ambient concentrations of ozone to very low levels. In addition, any variation in test specimen thickness greatly affects results. For crack initiation tests, cracks are started from natural surface flaws, which can be quite small, variable from one test series to another. Poorly dispersed compounding ingredients can cause stress points and affect test results. Cut growth tests require a cut. From Section 3.12 (tear), it can be seen that how the cut is made is quite important and a potential source of variation. Generally, fatigue tests under tension cycling may be more reliable than standard fatigue bending tests.

Fatigue testing can also be affected by the cured modulus of the specimen because the compound modulus determines the energy available for crack propagation. For example, a higher modulus compound allows more energy for crack propagation than a lower modulus compound provides under conditions of a *controlled strain* cycle. However, under fatigue test conditions of a *controlled stress* cycle, a higher modulus specimen has a lower resulting deformation per cycle compared to a test specimen with a lower modulus. Therefore, it is important to know whether the fatigue test is under controlled stress or controlled strain *and* the deformational history of the actual rubber product in service.

ASTM D430 describes three different bending flex fatigue tests: the *Scott* Flexer (Method A), the unpiered (uncut) *DeMattia* tester (Method B), and the *DuPont* Flexing Machine (Method C). The Scott Flexer was discussed earlier in Section 3.11 (Cured Adhesion). It is a dynamic adhesion tester designed to test for ply separations in specimens prepared from tires and belts. The DeMattia tester is the most commonly used flex fatigue tester. This method is used to measure crack initiation of a cured molded rubber strip with a transverse groove. The test piece is bent by the two end clamps moving toward and away from each other, bending this test piece in a loop-like pattern and placing a maximum strain at the surface in the groove. Lastly, the DuPont Flexer works on the principle of test pieces coupled together to make an endless belt configuration that goes over specified pulleys. ISO 132 is also an unpierced DeMattia test; however, it differs from ASTM D430 Method B.

ASTM D813 refers to the use of the DeMattia Tester for measuring cut growth from a small, specified cut applied before flexing. As discussed earlier, crack initiation and crack growth are two very different properties. ISO 133 also describes the use of the DeMattia tester for measuring crack growth; however, this ISO standard is different. For example, it calls for the test results to be reported in a different manner than the ASTM procedure.

ASTM D1052 describes measuring cut growth with the *Ross* Flexing Machine. This method uses pierced strip test specimens. The Ross is another bending test which can more precisely control the strain. It is primarily used to test rubber shoe sole compounds. This method does not have an ISO equivalent.

ASTM D3629 describes the TEXUS® Flex Tester, which is also a cut growth bending test method. It is also referred to as the "flipper tester." This method calls for the curing and molding of test specimens in the shape of a 'T'. These specimens are mounted on a circular platform and are struck repeatedly against deflector bars while being rotated. These specimens are also carefully pierced before testing. There is no equivalent ISO method.

ASTM D4482 refers to fatigue testing under tension, a method also known as the Fatigue-to-Failure test. Dumbbell specimens are cyclically strained at a fixed frequency and strain (extension ratio). This cyclical strain passes through zero. During this cyclical strain, cracks are initiated from the naturally occurring flaws for each specimen. No deliberate flaws (cuts) are made on the dumbbell specimens. The extension ratio is calculated as follows:

$$\lambda = L/L_0$$

where λ is the extension ratio, L_0 is the unextended length, and L is the extended length of the specimen.

This standard allows tests to be performed at one extension ratio or at a number of different extension ratios. The log of fatigue life may be plotted against the maximum strain (or the log of strain energy defined in the Annex of the procedure). ISO 6943 is somewhat similar to D4482; one difference is that the ISO method allows not only dumbbells but also ring test pieces.

At this time, there is *no* agreed upon standard method for measuring "tearing energy". The concept of tearing energy with rubber was promoted in the 1950s by R. S. Rivlin and A. G. Thomas at the Malaysian Rubber Producer's Association (MRPRA) [31].Tearing energy calculations come from the theory of fracture mechanics. These measurements can be made under the proper conditions on a servo-hydraulic dynamic testing machine. From these studies, plots of log of crack growth rate vs. log of tearing energy can be constructed. From these plots, it can be found that certain compounds have comparatively better fatigue life under different tearing energy conditions. Their "slopes" may be quite different. Compound A may have lower crack growth than Compound B at low tearing energies; however, at higher tearing energies the reverse may be true. Also, at *very* low tearing energies, compounds generally can have an extremely slow or virtually no crack growth rate.

ASTM D623 is the heat build-up test from compressive cyclic deformation of a bulky rubber specimen. There are two different methods described in this standard. Method A describes the more commonly used *Goodrich* Flexometer (which tests under oscillating compression) while Method B utilizes the *Firestone* Flexometer (which tests with a rotary motion superimposed on a constant compression load). For the Goodrich Flexometer, the length of stroke, applied load, and base temperature can be set for the appropriate severity of test conditions for the compound in question. Properties such as heat build-up and permanent set can be recorded. Special conditions, such as a higher stroke under a higher load and elevated base temperature can be set up to run the sample to failure (run to "blow out"). Under these severe conditions, sometimes very subtle differences in a compound can be detected while they might not be seen with other traditional physical tests. Therefore, these tests make good quality control methods and are commonly used in the tire industry.

ISO 4666 is different from ASTM D623. Part 3 of ISO 4666 is somewhat related to the Goodrich Flexometer Method A of ASTM D623. However, Part 2 of this ISO standard describes a rotary flexometer known as the St. Joe Flexometer. This used to be described in an ASTM Standard but was discontinued in the early 1960s.

3.13.2   Heat Resistance

Rubber compound properties change with time at ambient temperatures or at a faster rate from exposure to heat. Therefore, heat aging tests are commonly performed to measure the change in physical properties for a rubber compound at higher temperatures which may approximate the actual product service temperatures or to use heat aging as "accelerated aging" tests to relate to actual product life at lower temperatures. However, there are several reasons heat aging tests do not correlate well with actual service life at lower temperatures. First, at elevated temperatures, oxidative effects are stronger. Second, oxygen diffusion rates are faster at elevated temperatures. Third, the geometry of the test specimen is usually different from the rubber product, so the exposed surface area-to-volume ratio is different. This means the quantity of oxygen that diffuses into the test specimen is different from the quantity that can diffuse into the product. This factor explains why there can be a poor correlation of a heat aging test not only to product performance at lower temperatures, but also to product performance at the heat aging test temperature itself.

ASTM D573 describes a standard method for aging of rubber specimens in air ovens. Multiple dumbbell specimens from each of the compounds tested are placed in the oven chamber. Here, dumbbell specimens are aged usually at either 70 or 100 °C. Some of the specimens from each compound (or batch) are tested for tensile strength, ultimate elongation, and tensile stress at different elongations, usually at intervals of 2, 4, 7, and 14 days. The percent change from the original unaged values can be tabulated. There is usually an optimal condition of temperature and time for the best test sensitivity to subtle compound differences. If these temperature and time conditions are too low or too high, the tested compounds may start yielding similar results. Also, in this method, multiple specimens from different compounds are placed in a single chambered oven. When different test specimens are in the same oven chamber, volatile compounding ingredients, such as plasticizers or antidegradants, can migrate from one test sample to another, affecting the aging test results. Lastly, it is important to note that the velocity of the air flow of the oven is very important. A higher air velocity imparts a higher aging severity than a lower air velocity imparts even at the same oven temperature. ISO 188 Method A is somewhat similar to ASTM D573, except that the ISO method allows for either a single chamber or a multi-cell oven. By using the multi-cell oven properly, migration effects can be avoided. Also, the ISO method is not restricted to only tensile testing.

ASTM D865 is generally called the Test Tube Aging Method. The idea here is to heat rubber dumbbell specimens in test tubes in a heat exchange medium to solve the problem of migration that may occur in a single chamber oven. The greatest problem with this method is that there is little air flow. So, test tube aging conditions are usually not as severe as air oven aging. There is no ISO for this method.

ASTM D572 is called the "oxygen bomb" method because it exposes the rubber dumbbell specimen to a higher specified pressure of oxygen at a specified elevated temperature for a predetermined time period in an appropriate chamber all in accordance with the ASTM standard. All the published ASTM safety precautions must be followed. The advantage of this method is that the higher oxygen pressure increases the oxygen diffusion rates into the specimens. There is a related method provided by ISO 188 Method B.

ASTM D454 is called the "air bomb" method. It is similar in principle to ASTM D572 which was just discussed, except it uses air instead of oxygen at different temperature and pressure conditions. Again, all published ASTM safety precautions must be followed. This method produces results somewhere between the "oxygen bomb" test and ambient pressure air aging tests. There is no corresponding ISO standard for the "air bomb" test.

International Standard ISO 6914 is titled "Determination of Aging Characteristics by Measurement of Stress at a Given Elongation." Unofficially, it has been called "continuous stress relaxation during aging" and "intermittent stress relaxation during aging." "Stress relaxation" here may be a misnomer because the stress, during aging, may actually rise, not fall.

3.13.3   Ozone Resistance

Compounds based on general purpose elastomers and some specialty elastomers can be very sensitive to ozone attack while in service under strain if not protected by an effective wax and/or antiozonant system. Ozone (O_3) tests cannot be performed on unstrained rubber specimens. However, sometimes a very small strain condition can cause the most harmful "deep cracking" to occur in an ozone environment. Usually, applying a higher strain causes the more frequent, smaller cracks to form. Also, very low concentrations of only 1 part per hundred million (pphm) can cause ozone cracking. Some rural areas of the world can have atmospheric concentrations of ozone between 1 to 10 pphm while in urban areas, levels can be much higher. Usually, standard ozone tests are carried out in a special ozone chamber where the O_3 concentration is set at a predetermined value in a range between 25 and 200 pphm.

Ozone attack occurs at the surface of the rubber compound. Many compounds contain antiozonants such as p-phenylene diamines and special blends of paraffin and microcrystalline waxes to improve resistance to ozone attack. Before performing ozone tests, compound specimens containing these antiozonants have to be conditioned for a given amount of time in darkness so these antiozonants have time to bloom (exude) to the rubber surface.

There are both static and dynamic ozone tests for measuring ozone resistance. Static tests do not destroy protective wax barriers, but the dynamic ozone tests which do cause protective barriers to weaken and rupture, relate better to dynamic product service. Therefore, dynamic test conditions are better for measuring the relative effectiveness of antiozonants such as the p-phenylene diamines. It is also important that a low frequency of deformation be applied for dynamic ozone testing so that cracks do not form because of a fatigue mechanism.

The temperature at which an ozone test is performed is very important. At very high temperatures, ozone becomes unstable. For example, 40 °C is a commonly selected ozone test temperature. If different waxes are being tested for their effectiveness in imparting ozone resistance, Jowett proposed that an alternate temperature of 0 °C be used as well [32]. Also, high humidity can affect ozone test results and should be part of the test report. It is advantageous to conduct these tests under low humidity conditions.

ASTM D1149 describes a standard static ozone test method that calls for either stretched or bent rectangular strips to be used. ASTM D1171 is another static ozone test using triangular cross-section specimens mounted on circular mandrels. ASTM D3395 covers standard methods for measuring dynamic ozone resistance. Method A discusses the tension cyclical flexing of rubber rectangular strips at a frequency of 0.5 Hz and a maximum amplitude tensile strain of 25%. Method B is an ozone belt flex test.

ISO 1431 ozone test method differs greatly from the ASTM methods. ISO 1431 has sections covering both static and dynamic ozone testing. The chamber gas velocities for

ISO 1431 vs. ASTM D1149 can be quite different, which can cause different results from these two methods. Also, the dynamic ozone section of ISO 1431 does not include ozone belt flex testing as described in ASTM D3395. On the other hand, the ISO 1431 Dynamic Ozone Testing section does allow T50 dumbbells as well as strips to be used. Other differences exist between the ISO and ASTM methods as well.

3.13.4   Weathering Resistance

The evaluation of weathering resistance can be complicated. Cured rubber products can deteriorate from exposure to bright sunlight, which includes effects from ultraviolet radiation; moisture from either rain or dew; oxygen; and ozone. This exposure can result in surface cracking, chalking, crazing, peeling, and discoloration. These surface defects can ultimately lead to product failure. We have already discussed test methods designed to directly measure the effects of oxygen and ozone on the vulcanizate, even though these tests may not correlate with actual product performance. Probably the best way to test weatherability resistance is to plan far enough in advance to arrange an independent testing agency to do outdoor comparative testing against a selected standard reference in a location such as Florida. The disadvantage is the unpredictability of weather. There are standard test methods available to simulate outdoor weathering under more controlled conditions. The carbon black in most compounds provides for very good ultraviolet protection. However, pigmented (non-black) rubber vulcanizates are subject to discoloration from weathering and radiation.

ASTM D518 describes rubber specimen preparation procedures for ''normal weathering'' or ozone testing (described in the previous section). ASTM D750 refers to the arc weathering apparatus which exposes rubber specimens to bright light and intermittently to water spray under controlled conditions. ISO 4665 differs somewhat from the ASTM methods. Part 1 of the ISO standard relates to assessment of property changes after exposure to natural weathering or artificial light. Part 2 addresses methods for testing for exposure to natural weathering, and Part 3 relates to artificial light.

3.13.5   Resistance to Liquids

When a rubber product has frequent contact with various liquids, which may be the case for many automotive rubber components, the resistance to the specific liquid (or group of liquids) becomes very important. Exposure to a liquid can affect a rubber vulcanizate in three ways. The most frequent effect is *swelling* from absorption of the liquid into the mass of the cured rubber. The second, less common, effect is *extraction*, where certain components of the rubber vulcanizate are solubilized and removed, actually reducing the rubber volume. The third possible effect is for a liquid to *chemically react* with the rubber itself. This may not cause a volume change; but it can result in a large loss in physical properties. In real world situations, all three effects can happen simultaneously to differing degrees. However, swelling remains the most common result.

ASTM D471 describes how to select standard test liquids, specimen conditioning requirements, test conditions for liquid exposure and to measure gravimetrically (or by dimensional changes) the specimen volume change from exposure to the selected liquid, determine percent soluble components which are extracted, and measure changes in hardness and stress–strain properties. ISO 1817 is somewhat similar to ASTM D471, but there are some differences. ASTM D1460 is specifically designed to measure the change in length of a given rubber specimen from immersion in a specific liquid. This method is sometimes applied to more volatile liquids.

3.13.6   Abrasion and Wear Resistance

Wear resistance is an important rubber compound property related to the useful product life for tires, belts, shoe soles, rubber rolls, and sandblasting hose, among other products. A wide variety of different abrasion testers have been developed over the years in an attempt to correlate to these product wear properties.

The rubber compound properties that affect wear behavior are very complicated. Wear is related to a rubber compound's cut resistance, tear resistance, fatigue resistance, hardness, resiliency, and thermal stability. Abrasion tests try to accelerate the process by applying more cutting-like conditions; however, this approach may not simulate actual wear. It is also important to try to match the severity of the abrasion test to the severity of the product wear conditions. For example, the severity of test conditions imparted by most abraders is usually greater than what the highway pavement may impart to a tire tread compound during normal driving.

Abrasion testers use an abradant to be applied to the surface of a rubber specimen. Tested compounds are usually compared on a "volume loss" basis which is calculated from the weight loss and density of the compound. Abrasion test results are known to be variable; therefore, it is important to control and standardize the abradant used in the test. Also, it is a good idea to relate test results to a standard reference vulcanizate(s).

ASTM D394, the *DuPont* Abrasion Test Method, consists of a pair of rubber test pieces pressed against a disk of a specified abrasive paper. Care should be taken with soft rubber compounds because "smearing" can occur, affecting test results.

ASTM D1630 describes what is commonly referred to as the *NBS* Abrader. This abrader is used on rubber compounds for shoe soles. This "rotating cylinder" test with a specified abrasive paper also calls for a specified standard reference compound to be used for the calculation of an abrasive index. Caution should be taken to assure that abraded compound particles do not smear or clog the abrasive paper.

ASTM D2228 describes the *Pico* Abrader. This unique test works on the principle of abrading the rubber surface by rotating a rubber specimen against a pair of tungsten carbide knives. A special dusting powder is used between the specimen and knives to "engulf" the rubber particles keeping the knives somewhat free of resin, oil, and other contaminants. This method specifies five calibration compounds, and provides their formulations and mixing cycles.

ASTM D3389 refers to the *Taber* Abrader, a method *not* originally from the rubber industry. This very general method uses two abrasive wheels against the rubber test piece (disk) attached to a rotating platform. The amount of slip cannot be changed; however, some other test conditions can be varied.

ISO 4649 refers to the *DIN* Abrader, so called because it is based on the German standard. The rubber test piece with a holder traverses a rotating cylinder covered with the specified abradant paper. By allowing the sample holder to move the test piece across the drum as it rotates, there is less chance of rubber buildup on the abradant paper. This method is used extensively in Europe.

British Standard BS903: Part A9 still describes the *Akron* Abrader, as well as three other abraders. The rubber test piece is a molded wheel which is positioned against an abrasive cylinder under controlled speeds. The Akron Abrader has the advantage of allowing variations in the degree of slip in the test.

References

1. ISO 2781 – Rubber,Vulcanized – Determination of Density, International Standard, 1988.
2. John Sommer, "Physical Properties and Their Meaning," *Rubber World*, August, 1996, p. 17.
3. L. Mullins, *J. of Rubber Research*, 15, 275 (1947).
4. A. N. Gent, "Chapter 1- Rubber Elasticity: Basic Concepts and Behavior," *Science and Technology of Rubber*, F. R. Eirich (ed.), Academic Press, New York, 1978.
5. E. G. Kinnich, *India Rubber World*, 103, 45 (1940).
6. John Sommer, "Physical Properties and Their Meaning" Part 4, *Rubber World*, October, 1996 (Figure 15).
7. A. N. Gent, "Rubber Elasticity: Basic Concepts and Behavior," Chapter 1, *Science and Technology of Rubber*, F. R. Eirich (ed.), Academic Press, New York, 1978.
8. John Sommer, "Physical Properties and Their Meaning," Part 5, *Rubber World*, December, 1996.
9. J. D. Ferry, (1970), *Viscoelastic Properties of Polymers*, Wiley.
10. Meng-Jiao Wang, William J. Patterson, George B. Ouyang, "Dynamic Stress-softening of Filled Vulcanizates," Paper No. 33 presented at the Rubber Div. ACS Meeting, May 5–8, 1996.
11. A. R. Payne, *J. Polymer Sci.*, 6, 57 (1962).
12. L. P. Smith, Chapter 11, Dynamic Stress–Strain, *The Language of Rubber*, Butterworth-Heinemann Ltd., 1993.
13. S. Futamura, "Critical Material Properties for Tire Traction," Presented at Akron Rubber Group, January 28, 1988.
14. Lou Gatti, "Applying Dynamic Mechanical Properties for Tire Development," a paper presented at the ITEC 1996 Meeting in Akron, Sept. 10, 1996 (sponsored by *Rubber and Plastics News.*).
15. L. F. Gatti, "Modeling Tractive Handling with Viscoelastic Properties," presented at the Akron Rubber Group, January 25, 1990.
16. Russell L. Warley, "Dynamic Properties of Elastomers as Related to Vibration Isolator Performance," presented at Rubber Div. ACS Meeting, May 2–5, 1995.
17. Donald S. Novotny, "Dynamic Rubber in Automotive Parts," *Rubber World*, Nov./Dec., 1989.
18. M. J. Gregory, "Dynamic Properties of Rubber in Automotive Engineering," *Elastomerics*, November, 1985, p. 19.
19. R. J. Schaefer, "Dynamic Properties of Rubber," *Rubber World*, January, 1995, p. 16.
20. R. J. Schaefer, "Dynamic Properties of Rubber," *Rubber World*, April, 1995, p. 22.
21. R. J. Schaefer, "Dynamic Properties of Rubber," *Rubber World*, June, 1995, p. 16.
22. R. J. Schaefer, "Dynamic Properties of Rubber," *Rubber World*, April, 1995, p23.

23. R. Schaefer, "Dynamic Properties of Rubber," *Rubber World*, January, 1995, p. 16.
24. L. E. Nielsen, *Mechanical Properties of Polymers*, Rheinhold, New York, 1962, p. 15.
25. L. A. Wood, *J. Polymer Science*, 28 (1958) p. 319.
26. L. P. Smith, Chapter 18, "Low Temperature Properties," *The Language of Rubber*, Butter-worth-Heinemann Ltd., 1993.
27. Lou Gatti, "Applying Dynamic Mechanical Properties for Tire Development," a paper presented at the ITEC 1996 Meeting in Akron, Sept. 10, 1996 (sponsored by *Rubber and Plastics News*.).
28. John D. Ferry, *Viscoelastic Properties of Polymers*, Third Edition, John Wiley & Sons, 1980, p. 15.
29. Roger Brown, *Physical Testing of Rubber*, Chapman & Hall, London, pp. 309–310.
30. R. S. Rivlin, A. G. Thomas, *J. Polym. Sci.*, **10**(3), 291.
31. *Ibid*.
32. M. L. Hill, F. Jowett, *Polymer Testing*, 1980, **1**(4).

4   Rubber Compound Economics

John M. Long

4.1   Introduction

Compound cost is not just the simple calculation of the cost per pound or kilogram of material; it must also take into account whether the application is volume-based or weight-based. The case of volume-based applications includes the concept of cost-volume relationships and calculations.

There is always the temptation to use the lowest cost compound. But one must be constantly vigilant that product quality is not compromised by less expensive ingredients. By the same token, over-design wastes money and can make products non-competitive. Often alternative materials can be used; however, careful evaluation, including appropriate cost calculations, are essential.

4.2   Compound Cost Calculations

Compound cost calculations requires three parameters:

1.   The amount of each ingredient
2.   The cost of each ingredient
3.   The specific gravity of each ingredient

The specific gravity is the specific gravity as it would appear in the polymer. See Section 4.3.

If your system is metric, the equivalent formulae can be used simply by substituting kg-vol for lb-vol in cost/lb-vol calculations. Cost can be any monetary unit as long as it remains the same for all calculations.

4.2.1   Specific Gravity

Specific gravity is calculated as follows:

$$\text{compound specific gravity} = \frac{\sum \left(\frac{\text{weight(PHR)i}}{\text{specific gravity i}} \right)}{\sum \text{weight(PHR)i}} \tag{4.1}$$

where i is each individual ingredient.

4.2.2   Cost/lb

Cost/lb is calculated as follows:

$$\text{compound cost/lb} = \sum \frac{\text{weight(PHR)i} \times \text{cost/lb i}}{\sum \text{weight(PHR)i}} \tag{4.2}$$

The cost of a fixed volume of compound is of particular importance when the end product has a fixed dimension, such as the tread of a specific tire, insulation block, motor mount, or golf ball core.

4.2.3   Lb-Volume Cost

Volume cost is generally calculated by following formula:

$$\text{lb-vol cost} = \text{cost/lb} \times \text{compound specific gravity} \tag{4.3}$$

4.2.4   Part Cost

Calculation of the actual cost of the compound used in a particular part requires information on the volume of compound in the part, the cost/lb of the compound, and the specific gravity of the compound. The formula follows

$$\text{part cost} = \text{compound volume} \times \text{cost/lb} \times \text{compound specific gravity}$$
$$\times \text{conversion factor} \tag{4.4}$$

$$\text{part cost} = \text{compound volume} \times \text{lb-vol. cost} \times \text{conversion factor} \tag{4.5}$$

It is important to note that when comparing costs between two compounds, the volume of the compound must be used, not the number of pounds of one compound or the other.

4.2.5   Conversion Factors for Calculating Part Cost

4.2.5.1 in^3 and cost/lb

When using compound volume in cubic inches and cost/lb the conversion factor is 0.036127 to find the total cost of the part. (This conversion is to put cubic inches in terms of lbs of compound at 1.0 specific gravity.)

4.2.5.2 cm^3 and cost/kg

When using cm^3 and cost/kg, the conversion factor is 0.001 to find the total cost of the part.

4.2.5.3 ft^3 and cost/lb

Should you be using ft^3, the conversion factor is 62.4259.

4.2.5.4 cm^3 and cost/lb

With cm^3 and cost/lb, use the conversion factor 0.0022046 to obtain the total cost of the part.

4.2.5.5   Relative Costs

If only the relative costs are of interest, conversion factors are not necessary. Part costs can be expressed as an index or percent of one versus another.

4.2.5.6   Developing Conversion Factors

The conversion factors were obtained as follows:

One kg of a substance with a 1.0 specific gravity occupies $1000\,cm^3$. The reciprocal would be 0.001, or

$$1 \bigg/ \frac{1000\,cm^3}{1\,(kg)} = 0.001\,kg/cm^3$$

One pound of a substance with a 1.0 specific gravity occupies $453.597\,cm^3$.

$$\frac{1\,liter\,(1000\,cm^3)\,or\,1\,kg}{\left(\dfrac{2.2046\,lbs}{kg}\right)} = 453.597\,cm^3/lb$$

The reciprocal then yields

$$\frac{1}{453.597} = 0.0022046\,lbs/cm^3$$

Likewise,

$$\frac{453.597\,cm^3/lb}{16.38706\,cm^3/in^3} = 27.680195\,in^3/lb$$

This reciprocal then yields

$$\frac{1}{27.680195\,in^3/lb} = 0.036127\,lb/in^3$$

In the case of ft^3, dividing in^3/lb^3 by ft^3/in^3 yields

$$\frac{27.680195\,in^3/lb}{1728\,in^3/ft^3} = 0.016019\,ft^3/lb$$

This reciprocal then yields

$$\frac{1}{0.016019\,ft^3/lb} = 62.4259\,lbs/ft^3$$

4.3   Measuring Specific Gravity (Density)

Before looking at practical examples, we need to understand the calculation of specific gravity as it relates to rubber compounds. In conventional measurements of solid materials, the material is put in a pycnometer and a suitable liquid introduced. The calculation is then made considering the weight of the material introduced and the specific gravity of the liquid. See ASTM D 1817 for details on performing this test.

In rubber, the incorporation (or voids present) may be more or less than with the conventional measurement. Although the two methods do not often yield significant differences, with high volume materials the impact on compound costs could be important. To determine the specific gravity of a component in rubber, place a known quantity (usually substantial, 50 PHR or more) of the material into a rubber mix that cures and calculate the specific gravity compared to the base compound. Obviously, some materials cannot be incorporated into the compound at these volumes without severely affecting the final compound. In those cases, a lesser quantity must be used; however, the accuracy of the determination may be reduced. The specific gravity is most often determined by weighing the cured sample in air and then immersing it in water. The calculation is made using the density of the water at the temperature during the test. This test must be done on a cured compound.

4.4   Cost Calculations

Having examined methods of determining densities, it is important to see how this impacts the costs of compounds used in applications.

4.4.1   Base Compound

Take, as an example, a rubber compound such as Model Compound I (Table 4.1), which includes the costs and densities as listed in Appendix A (Table 4.16). Examine the impact

Table 4.1   Model Compound I

Ingredient	PHR	Price/lb	Sp. Gr.	Volume	Price
SBR 1500	100.000	0.7200	0.94	106.38	72.00
N660	50.000	0.3275	1.80	27.78	16.38
Aromatic oil	15.000	0.0838	1.00	15.00	1.26
Zinc oxide	3.000	0.6000	5.57	0.54	1.80
Stearic acid	2.000	0.5100	0.85	2.35	1.02
TBBS	1.500	3.2900	1.28	1.17	4.94
Oil treated sulfur	2.000	0.1590	2.00	1.00	0.32
Total	173.500			154.22	97.71
Specific gravity	1.125				
Cost/lb	$0.563				
Cost/lb-vol.	$0.634				

of varying the fillers and other ingredients on the actual costs of the compound both in terms of equivalent volumes and costs where only the cost per pound is involved.

4.4.2   Same Ingredient Volume and Equal Cost

If you are manufacturing a part in a fixed volume mold that requires 1 lb of Model Compound I, the materials cost is $0.563. Substituting silica and treated clay for the carbon black, to yield equal volumes of these filler ingredients and the same cost/lb, leads to Model Compound II (Table 4.2). This is for purposes of illustrating costs only.

Table 4.2   Model Compound II

Ingredient	PHR	Price/lb	Sp. Gr.	Volume	Price
SBR 1500	100.000	0.7200	0.94	106.38	72.00
Silica	32.000	0.5950	2.00	16.00	19.04
Treated clay	31.000	0.1470	2.62	11.83	4.56
Aromatic oil	15.000	0.0838	1.00	15.00	1.26
Zinc oxide	3.000	0.6000	5.57	0.54	1.80
Stearic acid	2.000	0.5100	0.85	2.35	1.02
TBBS	1.500	3.2900	1.28	1.17	4.94
Oil treated sulfur	2.000	0.1590	2.00	1.00	0.32
Total	186.500			154.28	104.93
Specific gravity	1.209				
Cost/lb	$0.563				
Cost/lb-vol.	$0.680				

In this illustration, a cost penalty of $0.041/part is incurred, even though the cost/lb is the same as with Model Compound I.

$$\text{part cost} = (\text{compound volume} = 1/1.125 = 0.889)$$

$$\times \text{ [lb vol cost (Model Compound I} - \text{Model Compound II)}$$

$$= (0.634 - 0.680)]$$

$$= -0.041$$

(4.6)

If one were pricing a product strictly in terms of weight, high loadings of an inexpensive, high specific gravity filler would be advantageous, such as seen in Model Compound III (Table 4.3).

4.4.3   Low Cost/lb

As can be seen from Table 4.3, the cost/lb is about one-half that of Model Compound I (Table 4.1), but the lb-vol. cost is closer to three quarters that of Model Compound I. The physical properties of Compound III are significantly inferior to those of Model Compound I; however, if the inferior properties are not important and the compound is priced strictly based on cost/lb, then the lb-vol. cost is not a factor. As noted, the cost/lb-vol. is lower for

Table 4.3   Model Compound III

Ingredient	PHR	Price/lb	Sp. Gr.	Volume	Price
SBR 1500	100.000	0.7200	0.94	106.38	72.00
Hard clay	200.000	0.0590	2.62	76.34	11.80
Aromatic oil	15.000	0.0838	1.00	15.00	1.26
Zinc oxide	3.000	0.6000	5.57	0.54	1.80
Stearic acid	2.000	0.5100	0.85	2.35	1.02
TBBS	1.500	3.2900	1.28	1.17	4.94
Oil treated sulfur	2.000	0.1590	2.00	1.00	0.32
Total	323.500			202.78	93.13
Specific gravity	1.595				
Cost/lb	$0.288				
Cost/lb-vol.	$0.459				

Model Compound III. However, in the Model Compound II example, where the cost/lb-vol. is higher than Model Compound I, there was some attempt at maintaining reasonable physical properties.

4.4.4   High Specific Gravity

In some cases where it is mandatory that a high specific gravity compound be used, such as in aircraft wheel chocks, the high weight is essential for product performance. The key is to use the most cost-effective filler system to obtain high specific gravity compounds consistent with other physical property restraints. Model Compounds IV (Table 4.4) and V (Table 4.5) illustrate two methods of obtaining the same specific gravity compound, but Model Compound IV is more cost-effective.

These models are illustrations of cost effectiveness, not of real world compounds. One obvious material that increases the compound's specific gravity is Litharge (PbO) which has a specific gravity of 9.35. Examples of this material are not included because of the

Table 4.4   Model Compound IV

Ingredient	PHR	Price/lb	Sp. Gr.	Volume	Price
SBR 1500	100.000	0.7200	0.94	106.38	72.00
Barytes	200.000	0.1270	4.45	44.94	25.40
Aromatic oil	15.000	0.0838	1.00	15.00	1.26
Zinc oxide	3.000	0.6000	5.57	0.54	1.80
Stearic acid	2.000	0.5100	0.85	2.35	1.02
TBBS	1.500	3.2900	1.28	1.17	4.94
Oil treated sulfur	2.000	0.1590	2.00	1.00	0.32
Total	323.500			171.39	106.73
Specific gravity	1.888				
Cost/lb	$0.330				
Cost/lb-vol.	$0.623				

Table 4.5   Model Compound V

Ingredient	PHR	Price/lb	Sp. Gr.	Volume	Price
SBR 1500	100.000	0.7200	0.94	106.38	72.00
Zinc oxide	177.320	0.6000	5.57	31.83	106.39
Aromatic oil	15.000	0.0838	1.00	15.00	1.26
Stearic acid	2.000	0.5100	0.85	2.35	1.02
TBBS	1.500	3.2900	1.28	1.17	4.94
Oil treated sulfur	2.000	0.1590	2.00	1.00	0.32
Total	297.820			157.74	185.92
Specific gravity	1.888				
Cost/lb	$0.624				
Cost/lb-vol.	$1.179				

strict regulatory restrictions placed on the use of lead in manufacturing and environmental considerations.

4.5   Compound Design and Cost

There is a natural tendency to over-design a product just to be "on the safe side". While it is necessary and very important to design a product with sufficient safety margins, one must be realistic. Setting unrealistic safety margins limits the choices of filler and polymer that can be used with the potential of much higher compound costs. It makes no sense to design a component product to last 100,000 hours, when the main product is designed to last for only 1,000 hours. There is also a tendency to specify particular physical characteristics as engineering requirements that are not important to product performance, but may be considered in the industry as measures of "quality." Tensile strength can be one of those properties, for example.

In the opposite context, for those products where failure results in severe economic consequences or potential loss of life or injury, efforts must be focused on zero failures. For example, if the failure of a part of a larger system results in the equipment shut down, loss of productivity, and additional labor costs to effect the repair, increasing safety margins in the part simply makes sense. When worker safety is involved, costs must take a back seat to assuring product performance. When safety is assured, there may be opportunities to reduce product costs, but they can be implemented only after thorough testing.

4.6   Reducing Compound Cost

This section will examine the manipulation of specific compounding ingredients, and their cost implications.

4.6.1   High-Structure Carbon Blacks

The use of higher structure carbon blacks with increased oil to maintain 300% modulus levels decreases cost/lb and cost/lb-vol by taking advantage of these carbon blacks' ability to increase modulus. Subsequently the addition of oil reduces the modulus to the original level. However, other properties that may be important to the product's performance may be compromised.

4.6.2   White Compounds

In compounds where high levels of titanium dioxide are used for maximum whiteness, it may be possible to substitute light colored clays, light colored treated clays, calcium carbonate, and optical whiteners, among others, for some of the pigment. The selection and quantities of the substitutes will depend on several factors including the required whiteness level, physical properties (tensile, modulus, hardness), grindability, and fatigue resistance, among others.

4.6.3   Antioxidants/Antiozonants

Antioxidants/antiozonants are another class of materials that can significantly impact costs. There are numerous examples of synergy between two or more ingredients that are not only more effective than a single ingredient, but reduce costs as well. One example which might work is substituting 4 PHR 6PPD with 2.5 PHR 6PPD, 2 PHR TMQ, and 1 PHR wax. Tables 4.6 and 4.7 (Model Compounds VI and VII) reveal a lb-vol. cost reduction of $ 0.013 under these circumstances. While this may seem a very small saving,

Table 4.6   Model Compound VI

Ingredient	PHR	Price/lb	Sp. Gr.	Volume	Price
SBR 1500	100.000	0.7200	0.94	106.38	72.00
N660	50.000	0.3275	1.80	27.78	16.38
Aromatic oil	15.000	0.0838	1.00	15.00	1.26
Zinc oxide	3.000	0.6000	5.57	0.54	1.80
Stearic acid	2.000	0.5100	0.85	2.35	1.02
6PPD	4.000	4.3800	0.99	4.04	17.52
TMQ	0.000	2.4700	1.06	0.00	0.00
Wax	0.000	0.5800	0.90	0.00	0.00
TBBS	1.500	3.2900	1.28	1.17	4.94
Oil treated sulfur	2.000	0.1590	2.00	1.00	0.32
Total	177.500			158.26	115.23
Specific gravity	1.122				
Cost/lb	$0.649				
Cost/lb-vol.	$0.728				

Table 4.7   Model Compound VII

Ingredient	PHR	Price/lb	Sp. Gr.	Volume	Price
SBR 1500	100.000	0.7200	0.94	106.38	72.00
N660	50.000	0.3275	1.80	27.78	16.38
Aromatic oil	15.000	0.0838	1.00	15.00	1.26
Zinc oxide	3.000	0.6000	5.57	0.54	1.80
Stearic acid	2.000	0.5100	0.85	2.35	1.02
6PPD	2.500	4.3800	0.99	2.53	10.95
TMQ	2.000	2.4700	1.06	1.89	4.94
Wax	1.000	0.5800	0.90	1.11	0.58
TBBS	1.500	3.2900	1.28	1.17	4.94
Oil treated sulfur	2.000	0.1590	2.00	1.00	0.32
Total	179.000			159.75	114.18
Specific gravity	1.121				
Cost/lb	$0.638				
Cost/lb-vol.	$0.715				

if 10,000,000 lbs of the original compound are produced per year, the total savings amount to $115,864.

$$
\text{Savings} = \left(\text{compound volume} = \frac{10,000,000}{1.122} = 8{,}912{,}656 \right)
$$

$$
\times \ [\text{cost/lb-vol. (Model Compound VI} - \text{Model Compound VII)} \quad (4.7)
$$

$$
= (0.728 - 0.715) = 0.013]
$$

$$
= 115{,}864
$$

4.6.4   Polymer Substitutions

Polymers are often the largest constituent of a recipe, providing an opportunity for reducing compound costs, including lb-vol. costs.

There are three major areas of polymer substitutions. First, high specific gravity polymers or high cost polymers can be replaced. Second, non-oil-extended polymers can be replaced by oil-extended types; third, free carbon black/oil mixed compounds can be changed to carbon black/oil masterbatches.

4.6.4.1   High Cost/High Specific Gravity Polymers

Replacing chloroprene with NBR eliminates not only a high specific gravity material, but a higher cost one. Both the cost/lb and the cost/lb-vol. are reduced, as seen in Model Compounds VIII (Table 4.8) and IX (Table 4.9).

The cost savings per part, assuming 1 lb of Model Compound VIII (Table 4.8), is $0.551.

Table 4.8   Model VIII

Ingredient	PHR	Price/lb	Sp. Gr.	Volume	Price
Chloroprene W type	100.000	1.9100	1.23	81.30	191.00
N762	50.000	0.3275	1.80	27.78	16.38
Zinc oxide	5.000	0.6000	5.57	0.90	3.00
MgO	4.000	0.8500	3.20	1.25	3.40
Stearic acid	2.000	0.5100	0.85	2.35	1.02
TMTD	1.000	2.0500	1.42	0.70	2.05
TMTM	1.000	3.7000	1.37	0.73	3.70
Oil treated sulfur	1.000	0.1590	2.00	0.50	0.16
Total	164.000			115.51	220.70
Specific gravity	1.420				
Cost/lb	$1.346				
Cost/lb-vol.	$1.911				

$$\text{part cost} = (\text{compound volume} = 1/1.420 = 0.704)$$
$$\times \left[\text{cost/lb-vol. (Model Compound VIII} - \text{Model Compound IX)}\right.$$
$$= (1.911 - 1.129) = 0.782\right] \tag{4.8}$$
$$= 0.551$$

If the cost of the chloroprene were equal to the price of the NBR, which means the only difference between the two materials is specific gravity, the savings is still $0.148.

$$\text{part cost} = (\text{compound volume} = 1/1.420 = 0.704)$$
$$\times \left[\text{cost/lb-vol. (Model Compound VIII(chloroprene cost modified)}\right.$$
$$- \text{Model Compound IX)} = (1.339 - 1.129) = 0.210\right] \tag{4.9}$$
$$= 0.148$$

Again, these compounds are used for illustration of the cost effect of replacing polymers and are not intended to describe real world compounds. The cure systems and properties will be different.

Table 4.9   Model IX

Ingredient	PHR	Price/lb	Sp. Gr.	Volume	Price
NBR (35% ACN)	100.000	1.2500	0.99	101.01	125.00
N762	50.000	0.3275	1.80	27.78	16.38
Zinc oxide	5.000	0.6000	5.57	0.90	3.00
Stearic acid	2.000	0.5100	0.85	2.35	1.02
TMTD	1.000	2.0500	1.42	0.70	2.05
TMTM	1.000	3.7000	1.37	0.73	3.70
Oil treated sulfur	1.000	0.1590	2.00	0.50	0.16
Total	160.000			133.97	151.30
Specific gravity	1.194				
Cost/lb	$0.946				
Cost/lb-vol.	$1.129				

Table 4.10   Model Compound X

Ingredient	PHR	Price/lb	Sp. Gr.	Volume	Price
SBR 1500	100.000	0.7200	0.94	106.38	72.00
N234	75.000	0.4425	1.80	41.67	33.19
Aromatic oil	30.000	0.0838	1.00	30.00	2.51
Zinc oxide	3.000	0.6000	5.57	0.54	1.80
Stearic acid	2.000	0.5100	0.85	2.35	1.02
TBBS	1.500	3.2900	1.28	1.17	4.94
Oil treated sulfur	2.000	0.1590	2.00	1.00	0.32
Total	213.500			183.11	115.78
Specific gravity	1.166				
Cost/lb	$0.542				
Cost/lb-vol.	$0.632				

4.6.4.2   Clear and Oil-Extended Polymer Replacements

The replacement of clear polymers with oil-extended grades usually permits the use of higher loadings and oil content. Model Compounds X (Table 4.10) and XI (Table 4.11) examine only a change in oil content. Improvements in processing will also be evident.

As seen in these examples, a modest decrease in cost can be obtained by using oil-extended polymers instead of clear polymers and free oil. In this case, a part utilizing 1 lb of Model Compound X has the following cost advantage if Model Compound XI were substituted.

$$\text{part cost} = (\text{compound volume} = 1/1.166 = 0.858)$$

$$\times \; [\text{lb-vol. cost (Model Compound X} \; - \text{Model Compound XI)}$$

$$= (0.632 - 0.623) = 0.009]$$

$$= 0.008$$

(4.10)

Table 4.11   Model Compound XI

Ingredient	PHR	Price/lb	Sp. Gr.	Volume	Price
SBR 1712	137.500	0.5600	0.96	143.23	77.00
N234	75.000	0.4425	1.80	41.67	33.19
Aromatic oil	0.000	0.0838	1.00	0.00	0.00
Zinc oxide	3.000	0.6000	5.57	0.54	1.80
Stearic acid	2.000	0.5100	0.85	2.35	1.02
TBBS	1.500	3.2900	1.28	1.17	4.94
Oil treated sulfur	2.000	0.1590	2.00	1.00	0.32
Total	221.000			189.96	118.26
Specific gravity	1.163				
Cost/lb	$0.535				
Cost/lb-vol.	$0.623				

Table 4.12   Model Compound XII

Ingredient	PHR	Price/lb	Sp. Gr.	Volume	Price
SBR 1500	100.000	0.7200	0.94	106.38	72.00
N330	52.000	0.3750	1.80	28.89	19.50
Aromatic oil	10.000	0.0838	1.00	10.00	0.84
Zinc oxide	3.000	0.6000	5.57	0.54	1.80
Stearic acid	2.000	0.5100	0.85	2.35	1.02
TBBS	1.500	3.2900	1.28	1.17	4.94
Oil treated sulfur	2.000	0.1590	2.00	1.00	0.32
Total	170.500			150.34	100.41
Specific gravity	1.134				
Cost/lb	$0.589				
Cost/lb-vol.	$0.668				

4.6.4.3   Carbon Black/Oil Masterbatches Replacing Free Mix Compounds

The next example demonstrates the use of carbon black/oil masterbatches to replace free black mixed compounds. SBR 1606 is a masterbatch that contains 100 PHR of SBR, 52 PHR N330 carbon black, and 10 PHR aromatic oil. Tables 4.12 and 4.13 (Model Compounds XII and XIII) compare these costs.

In this example it appears that there is a cost penalty for using a carbon black/oil masterbatch of $0.035, based on a part using 1 lb of Model Compound XII.

$$\text{part cost} = [\text{compound volume}$$

$$= (1 \text{ lb } / \text{ specific gravity Model Compound XII}) = 1/1.134 = 0.882]$$

$$\times [\text{cost/lb-vol. (Model Compound XII} - \text{Model Compound XIII)} \quad (4.11)$$

$$= (0.668 - 0.708) = -0.040]$$

$$= -0.035$$

However, there is more to the story. Previously, we discussed only materials costs. Other costs relate to processing the compound from mixing, through calendering or extrusion,

Table 4.13   Model Compound XIII

Ingredient	PHR	Price/lb	Sp. Gr.	Volume	Price
SBR 1606	162.000	0.6068	1.12	145.29	98.31
Zinc oxide	3.000	0.6000	5.57	0.54	1.80
Stearic acid	2.000	0.5100	0.85	2.35	1.02
TBBS	1.500	3.2900	1.28	1.17	4.94
Oil treated sulfur	2.000	0.1590	2.00	1.00	0.32
Total	170.500			150.35	106.38
Specific gravity	1.134				
Cost/lb	$0.624				
Cost/lb-vol.	$0.708				

and finally vulcanization. In this example, mixing costs must be included to compare the total cost of these two formulations.

Assume that Model Compound XII needs to be mixed in two stages (the polymer, carbon black, oil, antioxidants, zinc oxide, and stearic acid in the first stage with the curatives added in the second stage). This two-stage process means added mixing costs. For the first stage, assume a mixer operating cost of about \$450/hour and a total mixing cycle time of about 4 min. for a 450 lb batch. The mixing costs for Model Compound XII then are as follows.

$$\frac{4\,\text{min}}{\left(\dfrac{60\,\text{min}}{h}\right)} = 0.067\,\text{h}; \quad \frac{0.067\,\text{h}}{450\,\text{lbs}} = 0.000148\,\text{h/lbs} \tag{4.12}$$

Multiplying the mixing time per lb and the mixing costs per hour, yields the mix cost per pound of \$0.067.

$$\frac{0.000148\,\text{h}}{\text{lb}} \times \frac{450}{\text{h}} = \frac{0.067}{\text{lb}} \tag{4.13}$$

Because the black/oil masterbatch compound can be mixed in one step, this first stage is completely eliminated, saving the \$0.067/lb of finished compound cost. So now, instead of incurring a cost penalty of \$0.035/lb, an actual savings of \$0.032/lb is realized.

$$\text{Mixing costs of } \frac{0.067}{\text{lb}} - \text{materials cost penalty of } \frac{0.035}{\text{lb}} = \frac{0.032}{\text{lb}} \text{ savings} \tag{4.14}$$

Besides the savings in mixing cost, there is the consideration of increased mixing capacity. By eliminating the first longer mixing stage, two additional final or one stage mixes can be accomplished, effectively increasing output by three times. This may be particularly important if the factory has limited mixing capacity. The cost calculations for Model Compounds XII (Table 4.12) and XIII (Table 4.13) can be done directly with the cost/lb because both compounds have the same specific gravity.

4.6.4.4   Extrusion Productivity

A relatively expensive processing aid such as that in Model Compound XIV (Table 4.14) increases the cost of the compound compared to Model Compound I (Table 4.1) and, on first glance, it would appear that an economic loss was incurred. However, if the processing aid improves extrusion output by 10%, the increased cost is more than offset. Assuming an extrusion operational cost of \$300/hour and an initial throughput of 10,000 lbs/hour, the following calculation can be made.

$$\begin{aligned} \text{part cost} &= (\text{compound volume} = \text{Model Compound I/specific gravity Model} \\ &\quad \text{Compound I} = 1/1.125 = 0.889) \\ &\quad \times [\text{lb-vol. cost difference} = (\text{Model Compound I} - \text{Model} \\ &\quad \text{Compound XIV}) = (0.634 - 0.635) = -0.001] \\ &= -0.001 \end{aligned} \tag{4.15}$$

Table 4.14   Model Compound XIV

Ingredient	PHR	Price/lb	Sp. Gr.	Volume	Price
SBR 1500	100.000	0.7200	0.94	106.38	72.00
N660	50.000	0.3275	1.80	27.78	16.38
Aromatic oil	15.000	0.0838	1.00	15.00	1.26
Processing aid	3.000	0.7000	1.00	3.00	2.10
Zinc oxide	3.000	0.6000	5.57	0.54	1.80
Stearic acid	2.000	0.5100	0.85	2.35	1.02
TBBS	1.500	3.2900	1.28	1.17	4.94
Oil treated sulfur	2.000	0.1590	2.00	1.00	0.32
Total	176.500			157.22	99.81
Specific gravity	1.123				
Cost/lb	$0.565				
Cost/lb-vol.	$0.635				

$$\text{extrusion cost savings} = \left[\text{extrusion costs Model Compound I} \right.$$

$$= \left(\frac{300}{\text{hour}} \right) \bigg/ \left(\frac{10{,}000}{\text{hour}} \right) = 0.030/\text{lb} \bigg]$$

$$- \left[\text{extrusion costs Model Compound XIV} \right. \tag{4.16}$$

$$= \left(\frac{300}{\text{hour}} \right) \bigg/ \left(\frac{11{,}000}{\text{hour}} \right) = 0.027/\text{lb} \bigg]$$

$$= 0.003/\text{lb}$$

If the compound cost penalty is combined with the extrusion cost savings, the result is a positive $0.002/lb. This number is small, but if the extruder operates 20 hours/day and 240 days/year, the throughput of Model Compound I (Table 4.1) is 48,000,000 lbs/year. A savings of $0.002/lb in the production of Model Compound I (Table 4.1) add up to $96,000/year under the processing conditions. In addition, output is increased by 4,800,000 lbs/year over levels prior to the addition of the processing aid.

4.6.4.5   Vulcanization Productivity

Secondary accelerators can increase compound cure rate and subsequently reduce part cure time. In Model Compound XV (Table 4.15) TMTM is added as a secondary accelerator and Model Compound I (Table 4.1) is the base. This addition reduces the cure time of a part from 10 min. at 347 °F (175 °C) to 9 min. at the same temperature.

In this example, the materials cost is higher for Model Compound XV compared to Model Compound I by $0.006/lb. Because the specific gravity is the same, the lb-vol. cost calculation is not needed. Assuming that each part is 1 lb, curing press costs are $50/hour, and 12 parts are produced per press cycle, the output for Model Compound I (Table 4.1) is 72 parts/hour with associated costs of $0.694/part. With Model Compound XV

Table 4.15   Model Compound XV

Ingredient	PHR	Price/lb	Sp. Gr.	Volume	Price
SBR 1500	100.000	0.7200	0.94	106.38	72.00
N660	50.000	0.3275	1.80	27.78	16.38
Aromatic oil	15.000	0.0838	1.00	15.00	1.26
Zinc oxide	3.000	0.6000	5.57	0.54	1.80
Stearic acid	2.000	0.5100	0.85	2.35	1.02
TBBS	1.500	3.2900	1.28	1.17	4.94
TMTM	0.350	3.7000	1.37	0.26	1.30
Oil treated sulfur	2.000	0.1590	2.00	1.00	0.32
Total	173.850			154.48	99.00
Specific gravity	1.125				
Cost/lb	$0.569				
Cost/lb-vol.	$0.641				

(Table 4.15), the output is 80 parts/hour and associated costs are $0.625/part. The savings in curing costs are $0.069. Reducing that savings by the increased compound cost of $0.006/lb yields net savings of $0.063/part. Output also increases by eight parts/hour, concurrent with the decrease in costs per part.

These are just a few examples of the substitutions and manipulations that can lead to savings in costs and increases in extrusion efficiencies. It is obviously important to examine all the ingredients in a formulation for possible substitution by more cost effective ingredients. Suppliers are usually very willing to assist compounders to find more cost effective solutions. Whenever making substitutions, whether of your own design or those suggested by a supplier, you should make sure to thoroughly test the resultant product to verify that unexpected, detrimental attributes have not been introduced. You probably know more about your product than anyone else so follow your instincts.

Appendix

Table 4.16   Cost and Densities of Common Materials (Circa 1997)

		US$/lb (1997)	Sp. Gr.
Primary accelerators			
MBTS	lb	2.3600	1.51
CBS	lb	3.2800	1.28
TBBS	lb	3.2900	1.28
MBS	lb	3.2900	1.37
OTOS	lb	3.4500	1.34
Secondary accelerators			
ZnDBC	lb	2.1900	1.21
MBT	lb	2.1900	1.52
TMTD	lb	2.0500	1.42
TMTM	lb	3.7000	1.37
DOTG	lb	4.0800	1.20
DPG	lb	3.5100	1.20
Sulfur donor			
DTDM	lb	4.4200	1.35
MBSS	lb	4.2500	1.51
Vulcanization activators			
Zinc oxide	lb	0.6000	5.57
Stearic acid	lb	0.5100	0.85
MgO	lb	0.8500	3.20
Antioxidant/antiozonants			
DPPD	lb	5.8900	1.28
TMQ	lb	2.4700	1.06
ODPA	lb	2.5300	0.99
6PPD	lb	4.3800	0.99
DMHPD	lb	5.2700	0.90
DTPD	lb	3.4800	1.18
Wax	lb	0.5800	0.90
Carbon blacks			
N234	lb	0.4425	1.80
N330	lb	0.3750	1.80
N339	lb	0.3750	1.80
N660	lb	0.3275	1.80
N762	lb	0.3275	1.80
N991	lb	0.4750	1.80
Fillers			
Titanium dioxide	lb	1.0200	3.88
Silica	lb	0.5950	2.00
Calcium carbonate	lb	0.0143	2.70
Hard clay	lb	0.0590	2.62
Barytes	lb	0.1270	4.45
Treated clay	lb	0.1470	2.62
Treated clay (light color)	lb	0.1605	2.62
Talc	lb	0.0485	2.85

Table 4.16 *Continued*

		US$/lb (1997)	Sp. Gr.
Ground coal	lb	0.0900	1.22
Mineral rubber	lb	0.1700	1.04
Water washed clay (light color)	lb	0.1575	2.62
Oil treated sulfur	lb	0.1590	2.00
Insoluble sulfur 20%OT	lb	0.7300	1.64[*]
Oil			
Aromatic	lb	0.0838	1.00[*]
Naphthenic	lb	0.1302	0.92[*]
Paraffinic	lb	0.1761	0.85[*]
Polymers			
SBR 1500	lb	0.7200	0.94[*]
SBR 1502	lb	0.7200	0.94[*]
SBR 1712	lb	0.5600	0.96[*]
SBR 1778	lb	0.6161	0.96[*]
SBR 1848	lb	0.5232	1.14[*]
SBR 1805	lb	0.5628	1.13[*]
SBR 1606	lb	0.6068	1.13[*]
Natural rubber	lb	0.7500	0.92
Chlorobutyl	lb	1.2250	0.92[*]
Solution SBR	lb	0.9800	0.94
NBR (35% ACN)	lb	1.2500	0.99[*]
Chloroprene	lb	1.9100	1.23[*]

[*] Cost determined by an independent survey of producers.
Costs courtesy of *Rubber World* except where asterisked.
Descriptions have been modified to the generic name or Abbreviated according to ASTM D-3853 where possible. This is an abstract of a full list of ingredient costs published in *Rubber World*.

5 The Technical Project Approach to Experimental Design and Compound Development

Alan G. Veith

5.1 Introduction

Rubber technology has two sources of complexity:

1. New materials and new processing methods continually appear on the technical horizon
2. Properties of rubber compounds depend in a complex way on the materials in the compound in addition to the processing operations.

These two sources of complication, combined with the intrinsic variation encountered in any performance testing, often make compound development a daunting task. Successful compound development requires a systematic and efficient style of experimentation embodied in what may be called a technical project, defined as an operation or process organized to solve a problem, answer a number of specific questions, or provide data to make technical decisions. A key feature of any technical project is the use of statistical experimental design.

Experimental design and statistical analysis can be conducted by a professional statistician working in conjunction with a technologist or by the technologist acting alone. Progress is greatest when the person planning the execution of a program is familiar with basic scientific concepts and testing procedures as well as the advantages and disadvantages of alternative options. Technologists have these capabilities and if they are knowledgeable about the basics of experimental design, the result is efficient problem solving. This chapter is devoted to assisting technologists to develop a rational approach to problem solving using experimental design. It assumes that elementary statistical concepts (mean, variance, tests of significance) are understood. Once the basic experimental design concepts have been mastered, technologists can address all but the most complex projects. If additional assistance is needed for advanced projects, technologists with solid experimental design experience are much better equipped to work with a professional statistician.

Information on experimental designs may be obtained from numerous texts or from dedicated computer programs that select experimental designs based on some initial input. Although the actual selection of a design is an important step in the solution of rubber technology problems, other equally important issues must be considered. Experimental designs must be supplemented by a number of additional procedures or steps that are part of a systems approach. Table 5.1 lists the steps in a technical project which may take one

Table 5.1   Steps in a Technical Project

Stage 1
1. Initial actions required
  1.1 Develop planning model
  1.2 Prepare work, time, and cost proposal
2. Experimental design
  2.1 Select potential or known factors or variables
    i. Select test instruments and procedures
  2.2 Propose a response model
  2.3 Select experimental design
    i. Screening or exploratory design
      (design must be consistent with model)
    ii. Select experimental design values
    iii. Obtain preliminary estimate of measurement uncertainty
    iv. Set up sampling procedure
    v. Set up replication and randomization procedures
3. Conduct measurements and obtain data
4. Conduct analysis and evaluate preliminary model
  4.1 Estimate factor effects or effect coefficients
  4.2 Estimate test error
  4.3 Determine significance of factor effects or model components
5. Prepare report (preliminary or final)

If model is not satisfactory and problem is not solved – GO TO STAGE 2

Stage 2
1. Re-define the technical project
  1.1 Revised work proposal
  1.2 Develop additional time schedule and estimated costs
2. New or augmented experimental design
  2.1 Develop revised or new model
    i. Select variables (original+new, entirely new)
    ii. If needed, select additional test instruments and procedures
  2.2 Select new experimental design
    i. New screening or exploratory design
      (design must be consistent with model)
    ii. Select design values consistent with previous values
    iii. Obtain new estimate of test error
    iv. Set up new sampling procedure
    v. Set up new replication and randomization procedures
3. Conduct additional measurements and obtain data
4. Conduct analysis and evaluate preliminary model
  4.1 Estimate factor effects or effect coefficients
  4.2 Estimate test error or uncertainty
  4.3 Determine significance of factor effects or model components
5. Prepare report

Repeat Stage 2 if required, for satisfactory solution

or more stages or iterations to reach a satisfactory conclusion. This chapter is divided into two parts:

Part 1: The Steps in a Technical Project is a discussion of the five steps in Stage 1 and 2 sections of Table 5.1. Elementary concepts of experimental design are provided here. After this section, the reader is better equipped to select appropriate experimental designs.

Part 2:   The Use of Experimental Designs is a more extended discussion, involving concepts not defined in Part 1, the analysis algorithms, and the individual designs listed in Appendix 1.

5.2   Part 1: Steps in a Technical Project

5.2.1   Initial Action Required

The technical program and models required to define the system are usually developed by a principle investigator in conjunction with specialists in allied technical disciplines. Topics that require attention are project objectives, resources and constraints, and decision procedures, as well as selecting the measurement methodology. A set of well designed and coordinated standard operating procedures (SOPs) must be specified. There are several different types of models: planning models, statistical measurement models, and direct response models.

5.2.1.1   Planning Model

The planning model is a generalized statement of the strategy to be used to solve a problem. It involves the selection of a set of SOPs and the required system elements to arrive at a solution. Planning models are descriptive in nature and not as rigorous as response and analysis models. For technical programs that emphasize performance, performance criteria must be clearly stated.

5.2.1.2   Work, Time, and Cost Proposal

Every technical project or program requires a work statement that includes the projected time and total cost needed to arrive at some set of conclusions. If unforeseen difficulties arise, conclusions may be tentative and further action may be required for a complete solution. Preliminary plans should address this highly probable outcome.

5.2.2   Experimental Design

5.2.2.1   Selecting Variables or Factors

A very important step is the selection of the factors or variables for the system under investigation. The system is defined as all elements that constitute any process and/or the performance of any compound or material. There are two variable categories:

1. the measured or observed variables called dependent variables or response variables

2. the independent variables or factors that describe the state of the system. The purpose of the experimental design is to specify the value or level for these independent variables or factors as well as selecting the appropriate response.

Independent variables may be continuous quantitative variables, such as percent concentration, temperature, pH, etc., or they may be discrete or categorical variables, usually at two values or levels, such as the presence or absence of a particular component or condition. Independent variables are usually selected at some specific levels, although some technical programs may require accepting whatever levels are available, such as those generated by the environment or not under control of the experimenter. All the experience and background of the technologist must be employed when selecting factors; these choices should be based on the technical relevance and the ease of setting that factor at certain specified levels. Prior experience, implicit assumptions, and underlying operational concepts of the system must be clearly understood.

5.2.2.2   Selecting Test Instruments and Procedures

The decision on responses usually dictates what test instruments and test conditions are used. Measurement and other operational conditions required to evaluate properties or performance are defined as a testing domain. Although the results of testing may be intended to have general applicability, they may apply only to the particular testing domain used. Thus if broad application is desired, standardized testing procedures and instruments that have demonstrated acceptance in the industry should be selected.

5.2.2.3   Developing a Response Model

The response model is a mathematical formulation or equation that describes a system with regard to response. There are two categories: empirical models, which use simple mathematical formulations that are linear in the independent variable terms of the model; and analytical or mechanistic models based on theoretical principles. Analytical models may be linear in the independent variables, but are more frequently non-linear, with multiplicative terms, and exponents. A second type of non-linearity may exist between the response and some or all of the independent variables. This behavior can be accommodated in a model linear in the independent variables by the use of squared (or higher power) independent variable terms. Only empirical models are considered in this chapter.

In attempting to develop a response model, the experimenter is usually faced with one of two situations:

1. The important independent variables are unknown or some variables may be suspected of being important.
2. The important independent variables are known, either from a previous investigation or from theoretical principles, but their functional relationship is not known.

For a system as described in (1), the purpose of experimental design is to discover and sort out the factors into a hierarchy of importance and give an approximate model. Once a group of factors are known to be important, further experimentation gives a more accurate model, one that better represents the actual physical system. Frequently, one or more test project stages, building on previous work, are required to generate an accurate model.

After the important variables have been identified, there are four important actions in constructing a more accurate model:

1. Selecting levels for the independent variables.
2. Developing an appropriate mathematical representation for these variables (building on any initial model from preliminary experiments).
3. Identifying any potential interaction among the independent variables as well as any non-linear response for the dependent variable vs. independent variables.
4. Validating the model, i.e., evaluating its usefulness for prediction.

Simple domains may require only a univariate model (one independent variable), while complex systems may require multivariate models with a number of independent variables. Some projects may have several response parameters, each of which may require a univariate or multivariate second stage model.

Types of Models

There are three types of response models. The simplest is a first-order model that gives an order of importance and approximate response effect for the independent variables of the system. One typical example that applies to continuous quantitative independent variables is

$$y = b_0 + b_1 x_1 + b_2 x_2 + b_3 x_3 \qquad (5.1)$$

where y is the response parameter and x_1, x_2, and x_3 are three independent variables or factors. The constants b_1, b_2 and b_3 are linear effect coefficients for the first order effects of the three respective variables; they indicate how much the response changes for a unit change in each respective x variable. The term b_0 is a constant that sets the general level for the measured response. The model is a linear combination of terms that reflects the influence of the three variables that are assumed to be independent; each contributes in a simple additive manner. A first-order model may be all that is needed to adequately describe a simple system.

A more complex first-order model is given by Eq. (5.2); it is designated as a first-order interactive model. This category can be used to represent a more complex system when the linear influence of one variable may depend on the level of one or more other variables. This is constructed for a three variable problem by using the first four terms of the first-order model described above, plus three two-factor interaction terms.

$$y = b_0 + b_1 x_1 + b_2 x_2 + b_3 x_3 + b_{12} x_1 x_2 + b_{13} x_1 x_3 + b_{23} x_2 x_3 \qquad (5.2)$$

An interaction effect exists between two independent variables, x_1 and x_2, when the effect of x_1 on the response y is different when x_2 is at a lower level compared to when x_2 is at a higher level. The interpretation of the interaction effects, may be understood by rearranging Eq. (5.2) as follows (only variables 1 and 2 are considered):

$$y = b_0 + b_1 x_1 + b_{12} x_1 x_2 + b_2 x_2 + \cdots \qquad (5.3)$$

Combining terms containing x_1 we obtain:

$$b_1x_1 + b_{12}x_1x_2 = (b_1 + b_{12}x_2)x_1 \tag{5.4}$$

The right side of Eq. (5.4) indicates how the value of the parenthetical term, the overall effect coefficient for x_1, depends on the value of x_2 in addition to b_1 and b_{12}. The other interactions may be explained on the same basis. For this type model, the relationship between y and any of the x variables is a linear one; however, if interaction exists, the slope (the parenthetical term as above) changes with the level of one or more of the other x variables.

The third type of model is second-order model. This is a linear combination of independent variable terms and in addition to interaction terms described above, some of the terms are second order (squared). The appearance of these allows for a curvilinear response for the dependent variable. A typical second order model with two independent variables, x_1 and x_2, is

$$y = b_0 + b_1x_1 + b_2x_2 + b_{12}x_1x_2 + b_{11}x_1^2 + b_{22}x_2^2 \tag{5.5}$$

In this model, the terms $b_{11}x_1^2$ and $b_{22}x_2^2$, combined with the other terms, allow for curvilinear behavior for the response vs. both x_1 and x_2; this gives a good representation of a much more complex system.

The empirical model for an experiment may be

1. a fixed-effects model, where only the selected levels of the factors in the experiment are of concern
2. a random-effects model, where the factor levels chosen represent a sample from a larger population
3. a mixed-effects model, where both random and fixed effects factors are included in the design.

Factors may be primary, i.e., the major factor(s) for evaluation potentially have a direct influence; or secondary, that is, factors that potentially have a less direct influence, e.g., environmental conditions.

5.2.2.4   Selecting an Experimental Design

Each experimental design consists of a specified series of values or combinations of values for the independent variables. This series is selected to permit the effect of each treatment and/or the model coefficients to be evaluated with a certain precision and sensitivity. A more complete discussion on selecting an experimental design is in Part 2 of this chapter. Some preliminary discussion is needed here to put the material on the succeeding topics in proper perspective.

5.2.2.4.1 Screening or Exploratory Designs

Experimental designs are basically of two types depending on the goal of the technical project:

1. Screening Designs, which are used when knowledge of the system is limited, i.e., where discovery of new information is desired; the goal is an approximate model; or first order or first order-interactive model.
2. Exploratory Designs, which are used when the important variables or factors have been identified; the goal is a second order model and a detailed description of how the important variables influence the system, especially the interaction and the curvilinear effects. These designs often seek optimum conditions.

For a simple system, a screening design and first order model may suffice. For more complex systems, the first action is usually to use a screening design followed by one or more exploratory designs. In some cases when initial knowledge is limited, more than one screening design is required to identify all the key factors used in the exploratory phase. If substantial background information on the important variables is available, exploratory designs may be used directly.

5.2.2.4.2 Selecting Experimental Design Values

The selection of the values for the suspected or known independent variables is important. The experimenter must confirm that:

1. The selected values are in the range of direct interest.
2. The lower and upper levels (and intermediate level if used) of all independent variables have a difference sufficient to detect any potential influence with good sensitivity.

The greater the test error or measurement uncertainty, the greater is the importance of both of these steps.

5.2.2.4.3 Preliminary Estimate of Measurement Test Error or Uncertainty

The selected testing domain(s) should be stable or in a state of statistical control with regard to test results on appropriate, homogeneous reference materials. If new or unstandardized testing instruments are required, their stable operation should be confirmed by a preliminary Statistical Quality Control program to provide an estimate of the measurement uncertainty or test error standard deviation on selected reference materials. For new testing domains, a preliminary evaluation of test error should be conducted on some uniform representative materials to be used in the technical project. The standard deviation so obtained can be used to select the number of replications for each individual run in the chosen experimental design, if this is required. It can also be used to evaluate the statistical significance of the measured effects and the model coefficients. Large inherent variation in a testing domain lowers the efficiency of any well designed experiment.

Test error or measurement uncertainty is evaluated using a statistical measurement model, which describes how test variation affects the measured or estimated value of a term designated as μ_R. A simple statistical measurement model for any response y is

$$y = \mu_R + \sum \varepsilon + \sum B \tag{5.6}$$

where μ_R is a reference value (often defined as a true value) equal to y without error or variation of any sort, for a system response for some selected combination of factor levels in any designed experiment. Under real world conditions, this reference value is perturbed by one or more potential random deviations, each generated by a specific cause system. The (algebraic) sum of deviations from all potential random cause systems is $\sum \varepsilon$. In addition, μ_R may also be perturbed by one or more potential systematic or bias deviations, each generated by a specific cause system. The (algebraic) sum of deviations from all potential bias cause systems is $\sum B$.

Random deviations are $+$ and $-$ differences about μ_R; each execution of the test gives a unique deviation and the mean value of ε is zero for a long run of test executions. Bias deviations are offsets or constant differences from μ_R; they may be $+$ or $-$, are frequently unique for specific test conditions and do not sum to zero in the long run. Biases may be due to a number of unique conditions; ambient temperature, skill level of different operators, type of equipment or purity of reagents.

There are two types of biases:

1. Biases that are truly constant, e.g., the misadjustment of an instrument output scale
2. Biases that are quasi-constant, i.e., always of a $+$ or $-$ nature with respect to μ_R for an intermediate time interval, but fluctuating over a longer time period.

For any testing domain in some selected time interval, each y measurement has associated with it a random ($+$ or $-$) value for each ε, and a fixed ($+$ or $-$) value for each B. The variance of y, Var(y), depends on the nature of the bias components and the time period for the measurements. If the biases are constant for any series of measurements, then

$$\text{Var}(y) = \sum \text{Var}(\varepsilon) \tag{5.7}$$

For a sufficiently long series of measurements at constant external conditions, the mean of the n values of y, defined as $\acute{Y}_n$, is (approximately)

$$\acute{Y}_n = \left[\mu_R + \sum B\right] \tag{5.8}$$

where all bias components B are constant. The term $\sum \varepsilon$ does not appear because it has been reduced to essentially zero by the high replication (long run) condition. In the ideal case with no biases (each $B = 0$), the long run value of y is

$$\acute{Y}_n = \mu_R \tag{5.9}$$

If measurements are now made at another time (day, week, or month later) and some unknown environmental test factor that biases the output has changed, a new bias term $\sum B^*$ perturbs the output and $\acute{Y}_n$ now estimates

$$\acute{Y}_n = \left[\mu_R + \sum B^*\right] \tag{5.10}$$

For this new Eq. (5.10) condition, Var(y) may be the same as for Eq. (5.9); however, $\acute{Y}_n$ in Eq. (5.10) now estimates a new value and a shift in output has occurred. Over a very extended time period for a testing domain, the simplest expression for measurement variance is given by Eq. (5.11):

$$\text{Var}(y) = \sum \text{Var}(\varepsilon) + \sum \text{Var}(B) \tag{5.11}$$

where $\sum \text{Var}(B)$ is the sum of the variances of individual biases resulting from the varying values for individual B terms over this extended time period. More complex expressions may also exist.

It is important to know if any serious biases exist in a testing domain because the total test error or uncertainty for any measurement y, is a function of both random and bias terms

$$\text{measurement uncertainty for } y = k\left[\sum \varepsilon + \sum B\right] \tag{5.12}$$

where k is determined by the measurement conditions. As noted above, replication can reduce the random uncertainty component(s) $\sum \varepsilon$, but it cannot reduce or eliminate $\sum B$. Such biased output can be reduced or eliminated by

1. Appropriate calibration procedures
2. Investigations of bias and elimination by appropriate modifications of the test
3. Testing standard reference materials and applying correction procedures to the data output.

5.2.2.4.4 Sampling Procedure

When simple comparative experiments (Designs C1, C2 – see Appendix 5.1) are employed, sampling procedures need attention. Two characteristics must be considered: the quality of the estimates of the properties of interest, and the cost of conducting the sampling. Increasing confidence in the estimates (more extensive sampling) with special emphasis on bias reduction, increases the cost of the sampling and testing. A sampling plan should ideally generate an objective estimate of any measured parameter by using strict probabilistic or statistical sampling. However there are situations where this type of plan would be excessively costly for the priority of the decisions to be made based on the data generated. In such cases, an alternative approach using subjective elements or technical judgment is usually employed. Because projects can range from the simple to the complex, three generic types of sampling plans may be applicable.

The first is intuitive sampling in which the skill and judgment of the technologist based on prior information is involved. Decisions made using such a plan are based on the experience of the tester combined with limited statistical conclusions. The second plan type is statistical sampling, which is based on strict statistical sampling procedures that provide authentic probabilistic conclusions. Usually, a large number of samples are needed if the significance of small differences is important. When the number of required samples is large, which may impose a testing burden, hybrid plans using some simplifying intuitive assumptions are frequently employed. The third plan type is protocol sampling, which is usually employed in producer-user situations, based on regulations that specify the type, size, frequency, and period of sampling as developed from previous relevant programs.

5.2.2.4.5 Replication and Randomization Procedures

The process of selecting the number of tests to evaluate the mean response for any combination of factor levels is called replication. The total uncertainty in the mean so obtained is a function of both the total test (measurement) error and the sampling error or uncertainty. For simple comparison designs, a good estimate of the testing error may be obtained by pooling the individual estimates from each treatment. To do this, it is necessary to assume equal variance among the treatments, even though the means among the treatments may be different. For factorial designs, the layout of the design ensures that each estimate of an effect is obtained as an average that utilizes all the response measurements in the design. Thus, a certain level of replication is inherent in essentially all designs; this is one advantage for using experimental designs.

Statistical inference and decisions derived from it are based on probability theory. Probability is defined as the long run relative frequency of a particular outcome for repeated trials for the event in question. For valid statistical inference, randomization is required. On a theoretical basis, randomization is defined as the operational process that ensures that every potential outcome has an equal chance of occurring. As an example, for a series of throws for a normal die, the numbers on each of the six faces have an equal chance of appearing on a given throw. The probability for each number is 1/6 or 0.167. This is a random process. On a practical basis for testing operations, randomization eliminates all known biases from any given measurement or other relevant process.

Each experimental design involves a specified number of measurements of the selected performance parameter, each measuring a treatment or combination of independent variable levels. In the list of factor combinations called the design matrix, the levels for each of the factors are given in a systematic pattern. If this specific design sequence is used for the series of measurements, the sequence may be correlated with a potential time trend or bias if some unknown environmental conditions change with time. Thus, for each selected experimental design, two important actions must be taken

1. Every effort must be made to eliminate all known or suspected biases
2. To counteract the lack of complete knowledge about biases, the individual runs of the design must be randomized

Randomization combines the potential bias effect(s) with the truly random error and prevents the bias effects from being confused or confounded with factor effects, thus misleading the experimenter. This combination of bias with random error avoids any confusion of bias effects with real factor effects; but, if trend-type bias is large, the influence of the factors is evaluated with poor precision and sensitivity because of the inflated test variance. This demonstrates the importance of knowledge about biases and the elimination of known biases for good testing operations.

5.2.3   Conduct Measurements and Obtain Data

Prior to the actual measurement phase of any project, the test or measuring system should be in a state of statistical control. The technologist should conduct testing operations carefully while alert for any unanticipated complications. Any unusual circumstances or outcome should be carefully documented. Standardized data forms should be used and all pertinent background information clearly recorded.

5.2.4   Conduct Analysis and Evaluate Preliminary Model

The analysis of experimental designs, which may be conducted with simple spreadsheet operations or through statistical analysis packages, is discussed in Section 5.3. There are three steps in this analysis:

1. Estimating the effect of treatments or model coefficients
2. Estimating the test error or uncertainty
3. Using the results of both operations to determine the statistical (and technical) significance of the treatments or model coefficient terms.

5.2.5   Prepare Report

The end product of any technical project is a full report. The report begins with a statement of the problem addressed or project undertaken, followed by a summary of the conclusions and/or technical decisions made as a result of the program. This can be followed by a detailed account of all the planning, work, and analysis conducted. If a complete solution is found for any project, no future action is needed. If, however, new questions arise, unanticipated technical problems are discovered, or only partial answers and/or decisions are attained, a recommendation should be made for either a Stage 2 iteration of all five steps or for a future work program. Any Stage 2 operation needs to be carefully organized so as to build on and take full advantage of the work conducted in Stage 1.

5.3   Part 2: Using Experimental Designs

Although technical objectives may vary, all experiments have the same operational objective – to provide maximum information of the highest quality possible for minimum cost. Experimental design is the process of efficiently planning and executing a series of experiments with this objective in mind. Many problems, especially for complex systems, should be approached in a sequential manner. The Stage 1 process should provide basic understanding along with an approximate solution. Subsequent stages build on what is learned in Stage 1, where unanticipated new information frequently leads to new ideas on how to proceed. This approach provides for more efficient and less costly problem resolution. Refer to Appendix 5.1 for all of the designs as discussed in Section 5.3.

5.3.1   Screening Designs – Simple Treatment Comparisons

The most elementary technical project is the comparison of the mean (or average) for two sets of measurements that represent potentially different populations. The sets are characterized by two different treatments. The word treatment implies some distinguishing feature, such as a specified material composition, a physical or chemical modification, or a processing operation to produce certain production parts. Typical examples are the use of two different antioxidants for the heat aging behavior of rubber or two surface treatments of rubber to reduce friction. Part 1 of Appendix 5.1 contains screening designs, the simplest of which are for simple experiments (C1 and C2) where two or more treatments are compared. Experimental design, C1, may be used in two ways:

1. All replicates for both treatments can be conducted under uniform test conditions
2. Testing conditions or some other operational factors are uniform for treatments 1 and 2 for each replicate set, but may be different among the replicate sets as a group.

5.3.1.1   Design C1 for Uniform Replication Conditions

For this situation a decision on the statistical significance for the means of two potentially different populations is determined by t-test. A typical testing project might consist of five replicates for each of the two treatments. The variance among the five replicates is used to evaluate test variation with the assumption that, although the means may be different, there is no difference in variance for the two treatments. The decision on the significance of the difference, $(\acute{y}_1 - \acute{y}_2)$, for the means of the five replicates for each treatment is based on the calculated t value, t_{calc}, obtained from

$$t_{calc} = (\acute{y}_1 - \acute{y}_2)/S_{dy} \tag{5.13}$$

where $\acute{y}_1$ = mean of 5 replicates for Treatment 1

$\acute{y}_2$ = mean of 5 replicates for Treatment 2

S_{dy} = standard deviation of difference of means (5 values each)

$S_{dy} = [(S_1^2/n) + (S_2^2/n)]^{1/2}$

S_1^2 = variance among 5 replicates for Treatment 1

S_2^2 = variance among 5 replicates for Treatment 2

n = number of replicates = 5

The expression for S_{dy}, a pooled value of both sets of replicates, is obtained from theorems on the propagation of error, i.e., the variance of a sum or difference of two values taken from potentially different populations is the sum of the variances of the two populations. If t_{calc}, with degrees of freedom, df $= 2(n-1) = 8$, is larger than the critical t value, t_{crit}, at some selected confidence level, $(1-\alpha)100$, or $P = \alpha$, then the null hypothesis, $H_0: \mu_1 = \mu_2$, is rejected and the alternative $H_0: \mu_1 \neq \mu_2$, is accepted. If α is 0.05, the confidence level is 95%. From standard t tables for df $= 8$, the double-sided value of $t_{crit} = 2.31$ for $\alpha = 0.05$.

5.3.1.2   Design C1 for Non-Uniform Replication Conditions

When it is not possible to have uniform conditions for all replicates, a paired comparison test can be used. The response parameter in this case is the difference in measured or observed behavior between the individual paired values.

A typical example is as follows. The performance of a rubber compound is determined by the loss of plasticizer. Two different plasticizers, designated A and B, may be used. It is desired to determine if one plasticizer is more fugitive than the other. The testing is a long term outdoor exposure evaluation in different weather conditions. A test specimen of each formulation (plasticizer A or B) is exposed for a fixed period at eight different locations. Data obtained are:

| | Plasticizer content (%) | | |
Exposure condition	After exposure: A	B	Difference, d (A − B)
3 months – Ohio	10.1	10.0	0.1
3 months – Maine	8.9	8.9	0.0
6 months – Georgia	8.9	8.8	0.1
3 months – Arizona	8.4	8.2	0.2
6 months – Texas	9.2	9.0	0.2
6 months – Florida	8.7	8.4	0.3
3 months – California	9.0	8.8	0.2
3 months – Oregon	8.8	8.7	0.1
Average	9.00	8.85	0.15
Std Dev (individual)	0.501	0.535	0.093
Std Dev (average 8)	0.177	0.189	0.0329

The difference across the range of locations, $d = 0.15\%$, is small but consistent. These eight differences are a sampling of the distribution of differences, d, and the standard deviation of these eight values is 0.093. The standard error (or deviation) of means of eight values is $0.093/(8)^{1/2} = 0.0329$, and t_{calc}, the ratio of the mean difference, 0.15, to the standard error of means, 0.0329, equals 4.56. The df for this is $8 - 1 = 7$ and the value of t_{crit} at the 99% confidence level or $P = 0.01$, is 3.499. Thus, the mean difference of 0.15 is highly significant.

5.3.1.3   Design C2 for Multi-Treatment Comparisons

When more than two treatments are compared, a design such as C2 may be used with a selected number of replicates for each treatment. When the testing process is in control, all treatments, 1 to j, may be evaluated under uniform conditions. Typically, an analysis of variance (ANOVA) is conducted to determine if the variation among treatments is significant compared to the pooled variance among all replicates. However, an ANOVA does not pinpoint which of the treatments among the j total treatments is different from other treatments. To sort the treatments into a hierarchy of values with indicated significant differences among the j $(j - 1)/2$ treatment pairs, a multi-comparison analysis is required. A number of analysis techniques have been proposed for this; see Duncan [1], Tukey [2], and Dunnet [3]. Computer statistical analysis programs usually include these multi-comparison analysis routines.

The Dunnet procedure uses a control and permits a decision on the statistical significance of differences between the control and the experimental treatments. The procedure ensures that the decision on the entire set of treatments is controlled at the selected $P = \alpha$ level rather than individual comparisons. The references mentioned above plus Winer [4]; Box, Hunter, and Hunter [5]; and Diamond [6] provide additional background on multi-comparison analysis. The last text is oriented toward manufacturing processes and the use of α and β errors. If it is not possible to have uniform testing conditions for all j treatments, the program may be conducted in two or more uniform condition blocks, with a control in each block. Using the Dunnet procedure for each block permits decisions on differences between the control and the treatments in that block.

5.3.2   Screening Designs – Multifactor Experiments

Much of the experimental effort in rubber compounding is devoted to evaluating the effect of variations in compound formulation and processing conditions. Experimental layouts with two or more factors, called factorial designs, are especially useful for this purpose, especially two-level factorial designs. Part 1 of Appendix 5.1 contains 11 designs for multi-factor screening experiments. Seven is a reasonable maximum number of factors to be evaluated in any program or project. Any system with more potential factors needs to be reviewed for ways to consolidate or otherwise reduce the number of operational factors on the basis of engineering or scientific analysis.

5.3.2.1   *Two-Level Factorial Designs*

These designs are characterized by special selected combinations of lower and upper levels across all the selected factors. When the factor levels are set at the values called for in a particular combination and a response measurement is made for this combination, this is called a (test) run. Each design has some number of specified runs and the total list of these runs is called the design matrix. A complete factorial design matrix is one where for each factor, all factor levels of the other factors appear equally at their lower and upper levels. Thus, for two factors investigated at two levels each, a complete factorial design requires four (2^2) response measurements or runs, each of which has a different combination of the two levels of the two factors. Design S1 in Appendix 5.1 is an example.

When the number of factors is large, a complete or full factorial requires too many test runs and designs called fractional factorials are used. In fractional factorials, a certain fraction of the full factorial number of runs is selected on the basis that the design is balanced with respect to the number of selected levels of each factor. The fractional factorials are designated as 1/2 fraction, 1/4 fraction, etc., of the full design. With the exceptions noted below, all of these designs allow for the evaluation of two-factor interactions that often are important in many technical investigations. Any design that allows for direct calculation of two-factor interactions is usually sufficient to give a good evaluation of any system. Usually three-factor interactions have no real significance.

All of the two-level screening designs (as well as the exploratory designs) are orthogonal in the independent variables, i.e., there is no correlation among these variables. Orthogonality permits the use of the matrix for easy analysis via a spreadsheet program. The designs are balanced; for any factor level for factor i, the levels of all other factors appear at their upper and lower values the same number of times. The designs and the model equations are set up using special coded units for the independent variables.

As discussed in Section 5.2, for any response variable y and two independent variables x_1 and x_2, a model equation that allows for the evaluation of any interaction between x_1 and x_2 is

$$y = b_0 + b_1 x_1 + b_2 x_2 + b_{12} x_1 x_2 \tag{5.14}$$

where $b_0 =$ a constant; in system of units chosen, it is the value of y when $x_1 = x_2 = 0$

$\quad\quad\quad b_1 =$ change in y per unit change in x_1

$\quad\quad\quad b_2 =$ change in y per unit change in x_2

$\quad\quad\quad b_{12} =$ an interaction term for specific effects of combinations of x_1 and x_2;
$\quad\quad\quad\quad\quad$ see discussion for Eqs. (5.3) and (5.4)

The coded units are obtained for each factor by selecting a value that constitutes a center of interest or a reference value and then selecting certain values that are below and above that center of interest by an equivalent amount. This is a straightforward process for quantitative, continuous variables or factors, but may not be possible for some qualitative or categorical factors which can exist at only two levels. In this case, the center of interest is considered theoretical or conceptual. If a system contains a number of categorical factors, the final expression for the analysis may be given in terms of main effects and their interactions. The coded units for any x_i are defined by:

$$x_i = (v_E - cv_E)/su \tag{5.15}$$

with $v_E =$ selected factor value for x_i, in physical (experimental) units

$\quad\quad\quad cv_E =$ center of interest value for x_i, in physical units

$\quad\quad\quad su =$ scaling unit, i.e., change in physical units equal to 1 coded unit

When v_E is higher than cv_E by an amount equal to su then $x_i = 1$; when v_E is less than cv_E by an equal amount, $x_i = -1$; and when $v_E = cv_E$, $x_i = 0$. The center of interest values for all factors constitute the central point in the multi-dimensional factor space for the experiment. The constant b_0 is the value for y at this center in the factor space; it is the (grand) average of all responses.

The design matrix for a full factorial 2^2 design is given in Design S1. An additional matrix called the interaction matrix is also listed, which for this simple design is only one column headed by $X_1 X_2$. Design S2, for three x-variables, has a more extensive three-column interaction matrix. For each row of the design, the entry for any column, $X_i X_j$, of the interaction matrix is obtained by multiplying the design matrix entries for column X_i by column X_j.

5.3.2.2   Analysis of the Designs

The use of coded units and spreadsheet calculations simplifies the analysis. Each design can also be analyzed by multiple regression analysis with computer software programs. The effects of the independent variables may be calculated in one of two ways:

1. As main effects, defined as the change in the response for a change in the x-variable from the lower to the upper level
2. As effect coefficients, defined as the change in the response for one positive scaling or coded unit change in the x-variable

On the basis of coded units there are two units of change in moving from the lower (-1) to the upper (1) level; thus, effect coefficients are one-half the numerical value of main effects and main effect interactions. All analysis algorithms in Appendix 5.1 are given in terms of effect coefficients. To express the results of any analysis in terms of main effects and their interactions, multiply each calculated linear or interaction effect coefficient by 2.

5.3.2.3   Calculating the Effect Coefficients

The design and interaction matrixes are used to calculate the effect coefficients of the independent variables. The generalized equations for a design with any number of factors are as follows:

The coefficient b_0 is evaluated from the first or Y column of the design

$$b_0 = \left(\sum y_1 \text{ to } y_i \right) \Big/ 2n \tag{5.16}$$

with y_1 to y_i = the measured responses; there are a total of i rows (four for S1); n = the number of runs with $X = 1$ (as well as $X = -1$), for each X variable; and $2n$ = total number of runs for the design.

Each b_i, or linear effect coefficient, is evaluated by using the column Y and column X_i in the design matrix:

$$b_i = \frac{1}{2} \left\{ \left[\sum y_i^*(+X_i) \Big/ n \right] - \left[\sum y_i^*(-X_i)/n \right] \right\} \tag{5.17}$$

where $y_i^*(+X_i)$ = the product of y_i and X_i for each of n rows where $X_i = 1$; the sum is taken over all $2n$ runs in the design and divided by n; and $y_i^*(-X_i)$ = product of y_i and X_i for each of n rows where $X_i = -1$; the sum is taken over all $2n$ runs in the design and divided by n.

Each b_{ij}, a two-factor interaction coefficient, is evaluated by using column Y and column X_iX_j in the interaction matrix:

$$b_{ij} = \frac{1}{2} \left\{ \left[\sum y_i^*(+X_iX_j) \Big/ n \right] - \left[\sum y_i^*(-X_iX_j) \Big/ n \right] \right\} \tag{5.18}$$

where $y_i^*(+X_iX_j)$ = the product of y_i and X_iX_j for each of n rows where $X_iX_j = 1$; the sum is taken over all $2n$ runs in the design and divided by n; $y_i^*(-X_iX_j)$ = the product of y_i and X_iX_j for each of n rows where $X_iX_j = -1$; the sum is taken over all $2n$ runs in the design and divided by n.

Each b_{ijk}, a three-factor interaction coefficient, is evaluated by using column Y and column $X_iX_jX_k$ in the interaction matrix

$$b_{ijk} = \frac{1}{2} \left\{ \left[\sum y_i^*(+X_iX_jX_k) \Big/ n \right] - \left[\sum y_i^*(-X_iX_jX_k) \Big/ n \right] \right\} \tag{5.19}$$

where $y_i^*(+X_iX_jX_k)$ = the product of y_i and X_iXjX_k for each of n rows where $X_iXjX_k = 1$; the sum is taken over all $2n$ runs in the design and divided by n; $y_i^*(-X_iX_jX_k)$ = the product of y_i and X_iXjX_k for each of n rows where $X_iX_jX_k = -1$; the sum is taken over all $2n$ runs in the design and divided by n.

These generalized effect coefficient equations are given in Appendix 5.1 as notes for Design S1. The specific coefficient equations for S1 and S2 are also given to illustrate the use of the generalized equations.

When fractional factorial designs are used, confounding may exist in the interpretation of the effect coefficients. This confounding, or dual meaning of output information, is called aliasing. An alias exists in fractionated designs when the exact sequence of -1, 1 values is the same for two or more columns among all the columns of the total matrix (both design and interaction). An alias also exists for any two or more columns with sign reversal for all entries. Thus, no unique evaluation of the individual coefficients represented by the columns is possible.

Aliases in the designs are indicated by an expression of the type given in Design S3. For Block I, the expression $b_3 = b_3 + b_{12}$ indicates that the calculation of the linear effect coeficient b_3 has the same set of 1 and -1 combinations (i.e., in column $X3$) as the calculation for the interaction coefficient b_{12} (i.e., column $X1X2$). The calculation actually evaluates the sum $b_3 + b_{12}$. If b_{12} has no real effect (gives a mean of zero in long run) the calculation actually evaluates only b_3.

Confounding also exists where a similar identical set of 1 and -1 values produces an unresolved dual interpretation for higher order coefficients with respect to block effects in the same sense as an alias. This is normally of lesser importance than aliases with two-factor interactions. All aliases and indications for main effects clear (no aliases) are given for each design.

5.3.2.4   Reviewing Designs S1 to S11

The simplest design S1 has been discussed above. Design S2 for three factors can be conducted in eight runs and evaluates a full first-order interactive model, i.e., all three interactions. Design S3 is also for three factors but this is divided into two blocks. This design can be used where, for example,

1. Each individual run requires substantial set-up and execution time and only four runs constitute a uniform condition test period
2. Any given batch of basic material involved in all runs is sufficient only for four runs and batch-to-batch variation may lead to biased response

If each set of four runs can be conducted under uniform test conditions, then this design will evaluate all main effects and two-factor interactions. Either block of this design can be used for a three-factor problem where only main effect coefficients are desired.

Design S4 can be used to evaluate a four-factor problem with all two-factor and three-factor interactions free of any aliases. An example using this design is given at the end of the chapter. If the evaluation of two-factor and higher interactions are not required for a four-factor project, Design S5 with only eight runs may be used. Design S6, for 5 factors, allows for the evaluation of main effects and all two-factor interactions in 16 runs.

If two-factor interactions are not important, Design S7 also for five factors may be conducted in eight runs. Design S8 for six factors allows for the evaluation of all main effect coefficients free of any two-factor (or higher) aliases; however, some of the two-factor interactions are aliased with each other. For seven factors, Designs S9 and S10 may be used. Design S9 is a 1/16 fraction in eight runs where the main effects are aliased with two-factor interactions. If this design is used and subsequent investigation requires that main effects be evaluated free of two-factor (and higher) aliases, a second set of eight runs as given by S10 can be conducted. Using the combined runs, as given in S11, permits the evaluation of main effects clear of any aliases.

The estimation of the response or y value test error in screening designs can be conducted by:

1. A repeat of the entire design with each run replicated once in random order
2. The use of some minimal number of replicated runs
3. The use of interaction effect estimates for the higher order interactions (three-factor) and/or two-factor interaction estimates that are small compared to the value of b_0 (of the order of 3–4%)
4. The use of error estimation based on previous testing on identical or similar materials or objects.

5.3.3   Exploratory Designs – Multifactor Experiments

Exploratory designs are used when the important factors or variables of a system are known and the purpose is to develop an accurate model of the system. These designs are typically used after a basic understanding of any system has been developed using screening designs.

Exploratory designs require that all factors be quantitative, continuous x-variables. All of these designs evaluate squared or second order terms ($b_{11}x_1^2$, $b_{22}x_2^2$, etc.), as well as interaction terms that jointly account for potential curvilinear response behavior. Part 2 of Appendix 5.1 lists the exploratory designs E1 (for two variables) to E6 (for five variables). For E1, the hexagonal design, equations for the calculation of specific effects are given. For the remaining exploratory designs generalized equations are given in terms of generic symbols, as identified in Design E2. Although generic symbols are used for any design, the exact equations for calculating the effect coefficients use certain constants that are different for each design. To illustrate the use of the generic symbols, the specific equations for E2 are also given.

The estimation of the response or y value test error in exploratory designs can be approached in several ways:

1. By use of the minimal number of replicated runs as indicated in each design
2. A repeat of the entire design with each run being replicated once in random order
3. The use of interaction effect estimates for the higher order interactions (three-factor) and/or two-factor interaction estimates that are small compared to the value of b_0 (of the order of 3–4%)
4. The use of error estimation based on previous testing on identical or similar materials or objects.

Because these designs frequently require many runs, several of the designs are given in two or more blocks. Blocking permits the evaluation of all main effect, two-factor and squared-term coefficients free of any bias attributable to block-to-block variation in testing conditions. In general, including many runs in exploratory experiments, with or without blocking, requires long term stability in the testing domain. Designs E2 to E6 give the relationship between the standard deviation for the response measurements, S_y, and the standard error of the effect coefficients, SE(b).

5.3.4   Evaluating the Statistical Significance of Effect Coefficients

After all effect coefficients have been evaluated for any given design, the next step is to decide on the statistical significance of these individual coefficients. Multiple regression and ANOVA techniques have been used for this purpose. However, it is more instructive to evaluate standard errors (standard deviations) of the effect coefficients and use t-tests to determine what coefficients are statistically significant.

5.3.4.1   Evaluating Standard Errors for Effect Coefficients: Screening Designs

There are two methods for evaluating SE(b), the standard error for an effect coefficient. The methods are:

1. Using replicated measured responses of the sort discussed above, with the response standard deviation S_y used to calculate SE(b)
2. Using interactions (of order two or higher) to estimate SE(b).

For method (1), the usual procedure is to conduct duplicate measurements for each run in the design. Each set of duplicates gives a 1 df estimate of the standard deviation of the response, or y measurement. The 1 df estimates are pooled to give an overall standard deviation estimate S_y with df $= N$, where $N = 2n$ or the total number of runs in the design; as defined previously, $n =$ number of runs at either the -1 or 1 level.

The key to obtaining a valid estimate of S_y is to ensure that randomization is genuine. If a number of operational adjustments have to be made in the testing system to obtain each measured value, back-to-back duplicates should not be obtained after any particular operational adjustment setting and used with other duplicates to estimate a pooled S_y. Such duplicates evaluate instrumental error only. Each replicate must be obtained in a randomized sequence after the required (new) set of adjustments have been made. This evaluates overall test error, which includes instrumental error as well as other components. It is assumed that the error as evaluated from pooled estimates is uniform (equal) throughout the entire factor space, i.e., over the y response range for all x-variable levels.

Each main or interaction effect (main or interaction effect $[i] = 2b_i$) is evaluated in general form by means of a difference. The variance of the main or interaction effect, in terms of this difference, is given by

$$\text{Var}(2b_i) = \text{Var}(\acute{y}_+ - \acute{y}_-) = [(S_{y+}^2/n) + (S_{y-}^2/n)] = (2/n)S_y^2 \qquad (5.20)$$

where $\text{Var}(2b_i) =$ variance of any main or interaction effect, i.e., for any column X_i; $\acute{y}_+ =$ sum of y-values where $X_i = 1$; there are n such values; $\acute{y}_- =$ sum of y-values where

$X_i = -1$; there are n such values; $S_y^2 =$ pooled value of variance of response or y measurements.

A pooled or common value for S_y^2 is used because the y response variance is assumed to be uniform (equal) over the entire factor space. On the basis of N total runs for any design, with $n = N/2$, substitution into Eq. (5.20) gives:

$$\text{Var}(2b_i) = S^2(2b_i) = \left(\frac{2}{N/2}\right)S_y^2 = (4/N)S_y^2 \tag{5.21}$$

The square root of $S^2(2b_i)$, the standard error for any main or interaction effect, is given by:

$$\text{SE(main effect)} = \text{SE}(2b_i) = [2/(N)^{1/2}]S_y \tag{5.22}$$

Since $\text{SE}(2b_i) = 2\text{SE}(b_i)$ or $\text{SE}(b_i) = SE(2b_i)/2$, substitution in Eq. (5.22) gives:

$$\text{SE}(b_i) = [2/(N)^{1/2}]S_y/2 = S_y/(N)^{1/2} \tag{5.23}$$

The estimation of the effect coefficient standard error by method (2) uses values obtained from calculations for the interaction matrix. Long term experience using factorial designs has demonstrated that real three-factor interactions are quite rare or not significant; frequently, a substantial number of the potential two-factor interactions are also not significant. The declaration "not significant" means that if an entire design were repeated many times, the long run average effect coefficient or main effect would be zero. When a finite value for an effect coefficient or an effect is obtained for a single execution of a design, this finite value appears because of variation in the response measurement. The finite value may be thought of as a difference from the expected or long run value of zero. When there are several three-factor (or perhaps two-factor) interaction term effect coefficients or main effects evaluated, the respective values (i.e., differences) may be squared and used in the usual manner for estimating a response variance.

Once an effect coefficient estimate and standard error have been obtained, each coefficient may be evaluated for significance by

$$t_{\text{calc}} = b_i/\text{SE}(b_i) \tag{5.24}$$

The significance of t_{calc} is evaluated by using the df for $\text{SE}(b_i)$ to obtain a t_{crit} from standard (two-sided) t-tables at a selected $(1 - \alpha)100$ or $P = \alpha$ significance level. If $t_{\text{calc}} > t_{\text{crit}}$, a declaration of statistical significance can be made at this level. The use of method (2) in evaluating a design is illustrated in the example below.

5.3.4.2   Four Factor Screening Design: An Example

An example of a typical screening design illustrates how the coefficients are evaluated, estimates of test errors are obtained, and the interpretation of the analysis. This same general approach may also be used for exploratory designs. Table 5.2 lists the essential features for a typical four-factor rubber compounding screening design. The purpose of the experiment is to determine the influence of four compounding variables:

1. Crosslink density
2. HAF content

Table 5.2   Four-Factor Screening Design – Effect of Crosslink Density, HAF, Oil, and Cure Temperature on Modulus and Cure Time*[*]

Run	Y_1 M100	Y_2 Cure time	X_1 Dxl	X_2 phr HAF	X_3 phr Oil	X_4 Temp.
1	0.93	100	−1	−1	−1	−1
2	1.21	61	1	−1	−1	−1
3	3.17	106	−1	1	−1	−1
4	4.69	72	1	1	−1	−1
5	0.68	95	−1	−1	1	−1
6	0.88	56	1	−1	1	−1
7	2.07	106	−1	1	1	−1
8	3.00	68	1	1	1	−1
9	0.91	42	−1	−1	−1	1
10	1.26	31	1	−1	−1	1
11	3.15	50	−1	1	−1	1
12	4.44	33	1	1	−1	1
13	0.68	43	−1	−1	1	1
14	0.92	33	1	−1	1	1
15	1.97	43	−1	1	1	1
16	3.00	40	1	1	1	1
Avg	2.06 (MPa)	61.2 (min)				

Coded Design Units vs. Actual Physical Units

Coded units	Physical units			
	Dxl**	HAF	Oil	Temp.
−1	3.8	20	5	144
0	4.4	40	10	149
1	4.9	60	15	154
	phr	phr		°C

[*] Cure times are corrected for thermal lag.
[**] Crosslinks per gram $\times 10^{19}$.

3. Process oil content
4. Cure temperature

There are two responses

1. Modulus at 100% elongation
2. Cure time needed to attain full development or maximum modulus

Design S4 was used with 16 runs executed in a random order..

At the bottom of Table 5.2 are data which give the relation between the coded (−1, 1) design units and the physical units for each variable. The design was centered on a compound with a crosslink density of 4.4×10^{19} crosslinks per gram, 40 phr of HAF, and 10 phr of oil, cured at 149 °C. The crosslink densities were evaluated by a separate, preliminary test program. The crosslink levels of 3.8 and 4.9 ($\times 10^{19}$) crosslinks per gram were achieved at full modulus development by using 0.83 phr of CBS accelerator with 1.53 phr of sulfur (the -1 level) and 1.14 phr CBS with 2.18 phr of sulfur (the 1 level) in combination with 5 phr ZnO and 2 phr stearic acid. Columns Y_1 and Y_2 list the measured modulus and cure time. The application of the generic equations for the calculation of the effect coefficients, as given in Design S1, yields the results in Tables 5.3 and 5.4. Coefficients b_1 to b_4 were obtained using the design matrix; coefficients b_{12} to b_{234} were obtained using the interaction matrix.

In Part 1 of Table 5.3 for modulus response, two sets of coefficient values are given, numerical calculated values and relative values, where each b_i is expressed relative to b_0 as 100. A review of the relative b_i values permits a quick appraisal of the approximate significance of the coefficients. Relative coefficients for b_{14}, b_{24}, and all three-factor

Table 5.3   Results of Effect Coefficient Analysis – Modulus at 100% (MPa)

Part 1

Coefficient	Numerical value	Relative value	t_{calc}
b_0	2.06	100.0	
b_1 (Dxl)	0.36	17.5	15.7
b_2 (HAF)	1.13	54.9	49.1
b_3 (Oil)	-0.41	-19.9	-17.8
b_4 (Temp)	-0.02	-1.0	-0.9
b_{12}	0.23	11.2	10.0
b_{13}	-0.07	-3.4	-3.0
b_{14}	0.00	0.0	0.0
b_{23}	-0.27	-13.1	-11.7
b_{24}	-0.03	-1.5	-1.3
b_{123}	-0.04	-1.9	-1.7
b_{124}	-0.01	-0.5	-0.4
b_{234}	0.01	0.5	0.4

$t_{calc} = b_i/(\text{std error } b_i)$
t_{crit} 95% (at df $= 5$) $= 2.57$
t_{crit} 99% (at df $= 5$) $= 4.03$

Part 2 – Estimation of Error

	b coefficient	(b coefficient)^2
b_{14}	0.00	0.00000
b_{24}	-0.03	0.00090
b_{123}	-0.04	0.00160
b_{124}	-0.01	0.00010
b_{234}	0.01	0.00010
	sum	0.00270
Var $=$	sum/5	0.00054
Std error	$b_i =$	0.023

Model equation:
$M100 = 2.06 + 0.36Dxl + 1.13HAF - 0.41Oil + 0.23Dxl^*HAF - 0.27HAF^*Oil - 0.07Dxl^*Oil$

Table 5.4   Results of Effect Coefficient Analysis – Cure Time (min)

Part 1

Coefficient	Numerical value	Relative value	t_{calc}
b_0	61.20	100.0	
b_1 (Dxl)	−11.90	−19.4	−24.4
b_2 (HAF)	3.60	5.9	7.4
b_3 (Oil)	−0.69	−1.1	−1.4
b_4 (Temp)	−21.80	−35.6	−44.7
b_{12}	0.44	0.7	0.9
b_{13}	0.69	1.1	1.4
b_{14}	6.81	11.1	14.0
b_{23}	0.19	0.3	0.4
b_{24}	−1.44	−2.4	−3.0
b_{123}	0.56	0.9	1.1
b_{124}	−0.31	−0.5	−0.6
b_{234}	−0.56	−0.9	−1.1

$t_{calc} = b_i / (\text{std error } b_i)$
t_{crit} 95% (at df = 6) = 2.45
t_{crit} 99% (at df = 6) = 3.71

Part 2 – Estimation of Error

	b coefficient	$(b \text{ coefficient})^2$
b_{12}	0.44	0.194
b_{13}	0.69	0.476
b_{23}	0.19	0.036
b_{123}	0.56	0.314
b_{124}	−0.31	0.096
b_{234}	−0.56	0.314
	sum	1.429
Var =	sum/6	0.238
Std error	b_i =	0.488

Model equation:
Cure time = 61.2 − 11.9Dxl + 3.60HAF − 21.8Temp + 6.81Dxl*Temp − 1.44HAF*Temp

interactions have magnitudes of less than 2%; these are not really technically significant and represent error. Part 2 of the table illustrates a calculation of the error from these five values with $SE(b_i) = 0.023$. Although the relative effect coefficient for temperature (variable 4) is less than 2%, it is not used in the error estimation because it is a main effect coefficient.

The $SE(b_i)$ estimate is used to generate the t_{calc} column which indicates that the significant effect coefficients (at the 99% confidence level) in order of importance are:

- b_2 for (HAF)
- b_3 for (Oil)
- b_1 for (Dxl)

- b_{23} for (HAF*Oil)
- b_{12} for (Dxl*HAF)
- b_{13} for (Dxl*Oil) [at 95% confidence level]

On a main effect basis, a unit increase in HAF (20 phr) has slightly more then 2.5 times the impact on modulus as a unit increase in either Dxl or oil level. Although in the experiment cure time was continued until full modulus was attained, it is theoretically possible that a difference in cure temperature affects the modulus. However, the analysis revealed that in this experiment cure temperature (in range 145 to 155 °C) has no effect on modulus.

There are highly significant interactions between Dxl and HAF and HAF and oil and a minor interaction between Dxl and oil level. The modulus interaction effects can be evaluated using the model equation given in Table 5.3. They reveal the following:

1. At low Dxl, a unit increase in HAF increases modulus by 0.90 MPa. At high Dxl, a unit increase in HAF increases modulus by 1.36 MPa, i.e., HAF is 1.5 times more efficient for absolute modulus enhancement at high crosslink densities (curative level)
2. At low HAF, a unit increase in oil level lowers modulus by 0.14 MPa. At high HAF, the same unit increase in oil lowers modulus by 0.68 MPa, i.e., unit increase in oil is roughly five times more efficient at at high HAF levels in reducing modulus.
3. At low Dxl, a unit increase in oil level lowers modulus by 0.34 MPa. At high Dxl, the same unit increase in oil lowers modulus by 0.48 MPa, i.e., oil is 1.4 times more efficient at high crosslink densities (curative level) in reducing modulus.

A review of the cure time relative coefficients in Table 5.4 indicates that there are three two-factor and three three-factor interaction relative coefficient values below 2%. These again are not technically significant and represent test error. Using these values to estimate $SE(b_i)$ gives a value of 0.488. The calculated t values indicate the order of importance for the significant (at 99% confidence level) effect coefficients as:

- b_4 for (Temp)
- b_1 for (Dxl)
- b_{14} for (Dxl*Temp)
- b_2 for (HAF)
- b_{24} for (HAF*Temp) [at 95% level]

On a main effect basis, the most important factor is the cure temperature. A unit temperature increase (5 °C) has roughly twice the effect of a unit change in crosslink density. There is a minor influence for HAF: increased HAF extends the cure time, although this may be a dilution effect rather than a carbon black surface chemistry effect. There is a strong interaction between temperature and Dxl, as well as a minor interaction between HAF and temperature.

Evaluating the cure time interaction effects with the model equation in Table 5.4 reveals the following:

1. At the low cure temperature, a unit increase in HAF increases cure time by 5 minutes while at high temperature the increase is 2.2 minutes, i.e., the effect of HAF at low temperatures is 2.3 times greater than its effect at high temperatures. A substantial part of this outcome is attributed to the enhanced rate of cure at the higher temperature, which reduces cure times.

2. At low cure temperatures, a unit increase in Dxl reduces cure time by 18.7 minutes, while at high cure temperatures the same Dxl increase reduces a cure time by only 5.1 minutes. Again part of this is the result of reduced overall cure times at higher cure temperatures.

References

1.   Duncan, D. B., *Biometrics*, (1955) **11**, 1.
2.   Tukey, J. W., *Biometrics*, (1949) **5**, 99.
3.   Dunnet, C. W., Biometrics, (1964) **20**, 482.
4.   Winer, B. J., *Statistical Principles in Experimental Design*, McGraw Hill (1971).
5.   Box, G. E. P., Hunter, W. G., Hunter, J. S. *Statistics for Experimenters*, J. Wiley & Sons (1978).
6.   Diamond, W. J., *Practical Experimental Designs – For Engineers and Scientists*, Van Nostrand Reinhold, New York (1989).

Appendix   A Catalog of Experimental Designs

Part 1   Screening Designs: Simple Treatment Comparisons and Multifactor Experiments

Designs for Simple Treatment Comparisons
Design C1: Two Treatment and/or Paired Comparison

Replicate or condition	Treatment no.		
	1	2	d
1	y_1	y_2	$(y_1 - y_2)$
2	y_1	y_2	$(y_1 - y_2)$
3	y_1	y_2	$(y_1 - y_2)$
4	y_1	y_2	$(y_1 - y_2)$
5	y_1	y_2	$(y_1 - y_2)$
j	y_1	y_2	$(y_1 - y_2)$

y_1 in Treatment 1 column = test value for each condition.
y_2 in Treatment 2 column = test value for each condition.
d = difference.

Design C2 – Multi-Treatment Comparisons

Replicate	Treatment no.				
	1	2	3	4	i
1	y_1	y_2	y_3	y_4	y_i
2	y_1	y_2	y_3	y_4	y_i
3	y_1	y_2	y_3	y_4	y_i
4	y_1	y_2	y_3	y_4	y_i
5	y_1	y_2	y_3	y_4	y_i
j	y_1	y_2	y_3	y_4	y_i

y_1, y_2, etc. for each replicate = test values for Treatment 1 to Treatment i.

Design and Interaction Matrix for Multifactor Screening Designs
Design S1: Two Factor – No Aliases

Run no.		Design matrix		
	Y	X_1	X_2	$X_1 X_2$
1	y_1	-1	-1	1
2	y_2	1	-1	-1
3	y_3	-1	1	-1
4	y_4	1	1	1

X_1, X_2, etc. = independent variables, $-1, 1$ coded values for independent variables.
See text for coded value, alias, confounding definitions.
b_1, b_2, b_{12}, etc. = effect coefficients for independent variables X_1, X_2, and interaction $X_1 * X_2$, etc.
Y = column of measured (observed) response (dependent variable).

Coefficient Calculations
General format for all calculations:

For b_0:

$$b_0 = \left(\sum y_1 \text{ to } y_i\right)\bigg/2n, \text{ where } 2n = \text{total number of rows or runs}$$

For each b_i:

$$b_i = \frac{1}{2}\left\{\left[\left(\sum y_i * (+X_i)\right)\bigg/n\right] - \left[\left(\sum y_i * (-X_i)\right)\bigg/n\right]\right\},$$

where $y_i * (+X_i)$ = product of y and X_i value for each row with a +1 value for X_i and $y_i * (-X_i)$ = product of y and X_i vaue for each row with a −1 value for X_i. The sums are for n values (all + or all −), taken over all rows or runs; n = number of runs with +1 (or −1).

For each b_{ij}:

$$b_{ij} = \frac{1}{2}\left\{\left[\left(\sum y_i * (+X_iX_j)\right)\bigg/n\right] - \left[\left(\sum y_i * (-X_iX_j)\right)\bigg/n\right]\right\},$$

using the same procedure as above; X_iX_j = product of column X_i and X_j for each row.

For each b_{ijk} (i.e. for three-factor or greater interactions, see designs below):

$$b_{ijk} = \frac{1}{2}\left\{\left[\left(\sum y_i * (+X_iX_jX_k)\right)\bigg/n\right] - \left[\left(\sum y_i * (-X_iX_jX_k)\right)\bigg/n\right]\right\},$$

where $X_iX_jX_k$ = three way product of columns X_i, X_j, X_K for each row.

For this specific design:

$$b_1 = \tfrac{1}{2}\{[(y_2 + y_4)/2] - [(y_1 + y_3)/2]\}$$
$$b_2 = \tfrac{1}{2}\{[(y_3 + y_4)/2] - [(y_1 + y_2)/2]\}$$
$$b_{12} = \tfrac{1}{2}\{[(y_1 + y_4)/2] - [(y_2 + y_3)/2]\}$$

Design S2: Three-Factor (One Block) – No Aliases

Run no.	Y	Design matrix			Interaction matrix		
		X_1	X_2	X_3	X_1X_2	X_1X_3	X_2X_3
1	y_1	−1	−1	−1	1	1	1
2	y_2	1	−1	−1	−1	−1	1
3	y_3	−1	1	−1	−1	1	−1
4	y_4	1	1	−1	1	−1	−1
5	y_5	−1	−1	1	1	−1	−1
6	y_6	1	−1	1	−1	1	−1
7	y_7	−1	1	1	−1	−1	1
8	y_8	1	1	1	1	1	1

Coefficient Calculations
For this specific design:

$$b_0 = \left(\sum_{i=1}^{8} y_i \right) \Big/ 8$$

$$b_1 = \tfrac{1}{2}\{[(y_2 + y_4 + y_6 + y_8)/4] - [(y_1 + y_3 + y_5 + y_7)/4]\}$$

$$b_2 = \tfrac{1}{2}\{[(y_3 + y_4 + y_7 + y_8)/4] - [(y_1 + y_2 + y_5 + y_6)/4]\}$$

$$b_3 = \tfrac{1}{2}\{[(y_5 + y_6 + y_7 + y_8)/4] - [(y_1 + y_2 + y_3 + y_4)/4]\}$$

$$b_{12} = \tfrac{1}{2}\{[(y_1 + y_4 + y_5 + y_8)/4] - [(y_2 + y_3 + y_6 + y_7)/4]\}$$

$$b_{13} = \tfrac{1}{2}\{[(y_1 + y_3 + y_6 + y_8)/4] - [(y_2 + y_4 + y_5 + y_7)/4]\}$$

$$b_{23} = \tfrac{1}{2}\{[(y_1 + y_2 + y_7 + y_8)/4] - [(y_3 + y_4 + y_5 + y_6)/4]\}$$

Design S3: Three-Factor (Two Blocks)

Block	Run no.	Y	Design matrix			Interaction matrix		
			X_1	X_2	X_3	X_1X_2	X_1X_3	X_2X_3
I	1	y_1	−1	−1	1	1	−1	−1
	2	y_2	1	−1	−1	−1	−1	1
	3	y_3	−1	1	−1	−1	1	−1
	4	y_4	1	1	1	1	1	1
II	5	y_5	−1	−1	−1	1	1	1
	6	y_6	1	−1	1	−1	1	−1
	7	y_7	−1	1	1	−1	−1	1
	8	y_8	1	1	−1	1	−1	−1

Aliases: Block I: $b_3 = b_3 + b_{12}$. Block II: $b_3 = b_3 - b_{12}$; b_{123} confounded with blocks.

The b_i to the left of the = symbol in all aliases indicates the value as evaluated by the specific coefficient calculation, i.e. one-half the main or interaction effect.

Coefficient Calculations
Same as for Design S2 above.

Design S4: Four-Factor – No Aliases

Run no.	Y	Design matrix				Interaction matrix								
		X_1	X_2	X_3	X_4	X_1X_2	X_1X_3	X_1X_4	X_2X_3	X_2X_4	X_3X_4	$X_1X_2X_3$	$X_1X_2X_4$	$X_2X_3X_4$
1	y_1	−1	−1	−1	−1	1	1	1	1	1	1	−1	−1	−1
2	y_2	1	−1	−1	−1	−1	−1	−1	1	1	1	1	1	−1
3	y_3	−1	1	−1	−1	−1	1	1	−1	−1	1	1	1	1
4	y_4	1	1	−1	−1	1	−1	−1	−1	−1	1	−1	−1	1
5	y_5	−1	−1	1	−1	1	−1	1	−1	1	−1	1	−1	1
6	y_6	1	−1	1	−1	−1	1	−1	−1	1	−1	−1	1	1
7	y_7	−1	1	1	−1	−1	−1	1	1	−1	−1	−1	1	−1
8	y_8	1	1	1	−1	1	1	−1	1	−1	−1	1	−1	−1
9	y_9	−1	−1	−1	1	1	1	−1	1	−1	−1	−1	1	1
10	y_{10}	1	−1	−1	1	−1	−1	1	1	−1	−1	1	−1	1
11	y_{11}	−1	1	−1	1	−1	1	−1	−1	1	−1	1	−1	−1
12	y_{12}	1	1	−1	1	1	−1	1	−1	1	−1	−1	1	−1
13	y_{13}	−1	−1	1	1	1	−1	−1	−1	−1	1	1	1	−1
14	y_{14}	1	−1	1	1	−1	1	1	−1	−1	1	−1	−1	−1
15	y_{15}	−1	1	1	1	−1	−1	−1	1	1	1	−1	−1	1
16	y_{16}	1	1	1	1	1	1	1	1	1	1	1	1	1

Coefficient Calculations
See general format equations as given for Design S1 for coefficients from X_1 to $X_2X_3X_4$.

Design S5: Four-Factor (1/2 Fraction)

Run no.	Y	Design matrix				Interaction matrix					
		X_1	X_2	X_3	X_4	X_1X_2	X_1X_3	X_1X_4	X_2X_3	X_2X_4	X_3X_4
1	y_1	−1	−1	−1	−1	1	1	1	1	1	1
2	y_2	1	−1	−1	1	−1	−1	1	1	−1	−1
3	y_3	−1	1	−1	1	−1	1	−1	−1	1	−1
4	y_4	1	1	−1	−1	1	−1	−1	−1	−1	1
5	y_5	−1	−1	1	1	1	−1	−1	−1	−1	1
6	y_6	1	−1	1	−1	−1	1	−1	−1	1	−1
7	y_7	−1	1	1	−1	−1	−1	1	1	−1	−1
8	y_8	1	1	1	1	1	1	1	1	1	1

Aliases: $b_4 = b_4 + b_{123}$, $b_{12} = b_{12} + b_{34}$, $b_{13} = b_{13} + b_{24}$, $b_{14} = b_{14} + b_{23}$, $b_{23} = b_{23} + b_{14}$, $b_{24} = b_{24} + b_{13}$, $b_{34} = b_{34} + b_{12}$, b_{1234} confounded with other fraction.

Coefficient Calculations
Same format, X_1 to X_3X_4, as for Design S1.

Design S6: Five-Factor (1/2 Fraction) – No Aliases

Run no.	Y	Design matrix					Interaction matrix									
		X_1	X_2	X_3	X_4	X_5	X_1X_2	X_1X_3	X_1X_4	X_1X_5	X_2X_3	X_2X_4	X_2X_5	X_3X_4	X_3X_5	X_4X_5
1	y_1	−1	−1	−1	−1	1	1	1	1	−1	1	1	−1	1	−1	−1
2	y_2	1	−1	−1	−1	−1	−1	−1	−1	−1	1	1	1	1	1	1
3	y_3	−1	1	−1	−1	−1	−1	1	1	1	−1	−1	−1	1	1	1
4	y_4	1	1	−1	−1	1	1	−1	−1	1	−1	−1	1	1	−1	−1
5	y_5	−1	−1	1	−1	−1	1	−1	1	1	−1	1	1	−1	−1	1
6	y_6	1	−1	1	−1	1	−1	1	−1	1	−1	1	−1	−1	1	−1
7	y_7	−1	1	1	−1	1	−1	−1	1	−1	1	−1	1	−1	1	−1
8	y_8	1	1	1	−1	−1	1	1	−1	−1	1	−1	−1	−1	−1	1
9	y_9	−1	−1	−1	1	−1	1	1	−1	1	1	−1	1	−1	1	−1
10	y_{10}	1	−1	−1	1	1	−1	−1	1	1	1	−1	−1	−1	−1	1
11	y_{11}	−1	1	−1	1	1	−1	1	−1	−1	−1	1	1	−1	−1	1
12	y_{12}	1	1	−1	1	−1	1	−1	1	−1	−1	1	−1	−1	1	−1
13	y_{13}	−1	−1	1	1	1	1	−1	−1	−1	−1	−1	−1	1	1	1
14	y_{14}	1	−1	1	1	−1	−1	1	1	−1	−1	−1	1	1	−1	−1
15	y_{15}	−1	1	1	1	−1	−1	−1	−1	1	1	1	−1	1	−1	−1
16	y_{16}	1	1	1	1	1	1	1	1	1	1	1	1	1	1	1

Coefficient Calculations
Same format, X_1 to X_4X_5, as for Design S1.

Design S7: Five-Factor (1/4 Fraction)

Run no.	Y	Design matrix X_1 X_2 X_3 X_4 X_5					Interaction matrix X_1X_2 X_1X_3 X_1X_4 X_1X_5 X_2X_3 X_2X_4 X_2X_5 X_3X_4 X_3X_5 X_4X_5									
1	y_1	−1	−1	−1	−1	−1	1	1	1	1	1	1	1	1	1	1
2	y_2	1	1	−1	−1	−1	1	−1	−1	−1	−1	−1	−1	1	1	1
3	y_3	−1	−1	1	1	−1	1	−1	−1	1	−1	−1	1	1	−1	−1
4	y_4	1	−1	1	−1	1	−1	1	−1	1	−1	1	−1	−1	1	−1
5	y_5	−1	1	1	−1	1	−1	−1	1	−1	1	−1	1	−1	1	−1
6	y_6	1	−1	−1	1	1	−1	−1	1	1	1	−1	−1	−1	−1	1
7	y_7	−1	1	−1	1	1	−1	1	−1	−1	−1	1	1	−1	−1	1
8	y_8	1	1	1	1	−1	1	1	1	−1	1	1	−1	1	−1	−1

Aliases: $b_1 = b_1 - b_{25}$, $b_2 = b_2 - b_{15}$, $b_3 = b_3 - b_{45}$, $b_4 = b_4 - b_{35}$, $b_5 = b_5 - b_{12}$.

Coefficient Calculations
Same format, X_1 to X_4X_5, as for Design S1.

Design S8: Six-Factor (1/4 Fraction)

Run no.	Y	Design matrix						Interaction matrix (2 Factor)														
		X_1	X_2	X_3	X_4	X_5	X_6	X_1X_2	X_1X_3	X_1X_4	X_1X_5	X_1X_6	X_2X_3	X_2X_4	X_2X_5	X_2X_6	X_3X_4	X_3X_5	X_3X_6	X_4X_5	X_4X_6	X_5X_6
1	y_1	-1	-1	-1	-1	-1	-1	1	1	1	1	1	1	1	1	1	1	1	1	1	1	1
2	y_2	1	-1	-1	-1	1	-1	-1	-1	-1	1	-1	1	1	-1	1	1	-1	1	-1	1	-1
3	y_3	-1	1	-1	-1	1	1	-1	1	1	-1	-1	-1	-1	1	1	1	-1	-1	-1	-1	1
4	y_4	1	1	-1	-1	-1	1	1	-1	-1	-1	1	-1	-1	-1	1	1	1	-1	1	-1	-1
5	y_5	-1	-1	1	-1	1	1	1	-1	1	-1	-1	-1	1	-1	-1	-1	1	1	-1	-1	1
6	y_6	1	-1	1	-1	-1	1	-1	1	-1	-1	1	-1	1	1	-1	-1	-1	1	1	-1	-1
7	y_7	-1	1	1	-1	-1	-1	-1	-1	1	1	1	1	-1	-1	-1	-1	-1	1	1	1	1
8	y_8	1	1	1	-1	1	-1	1	1	-1	1	-1	1	-1	1	-1	-1	1	-1	-1	1	-1
9	y_9	-1	-1	-1	1	-1	1	1	1	-1	1	-1	1	-1	1	-1	-1	1	-1	-1	1	-1
10	y_{10}	1	-1	-1	1	1	1	-1	-1	1	1	1	1	-1	-1	-1	-1	-1	1	1	1	1
11	y_{11}	-1	1	-1	1	1	-1	-1	1	-1	-1	1	-1	1	1	-1	-1	-1	1	1	-1	-1
12	y_{12}	1	1	-1	1	-1	-1	1	-1	1	-1	-1	-1	1	-1	-1	-1	1	1	-1	-1	1
13	y_{13}	-1	-1	1	1	1	-1	1	-1	-1	-1	1	-1	-1	-1	1	1	1	-1	1	-1	-1
14	y_{14}	1	-1	1	1	-1	-1	-1	1	1	-1	-1	-1	-1	1	1	1	-1	-1	-1	-1	1
15	y_{15}	-1	1	1	1	-1	1	-1	-1	-1	1	-1	1	1	-1	1	1	-1	1	-1	1	-1
16	y_{16}	1	1	1	1	1	1	1	1	1	1	1	1	1	1	1	1	1	1	1	1	1

Aliases: $b_1 = b_1 + b_{235} + b_{456}$, $b_2 = b_2 + b_{135} + b_{346}$, $b_3 = b_3 + b_{125} + b_{246}$, $b_4 = b_4 + b_{236} + b_{156}$, $b_5 = b_5 + b_{123} + b_{146}$, $b_6 = b_6 + b_{234} + b_{145}$, $b_{12} = b_{12} + b_{35}$, $b_{13} = b_{13} + b_{25}$, $b_{14} = b_{14} + b_{56}$, $b_{15} = b_{15} + b_{46}$, $b_{16} = b_{16} + b_{45}$, $b_{23} = b_{23} + b_{46}$, $b_{24} = b_{24} + b_{36}$, $b_{25} = b_{25} + b_{13}$, $b_{26} = b_{26} + b_{34}$.

Coefficient Calculations
Same format, X_1 to X_5X_6, as for Design S1.

Design S9: Seven-Factor (1/16 Fraction)

Run no.	Y	Design matrix						
		X_1	X_2	X_3	X_4	X_5	X_6	X_7
1	y_1	−1	−1	−1	1	1	1	−1
2	y_2	1	−1	−1	−1	−1	1	1
3	y_3	−1	1	−1	−1	1	−1	1
4	y_4	1	1	−1	1	−1	−1	−1
5	y_5	−1	−1	1	1	−1	−1	1
6	y_6	1	−1	1	−1	1	−1	−1
7	y_7	−1	1	1	−1	−1	1	−1
8	y_8	1	1	1	1	1	1	1

Main effects only are evaluated.
Aliases: $b_1 = b_1 + b_{24} + b_{35} + b_{67}, b_2 = b_2 + b_{14} + b_{36} + b_{57}, b_3 = b_3 + b_{15} + b_{26} + b_{47},$
$b_4 = b_4 + b_{12} + b_{56} + b_{37}, b_5 = b_5 + b_{13} + b_{46} + b_{27}, b_6 = b_6 + b_{23} + b_{45} + b_{17},$
$b_7 = b_7 + b_{34} + b_{25} + b_{16}.$

Coefficient Calculations
Same format as for Design S1, for X_1 to X_7 only.

Design S10: Seven-Factor (1/16 Fraction B)

Run no.	Y	Design matrix						
		X_1	X_2	X_3	X_4	X_5	X_6	X_7
9	y_1	1	1	1	−1	−1	−1	1
10	y_2	−1	1	1	1	1	−1	−1
11	y_3	1	−1	1	1	−1	1	−1
12	y_4	−1	−1	1	−1	1	1	1
13	y_5	1	1	−1	−1	1	1	−1
14	y_6	−1	1	−1	1	−1	1	1
15	y_7	1	−1	−1	1	1	−1	1
16	y_8	−1	−1	−1	−1	−1	−1	−1

Main effects only are evaluated
Aliases: Same as Design S9.

Coefficient Calculations
Same format as for Design S1, for X_1 to X_7 only.

Design S11: Seven-Factor (Two 1/16 Fractions Combined)

Run no.	Y	X_1	X_2	X_3	X_4	X_5	X_6	X_7	X_1X_2	X_1X_3	X_1X_4	X_1X_5	X_1X_6	X_1X_7	X_2X_4
		\multicolumn{7}{c}{Design matrix}							\multicolumn{7}{c}{Interaction matrix}						
1	y_1	−1	−1	−1	1	1	1	−1	1	1	−1	−1	−1	1	−1
2	y_2	1	−1	−1	−1	−1	1	1	−1	−1	−1	−1	1	1	1
3	y_3	−1	1	−1	−1	1	−1	1	−1	1	1	−1	1	−1	−1
4	y_4	1	1	−1	1	−1	−1	−1	1	−1	1	−1	−1	−1	1
5	y_5	−1	−1	1	1	−1	−1	1	1	−1	−1	1	1	−1	−1
6	y_6	1	−1	1	−1	1	−1	−1	−1	1	−1	1	−1	−1	1
7	y_7	−1	1	1	−1	−1	1	−1	−1	−1	1	1	−1	1	−1
8	y_8	1	1	1	1	1	1	1	1	1	1	1	1	1	1
9	y_9	1	1	1	−1	−1	−1	1	1	1	−1	−1	−1	1	−1
10	y_{10}	−1	1	1	1	1	−1	−1	−1	−1	−1	−1	1	1	1
11	y_{11}	1	−1	1	1	−1	1	−1	−1	1	1	−1	1	−1	−1
12	y_{12}	−1	−1	1	−1	1	1	1	1	−1	1	−1	−1	−1	1
13	y_{13}	1	1	−1	−1	1	1	−1	1	−1	−1	1	1	−1	−1
14	y_{14}	−1	1	−1	1	−1	1	1	−1	1	−1	1	−1	−1	1
15	y_{15}	1	−1	−1	1	1	−1	1	−1	−1	1	1	−1	1	−1
16	y_{16}	−1	−1	−1	−1	−1	−1	−1	1	1	1	1	1	1	1

Aliases: All main effects clear.
Two-factor interactions:

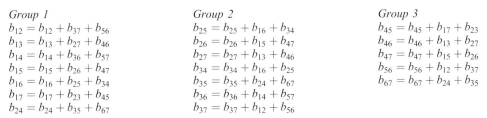

Group 1
$b_{12} = b_{12} + b_{37} + b_{56}$
$b_{13} = b_{13} + b_{27} + b_{46}$
$b_{14} = b_{14} + b_{36} + b_{57}$
$b_{15} = b_{15} + b_{26} + b_{47}$
$b_{16} = b_{16} + b_{25} + b_{34}$
$b_{17} = b_{17} + b_{23} + b_{45}$
$b_{24} = b_{24} + b_{35} + b_{67}$

Group 2
$b_{25} = b_{25} + b_{16} + b_{34}$
$b_{26} = b_{26} + b_{15} + b_{47}$
$b_{27} = b_{27} + b_{13} + b_{46}$
$b_{34} = b_{34} + b_{16} + b_{25}$
$b_{35} = b_{35} + b_{24} + b_{67}$
$b_{36} = b_{36} + b_{14} + b_{57}$
$b_{37} = b_{37} + b_{12} + b_{56}$

Group 3
$b_{45} = b_{45} + b_{17} + b_{23}$
$b_{46} = b_{46} + b_{13} + b_{27}$
$b_{47} = b_{47} + b_{15} + b_{26}$
$b_{56} = b_{56} + b_{12} + b_{37}$
$b_{67} = b_{67} + b_{24} + b_{35}$

Coefficient Calculations
Same format as for Design S1, for X_1 to X_7 and X_1X_2 to X_2X_4 in Group 1.
b_{25} to $b_{67} = b_{16}$ to b_{24} respectively, in sequence for middle b values as given by Groups 2 and 3.

Part 2   Exploratory Designs: Multifactor Experiments with Continuous Variables

Design E1: Two-Factor (Hexagonal Design – 8 Runs)

Run no.	Y	X_1	X_2
1	y_1	0	1
2	y_2	0.87	0.50
3	y_3	0.87	−0.50
4	y_4	0	−1
5	y_5	−0.87	−0.50
6	y_6	−0.87	0.50
7	y_7	0	0
8	y_8	0	0

Coefficient Calculations

$b_0 = \frac{1}{2}(y_7 + y_8)$

$b_1 = 0.289(y_2 + y_3) - 0.289(y_5 + y_6)$

$b_2 = 0.333(y_1 - y_4) + 0.167(y_2 + y_6 - y_3 - y_5)$

$b_{12} = 0.577(y_2 + y_5) - 0.577(y_3 + y_6)$

$b_{11} = 0.333(y_2 + y_3 + y_5 + y_6) - 0.167(y_1 + y_4) - 0.50(y_7 + y_8)$

$b_{22} = 0.500(y_1 + y_4) - 0.500(y_7 + y_8)$

Design E2: Two-Factor (Square + Star – 13 Runs)

Run no.	Y	X_1	X_2
1	y_1	−1	−1
2	y_2	1	−1
3	y_3	−1	1
4	y_4	1	1
5	y_5	−1.41	0
6	y_6	1.41	0
7	y_7	0	−1.41
8	y_8	0	1.41
9	y_9	0	0
10	y_{10}	0	0
11	y_{11}	0	0
12	y_{12}	0	0
13	y_{13}	0	0

SE $(b_i) = 0.354$ Sy

SE $(b_{ij}) = 0.50$ Sy

SE $(b_{ii}) = 0.379$ Sy

SE $(b_i) =$ standard deviation (error) for linear term coefficients

SE $(b_{ij}) =$ standard deviation (error) for interaction term coefficients

SE $(b_{ii}) =$ standard deviation (error) for squared term coefficients

Sy = standard deviation of (individual) y values

Coefficient Calculations

$b_0 = 0.20(oy) - 0.10[\sum(iiy)]$

$b_1 = 0.125(1y)$

$b_2 = 0.125(2y)$
$b_{12} = 0.25(12y)$
$b_{11} = 0.125(11y) + 0.01875[\sum(iiy)] - 0.01(oy)$
$b_{22} = 0.125(22y) + 0.01875[\sum(iiy)] - 0.10(oy)$
Generic symbols defined as follows where sums are taken over all rows of matrix:
(oy) = sum y_1 to y_i in column Y
(iy) = sum of products of column X_i and y_i values, of column Y
(ijy) = sum of products of columns X_iX_j and y_i values
For this design:
$(1y)$ = sum of products of column X_1 and y_i (summed over all rows of matrix)
$(2y)$ = sum of products of column X_2 and y_i (summed over all rows of matrix)
$(12y)$ = sum of products of columns X_1, X_2 and y_i (summed over all rows of matrix)
$(11y)$ = sum of products of column X_1 squared and y_i (summed over all rows of matrix)
$(22y)$ = sum of products of column X_2 squared and y_i (summed over all rows of matrix)
Note: This procedure is continued on to total number of X-variables, for subsequent designs and $[\sum(iiy)]$ indicates "sum of sums."

Design E3: Three-Factor (18 Runs)

Run no.	Y	X_1	X_2	X_3
1	y_1	−1	−1	−1
2	y_2	1	−1	−1
3	y_3	−1	1	−1
4	y_4	1	1	−1
5	y_5	−1	−1	1
6	y_6	1	−1	1
7	y_7	−1	1	1
8	y_8	1	1	1
9	y_9	−1.68	0	0
10	y_{10}	1.68	0	0
11	y_{11}	0	−1.68	0
12	y_{12}	0	1.68	0
13	y_{13}	0	0	−1.68
14	y_{14}	0	0	1.68
15	y_{15}	0	0	0
16	y_{16}	0	0	0
17	y_{17}	0	0	0
18	y_{18}	0	0	0

SE $(b_i) = 0.271$ Sy
SE $(b_{ij}) = 0.354$ Sy
SE $(b_{ii}) = 0.263$ Sy

Coefficient Calculations
$b_0 = 0.166(oy) - 0.0568[\sum(iiy)]$
$b_i = 0.0732(iy)$
$b_{ij} = 0.125(ijy)$
$b_{ii} = 0.0625(iiy) + 0.00689[\sum(iiy)] - 0.0568(oy)$

Design E4: Three-Factor (3 Blocks – 20 Runs)

Block 1

Run no.	Y	X_1	X_2	X_3
1	y_1	−1	−1	1
2	y_2	1	−1	−1
3	y_3	−1	1	−1
4	y_4	1	1	1
5	y_5	0	0	0
6	y_6	0	0	0

Block 2

Run no.	Y	X_1	X_2	X_3
7	y_7	−1	−1	−1
8	y_8	1	−1	1
9	y_9	−1	1	1
10	y_{10}	1	1	−1
11	y_{11}	0	0	0
12	y_{12}	0	0	0

Block 3

Run no.	Y	X_1	X_2	X_3
13	y_{13}	−1.63	0	0
14	y_{14}	1.63	0	0
15	y_{15}	0	−1.63	0
16	y_{16}	0	1.63	0
17	y_{17}	0	0	−1.63
18	y_{18}	0	0	1.63
19	y_{19}	0	0	0
20	y_{20}	0	0	0

SE $(b_i) = 0.274$ Sy
SE $(b_{ij}) = 0.354$ Sy
SE $(b_{ii}) = 0.275$ Sy

Coefficient Calculations
$b_0 = 0.165(oy) - 0.0577[\sum(iiy)]$
$b_i = 0.075(iy)$
$b_{ij} = 0.125(ijy)$
$b_{ii} = 0.0703(iiy) + 0.00540[\sum(iiy)] - 0.0577(oy)$

Design E5: Four-Factor (3 Blocks – 30 runs)

Run no.	Y	X_1	X_2	X_3	X_4
1	y_1	-1	-1	-1	-1
2	y_2	1	-1	-1	1
3	y_3	-1	1	-1	1
4	y_4	1	1	-1	-1
5	y_5	-1	-1	1	1
6	y_6	1	-1	1	-1
7	y_7	-1	1	1	-1
8	y_8	1	1	1	1
9	y_9	0	0	0	0
10	y_{10}	0	0	0	0

Block 2

Run no.	Y	X_1	X_2	X_3	X_4
11	y_{11}	-1	-1	-1	1
12	y_{12}	1	-1	-1	-1
13	y_{13}	-1	1	-1	-1
14	y_{14}	1	1	-1	1
15	y_{15}	-1	-1	1	-1
16	y_{16}	1	-1	1	1
17	y_{17}	-1	1	1	1
18	y_{18}	1	1	1	-1
19	y_{19}	0	0	0	0
20	y_{20}	0	0	0	0

Block 3

Run no.	Y	X_1	X_2	X_3	X_4
21	y_{21}	-2	0	0	0
22	y_{22}	2	0	0	0
23	y_{23}	0	-2	0	0
24	y_{24}	0	2	0	0
25	y_{25}	0	0	-2	0
26	y_{26}	0	0	2	0
27	y_{27}	0	0	0	-2
28	y_{28}	0	0	0	2
29	y_{29}	0	0	0	0
30	y_{30}	0	0	0	0

Block 1

SE $(b_i) = 0.204$ Sy   SE $(b_{ij}) = 0.250$ Sy   SE $(b_{ii}) = 0.191$ Sy

Coefficient Calculations
$b_0 = 0.166(oy) - 0.0417[\sum(iiy)]$
$b_i = 0.0417(iy)$
$b_{ij} = 0.0625(ijy)$
$b_{ii} = 0.0313(iiy) + 0.00521[\sum(iiy)] - 0.0417(oy)$

Design E6: Five-Factor (2 Blocks – 33 Runs)

Run no.	Y	X_1	X_2	X_3	X_4	X_5
1	y_1	−1	−1	−1	−1	1
2	y_2	1	−1	−1	−1	−1
3	y_3	−1	1	−1	−1	−1
4	y_4	1	1	−1	−1	1
5	y_5	−1	−1	1	−1	−1
6	y_6	1	−1	1	−1	1
7	y_7	−1	1	1	−1	1
8	y_8	1	1	1	−1	−1
9	y_9	−1	−1	−1	1	−1
10	y_{10}	1	−1	−1	1	1
11	y_{11}	−1	1	−1	1	1
12	y_{12}	1	1	−1	1	−1
13	y_{13}	−1	−1	1	1	1
14	y_{14}	1	−1	1	1	−1
15	y_{15}	−1	1	1	1	−1
16	y_{16}	1	1	1	1	1
17	y_{17}	0	0	0	0	0
18	y_{18}	0	0	0	0	0
19	y_{19}	0	0	0	0	0
20	y_{20}	0	0	0	0	0
21	y_{21}	0	0	0	0	0
22	y_{22}	0	0	0	0	0

Block 2

Run no.	Y	X_1	X_2	X_3	X_4	X_5
23	y_{23}	−2	0	0	0	0
24	y_{24}	2	0	0	0	0
25	y_{25}	0	−2	0	0	0
26	y_{26}	0	2	0	0	0
27	y_{27}	0	0	−2	0	0
28	y_{28}	0	0	2	0	0
29	y_{29}	0	0	0	−2	0
30	y_{30}	0	0	0	2	0
31	y_{31}	0	0	0	0	−2
32	y_{32}	0	0	0	0	2
33	y_{33}	0	0	0	0	0

Block 1

SE (b_i) = 0.204 Sy   SE (b_{ij}) = 0.250 Sy   SE (b_{ii}) = 0.182 Sy

Coefficient Calculations
$b_0 = 0.137(oy) − 0.0294[\sum(iiy)]$
$b_i = 0.0417(iy)$
$b_{ij} = 0.0625(ijy)$
$b_{ii} + 0.0313(iiy) + 0.00184[\sum(iiy)] − 0.0294(oy)$

6   Elastomer Selection

Rudy School

6.1   Overview

The selection of a particular rubber or elastomer depends on the intended job for the end product. Once a series of elastomers is selected for a project, the compounder then choose a polymer based on its properties, costs, or perhaps a combination of the two.

This chapter examines compounders' choices and attempts to assist decision-making in the area. For simplicity, this chapter uses the word ''elastomer'' in all references to rubber or elastomeric polymers, even though the two are defined differently by ASTM [1]. This chapter examines these materials in terms of their classifications by Fusco [2] in 1996:

- Commodity or general-purpose elastomers, such as natural rubber and SBR
- High-volume specialty elastomers, such as EPDM and polychloroprene
- Low-volume specialty elastomers, such as polyurethanes, silicones, and fluoro-elastomers

There now are more than two dozen major elastomers – both thermosetting types and thermoplastic types – available, but it is unlikely that any one elastomer meets all the properties required in an end use compound.

While the tire industry consumes about 70% of the 16 million metric tons of elastomers used annually, automotive- and non-automotive applications account for another 30%, making mechanical rubber goods the end use for much of the low- and high-volume specialty elastomers produced today.

Table 6.1 shows the various types of rubbers and elastomers and their abbreviations. Figure 6.1 shows a comparison of maximum tensile strengths that are achievable with various elastomers. Figure 6.2 compares polymers' brittleness points. Figure 6.3 is a price index of various elastomers compared to natural rubber (1.0 in the index). Caution: There are many specific polymers with different properties under each of the ''Types of Elastomers'' shown. Therefore it is very difficult for one set of numbers to represent what is actually many different polymers.

6.1.1   Commodity and General Purpose Elastomers

General purpose elastomers, often referred to as commodity elastomers, are recognized easily in the market because of their low selling prices relative to the specialty elastomers and their big volumes of usage. These elastomers usually provide high strength and good abrasion resistance along with low hysteresis and high resilience.

These elastomers require antidegradants in the mixed compound because they generally have poor resistance to both heat and ozone.

Table 6.1   Polymer Common Names and ASTM Designations

Polymer name	ASTM designation
Ethylene propylene diene	EPDM
Nitrile rubber	NBR
Hydrogenated nitrile rubber	HNBR
Polychloroprene	CR
Natural rubber	NR
Polyisoprene	IR
Styrene butadiene rubber	SBR
Butyl rubber	IIR
Bromobutyl rubber	BIIR
Chlorobutyl rubber	CIIR
Polybutadiene	BR
Silicone	MQ, VMQ, PMQ
Fluorosilicone	FMQ
Fluoroelastomer	FKM
Fluoroethylene/propylene	FEPM
Perfluoroelastomer	FFKM
Polyester urethane	AU
Polyether urethane	EU
Ethylene acrylic	AEM
Polysulfide	T
Epichlorohydrin	CO
Chlorosulfonated polyethylene	CSM
Chlorinated polyethylene	CPE
Polyacrylate	ACM
Polyoctene	POE
Thermoplastic elastomer	TPE

6.1.1.1   Natural Rubber (NR)

Natural rubber is the only non-synthetic elastomer in wide use, although some variations such as guayule continue to be researched for applications in tire and non-tire end uses. Derived from the milk-like liquid of the *Hevea* tree, natural rubber latex contains many organic and inorganic impurities.

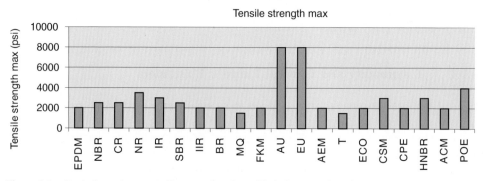

Figure 6.1   Typical maximum tensile strengths of specified elastomer-based, cured compounds.

Table 6.2   Polymer Chart

Polymer	Durometer min	Durometer max	Tensile strength max ('000)	Brittle pt °C	Average price ($)	Index
NR	30	90	3500	−58	0.45	1.00
SBR	40	90	2500	−58	0.50	1.11
BR	40	80	2000	−73	0.58	1.29
NBR	40	90	2500	−51	1.00	2.22
IR	30	90	3000	−58	1.00	2.22
POE	65	95	4000	−76	1.05	2.33
EPDM	40	90	2000	−58	1.10	2.44
IIR	40	80	2000	−58	1.26	2.80
CPE	50	90	2000	−40	1.80	4.00
CR	40	90	2500	−54	1.95	4.33
CSM	40	90	3000	−55	2.20	4.89
T	20	80	1500	−51	2.50	5.56
AU	60	95	8000	−51	3.50	7.78
EU	60	95	8000	−51	3.50	7.78
AEM	40	90	2000	−55	3.50	7.78
MQ	40	80	1500	−118	6.00	13.33
ECO	40	80	2000	−40	9.00	20.00
ACM	60	95	2000	−34	10.00	22.22
HNBR	45	90	3000	−55	13.00	28.89
FKM	55	90	2000	−40	19.00	42.22

Generally, NR is the material used to rate the properties of synthetic elastomers. It holds its strength during deformation. It is highly resilient and, because of its hysteretic properties, experiences little heat buildup during flexing, which makes it a material of choice when shock and dynamic load requirements are important. Heat buildup can cause thermal degradation in vibration-absorbing devices.

Natural rubber also has its shortcomings, such as poor resistance to ozone, high temperatures, weathering, oxidation, oils, and concentrated acids and bases. Often a synthetic rubber is blended with NR to help alleviate these problems, and antidegradants and other chemicals help to produce a usable compound for tires, vibration dampers, seals, isolators, shock mounts, and many other static and dynamic applications.

High green strength and tack are two important characteristics of natural rubber. The green strength allows an uncured component, such as a tire, to hold its shape during tire building and "green" tire storage, while tack is useful when building the different components of a tire. As a rule of thumb, compounds with a durometer between 30 and 95 can be produced by using NR.

6.1.1.2   Styrene-Butadiene Rubber (SBR)

Roughly 75% of all SBR used in North America goes into tires because SBR provides good traction properties and abrasion resistance. SBR can be produced by an emulsion process and a solution process; the tire industry now favors solution SBR because the

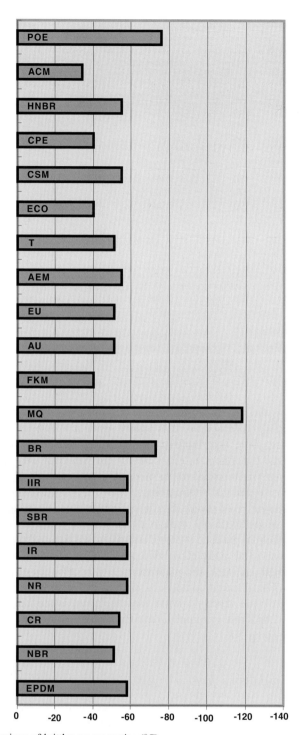

Figure 6.2   Comparison of brittleness properties (°C).

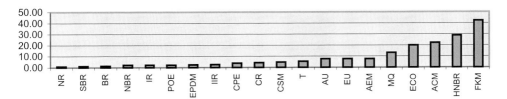

Figure 6.3   Price index of natural rubber (NR = 1.0). (Note: Price of natural rubber fluctuates greatly compared to prices of synthetic elastomers.)

material offers improvements in both abrasion and traction, as well as in hysteresis properties.

In emulsion polymerization, the monomers – styrene and butadiene – are emulsified in water by using a soap as an emulsifying agent [4]. High or low temperature processes are used, which leads to the designations of Hot SBR and Cold SBR. The cold polymers provide better abrasion resistance and also improved dynamic properties compared to the hot polymers. Cold SBRs also accept more fillers and other extending materials than do the hot SBRs.

The solution SBRs have higher purity because they do not contain the soap residues common to all emulsion SBRs. Such residues can be as much as 7% by weight of the polymer bale. Solution SBRs are made in a solution of hydrocarbons and yield products that can be tailored to a greater degree for stereospecificity and for molecular weight distribution. Alkyl-lithium-based catalyst systems are used because they are the only stereo-specific catalysts that copolymerize styrene and butadiene [5].

Regardless of type, SBR must be reinforced to provide acceptable tensile strength and tear resistance. Polymer processors provide a masterbatch of SBR and carbon black – also in an oil-extended version – for use by companies who have limited mixing capacity or prefer not to use loose carbon black in their manufacturing plants.

Extrusions made from SBR tend to be less scorchy during processing than those made from natural rubber. Hot SBRs are usable when a lower level of dynamic properties is required. One such application is solvent-based adhesives.

SBR can be mixed with other compounding ingredients either in an internal mixer or on an open mill. The resulting compounds are suitable for any type of processing, including calendering, molding (compression, transfer, or injection methods), and extruding.

6.1.1.3   Polybutadiene Rubber (BR)

Polybutadiene also is produced by either an emulsion process or by a solution process. The material offers the best low-temperature resistance of any organic elastomer; only silicone rubber is better. The polymer, however, suffers extensive cold flow at room temperature, which makes it difficult to store bales of the material in containers that are not specially reinforced.

The resistance to cold and also excellent flex properties make polybutadiene a useful synthetic rubber. Although more resilient than natural rubber, because of processing problems, BR is used in lower proportions in blends of two or more polymers.

Polybutadiene, which exhibits less dynamic heat buildup than NR, lacks the durability, toughness, and cut-growth resistance of the latter. It is blended with NR or with SBR to improve low temperature flexibility of those compounds. Internal mixing is the process of choice when compounds contain high proportions of polybutadiene.

The structures of polybutadiene vary by supplier and include high-*cis* (i.e., greater than 90% *cis* 1,4); low-*cis* (less than 40% *cis* 1,4); and vinyl. The solution BRs may contain a branching agent such as divinyl benzene to help improve cold-flow. Catalyst systems used in manufacture of the polymer also affect the choice of the compounder, with the neodymium- types banding on mill rolls more readily than the lithium-alkyl-catalyzed types.

Polybutadiene-based compounds can be sulfur-cured with systems activated by zinc oxide and fatty acids or crosslinked with peroxides. Peroxides can be used with high-*cis* BRs. The polymers generally should not be blended with polar elastomers such as nitrile rubber (NBR).

The choice of a high-vinyl BR results in a slower cure rate for the compound compared with conventional BRs [7]. The use of a primary and secondary accelerator together is suggested, and recommendations include sulfenamide- and possibly thiuram-type accelerators to "kick" the cure.

Polybutadienes have a fair resistance to oxidation despite their unsaturation, although most compounders prefer to add antidegradants to improve the ozone resistance.

6.1.2   High Volume Specialty Elastomers

While tires consume a small amount of these materials, the vast majority are used in automotive applications, industrial rubber goods, wire and cable, pharmaceuticals, and adhesives and sealants. Among the specialties, butyl and halogenated butyl polymers find most of their applications in tires.

High volume specialty elastomers generally are priced higher than commodity elastomers, but lower than low volume specialty elastomers.

6.1.2.1   Polyisoprene (IR)

These "synthetic natural rubber" products could be grouped as commodity rubbers for ease of comparison, but in reality are specialty elastomers. Polyisoprene is an excellent choice for applications that generally use natural rubber, especially where processing is important and tack and green strength are not as critical. Polyisoprene is gaining favor in medical applications as a replacement for natural rubber to eliminate the dangers of patients suffering allergic reactions to the natural proteins and impurities found in NR.

Polyisoprene also produces less odor than NR, but at a cost: IR costs generally are two to three times more than natural rubber. It also has a lower tear resistance, but is similar in chemical resistance, hysteresis, and oxidation resistance.

Polyisoprene processes better than NR, with extrusion speeds increased and extrudates smoother and less affected by viscous heating. Processing improvement results from tight manufacturing controls compared with the variability of the naturally occurring rubber. Polyisoprene has less nerve than NR, which reduces the die swell compared with NRs of similar plasticity.

The absence of impurities that affect cure also improves cure consistency, although slower cures generally occur with the synthetic. The reduction in *cis* content of polyisoprene, however, reduces its performance in applications that involve flex fatigue.

Finally, polyisoprene-based compounds can contain higher loadings of extenders than can NR in blends with SBR or EPDM.

Generally the compounder can offset the initial increased costs of IR at least partially by eliminating the need for pre-mastication of the polymer and reducing the residence time of the batch in an internal mixer.

6.1.2.2   Nitrile Rubber (NBR)

This highly polar elastomer is a copolymer of butadiene and acrylonitrile. The acrylonitrile content provides oil resistance to the elastomer, which improves as the level of acrylonitrile is increased. Unfortunately, an increase in acrylonitrile content harms low temperature flexibility and increases compound hardness. Typical ACN content ranges between 18 and 45%.

On a cost basis, NBR is the least expensive of the oil-resistant elastomers. Its resistance to many hydrocarbons is superior to that of polychloroprene but inferior to that of polysulfide elastomers. It also exhibits somewhat better heat-aging characteristics than natural rubber. NBR does not crystallize on stretching, so reinforcing agents are required to improve strength. Tear resistance is less than that of natural rubber.

The compounder has numerous choices for cure systems. High temperature applications usually employ peroxides and thiurams, as well as the use of a thiazole or sulfenamide with a secondary accelerator such as thiuram. The peroxides can be combined with coagents and blends of sulfur with thiazoles or sulfenamides.

The NBR-based compounds accept many antidegradants to improve compound performance, and some polymer makers permanently bind an antioxidant into the polymer molecule to improve resistance to heat. Such bound antioxidants are more difficult to extract than those added at the compound-mixing stage.

Compounders seeking a significant improvement in heat resistance should consider the more costly hydrogenated nitrile rubber (HNBR). The hydrogenation process removes most of the residual unsaturation in the polymer to make it far less vulnerable to attack by heat, ozone, and oxygen.

HNBR is cured with sulfur- or with peroxide-based systems with compound durometers ranging from 30 to greater than 90. HNBR has an excellent flex-fatigue resistance. It can be compounded for both high and low temperature uses. Some end use markets include synchronous timing belts, power-steering systems, and other automotive applications.

Carboxylated nitrile elastomers, often referred to as XNBR or CNBR, have improved strength properties and better abrasion resistance than conventional NBRs. The improvements are derived by the addition of carboxylic acid groups to the polymer backbone during the manufacturing process. These groups then provide additional crosslinking sites during the curing step.

Ozone resistance is improved by using a blend of NBR with polyvinyl chloride (PVC). The NBR is added to the plasticized PVC and fluxed in an internal mixer before being pelletized. The blend produces a PVC-based compound with a rubber-like feel but with better abrasion resistance and compression set than typical PVC compounds.

6.1.2.3   Ethylene-Propylene-Diene (EPDM)

EPDM is one of the most widely used "specialty" elastomers, finding application in single-ply roofing, automotive parts, wire-and-cable covers, and many other fields. One of its major attributes is superior resistance to ozone.

Compounders seeking a low-cost compound can rely on EPDM because it can be highly loaded with low-cost fillers, including clays, silicas, carbon black, and talcs.

EPDM can be cured with sulfur or with peroxides, although applications with high heat requirements should use peroxide-cured compounds. The peroxide also produces compounds with compression set properties that are superior to those of sulfur-cured EPDM compounds. A copolymer of ethylene and propylene (EPM) that lacks the diene segments also is available but can be cured only with peroxides because it is totally saturated.

Resistance of EPDM to aliphatic, aromatic, and chlorinated hydrocarbons is poor, but the polymer can be resistant to polar solvents such as ketones and alcohols. The polymer's non-polarity also gives it excellent electrical properties. Both the copolymer and terpolymer are highly resistant to ozone and oxygen. They offer superior weatherability without the use of antioxidants and antiozonants.

Reinforcing agents are required with EPDM because it lacks gum strength. High tensile and tear properties are achieved through high loading. Compression set resistance at high temperatures is good for both sulfur-donor and peroxide cures.

6.1.2.4   Polychloroprene (CR)

Polychloroprene, commonly referred to as Neoprene[*], is one of the oldest synthetic elastomers. Introduced in 1931, this elastomer's use is declining as many applications are switching to other elastomers, including EPDM-based thermoplastic vulcanizates.

Some compounders have called neoprene "chlorinated natural rubber" because its structure is similar to that of NR except that chlorine has replaced the methyl groups. The material actually is made from the emulsion polymerization of 2-chloro-1,3-butadiene. The addition of chlorine also increases the polarity of CR relative to that of NR. The polarity improves its resistance to oils, solvents, ozone, and heat.

Polychloroprene is not rated superior in any property category when compared to other elastomers, but it does offer an excellent balance of overall properties. End use

---

[*] Neoprene is a registered trademark of DuPont Dow Elastomers.

properties are dictated by grade used and how it is modified at polymerization, typically by sulfur or 2,3-dichloro-1,3-butadiene or both.

CR ages better than natural rubber and does not soften during heat exposure. High temperature tensile strength, however, often is lower than the comparable NR-based compound. The compounder choosing CR as a base should be aware that it is a dense synthetic rubber – CR density of about $1.24\,Mg/m^3$ vs. NR with a density of $0.92\,Mg/m^3$ – and also costs roughly three to five times as much as NR on a per-pound basis.

Because roughly 75% of all CR capacity is owned by DuPont Dow Elastomers, it is easiest to break down the various types of polychloroprenes by using DDE's classifications.

- "G" types are sulfur modified. They have limited shelf life because they are reactive and increase in viscosity if not stored in a cool (65 °F) environment. They can be peptized to some degree. Additional organic accelerators are not needed to cure these products, and the polymers provide good adhesion to fibers, excellent flex resistance and resilience. They are ideal for dynamic applications such as belts [8].
- "W" type polychloroprenes have excellent bin storage stability and can be milled without sticking to the rolls. These products require organic accelerators for curing and produce compounds with excellent resistance to heat and to compression set.
- "T" polychloroprenes are similar to the W types, but contain a highly crosslinked microscopic "gel" form of polychloroprene that acts as an internal processing aid [9]. These materials process better than both the G and W type CRs.

In addition to these three types, CRs are available for adhesive applications. They contain a high degree of crystallinity and are available in numerous grades.

Polychloroprene is processed as other elastomers, but the compounder must guard against precure or scorching. This generally involves low temperatures and short mixing cycles. High filler loading can improve CR calendering and extruding.

6.1.2.5 Butyl and Halogenated Butyl Elastomers

Butyl rubber was first commercialized in 1942, and there have been three significant milestones in variations of these products [10]:

- Commercialization of butyl rubber during World War II
- Commercialization of halogenated forms in the early 1960s
- Introduction of polymers based on polyisobutylene and paramethylstyrene in the late 1980s

Butyl rubber (IIR) is ideal for applications where low gas permeability is desired. It also is available in halogenated forms as bromobutyl (BIIR) and chlorobutyl (CIIR). The halogenation improves processing characteristics and some final properties.

Chemically, butyl rubber is polyisobutylene with isoprene units in the polymer chain. Now in its sixth decade of commercial manufacture, this elastomer is made via low temperature cationic polymerization. The halogenated forms, which are produced in hexane solution by using elemental chlorine or bromine, provide the compounder with greater flexibility in crosslinking and enhanced cure compatibility with general purpose elastomers.

These elastomers are chemically very inert and offer good resistance to oxidation and ozone. Compounds typically cover a durometer range of 30 to 80. As with other synthetic rubbers, there are many grades available based on Mooney viscosity and/or degree of unsaturation.

The compounder must use the more-active accelerators such as dithiocarbamates and thiurams in addition to sulfur with IIR because of the lower amount of available crosslinking sites compared to other elastomers. Resin cures also are an option with such materials as polymethylol-phenol resins.

Applications include tires, curing bladders, steam hose, o-rings, caulks, sealants, and roof coatings. The choice of accelerator system alters end use properties as well as the time required to crosslink the compound after molding or extrusion. This is one method a compounder can use to tailor-make end compounds. For example, end use compounds that require high thermal stability must be resin-cured.

Because of its chemical inertness, butyl polymers tend not to experience molecular-weight breakdown during processing. This allows operations such as heat treatment or high temperature mixing to affect the vulcanizate characteristics of a compound [11].

In addition, flexibility is increased by the addition of some mineral fillers in the clay, talc, and silica families along with promoters. Such promoters improve resilience and processing and can also increase compound modulus.

Another major end use for butyl rubber is vibration damping. According to Fusco [12], the viscoelastic properties of butyl rubber result from the molecular structure of the polyisobutylene chain because the two methyl groups on every other chain carbon atom of the molecular backbone results in greater delayed elastic response to deformation. As a result, compounds based on butyl rubber absorb and dampen shock, making the material superior in numerous automotive applications. The molecule also controls vibrational forces because of these properties.

Butyl-based compounds are resistant to moisture, making it a good electrical insulator. Butyl is affected a small amount by oxygenated solvents and other polar liquids but swells extensively when exposed to hydrocarbon solvents and oils. In phosphate ester fluids, for example, butyl swells only 25% as much as natural rubber and shows an 8-fold improvement over polychloroprene elastomers in percent volume increase after aging 70 hours at 156 °F.

Meanwhile, the halogenated butyls – bromobutyl and chlorobutyl – also have a very saturated backbone and have properties similar to butyl: impermeability to the passage of air, gases, and moisture; good vibration damping; and resistance to aging and weathering.

The halogenated varieties, however, have greater crosslinking activity because of the halogen component. The halogenation also allows the polymer to be blended with general purpose, highly unsaturated polymers. The compounder can use chlorobutyl where longer scorch time is needed and bromobutyl when greater reactivity is needed in multiple-polymer compounds.

While the addition of carbon black provides the same changes to a halobutyl-based compound as it does for most other synthetic elastomers, the choice of mineral filler can affect cure characteristics. Acidic clays produce very fast cures that can be tempered by the addition of magnesium oxide or other scorch retarders. Cures are slowed when the compound contains highly alkaline fillers.

Talc has little effect on cure but is only semi-reinforcing. A combination of two or three mineral fillers such as clay, talc, and silica can be used to fine-tune properties [13]. For example, hot tear and hardness increase with an increase in silica content, while a reduction in silica improves compression set. Compounders wishing to use plasticizers can select from both aliphatic and naphthenic mineral oils, and paraffin waxes.

6.1.2.6   Chlorinated and Chlorosulfonated Polyethylene

These two polymers have some similarities, but generally the chlorinated polyethylene (CPE) can be used in applications for which the chlorosulfonated polyethylene, or CSM, is over-specified for the end use requirements.

CPE has a saturated backbone and good resistance to petroleum oils. It also offers good resistance to oxidation. The chlorination is a random process in the manufacture of the CPE polymer, and the level of chlorination can be varied to achieve greater oil resistance (increased chlorine content), or for improved low temperature flexibility (decreased chlorine content). Compounders seeking improvements in flame retardation or in impact resistance can add CPE to other polymers in the compound. CSM finds applications over a broad temperature range of -65 to $250\,°F$ (-53.8 to $121.1\,°C$). It also provides good resistance to acids.

6.1.3   Low Volume Specialty Elastomers

Low volume specialty elastomers are generally priced much higher than the commodity and high volume specialties and have very narrow application ranges when compared to the other types. While there are many polymers in this category, the most important groups are:

* Fluoroelastomers
* Silicone rubber
* Fluorosilicone elastomers
* Polyurethane
* Ethylene acrylic
* Polyacrylate
* Epichlorohydrin
* Polyolefin elastomers
* Polysulfide rubber

6.1.3.1   Fluoroelastomers

Fluoroelastomers, denoted as FKM, typically are used in high temperature applications or in products that must be resistant to chemicals or oils. They are superior to most all elastomers in high temperature properties.

There are numerous types of fluoroelastomers, including:

* Tetra-fluoroethylene/propylene
* Vinylidene fluoride and hexafluoropropylene copolymers, which were among the first stable fluoroelastomers

- Terpolymers of vinylidene fluoride, tetrafluoroethylene, and perfluoro (methyl vinyl) ether, or PMVE
- Terpolymers of vinylidene fluoride, hexafluoropropylene, and tetrafluoroethylene

The primary suppliers of fluoroelastomers in North America are DuPont Dow Elastomers L.L.C. (Viton); Dyneon (Fluorel and Aflas); and Ausimont USA (Tecnoflon). The materials contain varying levels of fluorine, and many grades also contain a curative package. The fluorine generally is found at levels of 66 to 70% except for the tetrafluoroethylene-propylene polymers, which are more suited than the other fluoroelastomers to oil field applications where amine inhibitors are used. Although fluoroelastomers with similar fluorine levels are grouped together, they still possess varying degrees of viscosity and different molecular weight distributions so the compounder can pinpoint the proper grade for mold flow or extrudability.

Typically, a compounder must choose between three cure packages. Diamine cures were the original cure systems used when fluoroelastomers were introduced in the mid-1950s. Diamine cures are best suited for applications where compression set is not critical and where moderate resistance to steam or acid is required.

Bisphenol cures offer improved scorch protection over diamines, as well as the best compression set resistance obtainable for fluoroelastomers. Bisphenols also improve the steam- and acid-resistance of fluoroelastomer-based compounds.

Peroxide cures first were used in the late 1970s and improved steam and acid resistance resulted, compared with the other cure systems. Peroxides, however, do not provide the material with as good resistance to compression set as do the bisphenol cures.

To maximize the high temperature performance of a compound, the acceptor best for the fluoroelastomer must be determined. Common types are low-activity magnesium oxide; high-activity magnesium oxide; calcium hydroxide; calcium oxide; lead oxide (litharge); and zinc oxide

Fluroelastomers can be processed by compression, transfer, or injection molding; by extrusion; and by calendering. Common applications for fluoroelastomers are:

- Automotive components such as shaft seals, valve stem seals, fuel hoses, fuel-injector o-ring seals, and seals in emission-control devices
- Industrial uses such as binders for flares and explosives, diaphragms, electrical connectors, flue-duct expansion joints, and oil field components
- Aerospace-related parts such as manifold gaskets, firewall seals, protective coatings, and fuel-tank bladders [14].

6.1.3.2   Silicone and Fluorosilicone Rubber

These inorganic elastomers are based on polydimethylsiloxane. Generally they are low-strength compounds that are useful over a temperature range of -80 to $450\,°F$ (-62.2 to $232.2\,°C$). While they offer excellent flame, weather, and ozone resistance, silicone rubbers generally have poor resistance to fluids. Fluid and acid resistance are improved by the introduction of fluorine in the fluorosilicone elastomers, although processing is more difficult.

Silicone rubber is suited for applications that require excellent heat resistance. The polymer's low glass transition temperature – the result of using polydimethylsiloxane with a glass transition temperature of $-197\,°F$ $(-127.2\,°C)$ – allows the compounder to use silicone rubber where low-temperature flexibility is desired [15]. As noted previously, compounds based on silicone rubber have very low tear strengths and generally are not suited for dynamic applications.

Some silicone rubber polymers can be cured by room temperature vulcanization, but most are heat-cured by using organic peroxides. More recently, the compounder and end use manufacturer have added liquid-injection-moldable silicone rubber to their production capabilities. This material increases flexibility in the molding process not available with most elastomers.

Crosslinked by peroxides, fluorosilicone elastomers, while expensive, are suitable for applications that require significant solvent resistance and/or excellent low temperature properties. Originally developed for o-ring seals, the material now is used in other areas, such as shaft seals and gaskets; duct hose, and oil-field applications.

6.1.3.3   Polyurethane Rubber

The two types of polyurethane rubber in use are polyester-based products designated as AU and polyether-based products designated as EU. Polyurethanes are on the opposite end of the tear strength list from silicone rubber. PU possess very high tear strength and extremely good wear and abrasion resistance.

However, these polymers are very intolerant to heat and generally are limited to applications where the maximum temperature is $175\,°F$ $(79.4\,°C)$. Millable PU, developed in the late 1940s, is cured via peroxides, sulfur, or diisocyanates. The compounder can choose the curative on the basis of the end-use properties being sought.

There also are some castable grades of PU that are poured or injected into molds. These products often are sold as prepolymers that consist of a polyol backbone reacted with a diisocyanate [16].

Compounders must be extremely careful not to use these materials in applications that involve a high moisture or extremely warm environment during manufacture. Moisture and heat both cause chemical reactions that alter the prepolymer's properties prior to the incorporation of chain-extending chemicals.

The strength and flexibility of the end use product are controlled by the choice of polyol and diisocyanate in the prepolymer.

6.1.3.4   Ethylene-Acrylic Rubber

Ethylene acrylic elastomers, designated as AEM, are terpolymers of ethylene, methyl acrylate, and a third monomer that is present in a very small amount to serve as cure sites in the resulting polymer.

AEM has very good heat resistance, good oil resistance, and fair low temperature characteristics. The polymer is a good choice for use in environments that contain petroleum-based engine lubricants [17], and has replaced polyacrylate rubber in some seal applications on the basis of its comparatively low brittle point. Hertz [18] also notes that ethylene acrylic compounds are limited in their dynamic abilities for such applications as shaft seals that operate at high rpms, where leakage could occur as a result of the polymer's performance.

6.1.3.5   Polyacrylate Rubber

Polyacrylate rubber (ACM) has outstanding resistance to petroleum-based oils and fuels because of the polar acrylate group in its structure. This property also makes it a material of choice for applications that involve sulfur-bearing lubricants, because the chemical structure is resistant to crosslinking by these materials, which are finding increased use in automotive applications. The polymer provides superior resistance to oxidation, ozone, and sunlight. It also has high flex-crack resistance.

Although polyacrylate has somewhat inferior cold temperature properties, it functions in hot oil environments up to about 350 °F. It also is a better choice over nitrile rubber for components exposed to hot air.

The acrylic-based elastomer has its weaknesses, however. Compared to many other polymers, it is inferior in strength and water resistance. While polyacrylate lost some applications to ethylene acrylic elastomers in the 1980s, the polymer still finds application in such automotive components as automatic transmissions, engine gaskets, and power-steering gears.

Compounders have a choice of products with varying cure-site monomers for development of compounds with specific end properties. These monomers can include chlorine, carboxyl, epoxide, or combinations of these groups. Many crosslinking systems are usable, including amines, triazine, soap-sulfur, and peroxide.

6.1.3.6   Epichlorohydrin Rubber

Homopolymers (CO) and copolymers (ECO) make up the epichlorohydrin family of products. Developed nearly 40 years ago, the homopolymer is a rubbery, predominantly amorphous polymer, while the copolymer also contains ethylene oxide.

The polymers exhibit excellent resistance to gasoline and thus are ideal for many automotive applications. In addition to fuel resistance, CO and ECO have good aging resistance, are operational over a wide temperature range, and are relatively inexpensive when compared with fluoroelastomers and some other polymers. Applications can include hose, tubing, seals, rubber roll covers, and oil field products.

Cure systems can vary for epichlorohydrin, and include sulfur and organic accelerators; peroxide and coagent; ethylene thiourea and lead oxide; and a guanidine and magnesium oxide [19].

Limitations associated with epichlorohydrin include moderate low temperature flexibility, poor electrical properties, and poor abrasion resistance.

6.1.3.7   Polyolefin Elastomers

Polyolefin elastomers represent a fairly recent addition to the list of synthetic elastomers, with Union Carbide Corp., Exxon Chemical Co., and DuPont Dow Elastomers offering products.

These materials can offer greater resistance to heat than EPDM as well as comparable resistance to oils. They also are usable in blends with EPDM to enhance heat resistance or to modify the viscosity of the elastomeric compound. Grades of these products are available in viscosities ranging from extremely fluid to very viscous.

In addition, these polymers can produce compounds with tensile strengths ranging from as low as 800 psi to more than 3,700 psi. Final compound hardnesses of 65 to 95 typically are attainable for POE-based compounds that do not contain other elastomers in the base.

The compounder generally can use peroxides, radiation, or, in some cases, moisture cures to crosslink polyolefin elastomers. Sulfur is not usable in the crosslinking process. The compounder can use fillers, oils, and antioxidants as additives for POE in the same manner as used for EP-type elastomers. However, if a totally amorphous compound is needed, then certain grades of EPDM should be used instead.

6.1.3.8   *Polysulfide Rubber*

Polysulfide rubber, designated as T, can contain as much as 80% sulfur in its structure. This high density polymer is resistant to ketones, esters, and many solvents. Specific grades are suitable for fuel-tank sealants, gaskets, and fuel-hose liners.

6.1.4   Thermoplastic Elastomers

Thermoplastic elastomers have grown immensely in popularity during the past two decades – often at the expense of thermoset elastomers. These polymers provide the flexibility and elasticity found with thermoset elastomers, and they also provide the processing ease and recyclability found with thermoplastics.

While TPEs have replaced a significant amount of thermoset polymers in end-use applications, the former have a major limitation in that their ability to withstand high temperatures is limited by the melting point of the thermoplastic component. Processing also must be done above the melting point of the TP portion.

The choice of TPE depends on need, dynamic application, and end use properties. Choices range from the low-cost, mechanically blended olefinics and the styrenics to the specialty, high performance copolyester and polyamide varieties. While there are specialty and/or experimental types of thermoplastic elastomers constantly under development, general types are as follows:

- Styrenic block copolymers
- Olefinic blends
- Elastomeric alloys, often called thermoplastic vulcanizates because the elastomeric phase is partially vulcanized
- Thermoplastic urethanes
- Copolyesters
- Polyamides

These are discussed in detail in Chapter 10.

References

1. "Engineering Rubbers in Contrast," Lord Corp. reprint of article in *Machine Design*, 1980.
2. Fusco, James V., "The World of Elastomers," Paper presented at the Rubber Division, American Chemical Society, Meeting, Montreal, May 1996.
3. Reisch, Marc S., "Rubber Gets More Traction," *Chemical & Engineering News*, Vol. 75, No. 36, pp. 18–21 (September 8, 1997).
4. Hibbs, J., "Styrene-Butadiene Rubbers," *The Vanderbilt Rubber Handbook*, 13th Edition, R. Ohm (Ed.) (1990), p. 57.
5. *Ibid.*, p. 61.
6. Del Vecchio, R. J., "A basic Survey of Rubber Materials," Paper presented at a meeting of the Philadelphia Rubber Group, Philadelphia, April 1996.
7. Brown, T, "Polybutadiene," *The Vanderbilt Rubber Handbook, op. cit.*, p. 88.
8. Kane, R.P., "The Neoprenes," *The Vanderbilt Rubber Handbook, op. cit.*, p. 158.
9. *Ibid.*
10. "Reinventing Butyl," Exxon Chemical Co., Internet website, www.exxon.com (November 1998).
11. Fusco, J.V., and Hous, P., *The Vanderbilt Rubber Handbook, op. cit.*, p. 100.
12. *Ibid.*, p. 104.
13. *Ibid.*, p. 109
14. Crenshaw, L.E., and Tabb, D.L., "Fluoroelastomers," *The Vanderbilt Rubber Handbook*, R. Ohm (Ed. 13th edition updated as part of *1998 Rubber World Blue Book* CD-ROM (1998).
15. *Engineering with Rubber: How to Design Rubber Components*, A.N. Gent (Ed.), Carl Hanser Verlag, Munich (1992), p. 19.
16. Sheard, E.A., "Specialty Elastomers: Standard Heat Resistance," *The Vanderbilt Rubber Handbook," op. cit.*, p. 259.
17. Hertz, D.L., "Specialty Synthetic Rubbers," *The Vanderbilt Rubber Handbook, op. cit.*, p. 229.
18. *Ibid.*, p. 230.
19. *Ibid.*, pp. 236–237.

7   General Purpose Elastomers and Blends

Gary Day

7.1   Introduction

General purpose elastomers are the work horses of the tire and mechanical rubber goods industry. Natural rubber (NR), polyisoprene (IR), polybutadiene (BR), and styrene-butadiene rubber (SBR) are always included in this classification; sometimes nitrile-butadiene (NBR) and ethylene propylene rubber (EPR and EPDM) are included as well. This chapter covers only natural rubber, polyisoprene, polybutadiene, and styrene-butadiene rubber. These rubbers have good physical properties, processability, and compatibility and are generally very economical. They are the largest volume rubbers in the industry.

Compounding, processing, and curing are similar for all three of these polymers. They all contain unsaturation in their backbones and can be vulcanized with sulfur or crosslinked with peroxides. In sulfur vulcanization, zinc oxide and stearic acid are almost always used to activate organic accelerators. These polymers all respond well to fillers such as carbon black, silica, and clay. They can be extended with aromatic, naphthenic, or (sometimes) paraffinic oils. Because of the unsaturation in the backbone, they are all susceptible to ozone and oxygen attack, so they perform best with the addition of antiozonants and antioxidants.

These polymers have good physical properties including resistance to abrasion and tear, traction, and good low heat build-up. They cannot be used in high temperature applications or where they come into contact with oils and solvents. Other polymers can give lower gas permeability (e.g., butyls and halobutyls), better ozone resistance (e.g., EPR and EPDM), better solvent resistance (e.g., NBR), and better heat resistance (e.g., silicones and fluoroelastomers), but none can beat the combination of overall performance and cost inherent in general purpose elastomers.

7.2   Natural Rubber and Polyisoprene

Natural rubber and polyisoprene share the same monomer chemistry. Isoprene is the building block of these polymers and can polymerize in four different configurations. Polyisoprene can be polymerized with either a coordination catalyst (Ziegler) or alkyl lithium catalyst. Coordination catalysts are usually trialkyl aluminum and titanium tetrachloride. At a ratio around 1:1, the Al/Ti system can produce *cis* contents from 96 to 98%. The alkyl lithium catalyst system produces *cis* contents between 90 and 93%.

Natural rubber has a *cis* content of almost 100%. These microstructures give natural rubber a glass transition temperature (T_g) of approximately $-75\,°C$ and polyisoprene a slightly higher T_g $(-70$ to $-72\,°C)$ because of the presence of *trans*, 1,2 and 3,4 configurations.

Isoprene monomer: $CH_2=C-CH=CH_2$
                            |
                           CH_2

Isoprene can polymerize in four different configurations:

Cis-1,4

$$-[H_2C\overset{\displaystyle H_3C}{\underset{}{\diagdown}}\overset{}{\underset{\diagup}{C}}=CH\overset{}{\underset{\diagdown}{}}CH_2-CH_2\overset{\displaystyle CH_3}{\underset{}{\diagdown}}C=CH\overset{}{\underset{\diagdown}{}}CH_2]_n-$$

Trans-1,4

$$-[H_2C\overset{\displaystyle H_3C}{\underset{}{\diagdown}}C=CH\overset{CH_2-\!\!-CH_2}{\underset{H_3C}{\diagup\quad\diagdown}}C=CH\overset{}{\underset{\diagdown}{}}CH_2]_n-$$

1,2

$$-[CH_2-\underset{\displaystyle\underset{\|}{\underset{CH_2}{CH}}}{\overset{\displaystyle CH_3}{C}}-CH_2-\underset{\displaystyle\underset{\|}{\underset{CH_2}{CH}}}{\overset{\displaystyle CH_3}{C}}]_n-$$

3,4

$$-[CH_2-\underset{\displaystyle\underset{\|}{\underset{CH_2}{C-CH_3}}}{CH}-CH_2-\underset{\displaystyle\underset{\|}{\underset{CH_2}{C-CH_3}}}{CH}]_n-$$

There are three general forms of polyisoprene commercially available: high *cis* content, high trans content and high 3,4 content. A significant advantage in high *cis* content polyisoprene lies in its ability to undergo strain-induced crystallization. This crystallization phenomenon gives *cis*-polyisoprene very high tear strength and excellent DeMattia cut-growth resistance. This exceptional tear strength translates into superb physical properties and performance in the end product.

Gutta-percha or balata is a high *trans* content polyisoprene and is very hard at room temperature. Upon heating to 80 °C, the crystallinity "melts" and the rubber becomes soft and workable. When cooled to room temperature, the crystallinity reforms and the rubber becomes hard and unworkable again.

A commercially available polyisoprene with 60% 3,4 content is available. It has a reported T_g of about $-5\,°C$ and is processable at normal operating temperatures. For most applications, neither the high *trans* nor the high 3,4 polymer can match the high *cis* polymer for performance.

There are a number of important differences between synthetic polyisoprene and natural rubber. Synthetic polyisoprene has the advantage of consisting of up to 99% rubber hydrocarbon (RHC), while natural rubber is usually around 93% RHC. Synthetic polyisoprene is usually lighter in color, and more consistent in chemical and physical properties. Its lower molecular weight leads to easier processing. On the other hand, natural rubber has higher green strength and modulus, especially at higher strain levels and temperatures.

The weight average molecular weight (M_w) of natural rubber ranges from 1 million to 2.5 million while a typical synthetic polyisoprene ranges between 755 thousand and 1.25 million. Molecular weight distribution (MWD) is defined as the ratio of weight average molecular weight (M_w) divided by number average molecular weight (M_n) or (M_w/M_n). Molecular weight distributions vary from less than 2.0 for lithium-catalyzed polyisoprene to almost 3.0 for Al/Ti-catalyzed polyisoprene to greater than 3.0 (and widely variable) for natural rubber.

The 13th edition of the *Synthetic Rubber Manual* [1] lists 11 different producers of polyisoprene. Only one manufacturer produces low *cis* (91%) polyisoprene, Shell Nederland Chemie B.V. There is one listing for high *trans* polyisoprene, TP-301 from Kuraray Company, Ltd., and one listing for high 3,4 content (60%) from Karbochem of South Africa. Kuraray also produces a liquid polyisoprene that is often used as a processing aid. Because it is a liquid, it does not increase the green compound viscosity, but upon vulcanization, it crosslinks into the compound and cannot be extracted. There are nine producers manufacturing *cis* content greater than 96%: Goodyear Tire and Rubber Co., Nippon Zeon Co. Ltd., Japan Synthetic Rubber Co. Ltd., Kuraray, Nizhnekamskneftechim, Kauchuk Co., SK Premyer Co., Togliattisyntezkauchuk, and Volzhski Kauchuk Co. Not listed in the IISRP manual is a high *trans* IR from NCHZ Sterlitamak Co. in Russia.

There are two generic classifications of natural rubber, the crepes and sheets, and the technically specified grades. The conventional crepes and sheets are classified by the International Standards of Quality and Packing for Natural Rubber Grades (otherwise known as *The Green Book* [2]). There are eight grades of crepes and sheets with 35 subdivisions. Grading is done on a visual basis. Table 7.1 lists these grades [3].

The technically specified grades of natural rubber (TSR) are different from crepes and sheets in several respects, most noticeably in that TSR rubber comes in 75 lb bales rather than 250 lb blocks. TSR is not graded visually, but by chemical tests. Some of the more important TSR grades are listed in Table 7.2 [3].

Table 7.1   Grades of Natural Rubber

Type	Source
1. Ribbed Smoked Sheet	Coagulated field latex
2. Pale Crepe	Coagulated field latex
3. Estate Brown Crepe	Estate cuplump, tree lace
4. Compo Crepe	Cuplump, tree lace, wet slab, RSS cuttings
5. Thin Brown Crepe	Cuplump, tree lace, wet slab, unsmoked sheet
6. Thick Blanket Crepe	Cuplump, tree lace, wet slab, unsmoked sheet
7. Flat Bark Crepe	Cuplump, tree lace, earth scrap
8. Pure Smoked Blanket Crepe	Remilled RSS and RSS cuttings

Table 7.2   Technical Grades of Natural Rubber

Property	SMR L, CV	SMR 5	SMR 10	SMR 20	SMR 50
Dirt content, %	0.03	0.05	0.10	0.20	0.50
Ash content, %	0.50	0.60	0.75	1.00	1.50
Nitrogen content, %	0.60	0.60	0.60	0.60	0.60
Volatile matter, %	0.80	0.80	0.80	0.80	0.80
Wallace plasticity	30.00[1]	30.00	30.00	30.00	30.00
PRI, %	60.00	60.00	50.00	40.00	30.00

[1] Does not apply to SMR CV.
SMR L: This is a very clean, light colored rubber.
SMR CV: This is referred to as constant viscosity rubber. It is produced by adding hydroxylamine neutral sulfate before coagulation. It comes in several viscosity grades. The CV rubbers have fewer Mooney viscosity variations between lots and change less with age.
SMR 5: SMR 5 and SMR 1 are produced from a factory-coagulated latex but do not go through the RSS process first. This is a very clean grade of rubber, but is darker than SMR L.
SMR 10, 20, and 50: These grades are produced from field coagulation but may contain some RSS.

There are several other forms and grades of natural rubber, such as Oil Extended Natural Rubber, which is made by adding either aromatic or naphthenic oil to the latex before coagulation, or by blending in an extruder with the dry rubber. One of the newer and more interesting variations of natural rubber is epoxidized natural rubber, called Epoxyprene*. Epoxyprene comes in two grades, ENR-20 and ENR-50. The double bonds in the backbone are epoxidized to 20 mole% and 50 mole%, respectively, to make these grades. Epoxidation changes several physical properties, including increasing the T_g of the polymer. These polymers have higher damping, lower permeability to gases, and increased polarity, which reduces swelling in non-polar oils and increases compatibility with polar polymers such as polyvinyl chloride. The increased damping can be put to use in footwear and acoustic devices; the increased T_g can be used to improve wet traction in tire treads [4].

7.3   Polybutadiene

Polybutadiene (BR) comes in a wide range of micro- and macrostructures. BR is produced from the butadiene monomer and can be polymerized in three configurations:

Cis-1,4

$$-[H_2C \underset{}{\overset{CH=CH}{\diagup \diagdown}} CH_2-CH_2 \overset{CH=CH}{\diagup \diagdown} CH_2]_n-$$

Trans-1,4

$$-[H_2C \overset{CH_2-CH_2}{\underset{CH=CH}{\diagup}} \overset{}{\diagdown} CH=CH \diagdown CH_2]_n-$$

* Epoxyprene is a registered trademark of Kumpulan Guthrie Berhard.

1,2-vinyl

$$-[CH_2-CH-CH_2-CH]_n-$$
$$\quad\quad | \quad\quad\quad |$$
$$\quad\quad CH \quad\quad CH$$
$$\quad\quad || \quad\quad\quad ||$$
$$\quad\quad CH_2 \quad\quad CH_2$$

There are currently four popular variations of BR: high *cis* content, low or medium *cis* content, vinyl BR, and emulsion BR. A fifth variation, high *trans* content BR, is now under evaluation.

Within each classification, there can be many variations. High *cis* content polymers are produced with the use of Ziegler-Natta catalysts; the low/medium *cis* BR is produced with alkyl lithium catalysts; vinyl BR is produced with alkyl lithium in conjunction with a polar additive. All three types are produced in a hydrocarbon solution. Emulsion BR is produced in water using free radical initiators.

Emulsion polymerization is the oldest technology used to produce BR. The technology was first developed in the early 1940s, but it was not considered satisfactory for tires at that time. Not until ''cold polymerization'' was developed was acceptable emulsion BR produced. Even then, it was not until the development of the solution polymerization techniques that truly acceptable BR was produced. These solution polybutadienes found a market niche that emulsion BR could never find. Today, BR is the second largest commercial synthetic elastomer in production; SBR is the first.

There are several different Ziegler-Natta coordination catalysts in commercial use. The highest *cis* content is produced with neodymium. It is reported to give *cis* contents as high as 98% and a T_g of $-102\,°C$ [5]. The cobalt BRs produce materials next highest with *cis* contents (97%) and a $-101\,°C$ T_g. Nickel gives *cis* contents from 94 to 96% and titanium runs from 92 to 94% [6]. Alkyl lithium is used in anionic polymerization and gives *cis* contents of 35%, *trans* contents of 55%, and vinyl contents of 10% with a T_g of $-94\,°C$. These polymers are referred to as medium or low *cis* solution BR. The addition of polar solvents such as diglyme, TMEDA, THF, or oxolanyl alkanes [7–9] can raise the vinyl content up to 90%. Commercial polymers with 70% vinyl content, which are often referred to as high vinyl BR, have a T_g as high as $-25\,°C$. Because of the added cost of the modifier and the need to remove it from the hydrocarbon solvent before unmodified BR is made, the cost of these types of polymers is much higher than that of conventional BR. Emulsion BR has a microstructure in the same ratio as in emulsion SBR (ESBR). The *cis* content is 14%, *trans* content is 69%, and the vinyl content is 17% with a T_g of $-75\,°C$.

The microstructure dictates the glass transition temperature of the polymer, which in turn controls some of the performance of the compounds. Table 7.3 [10] lists typical T_g and T_m values for various microstructures.

Along with controlling microstructure, different catalysts create a wide range of macrostructures. The macrostructure is very important because it controls the processability of the polymer and the compound and the physical properties of the compound. Polybutadiene is rarely used as the sole polymer in a compound because it tends to bag on mills and has very rough extrudates. Linear polymers that have little long chain branching undergo cold flow. Monodispersed, linear polymers exhibit lower hysteresis and heat build-up.

The molecular weight distributions and relative branching that are reported in the literature vary widely. As a generalization, MWD increases in the high *cis* types from

Table 7.3   Microstructure Effect on Polybutadiene T_g and T_m

	T_g (°C)	T_m (°C)
cis	−106	2
trans	−107	97/125
syndiotactic 1,2	−28	156
isotactic 1,2	−15	126
atactic 1,2	−4	none

those produced with titanium catalyst to those produced with cobalt catalyst to those produced with nickel-based catalysts to those produced with the rare earth type catalysts. The manufacturer has a great deal of control over the macrostructure. For example, neodymium-catalyzed products can be very linear and have a narrow MWD. At least one manufacturer uses carbon disulfide to couple some of their neodymium-catalyzed grades to increase the MWD into the 2.5 range. Most high cis types have MWDs in the 2.0 to 4.0 range and a relatively high branching level. Cobalt-catalyzed materials usually have the highest degree of branching. Emulsion polybutadiene has a MWD in the 4 to 5 range and high branching levels. Alkyl lithium catalysts produce BR with the most linearity and narrowest molecular weight distribution. Anionic polymerization can be carried out in either batch or continuous reactors. In continuous reactors, the MWD can be pushed up to around 1.8 or higher if back flow is induced. In batch reactors, the MWD can be as low as 1.3 with very low branching. When lithium catalysts are used in the presence of polar modifiers, not only does the 1,2 content increase, but polymers with even less branching and a narrower MWD than the normal medium cis (35%) can be produced. These polymers can have severe cold flow problems. It is best to uncrate the bales only as they are used at the mixer.

Anionic polymerization is also called living polymerization. The polymer chain can continue to add additional monomer units until it is deliberately terminated. In practice, however, many of the polymer chains are terminated by impurities in the solvents or monomer feed streams. A living polymer chain can be terminated with specific molecules and coupling can be achieved by using multi-functional terminators such as tin tetrachloride or silicon tetrachloride. In addition, anionic polymerization can also be started with multi-lithium initiators. Polymer chains can be grown from two opposite ends in the case of bi-lithio initiators. Some initiators have three or more lithium initiation points, which can lead to very long chain branching architectures [11]. Multi-lithio initiators, in conjunction with multi-functional terminators, can produce materials with very high molecular weights, broad distributions, and sometimes, gelled polymer.

Another method of adding long chain branching is to co-polymerize with small amounts of divinylbenzene [12]. Also, molecules with special bonding abilities can be produced. Aromatic ketones can be used to attach the polymer chain end to carbon black [13]. These types of additions can have a large impact on MWD, cold flow, processing, and ultimate performance of the compound. This also shows the inherent flexibility of anionic polymerization. These same polymerization options are available in copolymers of butadiene, styrene and isoprene, which explains the interest that tire companies around the world have in this technology.

Table 7.4   IISRP List of Polybutadiene Manufacturers

Company	Location	Tradename
American Synthetic Rubber Corp.	USA	CISDENE
Ameripol-Synpol Co.	USA	BR
Asahi Chemical Industry Co., Ltd.	Japan	ASADENE
Bayer Buner GmbH, & Bayer AG	Germany	BUNA CB
Bayer Corp.	Canada/USA	TAKTENE
Buna AG	Germany	BUNA
COPERBO	Brazil	COPERFLEX
Efremov Synthetic Rubber Enterprise	Russia	—
Enichem Elastomeri S.r.l.	Italy and England	EUROPRENE NEOCIS
		EUROPRENE CIS
Firestone Synthetic & Latex Co.	USA	DIENE
Goodyear Rubber & Tire Co.	USA	BUDENE
Industrias Negromex, S.A de C.V.	Mexico	SOLPRENE
Japan Elastomer Co.	Japan	ASAPRENE
Japan Synthetic Rubber Co., Ltd.	Japan	JSR
Karbochem Co.	South Africa	AFDENE
Kauchuk a.s. Elastomers Division	Czech Republic	KRASOL
Kemcor Australian	Australia	AUSTRAPOL
Korea Kumho Petrochemical Co.	South Korea	KOSYN
Nippon Zeon Co., Ltd.	Japan	NIPOL
OMSK Kauchuk Co.	Russia	—
Repsol Quimica S.A.	Spain	CALPRENE
Ricon Resins, Inc.	USA	RICON
Shell Chemicals Europe	France and Germany	CARIFLEX BR
Taiwan Synthetic Rubber Corp.	Taiwan	TAIPOL
UBE Industries, Ltd.	Japan	UBEPOL
Voronezhsyntezkachuk Co.	Russia	—
Zaklady Chemiczne Oswiecim	Poland	KER

It is important to understand all of the parameters that the manufacturer can control. Generalizations between one type of polybutadiene and another or between the "same" product from different suppliers can lead to processing surprises and poor performance. Always evaluate the polymer provided by the supplier who you intend to use. Table 7.4 lists the BR manufacturers in the 13th Edition of the *Synthetic Rubber Manual* which is published by the International Institute of Synthetic Rubber Producers, Inc. (IISRP) [1].

7.4   Copolymers and Terpolymers of Styrene, Butadiene, and Isoprene

Styrene, butadiene, and isoprene copolymers have a wide range of properties and applications. These polymers can be rubbery or glassy at operating temperatures. They come in a wide array of micro- and macrostructures and are often specifically tailored to an end use or application.

The most common synthetic rubber in the world is styrene-butadiene copolymer (SBR). A new addition to the list of these polymers is styrene-isoprene-butadiene (SIBR). This material was first made available for general use in 1997. Styrene-isoprene copolymer (SIR) has been produced commercially for many years, but its use has been primarily in the adhesives industry. SIR is usually produced as a block copolymer, rather than a random copolymer. In a block copolymer, a sequence of isoprene monomers is sandwiched between end blocks of styrene, a configuration commonly referred to as SIS. Kraton polymers from Shell Chemical Co. are a major example of these types. SBR can also be produced as a block copolymer, and is used in plastics modification and adhesive applications.

Block copolymers are produced by alkyl lithium anionic polymerization, in which one monomer feed stream at a time is injected into a batch reactor. The basic configuration is a diblock consisting of one block of styrene and one block of butadiene. The feed streams can be continually alternated to make an A-B-A type of architecture, or a diblock can be coupled to produce an A-B-B-A type. Multi-functional coupling agents can be used to produce many additional architectures.

Block copolymers are used only for special rubber applications. High hardness can be achieved with SBS types, but the operating temperature for the application must remain below the T_g of the polystyrene blocks, i.e., $<100\,°C$. Small amounts of block styrene in otherwise random SBR can improve green strength and can provide thermoplasticity for better processing and extrusion in highly loaded stocks. Floor tile and cove molding are examples of some applications. Solprene 1205* is typically used in these applications. The most significant problem with block polystyrene is the increase in hysteresis and heat build-up [14]. Most applications that undergo dynamic flexing cannot tolerate the higher temperatures caused by the block polystyrene.

A new variation on block copolymers involves changing the microstructure within the different monomer blocks while maintaining a rubbery polymer. One example starts by adjusting the vinyl content of the butadiene monomers so that one end of the polymer chain has a high concentration of vinyl while the opposite end has a low concentration of vinyl butadiene. The same can be done with low and high concentrations of styrene, but without creating block styrene. The entire process can yield relatively clean blocks, resulting in multiple and distinct glass transition temperatures in the same polymer. Alternatively, the blocks can be tapered to give a single, but very wide T_g [15].

A new form of solution polymerization is called dispersion polymerization. As styrene levels in the copolymer are increased, the solubility parameter of the polymer changes. Depending on the hydrocarbon solvent, it is possible for high styrene copolymers to fall out of solution. Once the polymer falls out of solution, handling and polymerization of the polymer become extremely difficult. This new method allows the addition of high levels of styrene to the polymer chain by dispersing the polymer in the hydrocarbon solvent using low molecular weight block copolymers as dispersing agents [16]. This keeps the polymer cement processable and eliminates the problems encountered when a polymer falls out of solution. The advantage of dispersion of polymerization is that it allows the production of polymers with a high styrene content while maintaining the macrostructural advantages of anionic solution polymers.

---

* Solprene is a registered trademark of Phillips Petroleum Company.

Table 7.5   Comparison of Emulsion Polymerization on ESBR Performance

Tire performance	Polymerization temperature		
	$+50\,°C$	$+5\,°C$	$-10\,°C^*$
Tread wear	100.00	119.00	126.00
Heat build-up	100.00	93.00	89.00
Crack growth	100.00	50.00	15.00
Tensile (MPa)	19.77	24.59	25.22
300% Modulus (MPa)	8.06	0.09	8.75
Elongation (%)	560.00	600.00	600.00
Rebound	49.00	53.00	54.00

* An antifreeze in the water was needed to achieve a temperature below $0\,°C$.

The most common method of producing SBR has been emulsion polymerization (ESBR), which uses emulsifying agents to emulsify the monomers in water. Free radicals then initiate the addition of the butadiene and styrene monomers to the polymer chain. Originally, the source of the free radicals was the decomposition of a peroxide or peroxydisulfate. Decomposition of these materials is temperature dependent and polymerization temperatures in the early days of ESBR needed to be around 50 °C. Development of new redox routes led to adequate generation of free radicals at much lower temperatures which, it was soon discovered, led to significantly improved polymers. Comparisons in the performance of bias passenger tire tread compounds produced at different polymerization in temperatures are listed in Table 7.5 [17].

The reason for the improvement in performance of cold over hot polymerized ESBR results from changes in macrostructure. The cold ESBR has a higher primary molecular weight, less branching, lower vinyl configuration in the butadiene content, and less gel. Anionic solution polymerization continues to take the macrostructure in the direction that lower ESBR polymerization temperatures started. Vinyl content in SSBR drops to 8 to 10% from 17% in ESBR. The MWD (M_w/M_n) can be as low as 1.4 to 1.8 in solution SBR (SSBR), from as high as 5.0 in ESBR. Also, the SSBR can be much more linear, which makes the SSBR perform better than ESBR in many dynamic applications. On the other hand, SSBR is more difficult to process than ESBR.

There are many standardized grades of ESBR because the ESBR polymerization process leaves room for few microstructural changes, so consistency among manufacturers is more likely. Even so, a review of the *Synthetic Rubber Manual* [1] shows a wide range of Mooney viscosity values for these standard grades. In addition, some companies have made other modifications to improve hysteresis which in turn has affected processing. To assure consistency in your end products, stay with similar manufacturing technologies or with companies licensing technology from the same source.

ESBR grades can be neat, oil-extended, or carbon black masterbatches. The major grades of ESBR listed in the *Synthetic Rubber Manual* are shown in Table 7.6. There are no standardized grades of SSBR. Manufacturers who use the same technology (such as in licensing agreements) can produce equivalent grades of SSBR. But grades produced by manufacturers using different technologies cannot be compared.

As mentioned in Section 7.3, anionic solution polymerization can produce ''living'' polymers. This feature enables the addition of unique molecules to the chain end. While

Table 7.6   Selected IISRP Grades of ESBR

Grade	Styrene (%)	ML/1 + 4 (100 °C)	Staining	Oil/carbon black extension
S1500	23.5	50	Yes	
S1502	23.5	50	No	
S1712	23.5	45	Yes	37.5 phr aromatic oil
S1721	40.0	55	Yes	37.5 phr aromatic oil
S1778	23.5	50	No	37.5 phr naphthenic oil

this technology can be used when producing BR, it is more often found in the production of SBR and SIBR. One purpose for chain end modification is to couple multiple polymer chains together, which can be accomplished with the addition of tin tetrachloride, silicon tetrachloride, or carbon dioxide, as well as many other materials. Chain coupling can have many purposes, most of which concern macrostructure modification. However, it has been shown that during mixing with carbon black, the tin-coupled polymer can de-couple and expose polymer chain ends that react with carbon black. This activity results in improved dispersion of the carbon black and lower hysteresis at ambient temperatures and above. What is particularly unique and useful about this reaction is that it does not lower the hysteresis at 0 °C, which means tire compounders can improve wet traction and rolling resistance simultaneously.

Other molecules that do not couple but allow the same type of reactivity with carbon black have been successfully used to modify polymer chain ends. One of these, 4,4-bis(diethylamino)-benzophenone (EAB), was discovered and commercialized by Nippon Zeon [18].

The newest styrene-containing polymer to be commercialized is styrene-isoprene-butadiene rubber (SIBR). SIBR shares all of the typical advantages of a material produced by anionic solution polymerization. A wide range of micro- and macrostructures can be polymerized, and the living chain ends can be modified by any of the methods discussed earlier. By itself, SIBR can provide unique morphologies not achievable from traditional blends of IR, BR and SBR [19]. SIBR is not compatible with natural rubber, and in blends with natural rubber, SIBR can broaden the tangent delta (tan δ) curve of the compound. The physical properties of SIBR are comparable to those of SSBR.

7.5   Compounding with General Purpose Polymers

To demonstrate the effect of micro- and macrostructure on physical properties, a series of natural rubbers, polyisoprenes, polybutadienes, and styrene butadiene rubbers were mixed into the two formulas listed in Table 7.7. The polymers were mixed into a formula at 100 phr and into a 50/50 blend with SIR 10. The 100 phr formulas give a clean comparison between polymers and make it easy to see the effects of the polymer composition. However, most polymers are usually used in a blend. Miscibility of the polymers has a significant impact on performance of the compound and the 50/50 formulation is used to demonstrate these effects.

Table 7.7   Test formulations

	Single polymer (phr)	Blend (phr)
First mixing stage:		
Test polymer	100.0	50.0
SIR10	0.0	50.0
IRB #6	25.0	25.0
Zinc oxide	3.0	3.0
Stearic acid	1.5	1.5
N-1,3-Dimethylbutyl-N-phenyl-p-phenylenediamine	1.0	1.0
Microcrystalline wax blend	2.0	2.0
Total phr	132.5	132.5
Second mixing stage:		
First masterbatch	132.5	132.5
IRB #6	25.0	25.0
Total	157.5	157.5
Final mixing stage:		
Second masterbatch	157.5	157.5
Sulfur	1.8	1.8
N-t-butyl-2-benzothiazyl sulfenamide	1.4	1.4
Total phr	160.7	160.7

Compounds were mixed in three stages in a KSBI 1.5 liter internal mixer. An oscillating disc rheometer (ODR) was used to measure the time to 95% of optimum cure at 160 °C (320 °F). Thin test samples less than 2.5 mm (0.100 in.) thick were cured to the ODR-determined time for 95% of optimum cure. Thicker test samples were cured for five additional minutes.

Table 7.8 lists the polymers evaluated and their sources. Each manufacturer produces a unique polymer. While microstructures may be the same among different manufacturers, the macrostructures invariably differ, which has a significant impact on physical properties. If manufacturers supplied several grades of the same polymer, the material with a raw Mooney viscosity closest to 45 was chosen.

7.5.1   Polymer Characterization and Effect on Mixing

The polymers in this study were characterized by gel permeation chromatography (GPC). GPC can determine weight average molecular weight (M_w), number average molecular weight (M_n), molecular weight distribution (MWD = M_w/M_n), and degree of branching. The results from GPC are very dependent on the chromatography columns used, which makes it difficult to correlate results between different laboratories.

GPC results are shown in Table 7.9. Of the polyisoprenes, RSS#1 has the highest molecular weight and the narrowest molecular weight distribution (MWD). The high molecular weight caused it to build up heat quickly in the mixer, giving it the shortest mix time of the IR group. It also had the highest compound viscosity of the group. The chemical treatment used to produce CV60 reduces its molecular weight and increases its

Table 7.8   List of Experimental Polymers and Manufacturers

SIR 10	Natural rubber	
RSS #1	Natural rubber	
CV 60	Constant viscosity natural rubber	
Epoxyprene 50	Epoxidized natural rubber	Guthrie Latex, Inc.
Natsyn 2200	*cis*-polyisoprene	Goodyear Tire & Rubber Co.
TP-301	*trans*-polyisoprene	Kuraray Isoprene Chemical Co., Ltd.
Isogrip	3,4-polyisoprene	Karbochem
Budene 1208	*cis*-polybutadiene (Ni)	Goodyear Tire & Rubber Co.
Buna CB 24	*cis*-polybutadiene (Nd)	Bayer Corp.
Taktene 1203	*cis*-polybutadiene (Co)	Bayer Corp.
Taktene 4510	10% 1,2-polybutadiene	Bayer Corp.
Buna VI 47-0	47% 1,2-polybutadiene	Bayer Corp.
Buna VI 70-0 HM	70% 1,2-polybutadiene	Bayer Corp.
Cisdene 1203	*cis*-polybutadiene (Ti)	American Synthetic Rubber Corp.
E-BR 8405	Emulsion polybutadiene	Ameripol Synpol
SBR 1500	Cold emulsion SBR (23.5% styrene)	Ameripol Synpol
SBR 1006	Hot emulsion SBR (23.5% styrene)	Ameripol Synpol
SBR 1013	Hot emulsion SBR (40% styrene)	Ameripol Synpol
Duradene 711	Solution SBR (18% styrene)	Firestone Tire & Rubber Co.

Table 7.9   GPC Results for Experimental Polymers

Polymer		M_n	M_w	MWD
SIR 10	Natural rubber	258710	1154711	4.46
RSS #1	Natural rubber	502483	1413135	2.81
CV 60	Constant viscosity natural rubber	267443	1350088	5.04
Epoxyprene 50	Epoxidized natural rubber	(would not dissolve in solvent)		
Natsyn 2200	*cis*-polyisoprene	372802	1285557	3.45
TP-301	*trans*-polyisoprene	46252	539506	11.66
Isogrip	3,4-polyisoprene	203328	709230	3.49
Budene 1208	*cis*-polybutadiene (Ni)	127970	586133	4.58
Buna CB 24	*cis*-polybutadiene (Nd)	198775	530372	2.67
Taktene 1203	*cis*-polybutadiene (Co)	170794	505961	2.96
Taktene 4510	10% 1,2-polybutadiene	149887	411761	2.75
Buna VI 47-0	47% 1,2-polybutadiene	1739319	469319	2.70
Buna VI 70-0 HM	70% 1,2-polybutadiene	229155	551967	2.41
Cisdene 1203	*cis*-polybutadiene (Ti)	167266	547193	3.27
E-BR 8405	Emulsion polybutadiene	43438	234291	5.39
SBR 1500	Cold emulsion SBR (23.5% styrene)	98269	468899	4.77
SBR 1006	Hot emulsion SBR (23.5% styrene)	82868	388794	4.69
SBR 1013	Hot emulsion SBR (40% styrene)	63878	375468	5.88
Duradene 711	Solution SBR (18% styrene)	166348	489298	2.94
Duradene 750	SSBR (18% styrene, 37.5 phr oil)	194375	638982	3.28
Duradene 751	SSBR (25% styrene, 37.5 phr oil)	194879	651724	3.34
SBR 1712	ESBR (23.5% styrene, 37.5 phr oil)	120625	532038	4.41
SBR 1721	ESBR (40% styrene, 37.5 phr oil)	158188	659802	4.17

MWD. The synthetic high *cis*-IR has a molecular weight comparable to natural rubber. The lower polymer molecular weights resulted in lower heat build-up in the mixer, longer mix times, and lower compound Mooney viscosities.

As a group, the polybutadienes have about one-half the molecular weight of the natural rubber, and tend to have narrow molecular weight distributions. The compound Mooney viscosity of this group is much higher than that of the IR group. The mixed BR compounds came out of the mixer as fine crumbs; they were difficult to mass on the mill; and, with the exception of the emulsion BR, they bagged on the mill. The lithium BR was able to build torque and heat faster than the high *cis*-BR and hence, mixed faster. The lithium BR showed torque reduction during mixing (similar to the natural rubber) but the high *cis*-BR did not. When blended with the SIR 10 in a 50/50 ratio, the BR group mixed quicker and easier than the SIR 10/synthetic IR blends. The compound Mooney viscosity of these blends still remained higher than the IR group.

The solution SBR had the highest M_n and narrowest MWD of the SBR group. It mixed the quickest of the SBR group, but it bagged on the mill and had the highest compound Mooney viscosity. The cold emulsion SBR had significantly higher molecular weight values than the hot emulsion SBR, but processed about the same as the hot emulsion SBR. The molecular weight of SBR can be increased significantly, but processing suffers when this is done. To offset processing difficulties, polymers are often extended with aromatic or naphthenic oil. The last four entries in Table 7.9 show typical GPC results for oil-extended SBR. Again, solution SBR has a narrower MWD and higher molecular weight (especially M_n) than the emulsion SBR.

Overall, narrow molecular weight distribution and high molecular weight lead to faster mixing but higher compound Mooney viscosities. The higher compound Mooney viscosity leads to many processing difficulties, including mill bagging and ragged extrusion edges and surfaces. These results show why BR is seldom used alone, but is commonly blended with other materials.

7.5.2   Polymer Effect on Cure Rate

The single parameter with the greatest effect on cure rate is polymer microstructure. The higher the *cis* and *trans* content, the faster the polymer cures with sulfur. Comparing *cis*-polyisoprene with *trans*-polyisoprene (Table 7.10) shows little difference in cure rate. The higher the 3,4 isoprene content, the slower the polymer cures with sulfur. The same slow down in cure is seen when 1,2-polybutadiene replaces *cis* and *trans* polybutadiene. When the *cis* and *trans* content is diluted by the addition of styrene to SBR, the cure rate slows down. ESBR 1013 with 40% styrene (50% *cis* and *trans*) cures slower than 1006-ESBR with 23.5% styrene (63% *cis* and *trans*).

Cold emulsion SBR (1500-ESBR) cures slower than hot emulsion SBR (1006-ESBR), which cures slower than solution SBR (711-SSBR). The fatty acids and rosin acids used in emulsion polymerization retard the sulfur curing of these polymers. The same is seen when comparing the emulsion BR (E-BR[*] 8405 with 83% *cis* and *trans*) against the solution

---

[*] E-BR is a registered trademark of Ameripol-Synpol.

[Refs. on p. 171]

Table 7.10 Physical Properties of Polymers Mixed in the Single Polymer Recipe

		SIR 10	RSS #1	CV60	ENR-50	IrCis	IrTrans	3,4-Ir	Ni	Nd	Co	Li 110%	Li 47%	Li 70%	Ti	E-BR	1500	1006	1013	711
ODR @ 160°C																				
T_{min}	in.-lb.	7.40	8.00	7.60	2.7	6.8	4.2	5.9	9.5	9.7	10.2	9.9	8.5	9.8	10.1	11.5	7.3	5.7	5.1	11.8
T_{max}	in.-lb.	41.10	40.50	40.10	43.4	38.7	36.2	34.1	47	51.3	50.8	49.2	43.9	42.3	49.8	45.1	41.8	39	33.6	49
ts1	(min)	3.20	3.29	3.27	1.61	3.63	3.6	4.39	3.94	3.28	3.47	3.56	4.4	4.31	3.38	3.61	6.61	4.95	5.99	4.61
ts2	(min)	3.56	3.69	3.66	1.87	4.02	4.09	4.96	4.5	3.79	4.09	4.26	5.03	5.02	3.98	4.1	7.72	5.71	6.78	5.34
$t'c(30)$	(min)	4.56	4.74	4.72	2.64	5	5.12	6.64	6.29	5.43	5.61	6.14	7.36	7.64	5.53	5.6	10.25	7.29	8.36	7.54
$t'c(50)$	(min)	4.87	5.13	5.09	2.95	5.36	5.43	7.43	6.92	5.99	6.11	6.82	8.32	8.96	6.06	6.42	11.18	8.03	9.39	8.24
$t'(90)$	(min)	5.82	6.21	5.99	4.3	6.39	6.18	10.57	8.51	7.48	7.47	8.51	11.55	13.45	7.55	10.55	15.36	11.83	14.62	9.98
$t'c(95)$	(min)	6.24	6.66	6.39	5.18	6.87	6.45	11.89	9.24	8.2	8.1	9.46	13.95	17.06	8.27	12.64	17.5	13.87	17.41	10.96
Stress-strain @ RT																				
Mod @ 100%	kgf/cm^2	28.83	29.38	25.50	44.51	29.41	83.94	27.51	27.90	32.61	33.07	28.26	30.57	32.68	28.18	27.37	28.98	31.66	42.03	34.54
Mod @ 200%	kgf/cm^2	81.46	80.86	70.74	102.59	72.09	160.23	68.56	63.11	74.86	77.96	68.04	77.38	85.85	63.13	78.36	76.93	89.26	108.20	88.72
Mod @ 300%	kgf/cm^2	147.47	146.36	131.26	163.90	124.26	203.44	116.83	125.53	130.09	134.60	123.94	135.23	151.12	111.23	150.55	138.50	159.13	177.00	106.17
Break stress	kgf/cm^2	289.33	310.16	299.14	234.06	297.25	231.80	173.09	141.40	153.40	132.93	138.64	171.38	168.62	123.69	165.50	258.45	228.28	227.29	187.48
Break strain	%	540	566	586	450	587	410	446	325	343	295	327	363	327	323	317	505	405	379	357
Durometer	Shore A	63	64	62	75	65	98	67	65	68	68	67	65	68	67	67	65	65	70	68
Stress-strain @ 100°C																				
Mod @ 100%	kgf/cm^2	28.62	28.62	28.69	30.37	26.08	22.15	22.36	32.55	35.36	36.21	36.21	32.48	33.33	33.82	32.55	29.81	30.23	30.02	36.49
Mod @ 200%	kgf/cm^2	62.64	61.80	63.84	67.99	54.00	45.49	53.36			76.78	82.33	77.06		70.38		77.48		82.12	
Mod @ 300%	kgf/cm^2	101.80	98.29	102.79	103.07	84.65														
Break stress	kgf/cm^2	164.52	158.89	163.82	151.16	138.51	72.42	77.34	75.93	80.15	85.07	85.77	84.37	75.93	87.88	67.49	94.21	68.90	93.51	82.96
Break strain	%	464	470	464	466	454	291	273	177	182	203	205	213	187	216	172	232	181	221	186
Compression set 22 hours @ 100°C, 25%, 1.129″ sample																				
Set	%	47.2	49.4	49	73.7	41.4	51	34.8	39	33.7	38.2	42.6	34.3	29.5	35	40.8	38.8	41.3	31.1	37.8
B. F. Goodrich Flexometer Model II																				
Compression set	%	5.6	6	6.6	19.5	4.3	6.7	4.6	3.6	3.1	2.9	4.1	4.4	3.2	3.5	3.9	6	6.4	5.2	2.8
Monsanto Fatigue to Failure																				
8 cam avg	(cycles)	154,000	156,700	220,300	160,700	110,700	11,800	168,950	748,400	889,000	873,900	1,240,000	1,035,600	908,200	942,700	349,142	6410700	551,800	121,700	945,200
std dev		24,500	26,100	18,000	19,600	3,400	2,600	31,675	112,600	188,400	310,500	235,900	326,100	508,400	89,900	171,889	117,300	110,000	70,100	138,800
14 cam avg	(cycles)	62,900	86,700	101,900	74,700	63,000	15,600	58,667	304,500	320,000	266,000	271,300	209,700	200,700	458,200	112,833	139,100	116,600	47,400	264,200
std dev		21,100	19,300	10,000	9,400	4,900	3,000	18,537	60,700	73,600	149,600	176,300	118,600	116,400	95,800	56,284	105,700	77,000	13,600	124,000
18 cam avg	(cycles)	38,200	55,700	58,200	37,700	39,700	7,500	25,700	95,300	116,600	65,000	90,900	44,900	28,900	115,200	97,617	42,000	45,800	24,400	214,200
std dev		6,200	12,100	9,500	3,600	4,500	2,500	16,866	37,200	58,100	69,300	81,700	29,40	17,300	92,900	73,975	22,600	22,600	6,200	133,700

Demattia (cycles)																				
100	mm	2.5	2.5	2.5	2.5	2.5	7.6	0.0	3.8	5.1	6.4	3.8	2.5	3.8	17.8	15.2	2.5	7.6	6.4	15.2
500	mm	2.5	2.5	2.5	3.8	2.5	8.9	0.0	16.5	17.8	14.0	14.0	7.6	11.4	22.9	21.6	6.4	12.7	8.9	22.9
1000	mm	2.5	2.5	2.5	3.8	2.5	10.2	5.1	21.6	22.9	19.1	17.8	10.2	16.5		25.4	8.9	15.2	10.2	24.1
2000	mm	2.5	2.5	2.5	5.1	2.5	10.2	7.6		25.4	22.9	22.9	19.1	21.6			8.9	17.8	11.4	
3000	mm	2.5	2.5	2.5	5.1	2.5	11.4	10.2			25.4	25.4	21.6	25.4			11.4	19.1	14.0	
4000	mm	2.5	2.5	2.5	5.1	2.5	14.0	11.4									17.8		19.1	
5000	mm	2.5	2.5	2.5	5.1	2.5	14.0	14.0									19.1		22.9	
10000	mm	6.4	7.6	3.8	8.9	2.5	19.1	17.1												
15000	mm	7.6	8.9	5.1	8.9	3.8	20.3	19.1												
25000	mm	11.4	11.4	7.6	12.7	6.4	24.1	19.1												
35000	mm	14.0	12.7	7.6	12.7	7.6		24.1												
50000	mm	15.2	14.0	8.9	15.2	10.2		25.4												
65000	mm	16.5	16.5	10.2	17.8	12.7														
80000	mm	19.1	17.8	14.0	17.8	17.8														
100000	mm	22.9	20.3	15.2	21.6	20.3														
125000	mm	25.4	20.3	17.8	24.1	22.9														
150000	mm		24.1	20.3	24.1	24.1														
175000	mm		24.1																	
200000	mm																			

polybutadiene (Li 10%-BR with 90% *cis* and *trans*). While the *cis* and *trans* contents are similar, the emulsion polymer cures considerably slower.

The addition of epoxy units to natural rubber increases the cure rate because the epoxy units crosslink through non-sulfur chemistry and enhance the cure rate and the state of cure. When the polymers are blended with 50 phr of natural rubber (Table 7.11) the magnitude in differences decreases, but the same trends hold true. When used alone, slower curing polymers often are combined with secondary accelerators such as DPG or MBT. Blending with natural rubber increases the cure rate and decreases or eliminates the need for these secondary accelerators.

When peroxide cures are used, the 3,4-IR and 1,2-BR microstructures increase the cure rate and state of cure. High vinyl BR is sometimes added as a coagent in other polymer systems to enhance peroxide cures [20]. Microwave curing also benefits from an increase in 1,2 BR content.

7.5.3   Polymer Effect on Stress-Strain

Polyisoprenes tend to yield the highest tensile and elongation values (Tables 7.10 and 7.11). SBR exhibits lower tensile strength and polybutadiene, the lowest of all. Relative modulus values among polymers varies with strain. These differences result from the macrostructural differences of the polymers. While high molecular weight has a positive effect on tensile strength and elongation, strain-induced crystallization in high *cis*-IR has a much stronger impact, making natural rubber and synthetic high *cis*-IR the best performing general purpose elastomers available. As would be expected, blending polymers with SIR 10 tends to reduce variations between the different polymer types (Tables 7.10 and 7.11).

Various grades of natural rubber show differences in their physical properties. Ribbed smoked sheet #1 has the highest tensile value; the peptized CV60 has the highest elongation; and they all exhibit similar modulus values. Synthetic polyisoprene has a lower modulus value, but good tensile and elongation properties.

Changing the predominant microstructure from *cis* to *trans* dramatically increases the room temperature modulus because of the high melting point of the *trans* configuration. When *cis* and *trans* configurations are compared at a temperature well above the T_m of both microstructures, such as 100 °C in Tables 7.10 and 7.11, the *trans*-IR modulus is lower than the *cis*-IR modulus. The high *trans*-IR cannot provide either the tensile or elongation values of the high *cis* microstructure at any temperature. When blended 50/50 with SIR 10, the high *trans*-IR is compatible enough with the high *cis*-IR that neither microstructure has large enough domains to have a significant impact on properties. The *trans* domains do not increase the modulus, and the *cis* domains do not undergo strain-induced crystallization and increase tensile values.

The 3,4 microstructure provides the poorest tensile strength and elongation for the polyisoprene family. It also has very low modulus, especially at the higher strain levels. The addition of epoxidation significantly increases the room temperature modulus and durometer, but drops the tensile and elongation values. However, unlike the *trans* and 3,4-IR polymers, the ENR-50[*] maintains reasonable properties at 100 °C. In the final analysis,

---

[*] ENR-50 is a registered trademark of Kumpulan Guthrie Berhad.

high *cis* polyisoprene and natural rubber have the best overall combination of physical properties at both room and elevated temperature for all of the polymers evaluated.

SBR room temperature modulus and durometer values increase with styrene content because polystyrene has a T_g of 100 °C and is very hard and reinforcing at temperatures below that. When styrene begins to form blocks of consecutive units, stress-strain values increase. Emulsion SBR provides higher tensile and elongation values than solution SBR. SBR has only average properties at 100 °C, but the addition of natural rubber in the 50/50 formulation greatly enhances the high temperature properties, while effecting low temperature stress-strain results only slightly (Table 7.11).

The polybutadienes as a group have very low tensile and elongation values. The room temperature modulus values are equivalent to the polyisoprenes and styrene-butadiene copolymers and can be increased by raising the 1,2 microstructure content. This improvement results from the higher polymer T_g associated with the 1,2 content. However, this approach concurrently reduces tear properties and decreases cure rates. When blended with SIR 10, the higher 1,2 content reduces all the physical properties including tensile and elongation. Because BR is almost never used alone, higher 1,2 content is usually avoided for most applications. One exception is tire treads where the higher T_g of the 1,2 BR can improve wet traction without significantly increasing rolling loss. The reason rolling loss does not suffer too much is because of the higher M_n of the 1,2-BR. However, incorporating 1,2-BR in tire tread formulations reduces wear and cut/chip properties.

Overall, the differences within the medium to high *cis* polybutadiene group are relatively small. The 300% modulus of the emulsion BR is the highest and the modulus of titanium-catalyzed BR (Ti-BR) is the lowest. Of the high *cis* polybutadienes, Nd-BR has the highest tensile strength and Co-BR the lowest. In SIR 10 blends, the cobalt and neodymium versions show very slight advantages over the other medium/high *cis* versions.

7.5.4   Hysteresis

Hysteresis is the energy lost when an object is deformed. In filled rubber articles, energy losses result from internal friction and breaking and reforming filler/polymer contact points. Dynamic tests directly measure the temperature rise caused by this hysteresis (e.g., Goodrich Flexometer) or the storage (G', E', K') and loss (G'', E'', K'') modulus, which can then be used to predict hysteresis. The ratio of loss modulus to storage modulus (G''/G') is called the tangent delta (tan δ) and is a common measurement of hysteresis.

In this study, hysteresis was measured using an RPA2000 Rubber Process Analyzer from Alpha Technologies, and the ARES from Rheometrics Scientific. The RPA was used for temperatures above room temperature. Samples were inserted and cured to 95% of their optimum cure as determined by an ODR at 160 °C. The samples were cured in a static mode so that filler agglomeration would not be effected. After curing, the samples were cooled to 40 °C. At a frequency of 20 Hz, they were taken through a 0.5 to 15% strain sweep twice. The second strain sweep was used to collect dynamic data, including tan δ. The tan δ in the 40 to 70 °C range is often used by tire compounders to predict the tire rolling loss contribution from the tread [21,22].

Table 7.11 Physical Properties of Polymers Mixed in the Two Polymer Blend Recipe

		ENR-50	IrCis	IrTrans	3,4-Ir	Ni	Nd	Co	Li10%	Li47%	Li70%	Ti	E-BR	1500	1006	1013	711
ODR @ 160°C																	
T_{min}	in.-lb.	6.7	6.6	6.5	6.8	9.1	9	9.1	9.3	8	8.3	9.4	7.3	6.7	6	5.8	9
T_{max}	in.-lb.	41.5	41.3	41.5	37.9	44.7	46.9	46.1	44.8	41.9	41.9	46.7	43.4	39.3	39	38.4	9
ts1	(min)	2.01	3.55	3.27	3.39	3.29	3.28	3.38	3.44	3.59	3.69	3.6	4.19	4.52	3.99	4.36	4.05
ts2	(min)	2.32	3.98	3.63	3.78	3.68	3.69	3.81	3.85	4.03	4.18	4.04	4.74	5.1	4.5	5.01	4.51
t'c(30)	(min)	3.32	5.04	4.54	4.87	4.74	4.85	4.93	4.99	5.17	5.44	5.28	6.19	6.7	5.68	6.48	5.79
t'c(50)	(min)	3.53	5.39	4.81	5.26	5.1	5.27	5.3	5.38	5.61	5.9	5.7	6.73	7.28	6.12	7.03	6.22
t'(90)	(min)	4.7	6.35	5.55	6.52	6.08	6.34	6.3	6.43	6.94	7.39	6.8	9.29	9.64	8.33	9.55	7.28
t'c(95)	(min)	5.21	6.8	5.86	7.1	6.63	6.81	6.75	7.01	7.92	8.41	7.33	10.88	10.93	9.84	11.17	7.89
Stress-strain @ RT																	
Mod @ 100%	kgf/cm²	47.27	29.90	40.50	38.57	39.98	35.16	33.61	29.69	29.09	29.25	31.66	32.46	30.98	31.78	34.84	32.62
Mod @ 200%	kgf/cm²	103.54	78.62	84.64	88.89	96.30	88.66	85.65	75.13	75.20	71.85	79.81	86.79	83.63	84.23	89.25	85.13
Mod @ 300%	kgf/cm²	155.66	139.32	133.57	138.62	160.31	154.97	150.38	135.17	132.02	124.83	139.11	157.84	149.66	147.96	149.66	150.81
Break stress	kgf/cm²	242.32	279.87	203.83	185.62	200.54	220.03	226.99	211.54	194.99	170.22	201.66	223.79	249.55	264.00	254.63	241.10
Break strain	%	497	530	476	404	366	402	422	424	415	386	408	393	452	488	491	441
Durometer	Shore A	72	65	85	67	65	70	67	65	65	65	65	65	65	64	68	68
Stress-strain @ 100°C																	
Mod @ 100%	kgf/cm²	36.91	27.91	27.28	31.85	29.67	34.38	36.62	29.67	29.25	29.04	30.09	32.55	28.54	27.98	28.33	31.85
Mod @ 200%	kgf/cm²	76.14	61.94	60.11	69.53	64.61	73.54	69.32	65.17	66.30	64.05	65.88	74.74	67.14	64.75	65.81	72.98
Mod @ 300%	kgf/cm²	108.55	97.52	93.30	100.96	105.32	118.54	109.68	106.30			105.88		107.71	103.21	103.07	116.78
Break stress	kgf/cm²	160.30	146.94	94.21	104.05	114.60	132.88	138.51	115.30	86.48	78.74	106.16	104.05	142.02	135.69	135.69	136.40
Break strain	%	465.00	473.00	303.00	304.00	330.00	323.00	363.00	322.00	249.00	283.00	300.00	260.00	392.00	387.00	393.00	348.00
Compression set 22 hours @ 100°C, 25%, 1.129″ sample																	
Set	%	52.75	44.6	46.9	44.45	41.2	39.4	41.5	42.15	36.35	38.7	41.6	42.15	43.15	42.4	40.85	41.2
B. F. Goodrich Flexometer Model II																	
Compression set	%	11.6	7	7.6	5.7	4.9	3.5	3.3	4.7	3.8	4.2	4.3	4	5.7	4	5.5	3.3

Monsanto Fatigue to Failure

8 cam (cycles)	avg	43,783	125,483	173,517	247,033	206,033	147,600	334,050	164,133	305,833	297,883	226,667	142,367	149,917	261,100	84,200	185,300
	std dev	6,108	19,054	99,668	78,225	52,204	21,509	76,328	67,440	196,434	203,214	83,503	98,939	55,517	52,848	18,362	104,929
14 cam (cycles)	avg	25,467	63,750	50,600	84,083	175,417	149,167	76,717	35,867	81,367	66,600	64,133	61,850	48,550	47,850	36,367	81,683
	std dev	3,581	4,885	19,168	41,181	54,844	68,425	30,765	24,201	21,411	25,742	24,068	19,581	18,384	16,066	9,748	25,626
18 cam (cycles)	avg	18,717	40,683	32,100	59,900	45,483	21,850	43,350	25,617	44,917	39,067	56,883	27,783	28,483	22,050	15,117	37,850
	std dev	2,932	11,397	12,938	14,067	13,050	14,881	41,431	20,534	15,179	19,027	20,129	14,164	12,382	7,304	6,431	19,548

Demattia (cycles) — values in mm

cycles																	
1000	2.5	2.5	2.5	2.5	2.5	6.5	2.5	3.8	2.5	5.1	3.8	2.5	2.5	2.5	2.5	2.5
2000	2.5	2.5	2.5	2.5	5.1	8.9	5.1	7.6	6.4	7.6	7.6	2.5	2.5	2.5	2.5	3.8
3000	2.5	2.5	5.1	2.5	7.6	11.4	7.6	8.9	6.4	10.2	11.4	3.8	2.5	3.8	3.8	5.1
4000	2.5	2.5	5.1	2.5	8.9	15.2	11.4	11.4	8.9	11.4	14.0	6.4	3.8	6.4	6.4	7.6
5000	3.8	2.5	5.1	2.5	11.4	15.2	14.0	14.0	8.9	14.0	17.8	8.9	5.1	7.6	6.4	10.2
7500	6.4	2.5	7.6	2.5	15.2	17.8	16.5	15.2	14.0	16.5	19.1	14.0	7.6	9.5	9.5	13.3
10000	7.6	2.5	10.2	205	15.2	20.3	19.1	16.5	14.0	17.8	21.6	16.5	10.2	12.7	14.0	16.5
15000	10.2	2.5	12.7	2.5	19.1	24.1	24.1	20.3	19.1	21.6	24.1	21.6	12.7	17.8	16.5	19.1
20000	12.7	2.5	15.2	2.5	22.9	25.4	25.4	24.1	24.1	24.1	25.4	24.1	17.8	20.3	20.3	22.9
25000	14.0	3.8	19.1	2.5	25.4			24.1	25.4	25.4		25.4	20.3	21.6	21.6	24.1
30000	14.0	3.8	21.6	2.5				24.1				0.0	21.6	22.9	24.1	24.1
40000	16.5	6.4	25.4	3.8										25.4	25.4	24.1
50000	19.1	7.6		6.4												
65000	21.6	8.9		7.6												
80000	25.4	10.2		10.2												
100000		10.2		12.7												
125000		12.7		15.2												
150000		19.1		19.1												
200000		22.9		19.1												
225000		25.4		21.6												
300000																

The ARES places a small strip of material under tension and then a sinusoidal torsional deformation determines dynamic properties. The samples were run from -110 to $+100\,°C$ using 0.5% strain and 10 Hz frequency. The ARES data primarily looked at low temperature hysteresis.

Tire compounders often use the loss modulus or the $\tan\delta$ at 0 °C to predict the wet traction of tread formulations [22]. The higher the hysteresis in this temperature range the better the wet traction. The complex modulus between -20 and 0 °C is often used to predict snow and ice traction; the lower the value, the better [23].

Using the ratio of G' at 15% strain divided by G' at 0.5% strain (G'15%/0.5%), the Payne Effect can be quantified [24,25]. In filled compounds and at low strain levels (0.1 to 15%), the modulus drops as the strain level is increased, referred to as the Payne Effect. When the sample is immediately retested, the modulus values are lower than when the sample was originally tested. This softening phenomenon is referred to as the Mullins Effect. It is hypothesized that fillers have some affinity for one another and tend to agglomerate within the polymer matrix. These agglomerates have a finite rigidity and strength that contributes to the compound modulus. When the compound sample is stretched, these agglomerates are physically pulled apart and are no longer available to add their strength to the compound. This results in the lower modulus values when the sample is immediately retested and the agglomerates are not given enough time to reform.

Filler dispersion can contribute to the Payne Effect. With good dispersion, the filler is broken into very small particles and randomly distributed throughout the polymer matrix. Small particles surrounded by polymer cannot come into contact with other filler particles and form a reinforcing network that can contribute to the modulus of the compound. With fewer agglomerates breaking apart, there is less difference between the 15% strain modulus value and the 0.5% strain modulus value. The ratio of the modulus at 15% strain to 0.5% strain is higher and the Payne Effect is lower.

Filler dispersion can be measured by several different methods and at different scales. For example, optical microscopes measure a larger scale dispersion than electron microscopes. Dispersion at different scales can have different effects on hysteresis, cut and chip, and fatigue and crack growth properties. The filler dispersion as measured by the Payne Effect and the ratio of G' at 15% strain divided by G' at 0.5% strain (G'15%/0.5%) has a good correlation with hysteresis.

Figures 7.1 and 7.2 compare the 40 °C $\tan\delta$ and the Payne Effect for the single polymer and dual polymer formulations. There is some correlation between these two figures. The polybutadienes as a group have the lowest $\tan\delta$ values with the SBR and polyisoprenes more similar. The G'15%/0.5% strain ratio shows the polybutadienes to have the lowest Payne Effect, followed by the SBR, and then the polyisoprenes. In the SIR 10 blends, the G'15%/0.5% ratio splits the difference between the test polymer and the SIR 10.

Within each polymer group, the micro- and macrostructures impact the Payne Effect and $\tan\delta$ values. In the BR group, the Payne Effect increases with increasing 1,2-BR content and hence, the $\tan\delta$ value increases. In the IR group, the different microstructures show little difference in Payne Effect, but large differences in $\tan\delta$ are apparent. In the IR case, the $\tan\delta$ differences result from T_g differences in the microstructure. As testing temperatures approach the T_g of the polymer, the $\tan\delta$ value increases, which is easily seen in a temperature sweep of some IR compound batches (Fig. 7.3). In the SBR group,

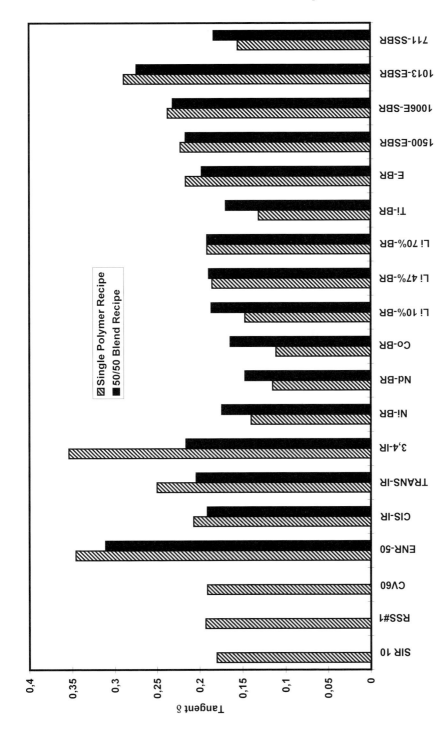

Figure 7.1   Tangent δ of selected polymers and blends.

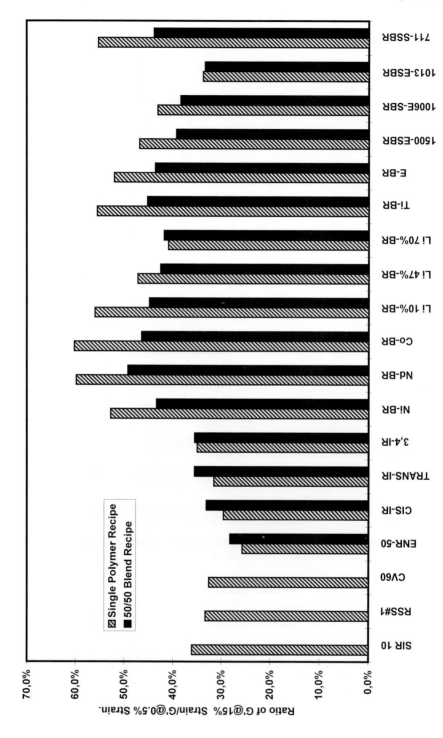

Figure 7.2   Payne effect of selected polymers and blends.

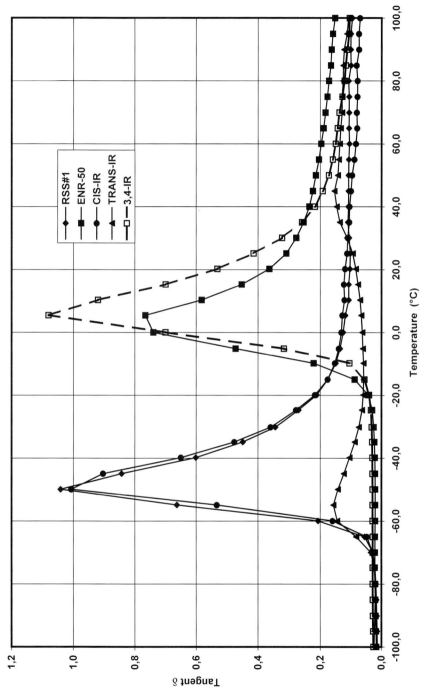

Figure 7.3 Temperature sweep of selected polyisoprene compounds.

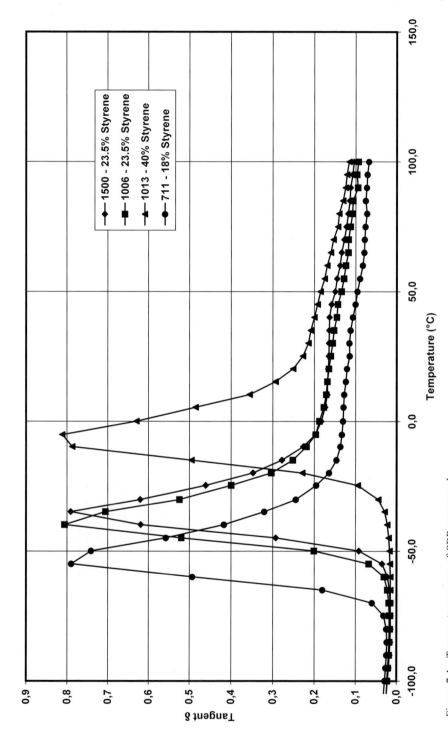

Figure 7.4   Temperature sweep of SBR compounds.

effects from both micro- and macrostructures can be seen. The higher styrene level increases the Payne Effect and increases tan δ values (1006-ESBR vs. 1013-ESBR). Part of that tan δ increase with increasing styrene content results from the styrene T_g (Fig. 7.4).

The macrostructure effect is best seen by comparing hot and cold emulsion SBRs with the same styrene content (1500-ESBR vs. 1006-ESBR). Both have the same microstructure, so all differences result from the macrostructure. The cold emulsion SBR is much better for Payne Effect and hence, tan δ values. The cold emulsion SBR has less long chain branching and a higher molecular weight than the hot ESBR. The solution polymerization is better than both the cold and hot emulsion polymerization. The solution SBR has the lowest level of long chain branching, the highest molecular weight, and the narrowest MWD (M_w/M_n).

The BR group provides another good review of the macrostructural effects on hysteresis. With the exception of the high 1,2-BR polymers, the other samples have similar enough T_g points to eliminate microstructural effects. The Nd- and Co-catalyzed polymers have the best Payne Effect and the lowest tan δ values. They also have the highest M_n and narrowest MWD. The emulsion BR has the lowest M_n and the widest MWD, resulting in the highest tan δ with only a slightly higher Payne Effect. The high tan δ value in the emulsion BR probably results from branching and molecular weight more than carbon black dispersion.

Hysteresis is influenced by the micro- and macrostructure along with their effects on filler dispersion. High M_w and narrow MWD tend to provide higher shear in the mixer, which results in better filler dispersion. The macrostructure also determines the polymer's contribution to hysteresis through polymer chain entanglements. Through determination of the T_g, the microstructure further impacts hysteresis and tan δ.

While compounders struggle to reduce hysteresis in most cases, it should be noted that hysteresis is critical for some applications such as traction in tires. Hysteresis is arguably the most important and unique feature of elastomers. In any dynamic application, it is always a critical factor.

7.5.5   Compatibility with SIR 10

The ARES temperature sweep data can also provide insight into polymer/polymer compatibility and its effect on low temperature hysteresis. If the glass transition temperatures (T_g) of the polymers are more than 30 °C apart, the loss modulus and tan δ curves show a broad peak or even two distinct peaks when the two polymers are not compatible. Non-compatibility allows the higher T_g polymer to maintain its impact on hysteresis in the wet traction range of the temperature curve. Compatibility pulls the high T_g polymer's impact to a lower temperature as the two polymers form a single loss modulus or tan δ peak at a compromise temperature between the two individual polymers' T_g. When the difference in T_g between the two polymers is less than 30 °C, this method cannot determine compatibility.

Blends of SIR 10 with most of the other polyisoprenes show single loss modulus and tan δ peaks. The 3,4-IR blend with SIR 10 appears to have a slightly wider G'' peak, but it is still a single peak. However, the tan δ curve shows a more clear separation for the 3,4-IR and SIR 10 blend (Figs. 7.5 and 7.6). This blend gives a very high tan δ value and also a

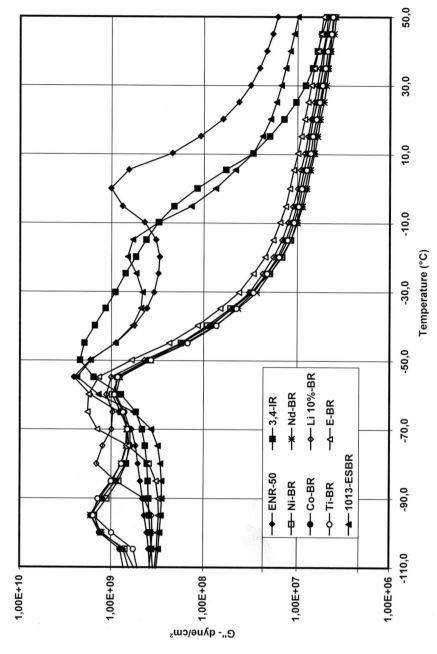

Figure 7.5   Loss modulus temperature sweep of two polymer blends.

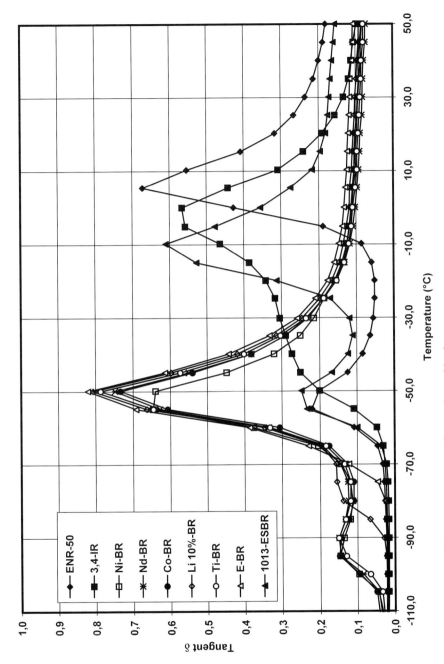

Figure 7.6   Tangent δ temperature sweep of two polymer blends.

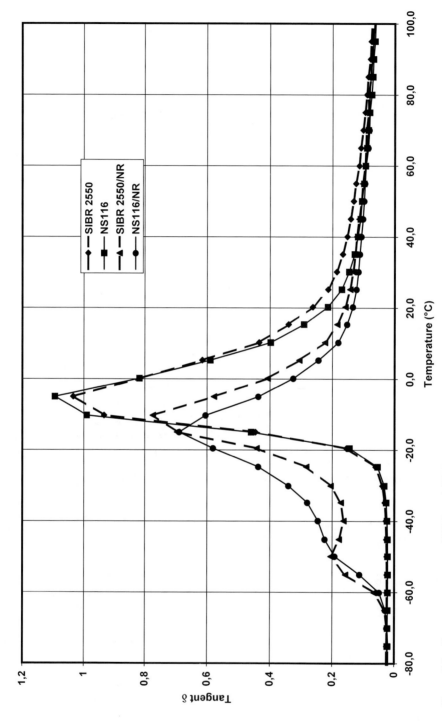

Figure 7.7   Temperature sweep of SIBR and SSBR.

relatively high G'' value at 0 °C, the temperature often used to predict wet traction in tires. The epoxidized natural rubber is not compatible with the SIR 10 and gives the highest G'' and tan δ peaks in the 0 to +5 °C range. The ENR-50 is expected to be an ideal candidate for blending with SIR 10 to improve wet traction in tires.

The high *cis*-BR samples are not compatible with SIR 10, including the 10% 1,2-BR. Higher 1,2 content polybutadienes have been reported by some researchers to be compatible with natural rubber. Because there is less than a 30 °C difference in T_g between the SIR 10 and the 47% and 70% 1,2-BR, this finding could not be confirmed in this study. It is difficult to determine if the emulsion BR is compatible with the SIR 10, because there is only one peak for both G'' and tan δ. But the peak appears a little broader than that for either polymer alone.

The high styrene 1013-ESBR was not compatible with the SIR 10, but the other SBR samples appeared to be. In the 1013-ESBR, there is 40% styrene, and the high styrene content is responsible for the incompatibility. Like the ENR-50 and the 3,4-IR, the 1013-ESBR combined with natural rubber would improve wet traction in tires.

Figure 7.7 shows the compatibility of SIBR with natural rubber. Using the same recipes listed in Table 7.7, SIBRFLEX 2550 from Goodyear Tire and Rubber Company and NS116 From Nippon Zeon were compared. SIBRFLEX 2550 is a SIBR with 25% styrene, 50% isoprene (40% of the IR is 3,4-IR), 25% butadiene (44% of the BR is 1,2-BR) and a microstructure that delivers a polymer T_g of −30 °C. The NS116 is a SSBR with 20% styrene, 80% butadiene (60% of the BR is 1,2-BR), and a similar T_g of −30 °C. It is interesting to see that the tan δ curves for the compounds containing only the synthetic polymers are almost identical. But when blended 50/50 with natural rubber, the SIBR broadens the curve and splits it into two distinct peaks, while the SSBR natural rubber blend has a narrower curve without two distinct peaks. In other words, SIBR shows less compatibility with natural rubber than an SSBR of comparable T_g.

Although these examples show that microstructure can determine compatibility in blends, it is known that molecular weight also has an influence on compatibility. Generally, the higher the molecular weight, the lower the compatibility [26].

7.5.6   Fatigue Properties

Fatigue properties are very important for predicting performance in dynamic applications. The best way to predict performance life is to test under dynamic conditions that are similar to the actual service conditions. Unfortunately, determination of those actual conditions can sometimes be very difficult. Also, testing time might be excessive (possibly years) if the exact service conditions are used. In the laboratory, higher strain levels or frequencies and pre-aging of the test sample might emulate service conditions. Some standard tests have been used for many years, two of which were used in this study: the Monsanto fatigue to failure (FTF) test and the De Mattia test.

The Monsanto FTF test looks at crack initiation and growth to failure. The total number of cycles to failure are usually reported. The dumbbell shaped samples can be stretched to any number of elongations by choosing different size cams. In this study the polymer compounds were evaluated with three different cams. Cam 8 corresponds to 80%

elongation; cam 14 corresponds to 100% elongation, and cam 18 corresponds to 120% elongation. The test was run according to ASTM D 4482.

The De Mattia test measures the growth of a pre-initiated crack. The crack is usually 2 mm long. This test was run according to ASTM D 813. There were two very clear patterns in the results from these two tests. The polyisoprene group was significantly better than the other compounds for De Mattia crack growth, and significantly worse for FTF (Tables 7.10 and 7.11). In most cases, the addition of SIR 10 improved the De Mattia crack growth for the sample polymers. Of the IR group, the CV60 and synthetic IR performed the best. They also have lower 300% modulus values than the other polyisoprenes. One exception is the 3,4-IR, which has a low modulus but did not show a good De Mattia crack growth result itself, but was very good when blended with SIR 10.

Much has been written about FTF for polybutadienes, although most of these reports disagree on which BR performs the best. In Table 7.10, the 10% 1,2-BR performs the best at 80% elongation; the Ti-BR performs the best at 100% elongation; and the Nd-BR and the Ti-BR tie for 120% elongation. As the 1,2 content in the lithium BR samples is increased, FTF is reduced. The high cis-BR samples are about equal to each other, with the Nd-BR and Ti-BR having just the slightest advantage over the others. The emulsion BR was very poor at low elongations, but was comparable at the higher elongations.

A study by Gargani and Bruzzone found somewhat similar results [27]. Using the Monsanto FTF test and Weibull mean life for comparison, they showed that higher 1,2 content was better than lower 1,2 content, and larger differences between high cis contents were also found. In addition, they showed that increasing weight average molecular weight (M_w) by 44% (457,000 to 657,000) increased fatigue life by 112%. The major reason offered for the excellent performance by the high cis polybutadienes was their crystallizing upon strain.

In still another study [28], results of high cis-BR samples were shown to vary with formulation. In a truck tread formulation, a Co-BR was best; in a light truck tread formulation, a Ti-BR was best; and in a sidewall formulation, an Nd-BR was best. This study presents the logical conclusion: always evaluate candidate polymers in the intended application recipe.

Table 7.11 shows the FTF results when blended with NR. The SIR 10 reduced the FTF values for all of the non-IR polymers. The SIR 10 reduced the FTF for the other polymers to a level comparable to the SIR 10 by itself.

7.5.7   Compression Set

Compression set is important in many applications, probably the most important being seals and O-rings. Compression set, ASTM D 395-B (22 h), and Goodrich Flexometer, ASTM D-623 A, were used to collect static and dynamic compression set values. Tables 7.10 and 7.11 list the compression sets for the two recipes used here. As a group, the polyisoprene samples have the highest compression set. The ENR-50 has the highest compression set of all polymers. The BR and SBR samples are all comparable. In the single polymer recipe, the 3,4-IR microstructure and the 1,2-BR microstructure improve their respective polymer types, which is somewhat surprising because these polymers have higher glass transition temperatures also.

7.6   Conclusion

The general purpose polymers represented by the polyisoprenes, polybutadienes, and SBR are widely accepted within the rubber industry. By carefully choosing the correct macro- and microstructures, a wide range of physical properties, performance, and processing parameters can be fine-tuned to fit most applications. Because macrostructures vary widely from one manufacturer to another, it is highly recommended that the candidate polymer(s) be evaluated in the application recipe. Switching from one manufacturer to another should always be preceded with adequate testing.

Technology for general purpose polymers is changing constantly. Alternative polymerization techniques, such as gas phase polymerization, are currently under study. Different microstructures, including high *trans*-BR, are being developed and commercialized, as are the styrene-isoprene-butadiene rubbers (SIBR). New comonomers such as p-methyl styrene are also under evaluation. New chain end modifiers are being studied, not only for carbon black reinforcement, but also for silica. Anionic polymerization allows the microstructure of SBR to be refined to the end application. While compounders have an ever-widening range of polymers from which to choose, it remains important to stay current on these new technologies and opportunities.

References

1. *The Synthetic Rubber Manual* 13th Edition, Copyright ©1995, International Institute of Synthetic Rubber Producers, Inc.
2. *The Green Book*, International Standards of Quality Packing for Natural Rubber Grades, Underwritten by Rubber Manufacturers' Association and Rubber Trade Association of North America.
3. Subramaniam, A., In *The Vanderbilt Rubber Handbook Thirteenth Edition,* 1990.
4. Reader, G., *Rubber Developments* (1994), **47**(3).
5. Miani, B., Lauretti, E., Mistrali, F., Paper presented at Rubber Division Meeting, Orlando, Florida, October (1993).
6. Sumner, A., Paper presented at ITEC September (1996).
7. Hall, J. E., United States Patent 5,112,929, May 12, 1992.
8. Hall, J. E., United States Patent 4,429,091, January 1, 1984.
9. Hall, J. E., United States Patent 4,429,090, January 31, 1984.
10. Brown, T., In *The Vanderbilt Rubber Handbook Thirteenth Edition,* 1990.
11. Kang, J. W., Takatsugu, H., Seaver, G. B., European Patent 476,665, September 9, 1995.
12. Sommer, N., Kautschuk + Gummi Kunststoffe (1975) **28**(3) pp. 131–135.
13. Namizuka, T., Koyama, Y., UK Patent GB 2,117,778, October, 13, 1983.
14. Day, G., Futamura, S., Paper presented at ACS Rubber Division Meeting, New York, New York, April 8–11, 1986.
15. Hall, J. E., Roggeman D. M., United States Patent 5,272,207, Dec. 21, 1993.
16. Hall, J. E., United States Patent 5,395,902, March 7, 1995.
17. White, *Industrial Engineering Chemistry* (1949) **41**, p. 1554.

18. Nagata, N., Kobatake, T., Watanabe, H., Ueda, A., Yoshioka, A., *Rubber Chemistry and Technology* (1987) **60**, pp. 837–855.
19. Halasa, A., *Rubber Chemistry and Technology* (1997) **70**, pp. 295–308.
20. Drake, R. E., Labriola, J. M., Paper presented at ACS Rubber Division Meeting, Pittsburgh, Pennsylvania, October (1994).
21. Hess, W. M., Klamp, W. K., *Rubber Chemistry and Technology* (1983) **56**, pp. 390–417.
22. Futamura, S., *Rubber Chemistry and Technology* (1991) **64**, pp. 57–64.
23. Futamura, S., *Rubber Chemistry and Technology* (1996) **69**, pp. 648–653.
24. Payne A. R., *Journal of Applied Polymer Science*, (1962) **6**, pp. 57–63.
25. Engelhardt, M. L., Day, G. L., Samples, R., Baranwal, K. C., Paper presented at ITEC, Akron, Ohio, September 10–12, 1996.
26. Gardiner, E. S., Educational symposium presented at ACS Rubber Division Meeting, Montreal, 1996.
27. Gargani, L., Bruzzone, M., *Advances in Elastomers and Rubber Elastomers and Rubber Elasticity*, (1986) Plenum Publishing Corp.
28. Kumar, N. R., Chandr, A. K., Mukhopadyay, R., *International Journal of Polymeric Materials* (1996), **34**, pp. 91–103.

8 Specialty Elastomers

8.1 Introduction

Specialty elastomers are very important to the rubber industry. They impart unique properties to a rubber compound, which cannot be matched by a general purpose elastomer. The specialty elastomers being discussed in this chapter are:

8.2 Butyl Rubber (IIR)
8.3 Chlorobutyl and Bromobutyl Rubbers (CIIR and BIIR)
8.4 Ethylene Propylene Rubber (EPM and EPDM)
8.5 Acrylonitrile Butadiene Rubber (NBR)
8.6 Hydrogenated Acrylonitrile Butadiene Rubber (HNBR)
8.7 Polyacrylate Rubber (ACM)
8.8 Polychloroprene (CR)
8.9 Chlorinated Polyethylene (CM)
8.10 Chlorosulfonated Polyethylene (CSM)
8.11 Polyepichlorohydrin (CO, GCO, ECO, GECO)
8.12 Ethylene-Acrylic Elastomers
8.13 Polynorbornene
8.14 Fluoroelastomer (FKM)
8.15 Silicone Elastomer (MQ, VMQ, PMQ, PVMQ, FVMQ)

8.2 Butyl Rubber

G.E. Jones, D.S. Tracey, A.L. Tisler

8.2.1 Introduction

Commercial butyl-rubber grades poly(methylpropene-co-2-methyl-1,3 butadiene) or poly(isobutylene-co-isoprene), are prepared by copolymcrizing 1 to 3% of the isoprene monomer feed with isobutylene, catalyzed by $AlCl_3$ dissolved in methyl chloride. Butyl rubber had its origin in the work of the researchers Gorianov and Butlerov (1870) and Otto (1927), who found that oily homopolymers of isobutylene could be produced in the presence of boron trifluoride. It was not until the 1930s that the I.G. Farben Company of Germany produced high molecular weight polyisobutylenes that possessed rubber-like properties, but these products could not be vulcanized by normal methods because of their saturated hydrocarbon structures. In subsequent research in the 1930s, W.J. Sparks and R. M. Thomas [1–5] of Exxon Research and Engineering Company advanced the state of the art of isobutylene polymerization. In 1937, they produced the first vulcanizable isobutylene-based elastomer by incorporating small amounts of a diolefin, particularly isoprene, into the polymer molecule. This finding introduced the concept of limited olefinic functionality for vulcanization in an otherwise saturated copolymer.

8.2.2   Butyl Rubber Physical Properties

The most widely used, commercially available, butyl rubbers are copolymers of isobutylene and isoprene and are described in Table 8.1. Grades are distinguished by molecular weight (Mooney viscosity) and mole % unsaturation, or the number of moles of isoprene per 100 moles of isobutylene.

Important modified product forms of butyl rubber are produced by adding a polymer branching agent, which allows the isobutylene and isoprene molecules to attach to another short molecule. These short initiator molecules, (styrene-isoprene copolymer or styrene-butadiene copolymer) attach to the butyl molecules via one of two different reactions to create a new arrangement of 10 to 40 polymer "arms" in a "star" format. These stars agglomerate in clusters of about 10 star molecule units. This structure contributes to the improved green strength seen in this product form. The star format is illustrated in Fig. 8.1.

If the molecular weight distributions are measured, the result is a bimodal distribution with a fraction of higher weights of about 13 to 15 wt%.

8.2.3   Butyl Rubber Properties, Vulcanization, and Applications

The molecular characteristics of low levels of unsaturation between long segments of polyisobutylene produce unique elastomeric qualities that find a wide variety of applications. These special properties are:

Table 8.1   Commercial Butyl Rubber Grades

Typical range for Mooney viscosity		Unsaturation [Ave. Mole%]	Supplier and typical grades		
ML @ 100 °C (1 + 8)	ML @ 125 °C [1 + 8]		Exxon Chem. Co.	Bayer Polysar (Canada)	Russian
41–49	29–35	0.7 to 1.1[a]	Exxon Butyl 065	Polysar Butyl 100	
43–51	30–36	2.0 to 2.6[a]	Exxon Butyl 365	Polysar Butyl 402	
41–49	29–35	1.1 to 1.5[a]	Exxon Butyl 165	–	
–	46–56	1.5 to 1.8[a]	Exxon Butyl 268	Polysar Butyl 301	BK–1675N
–	43–53	0.9 to 1.3[a]	Exxon Butyl 077	–	
–	46–56	~1.7	Exxon Butyl 007	Polysar Butyl 101–3	
–	34–44	1.4[a]	Exxon SB Butyl 4266		
	46–56	~1.4[a]	Exxon SB Butyl 4268		
	46–56	~1.0[a]	Exxon Butyl 068		

[a] Grade stabilized with non-staining antioxidant.

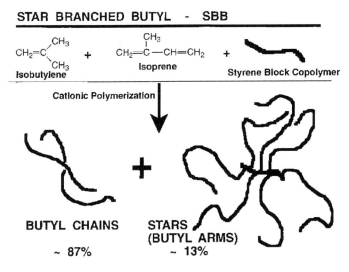

Figure 8.1   Star branched butyl arms.

1. Low rates of gas permeability
2. Ozone and weathering resistance
3. Thermal stability
4. Vibration damping and higher coefficients of friction
5. Chemical resistance
6. Moisture resistance

The chemically inert nature of butyl rubber is reflected in the lack of significant molecular weight breakdown during processing. This property allows operations such as heat treatment or high temperature mixing to alter the vulcanizate characteristics of a compound to occur. With carbon black-containing compounds, hot mixing techniques promote pigment polymer interaction [6], which alters the stress-strain behavior of a vulcanizate. The shape of the stress-strain curve of the vulcanizate from a heat-treated mixture is a reflection of a more elastic network. The heat treatment has resulted in vulcanized compounds with more flexibility for a given level of a carbon black type. Butyl rubber compositions with greater flexibility have also been prepared with certain types of mineral fillers, such as reinforcing clays, talcs, and silicas that contain appropriately placed OH groups in the lattices. This enhancement in pigment-polymer association usually occurs in the presence of chemical promoters [7,8].

8.2.4   Gas Permeability

The permeability of elastomeric films to the passage of gas is a function of the diffusion of gas molecules through the membrane and the solubility of the gas in the elastomer [9]. The polyisobutylene portion of the butyl rubber molecule provides a low degree of permeability to gases and is a familiar property, leading to an almost exclusive use in inner tubes. The difference in air retention between a natural rubber and a butyl inner tube can be

Table 8.2   Air Loss of Inner Tubes During Driving Tests

Inner tube	Original pressure (psi)	Air pressure loss (psi)		
		1 Week	2 Weeks	3 Weeks
Natural rubber	28	4.0	8.0	16.5
Butyl	28	0.5	1.0	2.0

Table 8.3   Composition, Butyl Inner Tube

Exxon Butyl 268 (1.4 to 1.8 mol% unsaturation)	100.0
N660 carbon black	70.0
Paraffinic process oil	25.0
Zinc oxide	5.0
Stearic acid	1.0
Sulfur	2.0
TMTDS (tetramethyl thiuram disulfide)	1.0
MBT (mercaptobenzothiazole)	0.5

Cure range 5 min at 175 °C to 8 min at 330 °F

demonstrated by data from controlled road tests on cars driven 60 mph for 100 miles per day. Under these conditions, it was shown (Table 8.2) that butyl is at least eight times better than natural rubber in air retention.

Other gases such as helium, hydrogen, nitrogen, and carbon dioxide are also well retained by a butyl bladder membrane. While the significance of these properties in inner tubes is waning, it is of importance in air barriers for tubeless tires, air cushions, pneumatic springs, accumulator bags, air bellows, and the like. A typical formulation for a butyl rubber passenger tire inner tube is given in Table 8.3.

8.2.5   Ozone and Weathering Resistance

The low level of chemical unsaturation in the polymer chain produces an elastomer with greatly improved resistance to ozone when compared to polydiene rubbers. Butyl rubber, with the lowest level of unsaturation (0.8 mol%), produces high levels of ozone resistance, which are also influenced by the type and concentration of vulcanizate crosslinks. For maximum ozone resistance, as in electrical insulation, and for weather resistance, as in rubber sheeting for roofs and water management application, the least unsaturated butyl is advantageously used. A typical butyl rubber sheeting compound in given in Table 8.4.

8.2.6   Butyl Rubber Vulcanization

Regular butyl rubber is commercially vulcanized by three basic methods [10]:

1. Accelerated sulfur vulcanization
2. Crosslinking with dioxime and related dinitroso compounds
3. The polymethylol-phenol resin cure

Table 8.4   Butyl Sheeting Compound

Exxon butyl 065 (0.7 to 1.4% mole unsaturation)	100.0
N330 carbon black	48.0
N762 carbon black	24.0
Zinc oxide	5.0
Petrolatum	3.0
Wax	4.0
Sulfur	1.0
TDEDC (tellurium diethyldithiocarbamate)	0.5
ZDMDC (zinc dimethyldithiocarbamate)	1.5
MBT (mercaptobenzothiazole)	0.5

8.2.6.1   Accelerated Sulfur Vulcanization

In common with more highly unsaturated rubbers, butyl rubber may be crosslinked with sulfur and activated by zinc oxide and organic accelerators. In contrast to the higher unsaturated varieties, however, adequate states of vulcanization can be obtained only with the very active thiuram and dithiocarbamate accelerators. Other less active accelerators, such as thiazole derivatives, may be used as modifying agents to improve the safety of processing scorch. Most curative formulations include the ranges of ingredients shown in Table 8.5.

8.2.6.2   The Dioxime Cure

The crosslinking of butyl rubber with p-quinone dioxime or p-quinone dioxime dibenzoate proceeds through an oxidation step that forms the active crosslinking agent, p-dinitroso-benzene. The use of PbO_2 as the oxidizing agent results in very rapid vulcanizations, which can produce room temperature cures for cement applications. In dry rubber processing, the dioxime cure is largely used in butyl rubber formulations for electrical insulation to provide maximum ozone resistance and moisture impermeability to the vulcanizate.

8.2.6.3   The Resin Cure

Crosslinking of butyl rubber (and other elastomers containing olefinic unsaturation) by this method is dependent upon the reactivity of the phenolmethylol groups of reactive phenol-formaldehyde resins, as shown in Fig. 8.2.

Table 8.5   Curative Formulations

	Parts by Wt.
Butyl elastomer	100
Zinc oxide	5
Sulfur	0.5 to 2.0
Thiuram or dithiocarbamate accelerator	1 to 3
Modifying thiazole accelerator	0.5 to 1

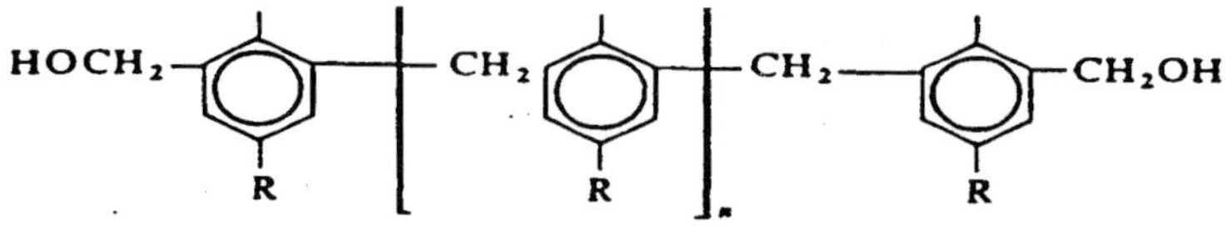

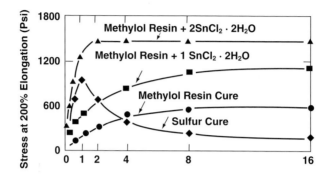

Figure 8.2   Reactivity of phenolmethylol.

Figure 8.3   Relative rates of cure and reversion of resin- and sulfur-butyl compounds [11].

The low levels of unsaturation in butyl rubber require resin cure activation by halogen-containing materials such as $SnCl_2$, or halogen-containing elastomers such as chloroprene [11]. A series of curves in Fig. 8.3 illustrates the activating effect of stannous chloride and the stability of the resultant crosslink to reversion upon prolonged heating. This feature of the resin cure is utilized in the fabrication of tire curing bladders, as previously mentioned.

A more reactive resin cure system that requires no external activator results if some of the hydroxyl groups of the methylol group are replaced by bromine. Such a resin commonly used is Schenectady Chemical's SP-1055.

8.3   Halogenated Butyl Rubber

G.E. Jones, D.S. Tracey, A.L. Tisler

8.3.1   Introduction

Using halogenation to provide more active functionality to the butyl rubber molecule was first introduced by B.F. Goodrich Tire and Rubber Company researchers in 1954–56 [12–16]. Their work emphasized the attributes of the brominated derivative. Goodrich

Table 8.6   Commercial Grades of Halobutyl Rubber

Supplier and Typical Grade[a]	Typical Mooney viscosities and halogen content	
	ML at 125 °C (1+8)	Typical wt.% halogen
Exxon chlorobutyl 1066[b]	33–43	1.2
Exxon chlorobutyl 1068	45–55	1.2
Exxon SB chlorobutyl 5066	35–45	1.5
Polysar chlorobutyl 1240	34–44	1.2
Polysar chlorobutyl 1255	45–55	1.2
Exxon bromobutyl 2222[c]	27–37	2.0
Exxon bromobutyl 2233	32–42	2.1
Exxon bromobutyl 2244	41–51	2.1
Exxon SB bromobutyl 6222	27–37	2.4
Polysar bromobutyl X2	41–51	2.0
Polysar bromobutyl 2030	28–36	2.0
Polysar bromobutyl 2040	34–44	2.0

[a] All grades contain non-staining antioxidants at about 0.1 wt%.
[b] Chlorobutyl contains Ca stearate to protect polymer from dehydrohalogenation.
[c] Bromobutyl contains Ca stearate and epoxidized soybean oil to protect polymer from dehydrohalogenation.

commercialized a brominated butyl rubber in 1954 (Hycar 2202), prepared from a bulk-batch halogenation, but withdrew it in 1969. In this same period, Exxon researchers originated the chlorobutyl product concept [17,18] and commercially introduced it in 1961. Bromobutyl rubber was again commercially introduced by Polysar Limited of Canada [19], now a Bayer company, in 1971 and by Exxon Chemical Company in 1980. Commercial grades of halobutyl are summarized in Table 8.6 along with some physical properties.

Star-branched butyl rubber is a bimodal polymer with a high molecular weight branched mode and a lower molecular weight linear component. The polymer is made by a conventional carbocationic copolymerization of isobutylene and isoprene in methyl chloride at temperatures below −80 °C with a Friedel-Crafts catalyst, e.g., $AlCl_3$, in the presence of a polymeric branching agent as shown in Fig. 8.1.

8.3.2   Compounding Halobutyl and Star-Branched Halobutyl Rubbers

The following section summarizes the effect of compounding ingredients on the processing and vulcanizate properties of halobutyl rubber.

8.3.2.1   Carbon Black

Carbon blacks affect the compound properties of halobutyl rubber similarly to the way they affect the compound properties of other rubbers. Particle size and structure determine the reinforcing power of the carbon black and hence, the final properties of the halobutyl rubber compounds, as described next:

- Increasing reinforcing strength (N660 to N330) raises viscosity, hardness, and cured modulus
- Cured modulus increases with the carbon black level up to 80 phr
- Tensile strength goes through a maximum at 50 to 60 phr carbon black level
- Star branched halobutyl rubbers have a lower Mooney (46 vs. 49) at mid levels of oil and black loading
- Higher green strength (~25%) for star branch halobutyl rubbers at the same carbon black loading levels

8.3.2.2   Mineral Fillers

Common mineral fillers may be used with halobutyl rubber. However, because they vary significantly both in particle size and chemical composition, fillers may alter vulcanization characteristics. As a general rule, highly alkaline fillers, such as calcium stearate, and hygroscopic fillers can strongly retard a halobutyl rubber cure. Acidic clays give very fast cures; therefore, extra scorch retarders such as magnesium oxide may be needed. Hydrated silicas should be used at moderate levels because they promote compound stiffness and strongly affect cure rates as a result of their tendency to absorb curatives. Talc is semi-reinforcing in halobutyl rubber without a major effect on cure. Calcium carbonate (whiting) gives relatively little reinforcement, but yields good elongation after heat aging. Calcium carbonates coated with calcium stearate strongly retard cure rates.

8.3.2.2   Plasticizers

Petroleum-based process oils are the most commonly used plasticizers for halobutyl rubber. They improve mixing and processing, soften stocks, improve flexibility at low temperatures, and reduce cost. Paraffinic/naphthenic oils are preferred for compatibility reasons. Other useful plasticizers are paraffin waxes and low molecular weight poly-ethylene. Adipates and sebacates improve flexibility at very low temperatures.

8.3.2.3   Processing Aids

The following are examples of processing aids:

- Struktol 40 MS and mineral rubber not only improve the processing characteristics of halobutyl rubber compounds by improving filler dispersion, but also enhance compatibility between halobutyl and highly unsaturated rubbers.
- Tackifying resins should be selected with care. Phenol-formaldehyde resins, even those with deactivated reactive methylol groups, react with halobutyl (especially bromobutyl) rubber, causing a decrease in scorch time. Partially aromatic resins such as Koresin have an intermediate shortening effect on scorch in bromobutyl rubber.
- Stearates, stearic acid: It should be noted that zinc stearate (which can also be formed via the zinc oxide and stearic acid reaction) is a strong dehydrohalogenation agent and a cure catalyst for halobutyl rubber. Similar effects are observed with other organic acids, such as oleic acid and naphthenic acid. Alkaline stearates such as calcium stearate, on the other hand, retard the halobutyl rubber cure.

- Anti-degradants. Amine-type antioxidants/antiozonants such as 1,2-dihydro-2,2,4-trimethylquinoline (TMQ), mercaptobenzimidazole, and especially p-phenylenediamines, react with halobutyl rubber. They should be added with the curatives, not in the master batch. Phenol derivative antioxidants are generally preferred.

8.3.3   Processing Halobutyl Rubber

The following recommended processing conditions are applicable to both chloro- and bromobutyl rubbers:

8.3.3.1   Mixing

Mixing is done in two stages. The first stage mixes all the ingredients except for zinc oxide and accelerators. The batch weight should be 10 to 20% higher than that used for a comparable compound based on general purpose rubbers. A typical mixing cycle (for internal mixers) and processing for a halobutyl rubber innerliner compound are as follows:

First Stage	0 min.	Halobutyl, retarder, process aids
	0.5 min	Fillers
	1.5 min	Plasticizers, stearic acid
	3.5 min	Dump at 120 to 140 °C
		Higher dump temperatures could result in scorching.
Second Stage	0 min	Master batch + curatives
	2 min	Dump at 100 °C

While mixing with the star-branched halobutyl polymers, it has been observed that:

- A faster mixing (5 to 10%) in internal mixers is seen
- High molecular weight fraction increases shear and incorporates the fillers more rapidly
- Polymer lumps are reduced or eliminated because of the increased shear

For Mill Mixes: A two-roll mill is best accomplished with a roll speed ratio of 1.25:1 and roll temperatures of 40 °C on the slow roll and 55 °C on the fast roll. The following sequence of addition is recommended:

- Part of the rubber together with a small amount of a previous mix as a seed
- 1/4 fillers plus retarder
- Remainder of polymer
- Rest of filler in small increments
- Plasticizers at the end
- Acceleration below 100 °C

8.3.3.2   Calendering

Feed preparation can be done either by mill or by extruder. Halobutyl rubber follows the cooler roll; therefore, a temperature differential of 10 °C between calender rolls is recommended.

Starting roll temperatures should be:

Cool roll   80 to 85 °C
Warm roll   85 to 95 °C

Normal calendering speeds for halobutyl rubber compounds are 25 to 30 meters/minute. Rapid cooling of the calendered sheet is beneficial for optimal processability (handling) and maximum tack retention.

Use of the star branch (SB) halobutyl rubber in calendering vs. regular halobutyl rubbers has resulted in:

- More uniform calendered sheets with the high green strength and faster stress relaxation of SB polymers
- Reduced air blisters because the high green strength reduces pooling of entrapped air
- Easier roll release because of the faster stress relaxation and more uniform mix

8.3.3.3   Extrusion

The feed temperature should be 75 to 80 °C, while the temperature of the extrudate is around 100 °C. During calendering and extrusion of halobutyl rubber compounds, the most significant problem is blister formation. The reason for this phenomenon is the low permeability of these polymers, which tend to retain entrapped air or moisture. Preventive actions should be taken at all stages of the process, such as:

- Ensure the stock is well mixed in a full mixer to prevent porosity
- Avoid moisture at all stages
- Keep all rolling banks on mills and calender nips to a minimum

When butyl polymers are extended, the result is higher output, lower die swell, and lower temperatures.

8.3.3.4   Molding

Halobutyl rubber can be formulated to have a fast-cure rate, good mold flow and mold release characteristics, and can, therefore, be molded into highly intricate designs with conventional molding equipment. Entrapped air can be removed by bumping the press during the early part of the molding cycle. Halobutyl is also very well suited for injection molding because of its easy flow and fast reversion-resistant cures. Low molecular weight polymer grades may be required for optimum flow and good scorch safety.

8.3.4   Halobutyl Rubber Vulcanization and Applications

The presence of both olefinic unsaturation and reactive chlorine/bromine in halobutyl rubber provides for a great variety of vulcanization techniques. The main difference between bromo- and chlorobutyl rubbers are in the higher reactivity of the C–Br bond compared to that of C–Cl. This manifests itself in:

1. A higher cure versatility, i.e.:
    - Greater choice in curatives
    - Lower curative level needs
    - Generally faster cures
    - Generally shorter scorch times
2. A higher tendency for covulcanizaiton with highly unsaturated rubbers

It should be emphasized that, in contrast to highly unsaturated rubber-sulfur cures, halobutyl cures are generally accelerated by acids (and retarded by bases), and usually result in very stable crosslinks, i.e., C–C or C–S–C bonds.

8.3.4.1   Straight Sulfur Cure

Bromobutyl rubber cures with sulfur as the sole curative in the absence of zinc oxide. This cure, which is of more academic than practical interest, is retarded by phenolic antioxidants pointing to a radical reaction mechanism. Especially with bromobutyl rubber, this cure efficiently uses the sulfur atom, which partially explains bromobutyl rubber's enhanced co-vulcanizability with high unsaturation rubbers.

8.3.4.2   Zinc Oxide Cure and Modifications

Zinc oxide, preferably with some stearic acid, can function as the sole curing agent for halobutyl rubber. After vulcanization, most of the halogen originally present in the polymer can be extracted as a zinc halide. A proposed mechanism by Baldwin is based on formation of stable C–C crosslinks through a cationic polymerization route [18]. When zinc oxide is used as the vulcanizing reagent, the necessary initiating amounts of zinc halide are likely formed as a result of thermal dissociation of some of the allylic halide to yield hydrogen halide. Subsequent reaction of the hydrogen halide with zinc oxide then provides a catalyst. It is not likely that the propagation step proceeds very far, but for vulcanization purposes, only one step is needed, particularly because both of the termination processes suggested result in the production of more catalyst. The attainment of the full crosslinking potential of halobutyl by the ZnO system is relatively slow. This situation can be remedied by inclusion of thiurams and thioureas in the curing recipe.

8.3.4.3   Zinc-Free Cures

Zinc-free compounds are required in some applications. Bromobutyl rubber is capable of vulcanizing without zinc oxide or any other zinc salt. Preferred curatives are diamines such as hexamethylene diamine carbamate (Diak No. 1).

8.3.4.4   Peroxide Cures

Although butyl polymers undergo molecular weight breakdown with peroxide cures, chlorobutyl, and even more so bromobutyl rubbers, are capable of crosslinking with peroxides. A good combination is dicumyl peroxide and bismaleimide (HVA-2) as a coagent. Studies have shown that bromobutyl rubber cures with bismaleimide alone without peroxide, giving low compression set and extremely high heat resistance.

8.3.4.5   Vulcanization through Bis-Alkylation

The other unique and valuable curing method for chlorobutyl rubber involves bis-alkylation reactions. This reaction is perhaps best illustrated by crosslinking with primary diamines, a vulcanization reaction that proceeds rapidly to yield good vulcanizates. Extensions of this bis-alkylation vulcanization technique are numerous, and in general any molecule with two active hydrogens can, under the proper catalytic conditions, crosslink the polymer. Typical examples are vulcanization with diihydroxy aromatics, such as resorcinol, and with dimercaptans.

8.3.4.6   Resin Cure

Both chlorobutyl and regular butyl rubbers are capable of vulcanization with heat reactive phenolic resins, which are usually characterized by consisting of 6 to 9 wt% methylol groups. Unlike conventional butyl rubber, no promoter or catalyst other than zinc oxide is needed for efficient vulcanization, and a fast, tight cure is obtained with considerably less reagent.

8.3.4.7   Scorch Control

The modified zinc oxide cures are very fast and tend to scorch. As a general rule, acidic materials activate the ZnO cure of halobutyl rubbers, while basic materials hamper or retard it. The retarding effects of some alkaline materials, such as magnesium oxide, can be used in a very practical way to provide processing safety. It can be seen in Table 8.7 that the addition of 0.25 phr of MgO increases the margin against incipient vulcanization during processing at 126.5 °C (260 °F), as judged by Mooney scorch measurements, from 5

Table 8.7   Effects of Magnesium Oxide Addition on Vulcanization

Compound No.	1	2	3	4
TMTD	1.0	1.00	1.0	0.5
Magnesium oxide	–	0.25	0.5	–
Mooney scorch measurement[a]				
Time to 5 point rise at 126.5 °C, min.	5	15	30	8
Room temperature tensile properties[a]				
Cured 45 min. at 153 °C				
Modulus @ 300%, psi	2500	2370	1850	2360
Tensile strength, psi	2600	2720	2410	2580
Ultimate elongation, %	315	335	380	320

[a]Formulation (phr): chlorobutyl – 100; antioxidant 2246 – 1; N330 carbon black – 50; stearic acid – 1; ZnO – 5; curatives as indicated.

to 15 minutes, without greatly affecting tensile properties at the vulcanization temperature of 153 °C (307 °F). However, if the concentration is increased to 0.5 phr, the cure rate is depressed significantly.

As a scorch retarder, MgO is effective with all cure systems except the amine cure. In this case, it has the reverse effect because it prevents the hydrogen halide, generated during crosslinking, from reacting with the curing agent. The choice of MgO type and concentration depends upon the compound used, as well as the particular application involved.

8.3.4.8   Stability of Halobutyl Crosslinks

Because halobutyl rubber has essentially the same structure as the butyl polymer, all the properties inherent to the butyl backbone are found in the halobutyl rubber. These properties include low gas and moisture permeability, high hysteresis, good resistance to ozone and oxygen, resistance to flex fatigue, and chemical resistance. Additionally, halobutyl (especially chlorobutyl) rubber offers appreciably higher heat resistance than regular butyl rubber cured with conventional sulfur vulcanization systems.

The combination of TMTD with zinc oxide for curing chlorobutyl rubber is effective in producing vulcanizates capable of withstanding temperatures up to 193 °C (380 °F). At that temperature most vulcanizates, other than chlorobutyl rubber, become excessively soft after a few hours of exposure, having lost all elastomeric behavior. Many chlorobutyl rubber cure systems, other than the ZnO-TMTD cure described above, can be selected for heat resistance, such as straight zinc oxide, ZnO-dithiocarbamate, and resin cure systems. Dithiocarbamates, such as the lead and zinc derivatives, have been successfully used in chlorobutyl-zinc oxide systems.

8.3.5   Halobutyl Rubber General Applications

Since the introduction in 1961 of chlorobutyl rubber (and re-introduction of bromobutyl rubber in 1971), these products have often proven to be highly useful in many commercial rubber products. Compounds with halobutyl rubbers generally exploit desirable character-istics generic to butyl polymers, such as the resistance to environmental attack and low permeability to gases. In addition, chlorobutyl rubber offers the superimposed advantages of cure versatility, highly heat stable crosslinks, and the ability to vulcanize in blends with highly unsaturated elastomers. The applications for halobutyl (chloro- and bromo-) rubber include tire white and black sidewalls, innerliners and tire tubes.

Halobutyl rubber is used in many rubber articles in addition to tires and tubes. In these applications, the cure versatility of halobutyl rubber and the stability of its vulcanizates are of particular importance. For example, chlorobutyl rubber can be cured with nontoxic cure systems, such as zinc oxide with stearic acid, for use in products in contact with food. Also, most cure systems provide fast, reversion-resistant cures with chlorobutyl rubber. This property helps thick articles cure uniformly. Because of its exceptional heat resistance and compression-set properties, properly compounded chlorobutyl rubber gives good service at temperatures up to 150 °C (302 °F).

Table 8.8   Air and Water Vapor Permeability Rates at 65 °C (Index)

Typical tire innerliner compound	Air	Moisture
All natural rubber	8.3	13.3
All SBR	6.8	11.0
Halobutyl/natural rubber (60/40)	3.1	3.0
All halobutyl rubber	1.0	1.0

Examples of typical halobutyl rubber applications are hose (steam, automotive), gaskets, conveyor belts, adhesives and sealants, tire-curing bags, tank linings, truck-cab mounts, aircraft engine mounts, rail pads, bridge bearing pads, pharmaceutical stoppers, appliance parts, curable contact cements for rubber-to-rubber adhesion, sheeting, and adhesives where room temperature cures are required.

8.3.6   Cured Properties

Bromobutyl, like chlorobutyl rubber, has several outstanding features related to both the polyisobutylene backbone and the cured network structure.

8.3.6.1   Permeability

Both chloro- and bromobutyl rubbers have outstandingly low permeability to air and water vapor, unsurpassed among rubbers, as further illustrated in Table 8.8.

8.3.6.2   Heat Resistance

For several high temperature applications, particularly those where low permeability is also required, bromobutyl rubber is preferred. The highest heat resistance is obtained when using bismaleimide (HVA-2) as a curative either with or without peroxide.

8.3.6.3   Resistance to Chemicals and Solvents

Bromobutyl rubber is a nonpolar hydrocarbon rubber with very low unsaturation. It therefore swells in the cured state in hydrocarbon solvents but resists polar solvents, much the same as chlorobutyl rubber. Moreover, chemical attack is minimal, i.e., attack by acids, bases, oxygen, and ozone. Except for special cases, halobutyl rubber compounds usually do not require the addition of antioxidants/antiozonants.

8.3.7   Flex Resistance/Dynamic Properties

Halobutyl rubbers have excellent flex crack resistance, both alone and in blends with other rubbers. Of course, with either bromobutyl or chlorobutyl rubbers, the formulation should be adjusted according to the type of deformation involved in a particular application. For example, in a tire innerliner compound, lower modulus compounds lead to higher flex crack resistance. Halobutyl rubber has the same vibration damping properties as the other rubbers of the butyl family.

8.3.8   Compatibility with Other Elastomers

Although halobutyl and natural rubbers have completely different backbone structures, different unsaturation (1.5 versus 100 mole percent), and different vulcanization chemistry, they can be blended in all proportions to attain the desired combination of properties. This is the result of the versatility of the halobutyl rubber vulcanization. Not only can halobutyl vulcanize independently of the NR sulfur cure, merely via crosslinking with zinc oxide, but it also uses sulfur very efficiently. Moreover, it has been shown that many cure systems result in chemical interaction between the two rubber phases, i.e., covulcanization. Recommended cure systems for optimal covulcanization are in Table 8.9.

Generally, bromobutyl rubber adheres better to polyisoprene than chlorobutyl does. Also, cured adhesion with other general purpose rubbers, such as polybutadiene and styrene–butadiene rubber, is possible, but adhesion levels are lower than with polyisoprene.

Table 8.9   Recommended Cure Systems for Covulcanization

ZnO – 3	Sulfur – 0.5	MBTS – 1.5	TMTD – 0 to 0.25
ZnO – 3	Sulfur – 0.5	TBBS – 1.0	TMTD – 0 to 0.25
ZnO – 3	Sulfur – 0.5	BNDS[a] – 1.5	MBTS – 1.5

[a] BNDS – betanaphtholdisulfide.

8.3.9   Halobutyl Rubber Compound Applications

8.3.9.1   Tire Innerliners

Because of its low permeability to gas and moisture vapor and its ability to adhere to high unsaturation rubbers, bromobutyl rubber is generally the preferred choice for 100% halobutyl innerliner formulations in tubeless tires. More than 80% of the bromobutyl rubber produced is currently used in tire innerliners. It is used in truck tires, which require the highest quality because of their severe service conditions. Its use in passenger tires in 100% halobutyl rubber compositions is also gaining importance and growing rapidly. A high quality halobutyl innerliner maintains the tire inflation pressure better, with subsequent improvements in rolling resistance, tread abrasion, tire duration, and most importantly, safety. The formulations shown in Table 8.10 are typical examples of halobutyl innerliner compounds.

8.3.9.2   Pharmaceutical Closures

Butyl and halobutyl rubbers are extensively used for pharmaceutical closures because of their inherent properties, including:

Table 8.10   Composition of Halobutyl Innerliners

100% Bromobutyl innerliner compound		100% Chlorobutyl innerliner compound	
Exxon bromobutyl 2255	100.0	Exxon chlorbutyl 1066	100.00
N660 carbon black	50.0	N660 carbon black	60.00
Paraffinic oil	8.0	Napthenic oil	8.00
Stearic acid	2.0	Stearic acid	2.00
Magnesium oxide	0.5	Magnesium oxide	0.15
Struktol 40 MS	7.0	Struktol 40 MS	7.00
Zinc oxide	3.0	Zinc oxide	3.00
Sulfur	0.5	Sulfur	0.50
MBTS	1.5	MBTS	1.50

- Low permeability to gases and H_2O
- Availability of special non-toxic vulcanization systems
- Resistance to heat, UV, and ozone
- Chemical/biological inertness

While butyl rubber still requires sulfur for vulcanization, halobutyl rubbers can be cured with sulfurless cure systems, resulting in very low extractables. A preferred cure system is based on zinc oxide combined with 1.5 to 2.5 phr phenol-formaldehyde resin.

Bromobutyl rubber has a further advantage in that it can be vulcanized without zinc, which is useful for some special closure types. A diamine such as Diak No. 1 is suitable for vulcanization in this case. Calcined clay is the preferred filler, but it can be partially replaced by talc, whiting, or even carbon black. For improved hot tear resistance, silica can be added, but only in low amounts, because it retards cure. All the ingredients should be of the purest grade possible. An example of a sulfurless zinc-free formulation is in Table 8.11.

This compound, cured for 10 minutes at 170 °C, gives a Shore A hardness of 45, a tensile strength of 6.0 Mpa, and a compression set of 16% (22 hours at 70 °C).

8.3.9.3   Heat Resistant Conveyor Belt

Several types of rubber are used for heat-resistant conveyor-belt covers and skim compounds: resin-cured butyl rubber, chlorobutyl rubber, EPDM and SBR. Bromobutyl rubber gives the best compromise between heat resistance and adhesion to textile. Moreover, it generally cures faster than the other rubbers just listed. The following is a

Table 8.11   Sulfurless Zinc-Free Formulation

Exxon bromobutyl 2244	100
Complex aluminum silicate	60
Primol 355 (white oil)	5
Polyethylene AC167A	3
Paraffin wax	2
Vanfre AP-2 (Processing Aid)	2
Stearic acid	1
Diak No. 1	1

Table 8.12   Bromobutyl Belt Cover Compound

Exxon bromobutyl 2244	100.0
N550 carbon black	40.0
Magnesium oxide	1.0
Carbowax 4000 (polyethylene glycol)	1.5
Stearic acid	1.0
MBT	1.0
Zinc oxide	3.0
Dicup 40C (peroxide)	2.0
HVA-2 (bismaleimide)	1.0

	Original	Heat aged 7 days at 150 °C
Hardness, Shore A	59.0	64.0
Tensile strength, MPa	11.2	9.1
Elongation at break, %	250.0	200.0

typical bromobutyl rubber belt-cover compound, cured 35 minutes at 160 °C; it exhibited the properties shown in Table 8.12.

References

1. R. M. Thomas and O.C. Slotterbeck, U.S. Pat. 2,243,658 (May 27, 1941).
2. R. M. Thomas and W. J. Sparks, U.S. Pat. 2,356,128 (Aug. 22, 1944).
3. W. J. Sparks and R. M. Thomas, U.S. Pat. 2,356,129 (Aug. 22, 1944).
4. R. M. Thomas, I. E. Lightbown, W. J. Sparks, P. K. Frolich and E. V. Murphree, *Ind. Eng. Chem.* (1940) **32**, 1283.
5. R. M. Thomas & W. J. Sparks, *Synthetic Rubber*, G. S. Whitby (ed.) (1963), John Wiley & Sons, New York, ch. 24.
6. A. M. Gessler, *Rubber Age* (1953) 74, 59
7. A. M. Gessler and F. P. Ford, *Rubber Age* (1953) 74, 243.
8. A. M. Gessler and J. Rehner, Jr., *Rubber Age* (1955) **77**, 875.
9. G. J. Van Amerongen, J. *Applied Phys.* (1946) **17**, 972; *J. Polym. Sci.* (1950) 5, 307.
10. C. J. Jankowski, K. W. Powers, and R. L. Zapp, *Rubber Age* (Aug., 1960) **87**, 833.
11. P. O. Tawney, J. R. Little, and P. Viohl, *Rubber Age* (1958) **83**, 101.
12. R. A. Crawford and R. T. Morrissey, U.S. Pat. 2,631,984 (Mar. 17, 1953).
13. R. T. Morrissey, *Rubber World* (1955) **138**, 725.
14. R. T. Morrissey, *Ind. Eng. Chem.* (1955) **47**, 1562.
15. R. T. Morrissey, U.S. Pat. 2,816,098 (Dec. 10, 1957).
16. V. L. Hallenbeck, U.S. Pat. 2,804,448 (Aug. 27, 1957).
17. F. P. Baldwin and R. M. Thomas, U.S. Pat. 2,964,489 (Dec. 13, 1960).
18. F. P. Baldwin, D. J. Buckley, I. Kuntz, and S. B. Robison, *Rubber and Plastics Age* (1961) **42**, 500.
19. J. Walker, R. H. Jones and G. Feniak, Philadelphia Rubber Group Tech. Meeting (Oct. 22, 1972).

8.4   EPM/EPDM

Rajan Vara and Janet Laird

8.4.1   Introduction

Polymers which are based on olefinic monomers represent a large commercial industry today, including polyethylene and polypropylene in their many forms. These polymers are considered to be thermoplastic, as they have significant levels of crystallinity present at operational temperatures to provide strength; yet this crystallinity can be melted at processing temperatures, allowing repeatable forming operations. In these thermoplastic materials, the crystallinity levels are high enough that the material has minimal elastic characteristics. However, copolymers of ethylene and propylene do exhibit a level of strength, flexibility, and elasticity that allow them to be classified as elastomers. These materials are commonly known as EPDM or EPM.

EPM represents copolymers of ethylene and propylene only. These polymers are completely saturated, and as such are crosslinked using peroxide cure systems. EPM grades are available with different molecular weights (MW), and also with different ethylene/propylene (E/P) ratios. These two factors represent the two variables that need to be considered when choosing the appropriate grade for a specific application.

EPDM, on the other hand, contains ethylene and propylene, and also a third monomer that has been incorporated into the polymer for the express purpose of enabling a sulfur crosslinking. The third monomer contains a double bond that can then be accessed by the sulfur. Therefore, when choosing a grade of EPDM, it is important to consider the MW and E/P ratio as mentioned above, but in addition the amount and type of third monomer must also be known.

8.4.2   Ethylene/Propylene Content

By definition, the classification of EPDMs would include any polymer that includes both propylene and ethylene monomers along a single polymer chain. Practically, commercial grades of EPDM today represent ethylene contents from 40 to 80 wt%. Therefore, in the majority of cases, there is a greater amount of ethylene than propylene. Some general property trends can be stated about the polymer when the ethylene propylene contents are within this range. As the ethylene content increases, the polymer crystallinity increases. Conversely, as the ethylene content decreases and the propylene content increases, the polymer is increasingly amorphous.

Typically, EPDM polymers are classified as either semi-crystalline or amorphous. Semicrystalline grades generally have ethylene contents of 62 wt% or greater, while amorphous grades generally have ethylene contains of less than 62 wt%. However, these two classifications are simply a convenient way to describe the morphology of the different grades, and it should not be construed that a polymer having 65 wt% ethylene has similar morphology to a grade having 72 wt% ethylene. Rather, as the ethylene content increases, the amount of crystllinity increases. The amount of crystallinity can be quantified by measuring the onset temperature of the crystalline melt (T_c) or the heat required to pass through the crystalline melt (J/g). The energy required to melt the crystallinity is typically recorded as a % crystallinity, and the J/g level is compared against a known highly crystalline material such as polyethylene.

Amorphous grades, which have minimal crystallinity, are more flexibile at low temperatures, lower in hardness, and more elastic. Semicrystalline grades, on the other hand, have several properties that are directly attributable to their crystallinity levels. When compared to the amorphous grades, they exhibit higher green strength, higher tensile and modulus, and higher hardness. They have less flexibility at low temperatures, and lower compression sets.

8.4.3   Diene Content

In EPDM, as opposed to EPM, a third monomer is incorporated to add unsaturation to the polymer. These are typically non-conjugated dienes. The unsaturation allows the use of sulfur crosslinking. Three different monomers are used commercially for this purpose. The most common is ethylidene norbornene (ENB), and the two other types are dicyclopentadiene (DCPD) and 1,4-hexadiene (HD). These monomers contain two double bonds, one of which is consumed during the polymerization reaction, while the other remains in the resulting polymer. In all three cases, the remaining diene is present on a side chain, and not on the backbone of the polymer. See Fig. 8.4.

In general, several comparisons can be made between the different dienes containing monomers. The fastest cure rates are achieved with ENB, followed by HD and DCPD. Actual rates vary greatly depending on the cure systems used. Conversely, ENB also has

I. Ethylene:

$H_2C = CH_2$

II. Propylene:

$H_3C - C = CH_2$

$|$

H

III. Diene Containing Monomer:

A. *1,4 Hexadiene:* $H_2C = CH - CH_2 - CH = CH - CH_3$

B. *Ethylidene Norbornene (ENB):*   **C.** *Dicyclopentadiene (DCPD):*

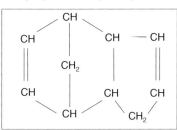

Figure 8.4   Commercially used diene containing monomers.

less scorch safety than the other dienes, provided cure systems are identical. Both of these trends result from the higher reactivity of the ENB.

The amount of diene incorporated varies depending on the grades of EPDM, as well. At one extreme, there is the diene containing no monomer, a material classified as EPM. At this extreme, the polymer cannot be sulfur cured, and it exhibits better heat resistance. When diene is added, it is typically between 0.5% (by weight) and approximately 12% (by weight). As the diene increases, the cure rate increases. Typical classifications for cure rates include medium, fast, very fast, and ultrafast. The industry has not standardized the meanings of these terms. A minimum of approximately 2% (by weight) diene is required for an effective sulfur cure. Vulcanization processes that utilize batch cures can typically use polymers containing 2 to 6% (by weight) diene. Continuous cure lines require faster cure rates; therefore, diene levels greater than 6% (by weight) are preferred. Sponge materials, in which the blowing rates must be matched with a very fast cure rate, typically use EPDM grades containing the highest levels of diene.

8.4.4   Rheology

The most common measure of the molecular weight is Mooney viscosity. Clear grades of EPDM can be obtained with Mooney viscosities from 20 to approximately 80. Measurements of Mooney viscosity values are standardized at 125 °C. Grades are also available that have higher Mooney viscosity levels, and these are sold as oil-extended grades. They contain process oils at amounts from 10 to 100 phr, so that the resulting Mooney viscosity of the blends is at or below 80. This allows the suppliers to provide a very high MW polymer that is still processable.

Another important measure of rheology of an EPDM is the MWD. Grades are produced today that have very narrow, medium, broad, and very broad MWDs. Again, the industry has

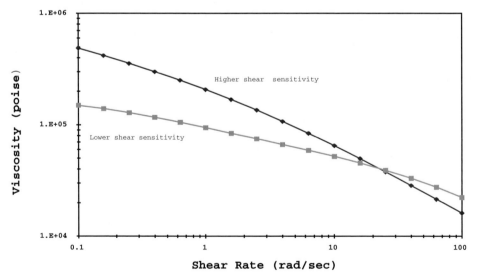

Figure 8.5   Viscosity vs. shear rate.

not standardized the meaning of these terms, but the designations can be used for comparative purposes.

Although Mooney viscosity is the common measure of the rheological properties of the EPDM polymer, its processing performance is significantly impacted by its MW and MWD. The impact of these variables can be seen in the shear viscosity curve that identifies the polymers, of which Mooney viscosity is but a single point. See Figure 8.5.

Reference

1.   Baldwin, F. P. and Ver Strata, G. "Polyolefin elastomers based on ethylene and propylene," *Rubber Chemistry and Technology*, 1972.

8.5   Acrylonitrile-Butadiene Rubber

Michael Gozdiff

8.5.1   Introduction

This section is an overview of acrylonitrile-butadiene rubber. Acrylonitrile-butadiene rubber is also known as nitrile, Buna-N and NBR. It is a complex elastomer family that like most others, requires formulating with added ingredients, and further processing to make useful articles. The additional ingredients include reinforcement fillers, plasticizers, protectants and vulcanization packages. The processing includes mixing, pre-forming to required shape, application to substrates and vulcanization to make the finished rubber article. Individual NBR suppliers should be consulted for specific formulary assistance, and detailed information concerning the use of their elastomers. "Elastomer" and "polymer" are used interchangeably when referring to NBR.

8.5.2   Chemical and Physical Properties – Relating to Application

Most NBR manufacturers make at least 20 conventional elastomer variations, with one global manufacturer now offering more than 100 from which to choose [1]. NBR producers vary polymerization temperatures to make "hot" and "cold" polymers. Acrylonitrile (ACN) and butadiene (BD) ratios are varied for specific oil and fuel resistance requirements. Molecular weight, emulsifier, stabilizer and coagulant systems are varied to provide other specific processing and end-use properties. NBR polymers containing carboxylated [2], divinyl benzene or amide terpolymers [3] are also offered. Some NBR elastomers are hydrogenated [4] to reduce the chemical reactivity of the polymer backbone, significantly improving heat resistance. Each modification contributes

Table 8.13   Common Properties of NBR Polymers

	Acrylonitrile content-dependent properties [9]	
NBR with higher acrylonitrile content		NBR with lower acrylonitrile content
Property improvement	← Processability	
	← Cure rate w/sulfur cure system	
	Cure rate w/peroxide cure system →	
	← Oil/fuel resistance	
	Resilience →	Property improvement
	Lower hysteresis →	
	← Compatibility w/polar polymers	
	Low temperature	
	Flexibility →	
	← Air/gas impermeability	
	← Tensile	
	← Abrasion resistance	
	← Heat-aging	

uniquely different properties. Saying "I want to use a nitrile" is much the same as saying, "I want to buy a car." There is a wide choice with many options.

8.5.2.1   Acrylonitrile Content (ACN)

The ACN content is one of the two primary criteria for defining every NBR. The ACN level, by reason of polarity, determines several basic properties, such as oil and solvent resistance [5,6], low temperature flexibility, glass transition temperature [7,8] (T_g) and abrasion resistance, to name a few. Higher ACN content provides improved solvent, oil and abrasion resistance, along with higher T_g. Table 8.13 summarizes most of the common properties for conventional NBR polymers.

8.5.2.2   Mooney Viscosity

Mooney viscosity is the other most commonly cited criterion for defining NBR. The Mooney test is reported in arbitrary units. It is the current standard measurement of the polymer's collective architectural and chemical composition. The Mooney viscosity provides data measured under narrowly defined conditions, with a specific instrument that is fixed at one shear rate. There is an almost unlimited number of ways of modifying polymers to derive a given Mooney number, with little assurance of uniformity in the polymer's architecture. More definitive polymer characterization can now be achieved using newer instruments and techniques that measure properties at shear rates pertinent to specific processing requirements. Using these newer instruments, such as the RPA2000, MPT, MDR2000, Capillary Die Rheometer and the newer Mooney machines, it is now also possible to rheologically measure elastic, as well as viscous characteristics. The RPA2000 and MDR2000 measure cure rates and cure states, as well.

Table 8.14 Influence of Emulsifier on Processing

	Rosin acid	Mixed rosin/ fatty acids	Fatty acid	Mixed fatty acid/synthetic	Synthetic
Cost	(3)	(2)	Lowest(1)	(4)	Highest (5)
Mixing	Best (1)	(3)	Worst (5)	(4)	(2)
Extrusion	Best (1)	(2)	Worst (5)	(4)	(3)
Calendering	Best (1)	(2)	Worst (5)	(4)	(3)
Building tack	Best (1)	(2)	Worst (5)	(4)	(3)
Adhesion	Best (1)	(3)	Worst (5)	(4)	(2)
Mold flow	(4)	(3)	Best (1)	(2)	Worst (5)
Mold fouling	Most (5)	(4)	(3)	(2)	Least (1)
Cure rate	Slowest (5)	(3)	Fastest (1)	(2)	(4)

8.5.2.3 Emulsifier

The emulsifier system [10,11] is one of the more important polymer attributes character-izing performance. It is usually next in importance after ACN content and Mooney. There are three basic systems used. They are fatty acid, rosin acid and synthetic-based soaps. These are often further subdivided into blended soap systems. Although the specifics of emulsifier systems are proprietary to polymer manufacturers, some generalities could be made. Each type of emulsifier system influences factory processing differently. The soap's influence includes mixing, calendering, building tack, extrusion, cure rate, mold fouling and adhesion.

Table 8.14 is a guide to indicate the general influence of emulsification systems on processing. As with all generalizations, this table does not account for other factors that influence processing.

8.5.2.4 Stabilizer

Stabilizer systems consist of chemical additives that provide elastomer stability [12] during storage and mixing. Staining and semi-staining amines, non-staining phenolics and non-staining complex phenyl phosphites are the antioxidants most commonly used. The correct elastomer/stabilizer selection is based on the final product's requirements, e.g., high temperature oil service, FDA requirements or color fastness.

8.5.2.4 Coagulation

Coagulants [13] used in making NBR can influence a mixed stock's cure rate and cure state. Coagulating systems can also influence cured performance properties, such as modulus, compression set, water swell and corrosion. Calcium chloride is one of the most commonly used coagulants. It has the benefits of fast cure rate, high modulus, and low compression set. However, if the polymer is not washed well, it may contribute towards water sensitivity and corrosion in some sealing applications. Other common coagulants include alum and magnesium sulfate. Alum's primary benefit is low water sensitivity and low corrosion contribution. Magnesium sulfate provides some benefit with heat resistance.

8.5.3   Polymer (Elastomer) Microstructure

Microstructure usually refers to how monomers are physically assembled within the polymer chains. This includes types of monomers, quantities of each and how they are dispersed throughout the polymer chains. At a given ACN content, the acrylonitrile can be "single charged," providing a benefit with low temperature impact brittleness; or "evenly distributed," providing an improved (lower) T_g. Each distribution is better for some applications, and deficiencient in others, depending on the specified requirements.

There are relatively standard proportions of 1,4-*cis*, 1,4-*trans* and 1,2-butadiene within the polymer backbone. Cold NBRs have a higher 1,4-*trans* content than is found in the hot varieties. The higher trans content is one of many factors which makes cold polymers processing more easily than their hot counterparts.

8.5.4   Polymer (Elastomer) Macrostructure

Macrostructure usually refers to the structural assembly of the polymer mass. It deals with the polymer's architecture. When NBR is polymerized, molecular chains do not all begin their formation at the same time. The rates of formation are also not uniform. For practical purposes, at any given instant, there is an enormous size differential in the population of constantly growing polymer chains. The physical character of the polymer mass is a summation of the positive and negative contributions of all these chain members as fixed when the reaction stops. There is a gradient of molecular chain lengths ranging from very short and plasticizer-like to very long and tough. This summation can be measured as molecular number, molecular weight, and distribution. Molecular weight distribution (MWD) is the attribute most compounders find of interest. Macrostructure also categorizes the extent to which polymer chains are linear, branched, or gelled.

8.5.5   Gel

Gel [14,15] is an often-discussed polymer attribute, with few individuals agreeing upon a single fully acceptable definition. It is usually defined as "crosslinked polymer." The premise is that cross-linking is a function of one polymer chain attaching to another, as an artifact of continued polymerization. Some would exclude intentional cross-linking formed during polymerization by using an additive, such as a difunctional monomer. Others would include it. Because test methods vary greatly, it is important to account for such things as solvent, filter size and sample preparation. A portion of what is often thought to be gel may actually be physically entangled polymer chains, and can be disentangled with minimal milling. True gel cannot be disentangled and is the cross-linked and solvent-insoluble material trapped by the filter. Irrespective of the test procedure, the possible presence of gel cannot be ignored, because it may be a significant part of the polymer content. The unexpected presence or absence of gel can significantly affect every aspect of factory processing.

A high gel content may be beneficial in applications requiring dimensional stability, but detrimental if good flow properties are needed. The same may be said for low gel: it all depends on the specific need. There are other good reasons for knowing the amount and types of gel in a polymer, including predicting factory processing and product performance. Because gel can sometimes be a significant portion of the product, it must also be included when analyzing molecular weight data. The molecular weight curve is actually a measure of the soluble polymer plus insoluble polymer smaller than the filter mesh size. The extent to which the curve accurately represents the polymer is inversely proportional to the gel content.

8.5.6   Molecular Weight

Molecular weight and its distribution (MWD) greatly affect how a polymer mixes, extrudes, molds and calenders in the factory. The molecular weight is proportional to viscosity, or toughness. A broad distribution means easier processing on wider varieties of equipment. The lower molecular weight (MW) fractions behave as processing aids. The medium to very high MW fractions are the strength members. A narrow MWD may not be quite so easy to process, but means higher green strength and less cold flow because of relative uniformity in the polymer chain lengths. This sometimes provides a benefit in some hose and molding operations.

The MWD also influences the physical properties of finished rubber articles. Narrow MWD polymers usually exhibit higher tensile, modulus and rebound. As with any other single set of properties, the MWD should not be the primary attribute considered for selecting a material. The balance of all properties and attributes fully defines an elastomer's suitability for an application.

8.5.7   Hot NBR

Hot-polymerized NBR polymers have many things in common, as well as a number of differences. They are all polymerized at the temperature range of 30 to 40 °C [16]. They are highly branched and prone to gel formation. Preventing gel formation with these types of elastomers requires extra precautions, not usually needed with cold polymers. For example, solvent-based adhesives, require a high degree of branching with a low gel content. The branching helps prevent mixed-in pigments from settling out and low gel promotes complete solubility. The combination of properties also encourages good tack and a strong bond. The physically entangled structure of this kind of polymer also provides a significant improvement in hot tear strength compared with a cold-polymerized counterpart. The hot polymers' natural resistance to flow makes them excellent candidates for compression molding and sponge. Other applications are thin-walled or complex extrusions where shape retention is important. Extrusion rates however, may be slow and power consumption may be high with hot NBR.

8.5.8   Crosslinked Hot NBR

Difunctional monomer crosslinked hot NBRs are excellent processing aids. Because these hot elastomers also are highly gelled through the difunctional monomer, tensile properties may be sometimes diminished. Rheological properties may also render them not suited for use as the sole polymer for use in many rubber recipes. They are generally used as a partial replacement at a 10 to 25 part level for other polymers, such as XNBR, SBR or cold NBR. This stabilizes extruded profiles for further processing or to control die swell. They are used in molded parts to provide sufficient molding forces, or "back-pressure," to eliminate trapped air. Another use is to provide increased dimensional stability of calendered goods and to improve release from the calender rolls. They provide dimensional stability, impact resistance, and flexibility for PVC modification.

8.5.9   Cold NBR

The current generation of cold NBRs spans a wide variety of compositions. Acrylonitrile contents range from 15 to 51%. Mooney values range from a very tough 110, to pourable liquids, with 20–25 as the lowest practical limit for solid material. They are made with a wide array of emulsion systems, coagulants, stabilizers, molecular weight distributions and monomer sequencing. Third monomers are added to the polymer backbone to provide advanced performance. Each variation provides a specific function.

Cold polymers are polymerized at a temperature range of 5 to 15 °C [17], depending on the balance of linear-to-branched configuration desired. The lower polymerization temperatures yield more-linear polymer chains. Reactions are conducted in processes universally known as continuous, semi-continuous and batch polymerization, with each process developing different MWDs.

Cold NBRs process relatively easily at lower temperatures than their hot counterparts. Linear polymer chains are less viscous and are able to slip free from their entanglements more easily than can the highly branched hot varieties. This increased mobility requires much less force to process the stock. The results are lower process temperatures and lower power consumption. The linearity contributes to easier incorporation of fillers and plasticizers. The reduced force required for stock flow improves the bottom line in extrusions and transfer and injection molding.

However, cold polymers cannot be categorically considered better than the hot varieties. The question must be asked – Better for what? The same linearity that allows stock to flow easily during mixing, extruding, calendering and molding, may be undesirable in other situations. The linearity may allow stock to cold flow or exhibit a shriveling. The usual remedies are to use a lower molecular weight elastomer or to blend in some crosslinked hot NBR.

8.5.10   Carboxylated Nitrile (XNBR)

Addition of carboxylic acid groups to the NBR polymer's backbone significantly alters processing and cured properties [18]. The primary reason for including carboxylation is to provide a network of ionic bonds to supplement conventional sulfur or carbon vulcanization

Table 8.15   XNBR Recipe Base

XNBR (7% carboxylation) [a]	100.0	100.0	100.0	100.0	100.0
Antioxidant[b]	2.0	2.0	2.0	2.0	2.0
Stearic acid	0.5	0.5	0.5	0.5	0.5
TMTD[c]	2.0	2.0	2.0	2.0	2.0
OBTS[d]	2.0	2.0	2.0	2.0	2.0
Sulfur[e]	0.4	0.4	0.4	0.4	0.4
Zinc oxide[f]	0.0	3.0	4.0	5.0	7.0

[a] Nipol NX775 (Zeon Chemicals L.P.)
[b] Wingstay 100 (Goodyear Chemical Division)
[c] Methyl Tuads (R.T. Vanderbilt)
[d] Amax (R.T. Vanderbilt)
[e] Sulfur, Spider Brand (C.P. Hall Company)
[f] Kadox 911C Zinc Oxide (Zinc Corporation of America)

Properties

Hardness, Shore A	48.0	57.0	64.0	67.0	68.0
Tensile (MPa)	3.8	7.2	10.7	14.0	16.9
Elongation,%	600.0	515.0	410.0	375.0	365.0
Rebound (23 °C), %	44.0	49.0	46.0	45.0	45.0
Rebound (100 °C), %	65.0	66.0	65.0	58.0	56.0
Compression set (70 h/100 °C), %	58.0	26.0	19.0	18.0	17.0

bonds. The additional ionic network is a series of metallic-carboxyl reactions. The result is a matrix with significantly increased strength, measured by improved tensile, tear, modulus and abrasion resistance. On the other side of the ledger, there are losses in water resistance, resilience and some low temperature properties. Another artifact of the carboxylation is usually a 12 to 16 point hardness increase in most recipes when there is a direct substitution of XNBR for NBR. Nitriles are fully compatible with XNBR for modifying properties [19] and cost. Table 8.15 shows the magnitude of the ionic bond's strength. A gum recipe is used for the purpose of focusing on the ionic bonding of the carboxylated nitrile vulcanizate [20].

Working with XNBR means working with multiple cure systems and applying inorganic chemistry to the organic chemistry usually used with rubber compounding. Conventional peroxide or sulfur cure systems that are common with conventional NBRs, react identically with the butadiene network in the carboxylated nitriles. However, there are now acid-base reactions to consider. Reactions of the carboxylation with metallic oxides, metallic salts, amines and a wide variety of other acid-reactive materials form a second network, which complicate matters. Water is the primary catalyst for the carboxyl-metallic reactions and must be treated as an "ultra" accelerator. The problem is that many of the ingredients contain water, including the polymer itself. Volatilizing the water through heat generated during mixing, is the best way for removing it. Desiccants such as calcium oxide do not work because they react with the carboxylic acid component.

8.5.11   Bound Antioxidant NBR

Nitriles are available with an antioxidant polymerized into the polymer chain [23–25]. The purpose is to provide non-extractable protection for the NBR during prolonged fluid service or in cyclic fluid and air exposure. These polymers work best with mineral filler systems and semi-ev cure systems. They have limited hot air aging capability when formulated with highly reinforcing furnace carbon black because of chemical reactivity between the amide terpolymer and the pigment. Abrasion resistance is improved when compared with conventional NBR, especially at elevated temperatures. They have also been found to exhibit excellent dynamic properties [26]. Although these elastomers have certainly found a firm place in the market, they cannot be viewed as low cost substitutes for HNBR. The HNBR elastomers are so superior in such properties as high temperature resistance, chemical resistance, abrasion resistance and dynamic properties that no comparison can be made.

When blending bound-antioxidant nitriles with XNBR, there are occasional disappointments. This elastomer blend places reactive amides of one polymer chain in close proximity with reactive carboxylic acid groups of another polymer chain. Unusually short elongation, coupled with high modulus is commonly seen.

Acknowledgment

The author gratefully acknowledges the resources of The Goodyear Tire & Rubber Company to enable the accomplishment of this work, and Zeon Chemicals L.P. as well as The Goodyear Tire & Rubber Company for permission to publish.

References

1. Table 14, Nitrile Dry Rubber (NBR), *IISRP Synthetic Rubber Manual, 13th Edition*, (1995)
2. *Op. cit.*, Table 21, Nitrile Carboxylated Rubber (XNBR)
3. *Op. cit.*, Table 14.
4. *Op. cit.*, Table 46, Highly Saturated Nitrile Rubber (HNBR)
5. Hofmann, W., *Nitrile Rubber*, (1964). Rubber Chemistry and Technology, A Rubber Review for 1963, 154–162
6. Jorgensen, A. H., Mackey, D., *Elastomers, Synthetic (Nitrile Rubber)*, Kirth-Othmer Encyclopedia of Chemical Technology (1993). 4th Edition, Vol. 8, 1005
7. Wiley, R. H. and Brauer, G. N., *Rubber Chemistry and Technology* (1949) **22**, 402–4
8. Hofmann, W., *Nitrile Rubber* (1964). Rubber Chemistry and Technology, A Rubber Review for 1963
9. *Ibid.*, 71.
10. *Ibid.*, 87–8.
11. Semon, W. L., *India Rubber World* (1947), **116**, 63–5, 132
12. Hoffman, W., *op. cit.*, 87–88
13. Gozdiff, M., *NBR Formulary Base – ACN vs. Properties*, Energy Rubber Group symposium, Jan. 16, 1992
14. Flory, P. J., *Journal American Chemical Society* (1947), **69**, 2893
15. Hoffman, W., *op. cit.*, 117.
16. Semon, W. L., *Nitrile Rubber, Synthetic Rubber* (1954), Division of Chemistry, American Chemical Society, 802–3

17. *Ibid.,* 803
18. Technical Brochure, *Chemigum NX775* (1991). No. 439500-10/91 Goodyear Chemical Division
19. Gozdiff, M., *Paper No. 59* (1997), Roll symposium, Meeting, Rubber Division, ACS, Anaheim, California
20. Technical Brochure, *Chemigum NX775,* (1991). No. 439500-10/91 Goodyear Chemical Division, p 10
21. Technical Brochure, *Chemigum HR* (1989). No. 819600-5/89 Goodyear Chemical Division
22. Horvath, J. W., Purdon, J. R., Mayer, G., Naples, F., Applied Polymer Symposium No. 25 (1974) 157–203
23. Kline, R. *Polymerizable Antioxidants in Elastomers* (1974). Presented Meeting, Rubber Division, ACS, Toronto, Canada
24. Horvath, J. W., Paper No. 18, *Dynamic Properties of Nitrile Rubbers,* (1990). Presented Meeting, Rubber Division, ACS, Las Vegas, Nevada

8.6   Hydrogenated Nitrile Butadiene Elastomers

Michael E. Wood

8.6.1   Introduction

Hydrogenated nitrile butadiene rubber, or HNBR as designated by ASTM 1418, was first commercialized in 1984 by Zeon Corporation (Nippon Zeon Co., Ltd.) under the trade name Zetpol. Later that same year, Bayer also announced the commercialization of their new HNBR, Therban. This family of elastomers stemmed from the hydrogenation of diene-containing polymers, a well-known and established reaction process, which starts with an emulsion-polymerized acrylonitrile-butadiene copolymer (NBR). The NBR base polymer is then dissolved in a pre-selected solvent. Then, through the addition of a particular catalyst in combination with a predetermined pressure and temperature, the polymer undergoes a "selective hydrogenation" process producing a "highly-saturated" nitrile elastomer. The subscripted "x-z" after the initial ethylene unit shows the variation in the ethylene unit concentration as a function of the degree of hydrogenation of the butadiene units.

$$-\left[(CH_2-CH_2)_{x-z}-(CH_2-\underset{\underset{C\equiv N}{|}}{CH})_y-(CH=CH)_z\right]_n$$

Through control of this reaction, many different grades of HNBR are currently produced. The available grades vary in hydrogenation level, acrylonitrile (ACN) content, and Mooney viscosity as illustrated in Table 8.16.

Table 8.16   Range of Attributes of HNBR

% Hydrogenation	% ACN content	Mooney viscosity: ML(1+4) at 100 °C
~85 to 99+	17 to 50	50 to 150

8.6.2   Applications

The largest single application for HNBR is in synchronous timing belts for the automotive industry. Other major applications within the automotive industry are in power steering, air conditioning, and fuel systems. HNBR can also be found in many different industrial, aerospace, oil field, and food contact applications.

8.6.3   Properties

HNBR elastomers are very tough. They provide outstanding stress/strain characteristics and abrasion resistance, along with an excellent balance of low temperature and fluid resistance properties. HNBRs are similar to NBR in that, as the ACN content increases, oil and fuel resistance improves (volume swell decreases). As the ACN content decreases, low temperature properties improve. The static heat resistance of HNBR elastomers is dependent on the level of saturation within a given polymer. As the level of saturation increases, the static heat resistance improves. However, as the saturation level decreases, the dynamic heat generated, as a result of flexing a given part, decreases. So, one has to determine the most appropriate saturation level depending on the desired end use in question. There is a conveniently wide grouping of Mooney viscosities available which enables a manufacturer to produce parts from HNBR via a number of different processing methods (e.g., compression, transfer, or injection molding, extrusion, calendering, etc.).

8.6.4   Formulating

Compounding HNBR is quite similar to NBR with some very notable exceptions. One can use a wide range of carbon blacks, white fillers, antioxidants, and processing aids. However, the most commonly used plasticizer in HNBR is TOTM (trimellitate), which provides the best balance of high- and low-temperature performance. Other monomeric plasticizers can be used to enhance the low temperature properties of an HNBR compound, but they exhibit poorer high temperature properties. Likewise, there is an assortment of polymeric plasticizers that improve high temperature capabilities of an HNBR compound, but they exhibit poorer low temperature properties.

Although a select group of HNBR elastomers is sulfur curable, most are peroxide cured because of the enhanced heat and ozone resistance attained via this cure chemistry. The most common peroxides used in HNBR compounds are α,α'-bis (t-butylperoxy) diisopropyl-benzene, dicumyl peroxide, and 2,5-dimethyl-2,5-di-(t-butylperoxy) hexane. The main difference between HNBR and NBR compounding is that HNBR uses a significantly higher quantity of peroxide. HNBR compounds commonly use 7–12 phr of peroxide while an NBR compound may only use 2.5–3.5 phr of the same peroxide. Although not used in all cases, co-agents may play a key role in both the cure chemistry and processability of HNBR compounds. Certain co-agents tend to reduce the uncured viscosity of an HNBR compound, making it easier to process, while also improving the cured compression set properties. A typical HNBR compounding recipe is listed in Table 8.17.

Table 8.17   HNBR Compounding Recipe

Ingredients	phr
Polymer	100
Carbon black (e.g., N 774)	60
Plasticizer (e.g., TOTM)	5
Metal oxide (e.g., ZnO or MgO)	3
Antioxidants	2
Process aids	2
Curative (e.g., peroxide)	8

8.6.5   Processing

Both internal mixers or mills can be used to mix HNBR compounds. The most common method is a 2-pass mix with an internal mixer. The first pass includes all of the ingredients except the curatives. This pass should be discharged from the internal mixer at approximately 140 °C. After the compound has cooled to room temperature, the second pass through the mixer includes the masterbatch from the first pass plus any curatives. It should be mixed to a dump temperature of approximately 110 °C, and then discharged onto a mill where the final blending and initial cooling of the compound takes place. Beyond this point, HNBR compounds can be extruded (cold-feed, hot-feed, Barwell, etc.); press cured (compression, transfer, or injection); autoclave cured; etc., the details of which are beyond the scope of this discussion. Further details about any of these specific areas can be obtained from your HNBR supplier.

Chemical Abbreviation Key

Abbreviation	Chemical Name
HNBR	hydrogenated nitrile butadiene rubber
NBR	nitrile butadiene rubber
ACN	acrylonitrile
FDA	Food & Drug Administration
TOTM	trioctyl trimellitate
ZnO	zinc oxide
MgO	magnesium oxide

8.7   Polyacrylate Elastomers

Paul Manley and Charles Smith

8.7.1   Polymer Composition

Polyacrylate polymers are classified as heat and oil resistant polymers (designated by ASTM as ACM). Typical monomers used in the manufacturing of the polymers are ethyl

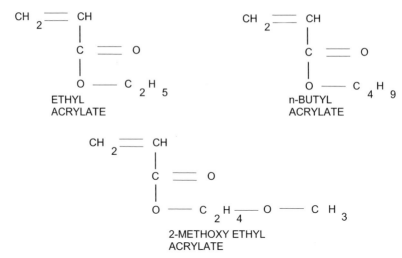

Figure 8.6   Major backbone monomers.

acrylate (EA), butyl acrylate (BA), and methoxy ethyl acrylate (MEA) and are shown in Fig. 8.6 [1].

The degree of heat and oil resistance exhibited by polyacrylates depends on the base polymer composition. Rubber compounds based on different grades of polyacrylate polymers can be formulated to meet application needs (e.g., engine sealing gaskets, hoses, and grommets) that range in temperature from −40 to 190 °C. The choice of base elastomer is a compromise of low temperature performance, heat resistance, and volume change in oil. The different base polymer properties can be seen in Table 8.18 [2].

Cure site monomers are added to the base polymer to enable the polymer to vulcanize quickly. There are different cure site monomers used in polyacrylate and a given set of cure systems for a given cure site. Different cure site monomers can be incorporated into the backbone of the polymer and are referred to by the functional pendant group. Table 8.19 shows typical cure systems used for a polyacrylate elastomer containing a carboxyl and chlorine dual cure site system. Table 8.20 shows typical cure systems used for a polyacrylate elastomers containing a chlorine cure site monomer. Table 8.21 shows cure systems used for an epoxy functional polyacrylate polymer.

Table 8.18   Comparison of Different ACM Polymers. ACM Dual Curesite

Product	Service temperature range, °C	IRM 903 oil swell %
High temperature ACM	−18 to 200	11
Intermediate low temperature ACM	−32 to 190	18
Low temperature ACM	−40 to 190	24
Low temperature ACM	−40 to 200	65

Table 8.19   Cure Systems. Chlorine/Carboxyl ACM

Ingredient	PHR	Purpose	Shelf stability
Sodium stearate	4 to 6	Curative	Excellent
Quaternary ammonium salt	2 to 3	Accelerator	
Urea compound	2 to 6	Retarder	
Stearic acid	1 to 3	Retarder	
Sodium stearate	4 to 6	Curative	Good
3,4-Dichlorophenyl-1.1-dimethylurea	2 to 6	Accelerator	
Stearic acid	1 to 3	Retarder	
2,5-Dimercapto-1,3,4-thiadiazole	0.4 to 1.0	Curative	Fair
Tetrabutylthiuram disulfide	2 to 4	Accelerator	
Zinc stearate	0 to 2	Retarder	

Table 8.20   Cure Systems. Chlorine Cure Site ACM

Ingredient	PHR	Purpose	Shelf stability
Sodium stearate	2 to 4	Curative	Good
Potassium stearate	0 to 1	Curative	
Sulfur	0.2 to 0.6	Accelerator	
Zinc dibutyldithiocarbamate	1 to 2	Accelerator	Excellent
2,4,6-Trimercapto-s-triazine	0.3 to 1	Curative	
Sulfonamide derivative	0.2 to 0.5	Shelf life	
N-(cyclohexylthiophthalimide)	0 to 0.5	Retarder	

Table 8.21   Cure Systems. Epoxy Cure Site ACM

Ingredient	PHR	Purpose	Shelf stability
Iso-cyanuric acid	0.5 to 1	Curative	Good
Trimethyl octa-decyl ammonium bromide	1 to 2.5	Accelerator	
N,N'-diphenyl urea	0.2 to 0.6	Retarder	
Ammonium benzoate	1 to 2	Curative	Fair

8.7.2   Basic Compounding of Polyacrylate Polymers

The typical recipe for a polyacrylate compound can be seen in Table 8.22. The base polymer or polymers are selected for the low temperature and oil swell properties the final product requires. Polymers with the same cure site functionality can also be blended to achieve the desired low temperature and oil swell balance. These blends can sometimes result in improved compound economy [3].

Fillers are needed in polyacrylate polymers to improve the tensile strength of the compound. Carbon blacks are most commonly used to reinforce polyacrylate polymers. As is the case for other synthetic polymers, the use of a smaller particle, lower structure

Table 8.22   Typical ACM Recipe. Compounding

Ingredient	Level (phr)	Purpose
ACM polymer	100	Oil swell/low temperature
Filler	25 to 150	Reinforcement
Retarder	1 to 4	Improves scorch safety
Process aids	1 to 4	Improves mill release
Antioxidants	0 to 4	Improves the thermal stability
Plasticizers	0 to 20	Lower the compound durometer
Curatives	1 to 10	Crosslink the polymer

carbon black results in higher tensile strength, higher ultimate elongation, and higher tear values, but also higher compression set values. Table 8.23 [4] shows how much of a given carbon black is needed to increase the hardness one durometer point in a given polymer.

Non-carbon black fillers can also be used to reinforce polyacrylate polymers. Care must be taken in the choice of the filler, as non-carbon black fillers must have a basic pH. Acidic fillers retard most of the cure systems. Table 8.24 [5] shows how much of a given filler is needed to increase the hardness one Shore A durometer point.

Plasticizers and processing aids are added in small amounts when required. Because most polyacrylate compounds require a post cure, permanent plasticizer must be used to prevent loss. Generally, 1 phr of plasticizer lowers the Gehman T100 value about 1 °C. Processing aids are added to prevent sticking in processes throughout the factory. Stearic acid, one of the best process aids for polyacrylate compounds, provides mold release and slows the cure rate but increases the compression set values.

8.7.3   Processing Guidelines

Polyacrylate polymers are processed by conventional rubber processing equipment. While no extraordinary equipment is required, it is advisable to observe several guidelines to ensure high quality, consistent mixes.

Table 8.23   The Amount of Carbon Black it Takes to Increase the Hardness by 1 Shore A Durometer Point. Calculated off a Base Hardness of 24

Polymer type	N774	N650	N550	N330
High temperature ACM	2.00	1.75	1.65	1.50
Low temperature ACM	3.65	2.60	2.20	1.80

Table 8.24   The Amount of Non-Black Filler it Takes to Increase the Hardness by 1 Durometer Point. Calculated off a Base Hardness of 24

Polymer type	Amino silane treated calcined clay	Low reinforcement precipitated silica	Talc
High temperature ACM	3.80	1.60	3.70

Compound sticking generally occurs from excessive heat build-up or a lack of effective processing aids. The use of cool mills and mixers usually stops batches from sticking during mixing [6].

Typical injection and compression molding cycles range from 1 to 4 minutes depending on the formulation at mold temperatures of 170 to 200 °C. A post cure of 4 to 8 hours at 170 to 190 °C may be required to achieve a low compression set value [7].

References

1. R. D. DeMarco, R. J., Flecksteiner "Polyacrylate Elastomers," Zeon Chemicals HyTemp Bulletin PA0901.4.
2. R. D. DeMarco, R. J., Flecksteiner "Polyacrylate Elastomers," Zeon Chemicals HyTemp Bulletin PA0901.5.
3. R. D. DeMarco, R. J., Flecksteiner "Polyacrylate Elastomers," Zeon Chemicals HyTemp Bulletin PA0901.5.
4. P. E. Manley, Zeon Chemicals Technical document TA95-018, "The effects of carbon blacks in polyacrylate formulations."
5. P. E. Manley, Zeon Chemicals Technical document TA94-035, "The effects of non-carbon black fillers in polyacrylate formulations."
6. Zeon Chemicals Processing Notes PA0600.1.
7. Zeon Chemicals Processing Notes PA0600.3.

8.8   Polychloroprene (Neoprene)

Leonard L. Outzs

8.8.1   Introduction

The basic chemical composition of Neoprene synthetic rubber is polychloroprene. The polymer structure can be modified by copolymerizing chloroprene with sulfur and/ or 2,3-dichloro-1,3-butadiene to yield a family of materials with a broad range of chemical and physical properties. By proper selection and formulation of these polymers, the compounder can achieve optimum performance for a given end use.

This section is a guide to selecting the appropriate type of polychloroprene in developing a compound for a particular application. In addition to achieving certain physical properties, choosing the Neoprene best adapted to processing and storage conditions is also considered.

8.8.2   Basic Characteristics of Polychloroprene

Polychloroprene synthetic rubber is a multi-purpose elastomer that yields a balanced combination of properties. All types have these inherent characteristics:

- Resistance to degradation from sun, ozone, and weather
- Good performance when in contact with oils and many chemicals
- Useability over a wide temperature range
- Outstanding physical toughness
- Better resistance to burning than exclusively hydrocarbon rubbers

8.8.3   Families of Neoprene

Neoprene for dry rubber applications is available in three different families: G-, W-, and T-types, with variations within each family. These elastomers offer a broad range of physical properties and processing conditions so that the user can formulate to his specific requirements. The characteristics of raw polymer and the vulcanizates of each family, in general terms, are displayed in Tables 8.25 and 8.26.

Frequently, available compounds of Neoprene can be modified to fit a new application by simply changing the type of Neoprene used. For example, substituting a crystallization-resistant type for a regular type can make the compound more suitable for low temperature applications.

8.8.4   Neoprene 'G' Family

Fundamental differences in behavior between the G family and other families of Neoprene types arise because G-types are made by copolymerizing chloroprene with sulfur and are stabilized, or modified, with thiuram disulfide. Most of the special characteristics of the G family can be predicted from the basic chemical structure of the polymer. G-type polymers have much wider MWD ranges than are found in the W- or T-types.

Table 8.25   Family Characteristics of Neoprene

	G-Types	W-Types	T-Types
Raw polymers	Limited storage stability Peptizable* to varying degrees No accelerators necessary Fast curing, but safe processing	Excellent storage stability Non-peptizable Accelerator required cure flexibility	Excellent storage stability Least nerve Accelerator required cure flexibility Best extrusion, calendering performance
Vulcanizates	Best tear strength Best flex Best resilience	Best compression set resistance, best heat aging	Properties similar to W-types

* Peptizable (i.e., reduction in viscosity) can be accomplished either mechanically or chemically.

Table 8.26 Comparsion Chart of the DuPont Neoprenes

Family characteristics	Type	Co-monomer[a]	Polymer stability	Mooney	Crystalliza-tion rate	Distinguishing features
G-TYPES **Raw Polymers:** limited storage stability; Peptizable to varying degrees; Fast curing, but safe processing; No accelerators necessary **Vulcanizates:** Best tear strength; Best flex; Best resilience	GW	Sulfur	Good	37 to 54	Slow	Best tear & flex; Best G-type for heat aging & c/s resistance; Approaches GRT crystallization resist; Very slightly peptizable; Contains thiuram
	GNA	Sulfur	Good	42 to 59	Medium	Staining stabilizer; Contains thiuram
	GRT	Sulfur/ACR	Good	36 to 55	Slow	Best tack; Crystallization resistance; Contains thiuram
	FB	Sulfur	Good	Liquid at 50 °C	Medium	Fluid at 50 °C; Contains thiuram
W-TYPES **Raw Polymers:** Excellent storage stability; Not peptizable **Vulcanizates:** Best compression set resistance; Best heat aging	W-M1	None	Excellent	32 to 42	Fast	General purpose – low viscosity
	W	None	Excellent	40 to 49	Fast	General purpose – std. viscosity
	WHV-100	None	Excellent	90 to 105	Fast	General purpose – med. high viscosity
	WHV	None	Excellent	106 to 125	Fast	General purpose – high viscosity
	WRT	ACR	Excellent	41 to 51	Very slow	Very slow crystallizing; Slower curing than w
	WRT-M1		Excellent	34 to 42	Very slow	Low viscosity WRT
	WD		Excellent	100 to 120	Very slow	High viscosity WRT; Contains high percentage of gel for processing; Vulcanizate properties reduced
	WB		Excellent	43 to 52	Medium	Generally used in blends
T-TYPES **Raw Polymers:** Excellent storage stability; Least nerve; Fast processing **Vulcanizates:** Properties similar to W-types	TW	None	Excellent	42 to 52	Fast	Gel-containing for good processing; Vulcanizate properties similar to W-types
	TW-100	None	Excellent	92 to 99	Fast	Higher viscosity form of TW
	TRT	ACR	Excellent	42 to 42	Very slow	Maximum resistance to crystallization; Contains gel for good processing

[a] Principal monomer is chloroprene; ACR = 2,3-dichloro-1,3-butadiene. ACR ( ) = 2,3-dichloro-1,3 butadiene.

As a group, members of the G family have these characteristics in common:

They are capable of peptization to a lower viscosity, mechanically or chemically, with thiurams or guanidines.

- They can be cured with metal oxides alone. Cure accelerators are not required, but can be used and may be desirable to develop certain properties.
- Their vulcanizates generally exhibit greater tear strength, slightly higher resilience and elongation, and a "snappier" feel, and adhere better to natural rubber and SBR substrates.
- Their compounds are more susceptible to reduction in scorch time by accumulated heat history than are those based on other types of Neoprene.
- Their raw polymer storage stability is more limited than that of W- or T-types.

Compounders normally choose G-types over other Neoprene where:

- Building tack is important in fabricating the final product.
- The end use involves severe flexing and/or other dynamic stresses, and minimum compression set is not required.
- The compound requires high loadings with a minimum of plasticizer. Because G-types break down, or soften, when subjected to shear during mixing, workable viscosities can be achieved in this type of compound.
- Organic accelerators cannot be used for some reason, such as in compounds requiring FDA approval. G-types cure with metal oxides alone.
- Good frictioning characteristics are desired. G-types, and particularly GRT, have found wide use in this application.
- A polymeric plasticizer for Neoprene, which does not lower hardness, is needed. In these cases, Neoprene FB reduces compound viscosity appreciably, and then cures along with the balance of the Neoprene polymer.

8.8.5   Neoprene 'W' Family

The W-types of DuPont Dow Elastomers Neoprene synthetic rubber are chloroprene homopolymers or copolymers of chloroprene and 2,3-dichloro-1,3-butadiene. They contain no elemental sulfur, thiuram disulfide, or other additives that are capable of decomposing to yield either free sulfur or a vulcanization accelerator. They also contain no staining stabilizer. They have a more uniform MWD than the G-types, with the greatest frequency occurring at about 200,000.

As a group, members of the W family have these characteristics in common (which are related to their chemical composition):

- They have excellent raw polymer storage stability – considerably better than that of the G-types.
- They do not decrease in molecular weight, either under mechanical shear or with chemical peptization. However, some reduction in viscosity may occur under high shear conditions. The degree of softening is greater for the high molecular weight polymers such as Neoprene WHV or WD.
- Their compounds mix faster, develop less heat during mixing, and are less tacky than those of Neoprene GNA or GRT.

- Their vulcanizates have good resistance to heat aging and compression set – substantially better than vulcanizates of Neoprene GNA or GRT.
- Cure accelerators must be added to achieve practical cure rates and acceptable vulcanizate properties. Therefore, the W-types offer considerable latitude in processing safety and cure rate through judicious choice of type and amount of accelerator.
- Before accelerators are added to the mix, Neoprene W compounds are much less susceptible to scorch resulting from heat history than are compounds based on members of the G family.
- Uncured extrusions resist collapse and distortion much better than do extrudates of compounds based on the G-types. However, the W-types are not as good as the T-types in this respect.
- If properly compounded with non-staining compounding ingredients, they can be used in applications where staining of finishes cannot be tolerated.

The W-types are chosen primarily where service conditions call for Neoprene with the best available heat and compression set resistance. Even when these properties are not the primary consideration, the W-types are frequently used because of their combination of excellent storage stability, uniform processability, broad compounding latitude, and all-around good vulcanizate properties.

8.8.6   Neoprene 'T' Family

In most of their chemical composition and performance characteristics, the T-types of DuPont Dow Elastomers' Neoprene synthetic rubber are similar to the W-types. They are chloroprene homopolymers or copolymers of chloroprene and 2,3-dichloro-1,3-butadiene; they contain no elemental sulfur, other curatives, or additives that can decompose into a curative; they contain no staining stabilizer; and, they have a much more uniform MWD than the G-types.

In addition, the T-types all contain a highly crosslinked ''gel'' form of polychloroprene, similar to that contained in Neoprene WB, which acts as an internal processing aid. As a result, the T-types process better than the W-types (except Neoprene WB). They offer the compounder a new dimension in smooth, fast extrusion and calendering with little or no loss in physical properties. As a group, members of the T family have the same characteristics as the W family.

8.9   Chlorinated Polyethylene (CM)

Laura Weaver

8.9.1   Introduction

Chlorinated polyethylene elastomers are abbreviated CM in accordance with ASTM nomenclature. CM resins are available in dry, flowable crumb, or powder form having an average particle size in the 12 to 60 mesh ranges, depending on product type. This form and particle size provides a significant advantage in blending and compounding, assuring more uniform and faster dispersion.

CM elastomers are chlorinated polyethylene polymers produced by combining chlorine and polyethylene in an aqueous slurry. This process produces a saturated, linear molecular backbone which gives CM elastomers many advantages in applications requiring heat, ozone, oil, weather, and ignition resistance.

Major applications and markets for CM elastomers include automotive under-the-hood air ducts; ignition wire jacketing; hose and tubing; wire and cable jacketing for power, instrumentation, control, industrial, and portable cord applications; as well as various high performance industrial applications such as hydraulic hose, gasketing, diaphragms, and many other industrial parts.

Chlorinated polyethylene (CM) is subject to dehydrohalogenation by Zn^{++} ions which can seriously degrade heat resistance and state of cure. It is strongly recommended to avoid incorporating any zinc-containing materials into a CM formulation. The use of zinc-containing coatings or dusting agents on the outside of these formulations is also discouraged.

The Tyrin (CM) elastomers produced by DuPont Dow Elastomers that are most often recommended for use in applications such as automotive hose and tubing, wire and cable jacketing, industrial hose, and molded goods are listed below.

- Tyrin CM0136 is a 36% chlorine, general purpose elastomer with the widest latitude for applications and processing; excellent colorability; good high and low temperature properties.
- Tyrin CM674 is a 25% chlorine elastomer with high crystallinity, high modulus, and extremely low sodium ion content for excellent moisture resistance.
- Tyrin CM0730 is a 30% chlorine elastomer offering superior low temperature performance, heat aging, and excellent processability.
- Tyrin CM0836 is a high molecular weight version of Tyrin CM0136.
- Tyrin CM2136P is Tyrin CM0136 with improved powder handling characteristics.
- Tyrin CM2348P is a 36% chlorine elastomer with high molecular weight and low sodium ion content.
- Tyrin 3611P is a 36% chlorine, low viscosity elastomer used primarily to improve processing characteristics in elastomer blends.
- Tyrin 4211P is a 42% chlorine, low viscosity elastomer used in blends where improved ignition, oil, and fuel resistance are required.

8.9.2   General Characteristics

Tyrin elastomers are produced by random and blocked chlorination of high density polyethylene. Several of the properties of Tyrin, such as resistance to heat, oxidation, and ozone can be attributed to the chemical saturation of the polymer chain.

Many of the properties contributed by Tyrin elastomers are also directly related to the chlorine content of the polymer. Compounding and blending techniques are used to develop specific performance attributes more fully for certain applications. General effects

Table 8.27   Effect of Chlorine Content on Properties of Tyrin Elastomers (Peroxide-and-Thiadiazole-Cured Compounds, Except where Noted)

Increasing % Cl	Decreasing % Cl
Oil and chemical resistance improves	Plasticizer acceptance increases
Hardness increases	Low temperature properties improve
Ignition resistance increases	Compression set resistance improves
Barrier properties improve	Heat aging improves
Tear resistance improves[1]	Dynamic properties improve
Thiadiazole-cured crosslinked density improves	Peroxide-cured crosslinked density improves

of increasing or decreasing chlorine content on cured properties of Tyrin elastomers are shown in Table 8.27.

8.10   Chlorosulfonated Polyethylene (CSM)

Charles Baddorf

8.10.1   Introduction

CSM is the ASTM-approved designation for a group of sulfur- and peroxide-curable elastomers based on chlorosulfonated polyethylene. These elastomers have a completely saturated backbone and pendant groups that are suitable for varied approaches to vulcanization. As a result of this configuration, vulcanizates of CSM are extremely resistant to attack by ozone, oxygen, and weather. In addition, properly prepared vulcanizates of CSM are outstanding in resistance to deterioration by heat, oils, and many chemicals and fluids.

CSM is used in the rubber industry because of its wide range of unique properties, including:

- Permanent bright colors
- Superb ozone and weather resistance, even in non-black products
- Heat resistance to 150 °C (302 °F)
- Resistance to a wide range of aggressive chemicals
- Intermediate oil and solvent resistance depending on the chlorine level
- Electrical insulating properties
- Low flammability characteristics (associated with chlorinated polymers)
- Excellent resistance to abrasion and mechanical abuse

CSM is used in a wide variety of applications, such as:

- Single-ply roofing
- Pond and reservoir lining

Table 8.28 Hypalon® Chlorosulfonated Type

	20	30	45	6525	40S	40	4085	48
Description								
Chlorine content, %	29	43	24	27	35	35	36	43
Sulfur content, %	1.4	1.1	1.0	1.0	1.0	1.0	1.0	1.0
Physical form	Chips	Chips	Chips	Chips	Chips	Chips	Chips	Chips
Color	White	White	White	White	White	White	White	White
Odor	None	None	None	None	None	None	None	None
Specific gravity	1.12	1.27	1.07	1.10	1.18	1.18	1.19	1.27
Mooney viscosity ML 1-4 at 100°C (212°F)	28	30	37	90	46	56	94	78
Storage stability	Excellent	Excellent	Excellent	Excellent	Excellent	Excellent	Excellent	Excellent
Distinguishing features	Readily soluble in common solvents. Good low-temperature flexibility.	Readily soluble in common solvents. Forms hard, glossy films.	High uncured strength. Good heat resistance. Good low-temperature flexibility.	High polymer viscosity. Good low-temperature and heat resistance. Good processing at high extensions.	Low polymer viscosity. Improves processing of dry, stiff stocks.	Medium polymer viscosity. Versatile, suitable for many applications.	High polymer viscosity. Good green strength. Improves processing of soft or highly extended stocks.	High polymer viscosity. Excellent oil and fluids resistance. High uncured strength.
Processing characteristics								
Extruding	Fair	Fair	Good	Excellent	Excellent	Excellent	Excellent	Good
Molding	Good	Fair	Excellent	Excellent	Excellent	Excellent	Excellent	Good
Calendering	Fair	Fair	Excellent	Good	Excellent	Excellent	Excellent	Good
Vulcanizate properties								
Hardness, Durometer A	45–95	60–95	65–98	40–95	40–95	40–95	40–95	60–95
Tensile strength, MPA								
Carbon black stocks	Up to 20	Up to 24	Up to 28	Up to 27	Up to 28	Up to 28	Up to 28	Up to 28
Gum stocks	Up to 8	Up to 17	Up to 28	Up to 28	Up to 28	Up to 28	Up to 28	Up to 24
Color stability	Excellent	Excellent	Excellent	Excellent	Excellent	Excellent	Excellent	Excellent
Low-temperature properties	Good	Poor	Good	Excellent	Good	Good	Good	Poor
Tear strength	Fair	Fair	Good	Good	Good	Good	Good	Good
Resistance to abrasion	Very good	Very good	Excellent	Excellent	Excellent	Excellent	Excellent	Excellent
Chemicals	Good	Excellent	Good	Very good	Excellent	Excellent	Excellent	Excellent
Compression set	Fair	Poor	Good	Very good	Good	Good	Excellent	Fair-good
Flame	Fair	Very good	Fair	Fair	Good	Good	Good	Very good
Heat aging	Very good	Fair	Very good	Very good	Very good	Very good	Very good	Good
Ozone	Excellent	Excellent	Excellent	Excellent	Excellent	Excellent	Excellent	Excellent
Petroleum oils	Fair	Excellent	Fair	Fair	Good	Good	Good	Excellent
Weathering	Excellent	Excellent	Excellent	Excellent	Excellent	Excellent	Excellent	Excellent

Hypalon® is a registered trademark of DuPont Dow Elastomers.

- Automotive and industrial hoses
- Wire and cable jackets
- Industrial roll covers
- Coating and adhesives
- Flexible magnets

Chlorosulfonated polyethylene (CSM) is subject to dehydrohalogenation by Zn^{++} ions, which can seriously degrade heat resistance. It is strongly recommended to avoid incorporating any zinc-containing materials into a CSM formulation. The use of zinc-containing coatings or dusting agents on the outside of these formulations is also discouraged.

Hypalon[R] chlorosulfonated polyethylene is a type of CSM produced by DuPont Dow Elastomers. There are several types of Hypalon. When properly compounded and cured, they all possess the basic characteristics of CSM. However, there are chemical and physical differences among the various types, which affect processability and properties. The compounder should select the type of Hypalon that best produces the desired results for the intended application. A general description of each of these types is shown in Table 8.28. They are more fully described in the text that follows.

8.10.2   General Purpose Types of Hypalon

Hypalon[R] 40 is considered the basic general purpose chlorosulfonated polyethylene elastomer, having a good balance of processability and vulcanizate properties. It provides good oil and chemical resistance and intermediate low temperature performance.

Higher and lower viscosity grades of Hypalon are available to provide a range of compounding and processing capability within the general purpose family. Hypalon 4085 is a higher viscosity grade that offers improved processability and dilutability for highly or extremely extended formulations or low durometer vulcanizates. On the lower end of the viscosity range is Hypalon 40S (soft). This polymer provides easier processing in lightly plasticized, higher durometer compounds.

8.10.3   Specialty Types of Hypalon

Hypalon 45 is more thermoplastic than Hypalon 40. It has a lower viscosity at processing temperatures, but gives better room temperature green strength and harder vulcanizates. Hypalon 45 has a lower chlorine content than Hypalon 40; therefore, it is flexible at lower temperatures, but has reduced oil and chemical resistance. It also provides better tear strength and slightly better heat resistance than the general purpose types. Hypalon 45 can be compounded to give very good mechanical properties in uncured stocks. It is frequently used uncured in sheeting applications because of its strength and ease of seaming.

Hypalon 6525 is a high viscosity, extensible polymer with characteristics similar to Hypalon 45. It has a lower chlorine content than Hypalon 40, which yields compounds with excellent low temperature properties and slightly improved oven aging performance. It can be compounded for lower cost because it can accept high filler and plasticizer loadings. Some decrease in oil and flame resistance is to be expected when compared to Hypalon 40.

Hypalon 48 is intermediate in thermoplasticity between Hypalon 40 and Hypalon 45. It provides better oil resistance, but poorer low temperature properties than Hypalon 40 or Hypalon 45. It also has very high resistance to permeation by some refrigerants and other gases. Hypalon 48 will impart greater oil and flame resistance than Hypalon 40 or Hypalon 45 in comparable positions, at a sacrifice in low temperature flexibility.

8.10.4   Unvulcanized Applications

Hypalon 45 can also be compounded to give remarkably good stress/strain properties in uncured stocks. These compounds may be used in applications such as cove base, magnetic door closures, roofing membranes, and pond and pit liners.

Uncured compounds based on Hypalon 48 have physical properties similar to those obtained with uncured compounds of Hypalon 45. In applications for uncured stocks, Hypalon 48 imparts greater oil and flame resistance than Hypalon 45 in comparable compositions, but at a sacrifice in low temperature flexibility.

8.11   Polyepichlorohydrin Elastomer

Clark Cable

8.11.1   Introduction

Polyether elastomers offer an excellent balance of properties for automotive and marine applications. The presence of the oxygen atoms in the polymer backbone imparts flexibility, and the polymer chain is saturated, providing superior ozone resistance (see Fig. 8.7). The commercially important elastomers are polyepichlorohydrin (ECH), ECH-ethylene oxide (EO) copolymer, and ECH-EO-allyl glycidyl ether (AGE) terpolymer. Table 8.29 lists the commercially available types from Zeon Chemicals L.P., with ASTM designations.

8.11.2   Properties

With about one-third of the polymer being chlorine (chloromethyl groups), the polymer is polar. Thus, CO\ECO\GECO have very good resistance to oils and fuels, but are not

Table 8.29   ASTM Designations of Polyepichlorohydrin Elastomers

Homopolymer of epichlorohydrin (ECH)	CO
Copolymer of ECH/allyl glycidyl ether (AGE)	GCO
Copolymer of ECH/ethylene oxide (EO)	ECO
Terpolymer of ECH/EO/AGE	GECO

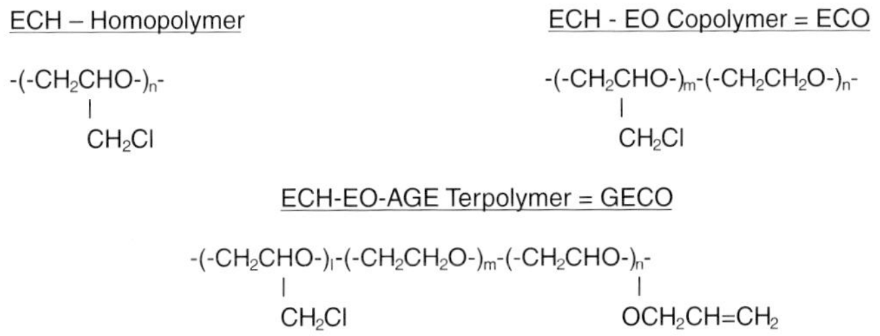

ECH – Homopolymer

-(-CH₂CHO-)ₙ-
    |
   CH₂Cl

ECH - EO Copolymer = ECO

-(-CH₂CHO-)ₘ-(-CH₂CH₂O-)ₙ-
    |
   CH₂Cl

ECH-EO-AGE Terpolymer = GECO

-(-CH₂CHO-)ₗ-(-CH₂CH₂O-)ₘ-(-CH₂CHO-)ₙ-
    | |
   CH₂Cl   OCH₂CH=CH₂

Figure 8.7   Commercially available elastomers.

resistant to (polar) brake fluid. Also, epichlorohydrin polymers have excellent permeation resistance because of the existence of the highly polar, and dense, chloromethyl groups.

Epichlorohydrin rubber is known for its balance of properties: heat resistance, cold temperature flexibility, ozone resistance, and oil/fuel resistance. They are classified by SAE J200 and ASTM D2000, to be in the CH (125 °C capable) category. Polyepichlorohydrin compounds can be formulated to withstand 70 hours at 150 °C in air, while being flexible at −40 °C.

The elastomer's volume resistivity values are as low as 1×10^8 ohms-centimeter. Making a compound with a high structure carbon black, i.e., N472, can give resistivity values in the conductive range, as low as 1.6×10^3 ohms-cm. The electrical properties of epichlorohydrin rubber can be used advantageously in a situation where the buildup of a static charge is to be reduced.

8.11.3   Formulating

Common carbon blacks, white fillers, and process aids can be used, along with various antioxidants (metal dithiocarbamates, amines, and imidiazoles). Plasticizers that are close to the same polarity as the polymer should be used, to a maximum of approximately 30 phr, depending on the filler level. These plasticizers are usually diesters and ethers. Examples of compatible plasticizers are DOP (dioctyl phthalate), DIDP (diisodecyl phthalate), TOTM (trioctyl trimellitate), and DBEEA [di(butoxy-ethoxy-ethyl) adipate].

Cure systems used for vulcanizing the CO and ECO elastomers are different than the sulfur and peroxide systems typically used in common elastomers. GECO can be cured with any of the just mentioned systems. Ethylene thiourea (ETU) and lead oxide has been the cure system of choice for epichlorohydrin compounds since the 1960s. (DTDM is the

Table 8.30   **Typical Formula for an Epichlorohydrin Elastomer**

Ingredients	phr	Range, phr
Polymer	100	
Carbon black (e.g., N762)	70	0 to 130
Plasticizer (e.g., DBEEA)	10	0 to 30
Antioxidant	1	0 to 2
Processing aid	1	0 to 3
Acid acceptor	5	3 to 10
Cure chemicals	1	0.3 to 2.0

Table 8.31   Nonlead / Alternative Cure Systems

Characteristic	Triazine	Thiadiazole	Bis-phenol
Disadvantages	Slightly lower tensile	Less scorch time – w/o barium carbonate (heavy metal)	Expensive
Advantages	Less toxic	Less toxic	Less toxic (use Dynamar RC-5251Q)
Adhesion to FKM	Fair to good	Fair to good	Excellent

best retarder for the ETU cure system, 0.5 phr is commonly used.) Recent laboratory work provides data that show nonlead containing cure systems to be effective alternatives to the ETU/lead system. Alternatives to the ETU/lead cure system are being pursued because the components are toxic at low levels, e.g., lead at 5 ppm.

8.11.4   Nonlead Cure Systems

The following three are viable cure systems for the CO/ECO/GECO elastomers: triazine (Zisnet F-PT), thiadiazole (Echo MPS), and bis-phenol (Dynamar 5157/5166). An overview of the systems is given in Table 8.31.

8.11.5   Adjustments

Table 8.32 shows how to adjust the cure rate of a Zisnet F-PT package.

DPG and CTP can be used together; the DPG seems to work more at curing temperatures, while the CTP works well at processing temperatures. One phr of CTP provides ample scorch safety and long shelf life. Levels of CTP above 0.5 phr may yield an obnoxious odor during and after curing. Vulkalent E/C (a sulfonamide derivative) is a very good replacement for the CTP, and does not have the odor problem.

8.11.6   Processing

Internal mixers (one or two pass procedures) and mills can be used to mix CO\ECO\GECO compounds. These compounds tend to stick to mill roll surfaces. This condition occurs erratically, and has been traced to residue on the mill roll surface. Not all residues cause sticking. Residue from a peroxide cured EPDM compound can cause severe sticking of a

Table 8.32   Adjustment of Cure Rate

Ingredient amounts, phr	Function	Initial cure rate	Faster cure rate	Slower cure rate
Calcium carbonate	acid acceptor	5.0	5.0	5.0
Zisnet F-PT	curative	0.5 to 1.0	0.8	0.8
DPG (diphenylguanidine)	accelerator		0.2 to 1.0	
CTP (Santogard PVI) or Vulkalent E/C	retarders			0.2 to 1.0

Table 8.33   Banbury Mix Procedure

Temperature °C (°F)	Approximate time, minutes	Actions
60 max (140)	0	Add polymer, carbon black, and all master batch and cure chemicals
85 (185)	1	Add all the plasticizer (if any)
90 (194)	2	Raise ram, scrape, lower ram
105 (220)	3	Drop batch onto the mill with clean roll surfaces (Maximum safe drop temperature 115 °C (240 °F)

following polyepichlorohydrin batch to the mill roll surface. To alleviate the sticking, use clean mill rolls, set roll temperatures at 60 to 80 °C, and/or use 1 phr of mono stearyl citrate (MSC).

8.11.7   Internal Mixer – Procedure

Polyepichlorohydrin compounds are typically mixed in a two-stage process. Upside-down and regular mix procedures have been used to make quality compounds, with good dispersion of the other ingredients. One pass mixing is also feasible. A one pass mix procedure follows in Table 8.33.

8.11.8   Extrusion

CO/ECO/GECO can be extruded successfully with the equipment typically used in rubber factories for such a process. Recommended temperatures are listed in Table 8.34.

Highly filled ECO compounds generally extrude better than rubber rich compounds. One reason is higher compound green strength, which is important in maintaining the extruded shape. Different carbon black types have been used successfully, such as N330, N550, N762, and N990.

8.11.9   Molding

CO/ECO/GECO compounds can be compression, transfer, or injection molded. Molding has included bonding to metal inserts.

Compounds for compression molding tend to have a higher viscosity, e.g., 90 to 120 Mooney units. This is usually accomplished with more reinforcement, and/or higher filler loading. Two to five phr of plasticizer should help the compound flow easier, to fill the cavity without it being too soft, which could lead to air entrapment.

Table 8.34   Recommended Extruder Profile Temperatures, °C

Barrel	Screw	Head	Die
70 to 90	70 to 90	90 to 100	100 to 110

Table 8.35   Commercially Available Elastomers from Zeon Chemicals L.P.

Series	Homopolymer ECH	Copolymer ECH/AGE	Copolymer ECH/EO	Terpolymer ECH/EO/AGE
Number	H-45 to H-75	H1100	C2000LL, C2000L & C2000	T3000, T3102, T3105 & T3100
Letter			C-55 to C-95	T-55 to T-95

Appendix 1   List of Chemicals

Names	Chemical	Source
Dynamar FC 5157	proprietary (bis-phenol) cross linker	Dyneon
DynamarFX 5166	proprietary (phosphonium salt) accelerator	Dyneon
Dynamar PPA-790	fluorochemical process aid	Dyneon
Dynamar RC-5251Q	proprietary	Dyneon
Millathane Glob	urethane + cleaning material	TSE Ind., Clearwater, FL
Naugard 445	4,4′-bis(a-dimethylbenzyl) diphenylamine	Uniroyal Chemical Co.
Santogard PVI- or CTP (PreVulcanization Inhibitor)	N-(cyclohexylthio) phthalimide	Flexsys America L.P.
Sulfasan R- or DTDM(also Vanax A)	4,4′-dithio-bis-morpholine	Elastochem, Inc.
Echo MPS (thiadiazole)	proprietary cross linker	Hercules, Inc.
TyPly BN	rubber to metal bonding agent	Lord Elastomer Prod.
Vanax A- or DTDM	4,4′ dithio-bis-morpholine	R.T. Vanderbilt Co. Inc.
Vanox MTI	2-mercaptotolulimidazole	R.T. Vanderbilt Co. Inc.
Vanox ZMTI	2-mercapto-4(5)-methylbenzimidazole	R.T. Vanderbilt Co. Inc.
Vulkanox MB-2/MG/C	Blend of 4 and 5 methylmercapto benzimidazole -oil treated	Bayer Fibers, Organics & Rubber Div.
Vulkalent E/C	(sulfonamide derivative plus 5% whiting and 1–2% oil)	Bayer Fibers, Organics & Rubber Div.
Zisnet F-PT (triazine)	2,4,6-trimercapto-s-triazine	Zeon Chemicals L.P.

Appendix 2   List of Plasticizers

Names	Chemical	Source
DOP	Dioctyl phthalate(or diethylhexyl phthalate)	C.P. Hall & various suppliers
DIDP	Diisodecyl phthalate	C.P. Hall & various suppliers
DBEEA (TP-95)	Di(butoxy-ethoxy-ethyl) adipate	C.P. Hall & various suppliers (Morton International)
TOTM	Trioctyl trimellitate	C.P. Hall & various suppliers

Compounds for transfer molding are likely to have lower viscosities, depending on the sprue/gate sizes and locations, and cavity shape. Waxy process aids are typically used in an attempt to allow the compound to flow even easier.

Compounds for injection molding can vary substantially in filler to plasticizer loading ratios, and therefore have a wide range of viscosity values. Here again, gate size and cavity shape are important factors in determining the requirements to successfully fill the mold set/cavity(s). Low viscosity compounds are usually wanted for injection molding. However, with the heating and mastication of the rubber compound by the barrel and screw, low viscosity may not be needed, or may not mold as well as a higher viscosity compound. A low viscosity compound may increase the occurrence of air entrapment.

Epichlorohydrin rubber compounds have a good balance of properties that make them an excellent choice for many under-the-hood applications. ECO and GECO are natural selections for electrostatic-dissipative (ESD) applications due to their electrical properties.

Polyether Elastomers Information

Polymers	ECH, wt.%	Chlorine wt.%	Ethylene oxide, wt.%	CAS Reg. No.	Specific gravity	ML Mooney at 100 °C	T_g, °C
ECH	100	38	0	[24969-06-0]	1.36	40–80	−22
ECO	68	26	32	[24969-10-6]	1.27	40—130	−40
ECH-AGE	93	35	0	[24969-09-3]	1.24	60	−25
GECO	65–76	24–29	13–31	[26587-37-1]	1.27	50–100	−38

These are produced by: (1) Zeon Chemicals L.P. in the USA under the trade name of Hydrin; (2) Nippon Zeon in Japan – trade name Gechron; (3) Daiso in Japan – trade name Epichlomer.

Reference

1. Kathryn Owens and Vernon Kyllingstad, *Elastomers Synthetic Polyether*, Kirk-Othmer Encyclopedia of Chemical Technology, 4th Ed., Volume 8, (1993), pp. 3–6, John Wiley & Sons, New York.

8.12   Ethylene-Acrylic Elastomers

Theresa M. Dobel

8.12.1   Introduction

Ethylene/acrylic elastomers, or AEM polymers according to its official ASTM designation, were first introduced commercially in 1975 by the DuPont Company under the trademark Vamac. The creation of this polymer was the result of an intensive research effort to

Table 8.36   Heat Resistance of Ethylene-Acrylic Elastomers

Temperature of continuous exposure	Approximate useful life >50% ultimate elongation[a]
121 °C (250 °F)	24 months
150 °C (302 °F)	6 months
170 °C (340 °F)	6 weeks
177 °C (350 °F)	4 weeks
191 °C (375 °F)	10 days
200 °C (392 °F)	7 days

[a] Per ASTM D412

develop an oil resistant polymer with heat resistance greater than a nitrile or chloroprene rubber at a moderate cost well below that of silicone or fluoroelastomers. Ethylene/acrylic elastomers are best known for their excellent heat and oil resistance. They also possess a good balance of compression set resistance, flex resistance, physical strength, low temperature flexibility, and weathering resistance. Some special attributes include temperature-stable, vibrational damping properties and the ability to produce flame-resistant compounds with combustion products with a low order of smoke density, without the presence of corrosive halogens.

The main features of ethylene/acrylic elastomers are heat (177 °C) and oil resistance. At elevated temperatures, ethylene-acrylic elastomers age by an oxidative crosslinking mechanism, resulting in eventual embrittlement rather than reversion. A general heat resistance profile is given in Table 8.36. If time taken to reach a final elongation of 50% is assumed to be a "useful life," then Vamac functions for up to 24 months at 121 °C, and up to 6 weeks at 170 °C continuous service temperature. Exposures up to 190 to 200 °C can be tolerated for short intervals.

Vamac compounds are well suited for applications requiring continuous exposure to hot aliphatic hydrocarbons, a class that includes most common automotive lubricants and hydraulic fluids. The higher methyl acrylic polymers exhibit lower oil swell than the standard polymers. Low, consistent swell in motor oils and transmission fluids indicates usefulness for service in various seals, gaskets, and hoses in both the transmission and engine, and seals for wheel and crankshaft bearings. The water and ethylene glycol resistance of ethylene acrylic elastomers is good, but some proprietary coolant packages can cause excessive stiffening of the terpolymer vulcanizates. Also, softening can occur after long term exposures above 100 °C in water and/or coolants.

8.12.2   Polymer Composition and Effect on Properties

Ethylene/acrylic terpolymers are primarily an addition copolymer of ethylene and methyl acrylate, with the incorporation of a third carboxylic monomer to provide cure sites for crosslinking with diamines (Fig. 8.8).

There is also a family of dipolymers that do not have the cure site monomer and are curable by peroxides. These polymers are non-crystalline, with the ethylene giving good low temperature properties, while the methyl acrylate content provides oil resistance. The polymer backbone is fully saturated in all types, making Vamac inherently resistant to ozone attack.

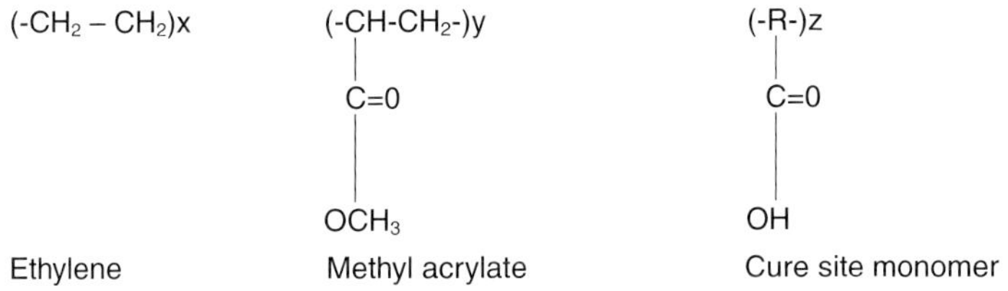

$$(-CH_2 - CH_2)x \qquad (-CH-CH_2-)y \qquad (-R-)z$$

	C=O	C=O
	OCH₃	OH
Ethylene	Methyl acrylate	Cure site monomer

Figure 8.8

8.12.3   Polymer Selection

Four different base polymers of Vamac ethylene/acrylic elastomers are commercially available (see Table 8.37). Until 1990, existing grades of ethylene/acrylic elastomers were based on a single gum polymer, Vamac G, defined as a terpolymer of average methyl acrylate content, ethylene, and a cure site monomer. In the 1990s, a higher methyl acrylate terpolymer, Vamac GLS, was introduced. The composition of this polymer was specifically chosen for significant improvements in oil resistance with minimal losses in low temperature flexibility.

A new family of peroxide-cured dipolymers was also introduced at about the same time. Compounds based on the dipolymer cure faster and exhibit good compression set properties without a postcure. The removal of the cure site has also made the polymer less susceptible to attack from amine-based additives. By varying the methyl acrylate level in

Table 8.37   Vamac Product Line Summary

Commercial name	Monomers	Methyl acrylate level	Product description	Applications
Vamac G	(MA/E/CS)	Average	Standard oil resistance Postcured required IRM 903 Oil Swell ~60% Best low temperature	General purpose seals, gaskets, and hose
Vamac GLS	(MA/E/CS)	High	Improved oil resistance Postcure required IRM 903 oil swell ~30%	Gaskets, seals, and EOC/TOC hose
Vamac D	(MA/E)	Average	Standard oil resistance No postcure required Improved amine resist (radiator fluids, oils) IRM 903 oil swell ~60% Best low temperature	NPC seals, gaskets, hose, and wire & cable
Vamac DLS	(MA/E)	High	Improved oil resistance No postcure required Improved amine resist (radiator fluids, oil) IRM 903 oil swell ~25%	NPC seals, gaskets WHV hose, and wire & cable

MA = methyl acrylate, E = ethylene, CS = amine active cure site monomer.

the dipolymer family, two materials were synthesized: Vamac D and its more oil-resistant counterpart, Vamac DLS. Note that the dipolymers are harder to process than their terpolymer counterparts. They tend to blister during processing and have the hot tear problems typically associated with a peroxide cure. They also are harder to bond to metal than their terpolymer counterparts. Therefore, while the dipolymers have found some niche applications where eliminating a postcure is advantageous, such as in automotive gaskets, the terpolymer is still the polymer of choice for most applications.

8.13   Polynorbornene

Clark Cable

8.13.1   Introduction

Polynorbornene is a high molecular weight unsaturated amorphous hydrocarbon (C_7H_{10}). Polynorbornene (PNR – not the official ASTM or ISO designation), sold by Zeon Chemicals L.P. under the trade name Norsorex, is a powder also available in a masterbatch form, with naphthenic oil [1]. PNR's unique characteristics yield extremely soft compounds, 15 to 20 Shore A hardness, while maintaining good mechanical properties. These highly filled compounds are readily processed on conventional rubber equipment.

8.13.2   Applications

Applications include seals, vibration isolators, vibration dampeners, shock absorbers, sound insulation sheets, machinery mounts, small motor and instrument mounts, bumpers, air intake hoses, hood stops, emergency brake boots, gear shift boots, grommets, diaphragms, gaskets, track pads, roll coverings, and racing tires.

PNR can be compounded to have very good vibration isolation and vibration damping characteristics. It is also an excellent insulator, with volume-resistivity values of 1×10^{16}

Table 8.38   Recipe for a 30 Shore A PNR Compound

Ingredients	phr (parts per hundred of rubber)
Norsorex 150 NA masterbatch	100
N762 carbon black	100
Whiting (calcium carbonate)	50
Sunthene 255 (naphthenic oil)	100
TMQ (antioxidant)	1
Stearic acid (processing aid)	1
Zinc oxide (activator)	5
Sulfur (vulcanizing agent)	2
TMTD (accelerator)	1
Total	360

ohms-cm. Adding 100 to 200 phr of N472 high structure carbon black to PNR makes it static-dissipative.

8.13.3   Compounding

The high binding power of PNR means the type and amount of fillers and plasticizers in the compounds can vary a great deal. A recipe for a 30 Shore A PNR compound is presented in Table 8.38.

8.13.4   Fillers

The fillers' contribution to compound properties is secondary to the oil type and amount. For example, at room temperature the oil provides the damping. The choice of oil type depends on the desired low temperature flexibility. An aromatic oil imparts more damping than a naphthenic one. For maximum cold resistance, a low viscosity naphthenic oil should be used. So, the oil choice is usually made first based on the desired low temperature flexibility and/or damping characteristics. Therefore, carbon black type and amount are the largest factors remaining to adjust the damping characteristics of a compound.

8.13.5   Oils/Plasticizers

PNR compounds typically contain 150 to 300 phr of plasticizer. PNR has a glass transition temperature above room temperature (35 °C), which makes it inflexible, like other plastics. However, through the addition of naphthenic oil or aromatic oil at very high loadings, these compounds can be made rubbery and flexible with T_g values in the range of -65 to -40 °C, respectively. As evidenced by the T_g values, naphthenic oils give significantly higher rebound than aromatic oils.

8.13.6   Cure System

Sulfur vulcanizes PNR. An example of a sulfur cure package used to vulcanize a PNR compound is presented.

Zinc oxide	5.0
TMTD (Methyl Tuads – R. T. Vanderbilt)	5.0
MBTS (Altax – R. T. Vanderbilt)	1.0
TeEDC (Tellurac – R. T. Vanderbilt)	0.5
TBTD (Butyl Tuads – R. T. Vanderbilt)	1.5
Sulfur	1.5

Polynorbenene can also be cured by peroxide. One example of a peroxide package for PNR is zinc oxide at 5 phr, and 6 phr of Di Cup 40 KE.

Nominal shrinkage for a typical, highly filled, PNR compound is 3.5%.

8.13.7   Rebound/Resilience

The lowest rebound values are obtained with aromatic oils. Results vary more after aging than with a naphthenic plasticize of the same viscosity. However, a high level of shock absorption at a low hardness cannot be obtained, even by changing the carbon black.

A minimum of 200 phr oil loading is needed to get small variations in resilience. Therefore, the rebound value is determined mostly by the type and amount of carbon black.

8.13.8   Vibration Damping

The oil/plasticizer is the largest contributor in determining the damping characteristics. The oil choice is usually determined by two factors: the desired low temperature flexibility, and damping characteristics. An aromatic oil imparts more damping than a naphthenic one. However, low viscosity naphthenic oils provide the best low temperature flexibility. So, a compromise will have to be made to get the best of both properties in one compound.

If staining is allowed, then a high viscosity aromatic oil can be used. If non-staining is a requirement, then use a high viscosity naphthenic oil.

Carbon black type and amount are the next largest factors remaining to adjust the damping characteristics of a compound. To increase damping use smaller particle size carbon blacks and/or higher structure blacks, and increase filler and/or oil loadings.

The ingredient(s) that have the third largest effect on damping are the cure chemical(s). Reducing the curative amount(s), for a lower cross link density, increases the damping.

8.13.9   Blends

PNR is compatible with other unsaturated polymers, i.e., NR, SBR, PBR, CR, and NBR. Even small amounts of PNR can markedly improve green strength and process ability of compounds using the elastomers just mentioned. Natural rubber can become dampening through blending with PNR. Such a compound would be adapted for low hardness engine or body mounts.

8.13.10   Mixing and Processing

Norsorex is a powder. It is also available in masterbatch form, with a naphthenic plasticizer and EPDM. Mixing a plasticizer into the powder is very difficult. A master-batch should be used, at least for 45 Shore A hardness or lower compounds.

8.13.10.1 Mill Mixing

The expanded structure of PNR allows the incorporation of huge amounts of oils. If this structure is destroyed, further oil incorporation is practically impossible. So, oils must be poured into the powder and stirred until absorption is complete.

Table 8.39   Internal Mixer Incorporation Procedure for PNR Powder

Time, minutes	Action
0	Add 10 phr PNR, then $\frac{1}{2}$ to $\frac{3}{4}$ of the oil, and the rest of the PNR
2	Introduce other elastomer type (if a blend)
3	Introduce $\frac{1}{2}$ to $\frac{3}{4}$ of the fillers
5	Introduce any remaining oil and fillers
8	Drop onto the mill (maximum safe drop temperature 130 °C)

Mill roll temperatures should be at approximately 80 °C to avoid bagging.

PNR tends to band to the hotter, faster mill roll. Mill roll temperatures should be 65 to 80 °C (150 to 180 °F), with the front roll hotter. If the PNR is not warm enough, it will lace (tear) badly. In such an instance, use a tight mill nip for about a minute.

8.13.10.2 Internal Mixers

PNR master batches can be mixed with common rubber ingredients, such as a typical elastomer. The fill factor should be from 1.1 to 1.4 for very soft PNR stocks, <40 Shore A. An overfill of 15% may be sufficient for soft PNR compounds, 40 to 60 Shore A. Compounds harder than 45 Shore A can be prepared by using a typical mix procedure. A special procedure is needed to mix the PNR powder (Table 8.39).

8.13.11   Calendering

PNR compounds can be prepared on a calender using standard industrial practices. Calender rolls should be kept warm, 65 to 80 °C (150 to 180 °F) for >40 Shore A stocks. Use standard procedures; thoroughly warm stock on the prep mill and use before it cools. Rough appearing stock is usually too cool. Heat the calender rolls to correct this problem.

8.13.12   Extrusion

Generally, extrusion rates for PNR compounds are faster than those of EPDM or SBR. Normally, higher die temperatures should be used than the ones used for other elastomer types at equal hardness. Extruded profiles can be cured in steam, hot air, salt baths, or microwaves. The cure system may need adjustment to work well with the curing equipment. Temperature profiles for 20, 40, and 60 Shore A PNR compounds are listed in Table 8.40.

Table 8.40   Suggested Temperature Profiles for 20, 40, & 60 Shore A PNR Compounds, °C (°F)

Compound	Feed	Body I	Vacuum	Body II	Head	Die
20 Shore A	40 (105)	40 (105)	50 (120)	70 (160)	80 (175)	90 (190)
40 Shore A	60 (140)	60 (140)	60 (140)	60 (140)	65 (150)	90 (190)
60 Shore A	100 (212)	80 (180)	65 (150)	80 (180)	100 (212)	100 (212)

Table 8.41   Commercial Norsorex® Products (Standard Grades)

	Description
Norsorex N	Powder
Norsorex® 150 NA	Masterbatch MB Norsorex® N 100 Naphthenic oil 150 Mooney ML1 + 4: 47 ± 55
Norsorex 150 NA/25EP	Norsorex® N 100 Naphthenic oil 150 EPDM 25 Mooney ML1 + 4: 45 ± 5
Norsorex 150 HNA	Norsorex® N 100 High viscosity naphthenic oil 150 EPDM 25 Mooney ML1 + 4: 55 ± 5

Table 8.42   Summary and Comparison of Mechanical Properties

Mechanical properties	Norsorex	butyl	EPDM	NR	NBR	CR	SBR
Hardness, Shore A	18–80	40–90	35–90	30–100	40–95	30–90	45–90
Tensile strength, MPa, max	23	15	15	26	23	23	21
Tear resistance	F	G	F	E	F	G	F
Abrasion resistance	G	G	G	E	G	G	G
Damping factor	M–H	H	M	L	M	M	M
Compression set	G	F	G	G	G	G	F
Electrical resistance	E	G	E	G	F	F	F–G

E = Excellent, G = Good, H = High, F = Fair, M = Medium, L = Low

Table 8.43   Summary and Comparison of General Properties

General properties	Norsorex	butyl	EPDM	NR	NBR	CR	SBR
Low temperature, °C	−45	−50	−50	−50	−40	−50	−45
Continuous high, °C	80	120	120	80	125	100	90
Intermittent high, °C	100	150	150	100	150	130	110
Ozone (protected)	G	G	E	F	G	G	F
Weather and sunlight	G	E	E	G	G	G	G
Gas permeation	F–G	E	G	F	G	G	F
Water resistance	E	E	E	G	F	G	G
Acid/base res., dil.	E	E	E	G	G	G	G
Acid/base res., conc.	F	G	G	F	G	G	F
Solvent res., aliphatic	F	L	L	L	G	G	L
Solvent res., aromatic	L	L	L	L	G	L	L

E = Excellent, G = Good, H = High, F = Fair, M = Medium, L = Low

8.13.13   Molding

Formulations based on PNR can be compression, transfer, or injection molded. Cavity temperature should be between 160 and 180 °C. The cure system may require some adjustment depending on the equipment being used.

8.13.14   Summary

Polynorbornene has the unique characteristic of good mechanical and green strength at extremely high loadings of oil/plasticizer, i.e., 250 phr. This allows for stocks with very low hardness, 15 Shore A, to be feasibly processed. The soft compounds have very good vibration damping and vibration isolation properties. These properties are useful in some mount and roll applications.

Reference

1.   Zeon Chemicals Literature on Norsorex compounds.

8.14   Fluoroelastomer (FKM)

Ronald D. Stevens

8.14.1   Introduction

Viton® is DuPont Dow Elastomer's tradename for its line of fluorohydrocarbon elastomers. Fluoroelastomers are a class of synthetic rubber that provide extraordinary levels of resistance to chemicals, oil, and heat, and useful service life at temperatures greater than 200 °C. The outstanding heat stability and excellent oil resistance of these materials result from the high ratio of fluorine to hydrogen, the strength of the carbon-fluorine bond, and the absence of unsaturation. Fluoroelastomers are referred to generically as FKM polymers, per the nomenclature noted in ASTM D1418. In the SAE J200/ASTM D2000 classification system for rubber materials, FKM is rated as a HK material as is seen in the Fig. 8.9.

8.14.2   Background

Viton® A, a copolymer of hexafluoropropylene (HFP) and vinylidene fluoride (VF_2), was developed by the DuPont Company in 1957 in response to high performance sealing needs in the aerospace industry. To provide even greater thermal stability and solvent resistance,

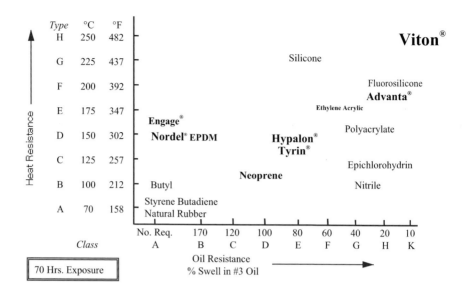

Figure 8.9   Differences in fluids resistance and low temperature between types of Viton®.

tetrafluoroethylene (TFE)-containing fluoroelastomer terpolymers called Viton® B were introduced in 1959. In the mid to late 1960s, lower viscosity versions of Viton® A and B were introduced. In the 1970s, a breakthrough in crosslinking occurred with the introduction of the Bisphenol AF cure system, which offered much improved heat and compression set resistance with better scorch safety and faster cure speeds. In the late 1970s and early 1980s, DuPont introduced a line of fluoroelastomers with improved low temperature flexibility by using perfluoromethylvinyl ether (PMVE) in place of HFP. These polymers, called Viton GLT and GFLT, require a peroxide cure.

In the late 1980s, DuPont introduced Viton® Improved Rheology Polymers (IRP). IRP are fluoroelastomers which, when bisphenol cured, show better flow properties, faster cures, and cleaner mold release. When the new IRP polymers are combined with a bisphenol curative, a precompound (polymer plus curative) is created, such as Viton A401C and Viton® B601C. The latest addition to the Viton line of polymers is Viton

Table 8.44   Applications for Viton® Fluoroelastomers

Aerospace	Automotive	Industrial
O-ring seals in fuels, lubricants, and hydraulic systems	Shaft seals	Hydraulic O-ring seals
Manifold gaskets	Valve stem seals	Check valve balls
Fuel tank bladders	Fuel injector O-rings	Military flare binders
Firewall seals	Fuel hoses	Diaphragms
Engine lube siphon hose	In tank and quick connect fuel	Electrical connectors
Clips for jet engines	System seals	Flue duct exp. joints
Tire valve stem seals	Gaskets (valve & manifold)	Valve liners
Lathe cut gaskets	Balls for check valves	Roll covers
	Sheet stock/cut gaskets	

Extreme ETP, introduced in 1997. ETP has a much broader fluids resistance profile than standard fluoroelastomers, and is able to withstand strong bases and ketones as well as aromatic hydrocarbons, oils, acids, and steam.

8.14.3   Applications

Viton[R] fluoroelastomers are used in a wide variety of high performance applications. FKM provides premium, long term reliability even in harsh environments. A partial listing of current end use applications for Viton[R] appear in Table 8.44.

8.14.4   Viton[R] Types

Viton[R] fluoroelastomers are classified into four groups; A, B, F, and Specialties. The A, B, and F types have increasing fluids resistance derived from increasing fluorine levels (66, 68, and 70%, respectively). Polymers with the "G" prefix are peroxide curable, and polymers in the Specialty types possess unique properties such as those with the "LT" suffix, which stands for improved low temperature flexibility.

One of the primary attributes of Viton[R] fluoroelastomers is their fluids resistance. Table 8.45 lists each type of Viton[R] fluoroelastomer and its performance in different fluids.

Table 8.46 shows the primary, but no all, types of Viton[R] fluoroelastomers commercially available. See DuPont Dow Elastomer's publication H-68143-02 for a complete listing.

Table 8.45 Fluids Resistance and Low Temperature Sealing Capability of Viton® Fluoroelastomers

	Viton® fluoroelastomer							
	Cure system							
	Bisphenol					Peroxide		
	A	B	F	GB	GF	GLT	GFLT	ETP
Aliphatic hydrocarbons, process fluids, chemicals	1	1	1	1	1	1	1	1
Aromatic hydrocarbons (toluene, etc.), process fluids, chemicals	2	1	1	1	1	2	1	1
Automotive and aviation fuels (pure hydrocarbons – no alcohol))	1	1	1	1	1	1	1	1
Automotive fuels containing legal levels (5-15%) of alcohols & ethers (methanol, ethanol, MTBE, TAME)	2	1	1	1	1	2	1	1
Automotive / methanol fuels blends up too 100% methanol (flex fuels)	NR	2	1	2	1	NR	1	1
Engine lubricating oils (SE-SF grades)	2	1	1	1	1	1	1	1
Engine lubricating oils (SG-SH grades)	3	2	2	1	1	1	1	1
Acid (H_2SO_4, HNO_3), hot water, and steam	3	2	2	1	1	1	1	1
Strong base, high pH, caustic, amines	NR	NR	NR	NR	NR	NR	NR	1-2
Low molecular weight carbonyls 100% concentration (MTBE, MEK, MIBK, etc.)	NR	NR	NR	NR	NR	NR	NR	1-2
Low temperature sealing capability TR-10 test results	−17 °C	−14 °C	−7 °C	−15 °C	−6 °C	−30 °C	−24 °C	−11 °C

1 = Excellent, minimal volume swell.
2 = Very good, small volume swell.
3 = Good, moderate volume swell
NR = Not recommended, excessive volume swell or change in physical properties.

Table 8.46 Partial Viton® Polymer Selection Guide

Application	Viton® product name	Mooney viscosity at 121°C	Fluorine %	Bisphenol curative included	Comp. set – 70 hrs at 200°C	Features and processing methods
O-rings, seals and gaskets needing low compression set						
A-type copolymers	A201C	20	66%	yes	16	A201C is designed for transfer or injection molded O-rings or seals.
	A401C	40	66%	yes	15	A401C is designed for compression molded O-rings or seals. Used to meet MIL-R-83248
	A601C	60	66%	yes	14	A401C is designed for compression molded o-rings or seals. Used to meet MIL-R-83248
	A200	20	66%	no – gum	20	Gum A200 needs curative (i.e., VC50 or VC20/VC30). Transfer molding.
	A500	50	66%	no – gum	15	Gum A500 needs curative (i.e., VC50 or VC20/VC30) or can be blended with curative containing type. Compression molding.
B-type terpolymers	B601C	60	68%	yes	20	B601C is designed for compression molded O-rings or seals
	B600	60	68%	no – gum	20	Gum B600 needs curative (i.e. VC50 or VC20/VC30). Compression molding
F-type terpolymers	F605C	60	70%	yes	30	For compression molding. Used to meet Ford's M2D401-A8
Custom molded goods and calendered sheet stock						
A-type copolymers	A331C	30	68%	yes	20	Higher elongation, better hot tear product
B-type terpolymers	B641C	60	68%	yes	30	For calendered sheet and flue duct expansion joints
Metal bonded seals – shaft seals						
A-type copolymers	A361C	30	66%	yes	20	For rubber-to-metal seals – includes adhesion promotor
B-type terpolymers	B435C	40	68%	yes	30	For rubber-to-metal seals – includes adhesion promotor
Speciality types of Viton						
Low fluorine	B70N	70	66%	no	30	Moderately improved low temp FKM; TR-10 = −19°C. Bisphenol cure with VC50
Terpolymer Peroxy cures	GBL-200	20	67%	no	40	Improved SF oil, water, and acids resistance. Transfer molding

Table 8.46 Continued

Application	Viton® product name	Mooney viscosity at 121 °C	Fluorine %	Bisphenol curative included	Comp. set – 70 hrs at 200 °C	Features and processing methods
	GBL-205LF	20	67%	no	35	Improved water, steam, and acid resistance with no lead (PbO) needed in cure system
	GF205NP	20	69%	no	35	High fluorine, peroxy cured type which needs no postcure for good c/s
Improved low temp	GF	70	70%	no	45	Improved fluids resistance
	GLT	90	64%	no	30	TR-10 = −30 °C. Used to meet MIL-R-83485 & M2D401-A2. Best low temp sealing.
	GFLT	75	67%	no	35	TR-10 = −24 °C. Used to meet Ford M2D401A3, A5. Best combination of fuel resistance and low temp sealing
Improved fluid resistance	*Extreme* ETP	90	67%	no	45	Broadest fluids resistance of any Viton® – no VF$_2$ in polymer backbone
Viton® curatives	VC 20 VC 30 VC 50					Accelerator in Viton® binder (33%) Bisphenol AF in Viton® binder (50%) Combination of accelerator with Bisphenol AF – melts at 80 °C

8.15   Silicone Elastomers

James R. Halladay

8.15.1   Introduction

Silicone elastomers can be divided into three broad classifications; polydimethylsiloxanes (abbreviated as MQ or VMQ), polydimethylsiloxanes with phenyl substituents (PMQ or PVMQ) and polydimethylsiloxanes with 1,1,1 trifluoropropyl substituents (FVMQ, better known as fluorosilicone). The "V" in the abbreviation refers to the small amount of vinylmethyl siloxane that is generally included in the polymerization process to add cure sites.

   Silicone rubber is rarely formulated directly from the raw gum, except by silicone gum manufacturers. Generally, a reinforced gum (gum with reinforcing fillers already incorporated) or a silicone base (completely formulated except for color and cure) is the starting point for makers of molded silicone rubber products. Formulating from a gum requires specialized manufacturing equipment such as a dough mixer, and the mixing cycles are measured in hours rather than in minutes.

8.15.2   Selection

Bases can be obtained in various ranges, typically imparting a cured 30 to 80 Shore A durometer hardness. Bases are generally preformulated for various properties or applications, i.e. general purpose, no post cure, high strength, high tear, or low compression set. Choosing the right polymer is an important step in successful formulating. The most common silicone compounds are made with the VMQ gums and these should be the first choice unless use requirements dictate otherwise. Linear polymers containing phenyl-methyl or diphenyl substituents (PVMQ gums) have enhanced resistance to stiffening at low temperatures. While the VMQ gums crystallize at temperatures below $-45\,°C$, the introduction of 5 to 7 mol% phenyl groups inhibits crystallization and extends the useful temperature range on down to $-90\,°C$. PVMQ gums also exhibit significantly improved resistance to radiation and somewhat improved heat resistance as compared to VMQ gums. The trade-off is in significantly higher cost and reduced oil resistance.

   Poly(trifluoropropyl)methylsiloxane or fluorosilicone is used when maximum resistance to oils and fuels is required. Fluorosilicone is much more expensive than VMQ and is somewhat less heat resistant. Any combination of the three types of silicone polymers can be blended together, but it is generally difficult to impossible to incorporate silicone into blends with organic elastomers or *vice versa*. Some liquid organic polymers such as 1,2 polybutadiene can be added to silicone rubber in small quantities, although they generally degrade the heat resistance of the elastomer.

8.15.3   Fillers

Extending fillers are generally used to reduce the cost of the compounded silicone. These are either non-reinforcing or semi-reinforcing and include calcium carbonate, barium sulfate, magnesium silicate, aluminum silicate, diatomaceous earth, ground quartz, and clay or kaolin. The most commonly used extending fillers include diatomaceous earth, which reduces tackiness in some compounds, and ground quartz, which can be used in large quantities with minimal loss in heat resistance. Ground quartz is quite detrimental to fatigue resistance, however.

Reinforcing fillers must be used in silicone elastomers because the tensile strength of an unreinforced, cured silicone gum is on the order of 0.15 MPa. When starting with a reinforced gum or a silicone base, reinforcing fillers have already been incorporated to an extent necessary to give the compound adequate strength. Further filler can be added to increase the modulus or hardness of the finished compound. The most commonly used reinforcing fillers are the pyrogenic or fume-process silicas. These silicas give the strongest cured elastomers but they have a pronounced tendency toward crepe hardening. They also provide the best moisture resistance, electrical resistance, and transparency. Untreated fumed silica can be added in only small quantities before crepe hardening or structuring makes the compound unprocessable. Hydrophobic-treated silica or an antistructuring agent should be added *in-situ* to reduce the crepe hardening phenomena. Precipitated silicas have a lower tendency to crepe harden and superior heat aging properties at temperatures above 200 °C, but are inferior in strength and compression set. High structure carbon blacks, particularly acetylene blacks, can be added to make conductive silicone elastomers.

8.15.4   Antistructuring Agents

Antistructuring agents or processing aids include any material that inhibits or eliminates the polymer filler interactions responsible for structuring or crepe hardening. Generally, these materials preferentially adsorb on the surface of the silica and tie up the reactive sites which would otherwise react with the silicone polymer. Effective antistructuring agents contain reactive groups (generally silanol) which promote strong adsorption with the silica surface. Two very effective materials are diphenylsilane diol and hexamethyldisilazane (HMDS). The need for an antistructuring agent is reduced or eliminated when using surface-treated hydrophobic silica. Surface-treated silicas that have been pre-reacted with HMDS or with dimethyldichlorosilane are commercially available.

8.15.5   Heat Stabilizers

Traditional antioxidants are not used in silicone elastomers, but heat stabilizers are generally incorporated. One of the best and cheapest is red iron oxide (1 to 2 phr). Other important stabilizers are barium zirconate (4 phr), which is especially useful in white or colored stocks, and thermal black, which is used at 0.1 to 0.5 phr. Most silicone bases already contain heat stabilizers and do not require further addition.

8.15.6   Peroxide Cures

Polymers that do not contain vinyl must be cured with peroxides, which are capable of hydrogen abstraction. Aryl peroxides such as benzoyl peroxide or dichlorobenzoyl peroxide are the most common examples of hydrogen abstracting peroxides. With vinyl monomer incorporated into the silicone polymer, alkyl and aralkyl peroxides may be used and are preferred for many applications such as seals because compression set is generally better. However, tear resistance is often poorer than when aryl peroxides are used.

The most commonly used cures for the high temperature vulcanization (HTV) of silicone involve free-radical abstraction, coupling, and addition reactions with the methyl or vinyl groups in the polymer chain. The cure reaction is generally initiated with organic peroxides and heat. Postvulcanization bakes or postcures do not complete the curing process, but they are often recommended for peroxide-cured silicones to rid the cured elastomer of low molecular weight silicone species and peroxide decomposition products. Peroxide decomposition products, particularly those from peroxides based on carboxylic acids which leave acid residues, can cause reversion of the elastomer over time.

The choice of peroxide is a very important decision. Some peroxides permit pressureless vulcanization without porosity. Others permit the use of carbon black as a filler. Because the decomposition of different peroxides varies with temperature, the choice of peroxide dictates to some degree the cure temperature range.

Diaryl peroxides have the lowest decomposition temperatures and the fastest rates of decomposition of the commonly used peroxides. They typically work at temperatures under 125 °C. Dichlorobenzoyl peroxide has both a very low decomposition temperature and decomposition products with low vapor pressures, which makes it ideal for curing without pressure, such as in continuous hot air. Diaryl peroxides cannot be used with compounds containing carbon black because the black interferes with the curing reaction. Dialkyl peroxides require higher curing temperatures (140 to 180 °C) and, because the decomposition products are volatile at curing temperatures, compounds must be cured under pressure to avoid porosity. They are particularly suitable for thick cross-section parts that must be cured slowly. Diaralkyl peroxides, such as dicumyl peroxide, are similar to the dialkyl peroxides in properties, although they tend to have a strong smell, especially if the post cure is omitted.

8.15.7   Platinum Cures

Another important cure mechanism involves hydrosilylation with platinum catalysts. Elastomers cured with platinum catalyzed addition cures show exceptional toughness and tensile strength, a tight surface cure, and non-yellowing translucency",4>translucency. The platinum-cured silicones require postcuring to complete the crosslinking process. One of the downsides to this type of cure system is it results in a much shorter shelf life of the catalyzed compound than when peroxides are used. Platinum cures are quite sensitive to cure inhibition from contamination with trace quantities of certain chemicals, such as sulfur or amines, commonly used in organic elastomers.

8.15.8   RTV Cures

RTV silicone rubber is crosslinked by either a condensation or an addition reaction. In the condensation reaction, a catalyst, typically dibutyl tin dilaurate or tin octoate, causes the terminal silanol (Si-OH) of a silanol-terminated polydimethyl siloxane to react with small quantities of multifunctional silanes. Addition reaction involves the addition of polyfunctional silicon hydride (Si-H) to unsaturated groups (usually vinyl groups) in the polysiloxane chain through a hydrosilylation reaction. This reaction is catalyzed using a platinum complex such as chloroplatinic acid.

9   Polyurethane Elastomers

Ronald W. Fuest

9.1   Introduction

Otto Bayer and his colleagues laid the foundation of present-day polyurethane chemistry and technology in the laboratories of I.G. Farbenindustrie in Germany in the late 1930s. Since that time, all types of polyurethanes, including high performance polyurethane elastomers, have continued to grow in many divergent areas to become the materials of choice in demanding applications.

The U.S. consumption of polyurethane resins grew to 2 million metric tons (4.4 billion pounds) in 1996. Of this, only a small fraction, approximately 72 thousand tons (159 million pounds), or less than 3.5%, can be classified as high performance elastomers used in rubber-like applications. The remainder of the 2 million tons of polyurethane chemicals is consumed in foam, coatings, RIM, shoe soling, thermoplastics, and adhesive and sealant applications [1]. Discussion of all of these types of polyurethane materials is beyond the scope of this chapter. Our focus is on liquid cast elastomers, with a brief discussion of thermoplastic polyurethane elastomers (TPU's) and millable gums.

While some description of the chemistry and physical properties of castable polyurethane elastomers is provided, detailed chemical structures and quantitative comparisons of physical properties are found in many reference publications. The purpose of this chapter is to provide an overview of the general processing and physical characteristics of castable polyurethane elastomers that distinguish them from other elastomers, and to encourage investigation into their application possibilities.

Applications for high performance polyurethane elastomers range from the familiar in-line roller skate wheels to heavy duty mining equipment. There are also many types of rolls, timing belts, industrial wheels, and countless miscellaneous machine parts made from cast polyurethanes. Polyurethane elastomers, sometimes simply called "urethane elastomers," are often chosen over many other rubbers because of their superior abrasion resistance, load-bearing ability, cut and tear resistance, and environmental resistance.

9.2   Polyurethane Chemistry and Morphology

With most diene elastomers (SBR, nitrile, natural rubber, etc.), desired elastomer properties are achieved by extensive compounding of a base stock material. A typical SBR compound may contain ten or more ingredients. With castable urethanes, compounding ingredients such as fillers, plasticizers and accelerators are generally not used except in

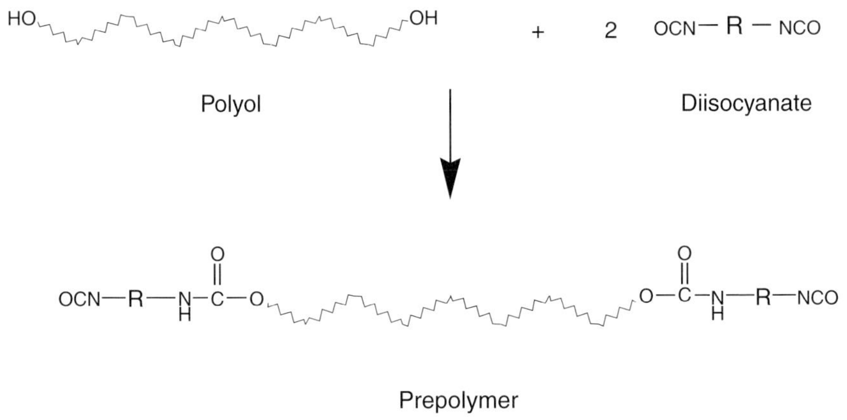

Figure 9.1   Formation of a polyurethane prepolymer.

special cases. When different properties are required, the usual course is to change to a different prepolymer and/or curative. A description of the chemical makeup and molecular structure of polyurethane elastomers answers some questions about why these compounding differences exist.

The prepolymer consists of a flexible molecular polyol chain "spine" or "soft segment" with reactive isocyanate groups on each end. This chain is produced by allowing the polyol to react with a diisocyanate according to the scheme shown in Fig. 9.1. It is called a "prepolymer" because it has a molecular weight of only a few thousand, but contains much of the "blueprint" of the final structure that the elastomer has when the polymerization reaction is completed by chain-extension (Fig. 9.2).

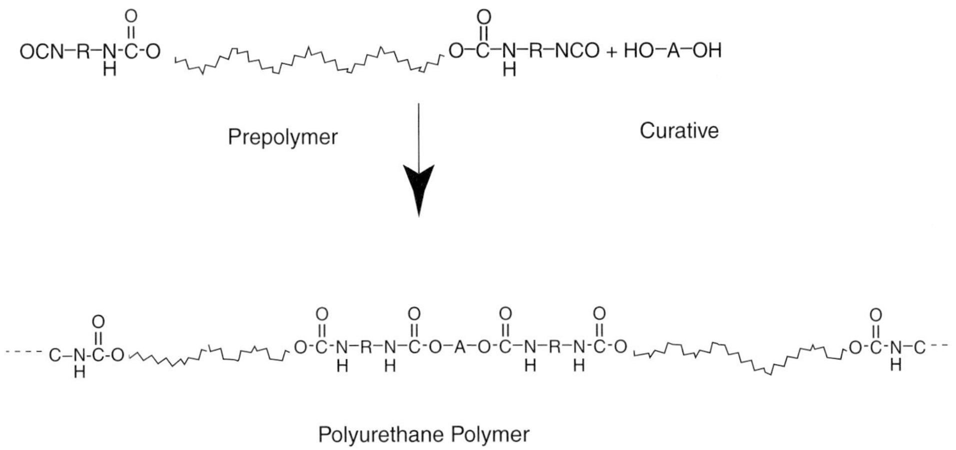

Figure 9.2   Chain-extension of a polyurethane polymer

The terms "chain extender" and "curative" are often used interchangeably in the cast polyurethane industry. Other terms, such as, "crosslinker" or even "catalyst," are sometimes used in this context, as well. "Crosslinker," while not totally incorrect, does not give an accurate description of the chemical reaction because often, little or no true crosslinking between polymer chains occurs. The use of the term, "catalyst," to refer to the curative is inaccurate and misleading. Catalysts are used in cast polyurethane processing to speed up the cure rate and do not become part of the structure of the final elastomer. Throughout this discussion, we use the term, "curative."

There is also the so-called "one-shot" approach to polyurethane elastomers. In this method, all of the components – polyol, diisocyanate and curative – are mixed together at the same time. This generally results in structures with a more random arrangement and more variability from batch to batch. It also requires the handling of pure isocyanates, which in the case of TDI (tolylene diisocyanate), especially, can be extremely hazardous.

Isocyanate groups are extremely reactive, and combine readily with many materials, especially those containing amino and hydroxyl groups (including water – more about water later). If the curative contains an amino or hydroxyl group on each end in the same way that the prepolymer has an isocyanate group on each end, the prepolymer and the curative can "link up" to produce molecules of very high molecular weight with high performance elastomeric properties.

Another approach is the "quasi" prepolymer approach. This is basically a half step between the true prepolymer and one-shot technologies. In this case, the diisocyanate is partly reacted with the polyol, and the remainder of the polyol is added along with the curative during processing. While both of these technologies have a place in the urethane elastomer industry, the true prepolymer approach is regarded as yielding the highest performance elastomers as well as the best reproducibility of properties. Additionally, prepolymers have the advantage of potential storage life of two years or more in unopened containers kept under normal conditions.

As noted in other chapters, the elastomeric properties in most rubbers are the result of crosslinking from vulcanization. This crosslinking allows the flexible segments to extend when the elastomer is stretched, but prevents the molecules from moving too far out of position. The material has "memory" and reverts to its original dimensions when the force is removed. In castable urethanes, there is relatively little crosslinking, and the elastomeric properties are the result of the association or crystallization of the so-called "hard blocks." For this reason, polyurethane elastomers are often referred to as "segmented block polymers." Under certain circumstances, the flexible spine can slowly crystallize as well, but this is usually not a desirable phenomenon, because it increases hardness and reduces flexibility of the elastomer over a period of months or years. Well-designed prepolymers are constructed to avoid soft-segment crystallization. Figure 9.3 is a conceptual drawing of a polyurethane elastomer in the relaxed and stretched states [2].

What is the source of these hard blocks? Earlier, we discussed the terminal isocyanate end groups on the prepolymer, and the terminal hydroxyl or amine groups on the chain extenders. The reaction of the prepolymer with a chain extender forms urethane links (in the case of a diol curative) or urea links (in the case of a diamine curative) in the chain. These highly polar urethane or urea groups have strong physical interactions, including hydrogen bonding, and thus, can associate in crystalline groups to serve the same function

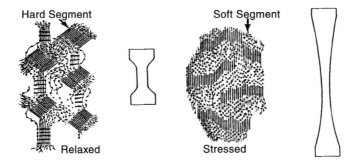

Figure 9.3   Polyurethane elastomer morphology.

as the crosslinks in a vulcanized material. Therefore, the soft segments allow flexibility and stretch, and the hard segments provide memory and contribute to the hardness and several other properties of the polyurethane elastomer. Although chemical crosslinking is not the primary force holding polyurethane elastomers together, some crosslinking is often introduced by curing with a slight ($\sim$5%) deficiency of curative. This allows the remaining isocyanate groups to react with already formed polymer chains to form biuret or allophanate linkages. Another way of introducing small amounts of crosslinking is by using a trifunctional curative in combination with the diamine or diol curative, which provides branching in the growing polymer chains.

   The soft segment can be of several types. One of the most common is the polyether type. Good hydrolysis resistance, high resilience, and good low temperature properties are the chief characteristics of polyethers. The other major class of soft segment materials is the polyesters. Higher cut, tear, and abrasion resistance and better oil and solvent resistance characterize the polyesters. The two most common diisocyanates are TDI and MDI (4,4'-diphenylmethane diisocyanate). Each of these diisocyanates can be combined with any polyol. With the various building blocks available, one can make hundreds of combinations of polyol backbones with various isocyanate/curative hard segments. Because of this, there are many trade-offs that can be made in choosing polyurethane elastomers for various applications. Each of these combinations has a unique set of properties, and one of them may be just the right solution to a specific engineering problem.

9.3   Polyurethane Products

Several of the most common polyurethane products are listed below:

- Castable elastomers
- Rigid and flexible foams
- Fibers and fabrics
- Adhesives
- Sealants

- Thermoplastics
- Millable gums
- Coatings

Castable elastomers are the materials of primary importance in this discussion. Other important polyurethane applications include both rigid and flexible foams. Rigid foams are used for insulation, while flexible foams are used in upholstery and cushioning. As indicated earlier, these are both very large markets. The familiar Spandex stretch fabrics are also based on polyurethane chemistry.

Many types of adhesives are also formulated from urethanes. Sealants based on polyurethane chemistry are widely used for applications such as sealing automobile windows. Thermoplastic urethanes are materials that are fully reacted and processed by melting and extrusion or injection molding.

Millable gums are urethane elastomers that are handled similarly to conventional rubbers, that is, they are compounded with fillers and other ingredients, processed on rubber machinery, and cured by vulcanization processes with peroxides or, in some cases, with sulfur. In this way, processors can achieve many of the unique properties of urethanes by the same methods and equipment used to process conventional rubbers.

Coatings, also widely based on polyurethane materials, cover a broad range: from varnishes and paints used on furniture and flooring to the sprayable elastomeric coatings for high abrasion or corrosion resistance. The latter not only protect surfaces from weather, but also provide tough, durable covers for sea buoys, truck beds, hoppers, bins and chutes, or added service life for costly mining machinery.

9.4   Cast Polyurethane Processing Overview

Cast polyurethane elastomers have become materials of choice in many applications because of two important attributes: the outstanding physical properties briefly mentioned above and versatile liquid processing. To discuss the second attribute first, the liquid "castability" of these materials means high performance articles can be manufactured with minimal capital investment in plant, equipment, and tooling. Most casting operations are carried out at relatively low temperatures, e.g., 25 to 120 °C (77 to 248 °F) and ambient or low pressures. Molds may be made inexpensively from many types of materials.

As shown in Fig. 9.4, the processing of cast polyurethanes consists of mixing two or more liquids, pouring the mixture into a mold, curing to a state where the object is self-supporting and can be demolded, then completing the cure during a post-cure operation. Finishing operations, such as grinding and lathe tooling, if needed, are similar to those performed on other elastomers. Processing operations can be done by handbatching methods to produce batches ranging from about 300 g (10.6 oz.) to 14 kg (30 lb.) or more, or by automatic, meter-mix machines. These machines can dispense materials continuously at rates ranging from less than 100 g (3.5 oz.) to more than 70 kg (150 lb.) per minute.

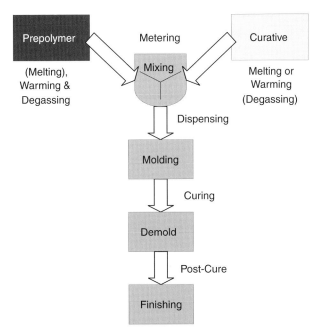

Figure 9.4   Polyurethane prepolymer processing.

Before the materials can be mixed and cast however, the prepolymer and chain-extender (curative) must be prepared. Each of the two materials (if solid) are melted and raised to the appropriate processing temperature, and then degassed under vacuum Vacuum degassing is critical because prepolymers can contain large amounts of dissolved air and carbon dioxide which, if not removed, cause bubbles in the finished casting. Often, the curative side does not require degassing because of the viscosity and gas solubility characteristics of many curatives. If handbatching is used, a second degassing is required to remove air entrained during mixing of the prepolymer and curative.

Because the curing, or chain-extension, step is a stoichiometric chemical reaction, calculation and close control of the ratio of the prepolymer and curative is necessary. In other words, if the amount of curative is not correct by as little as 2% (+ or −), there may be a significant effect on the physical properties and performance of the resulting elastomer. The mixed material is then dispensed into a mold or applied to some surface within its pot life. Pot life is the period of time between mixing the components and when the mixture becomes too viscous to be poured. Depending upon the particular system, this time may range from several seconds to several hours.

The list of available processing techniques is extensive, allowing the fabrication of intricate small articles weighing less than a gram (0.04 ounce) to massive items containing 4,000 kg (8800 lb.) or more of elastomer. Several of these techniques are described in a later section. Some of them are familiar to processors of gum rubber stocks; others are specific to liquids.

After it is cured to a self-supporting state in the mold, the part is removed from the mold and allowed to complete the cure reaction (''post-cure'') in an oven. The mold returns to the system to repeat the cycle. With some articles, such as roller skate wheels, demolding occurs in as little as 3 to 5 minutes. With other parts, such as large printing rolls, overnight oven curing in the mold is necessary.

A typical post-cure may take 16 hours in a 100 °C (212 °F) oven. Depending on the needs of the processor, the nature of particular polyurethane system used, and the end-use application, this schedule can vary considerably. In some cases, several days at room temperature, instead of an oven cure, are required to reach adequate properties. Before deviating from the manufacturer's recommendations however, the effect on properties of any modified cure schedule should be investigated [3].

Finishing operations may range from simple trimming of mold flash to elaborate machining to produce precision parts. When only flash trimming is required, finishing is often easiest just before or after demolding, while the material is still ''green'' and easily cut.

Although the process is simple in principle, it does require close attention to detail and involves handling toxic chemicals. There are many custom molders of high-performance polyurethane elastomers worldwide, and in many cases, the new user of polyurethane elastomers might consider sub-contracting a project to such a shop, thereby benefiting from the experience, established techniques, and plant equipment available.

9.5   Molding Methods

Because castable polyurethanes are processed as mobile liquids, many molding techniques may be used. As one might imagine, almost any process where a liquid can be held for a period of time in the shape of a desired final part is a potential molding method. Some of the more common ones are described below.

9.5.1   Open Casting

Open casting is the most common and usually the most economical method because there is little equipment needed other than an oven or hot table to maintain processing temperature. Additionally, there is no pressure applied to the mold. Molds can be made from many materials, including polished steel, aluminum, fiberglass/resin composites, silicone rubber, hard plastics, and castable polyurethane itself. In short, almost any material that can withstand modest molding temperatures of 100 to 120 °C (212 to 248 °F), is not porous, and does not retain moisture, can be considered as a potential mold material.

9.5.2   Centrifugal Molding

This technique consists of using centrifugal force to move the material into place and hold it there until sufficient cure has occurred. It is an excellent method for lining metal, fiberglass or other types of rigid pipe with an abrasion-resistant polyurethane interior surface. The process consists of introducing the liquid material into a pipe that has been prepared with a suitable adhesive and spinning it at a speed sufficient to distribute the material uniformly on its walls. Spinning continues until the material is cured enough to stay in place, usually 30 to 60 minutes.

9.5.3   Vacuum Casting

Vacuum casting is used when air entrapment is a problem, especially where intricate detail or undercuts in the mold make air removal difficult. Another important use is when the casting contains fiber reinforcement or fine wire windings. The entire mold is placed in a chamber, which is then evacuated. The material is dispensed from the outside directly from a meter-mix machine or from a supply vessel containing a pre-mixed batch. Degassing is accomplished during the molding cycle.

9.5.4   Compression Molding

This technique is borrowed from conventional rubber molding. Compression molding is used to make parts to close tolerances that must have finished surfaces on all sides. Sometimes prototype parts or small production runs are made from polyurethane elastomers using existing rubber molds with this method. The technique consists of pouring the mixed material into the mold and allowing it to gel to a high viscosity state, such that internal pressure is generated in the mass when the press is closed. With a little experience, an operator can learn when to close the press to get high quality parts every time. Disadvantages of this method are that it requires a press with heated platens; it does not lend itself to continuous processing because of the press cycle time; and molds must be of heavy metal construction to withstand the pressures generated.

9.5.5   Transfer Molding

This is another method borrowed from rubber processing. Transfer molding can be used for volume production of small precision parts. It is a variation on compression molding where, instead of being placed directly in the mold cavity, the mixed material is deposited in a cylindrical transfer cavity and allowed to thicken. Pressure is then applied to the cavity through a piston, and the viscous material is forced through sprues and runners into a single mold or series of mold cavities. Many of the same considerations apply here as in compression molding.

9.5.6   Liquid Injection Molding (LIM)

In this technique, liquid material is forced under moderate pressure into the mold directly from a meter-mix machine or from a pressure pot. The mold is usually filled from the bottom and the distance the mix flows depends upon the pot life of the material and the rate of fill. This method is particularly useful for making thin-walled tubing, rolls, and similar parts, because the air in the mold is displaced ahead of the liquid. The materials used and properties obtained by this process are the same as in other casting methods. In contrast, reaction injection molding (RIM) is a high-pressure, high-speed operation using one-shot chemistry. RIM has found considerable use in automotive fascia and other high-volume, medium performance applications.

9.5.7   Spraying

High quality polyurethane elastomers may be applied to surfaces requiring abrasion resistance that cannot be conveniently coated by other casting methods. Such applications include hoppers, bins, and chutes used in the mining industry and large buoys for marine applications.

9.5.8   Moldless Rotational Casting

One of the newest techniques for making rolls or similar articles is the rotational casting [4, 5], or Ribbon Flow* process. This method is a combination of special polyurethane systems and mechanical processing for producing high performance polyurethane rolls or other cylindrical items without the need for molds. In this process, short pot life (15 to 60 s) urethane systems are dispensed from a meter-mix machine directly onto a rotating roll core or other cylindrical object. All operations are conducted at or near ambient temperatures, avoiding the need to heat materials or metal cores. Roll covers in hardnesses of 50 A to 70 D durometer can be made. This process has the advantages of allowing rapid turn-around of customers' cores; requiring less overbuild and waste of materials, energy savings, and decreasing set-up time for large rolls. The process requires investment in a meter-mix casting machine and a lathe or similar turning device. In addition, the process is usually managed by a computer, which controls various key parameters, including rotational speed, dispensing nozzle traverse speed, height and angle, and machine output.

9.6   How to Select a Polyurethane Elastomer

It was mentioned earlier that many possible combinations of polyols and isocyanates, as well as many curatives or chain extenders, can be used to make polyurethane elastomers.

---

* Ribbon Flow is a registered trademark of *Crompton Corporation*.

Table 9.1   Basis of Selection of Polyurethane Elastomer for a Specific Application

Properties needed for the job	Processing characteristics
Pot life	–
Viscosity	–
Ratio control	–
Demold time	–
Process temperature required	–

With this wide array of materials available to the molder and design engineer, how should we choose a specific polyurethane elastomer for a particular application? Table 9.1 provides some direction.

There are two major considerations. First, what is required for the job in terms of physical properties and environmental resistance? Second, what are the processing characteristics of the polyurethane system chosen? The importance of the processing characteristics should not be overlooked. These include pot life, viscosity, ratio control, demold time (how many times one must turn each mold per day to have a cost-effective production system), and the process temperature required. Molding with some systems is feasible only with meter-mix equipment, while others can be easily processed by handbatching techniques. What controls the properties of these polyurethane elastomers? In part, properties are controlled by the chemical nature of the material and by the way it is handled. Table 9.2 lists factors controlling properties.

9.6.1   Types of Prepolymers

As we discussed in Section 9.2, polyurethane prepolymers consist of two major chemical structures. One is the diisocyanate. Most of today's commercial materials are based either on MDI or TDI. A recent improvement is the introduction of low free TDI monomer prepolymers. Traditionally, TDI prepolymers, depending on the type, contained up to 3% free TDI. Where ventilation and air extraction are inadequate, this free monomer can constitute a health hazard, because significant amounts of the TDI monomer can be evolved into the plant air during processing. Low free TDI prepolymers, with less than 0.1% free TDI, mean that ventilation and engineering requirements are significantly easier for processors to meet. In addition, low free TDI prepolymers usually have significantly lower process viscosity and better physical properties than their higher free TDI counterparts [6, 7]. Many processors purchase low free TDI prepolymers for the last two reasons alone. MDI monomer has much lower volatility than TDI; therefore little of the monomer escapes into the atmosphere during processing. For this reason, there has not been as much emphasis on lowering the free monomer content of MDI prepolymers. Each of these diisocyanates gives different properties to the final prepolymer; each requires different curing systems; and, in many cases, different processing conditions as well.

There are several other diisocyanates used in the casting industry, such as the aliphatics, the recently commercialized PPDI (para-phenylenediisocyanate), and NDI

Table 9.2   What Controls Properties?

Type of prepolymer

Isocyanate type
  MDI
  TDI
  Other (PPDI, aliphatic, etc.)

Polyol type
  PTMEG premium polyether
  PPG low-cost polyether
  Polyester
  Other (polycaprolactone, etc.)

Type of curative
  Diamine
  Diol
  Triol

Processing conditions
  Curative ratio
  Temperatures

Additives
  Plasticizers
  Fillers
  Protectants
  Colorants

(naphthylene diisocyanate). However, these materials constitute a relatively minor segment of the industry or, in the case of NDI, require special handling and processing and are not a subject of this discussion. PPDI prepolymers do merit special mention however, because they are the most rapidly growing class of polyurethane elastomers on the market.

One of the characteristics of PPDI materials is their outstanding dynamic performance. Their extremely low hysteresis properties lead to low heat build-up, even under high speed and load conditions. This makes them the materials of choice for amusement park ride wheels, flexible couplings, and extreme-service industrial wheels. The other major attributes of PPDI prepolymers are hydrolytic stability and environmental heat resistance. This makes them attractive for use in hydraulic seals and hydrocyclones under punishing conditions. With the recent introduction of low free-monomer PPDI prepolymers, they have become more "user-friendly" and promise to push the envelope of polyurethane application further despite their relatively high cost (3 to 4 times that of TDI and MDI prepolymers) [8].

The other component of a prepolymer is the polyol. There are three major types: PTMEG (polytetramethyleneether glycol), the so-called premium type of polyether spine; PPG (polypropylene glycol), a lower-cost polyether type; and a variety of adipic acid-based polyesters. Again, there are other polyols, such as polycaprolactones, which are increasing in importance in cast polyurethane elastomers. Polycaprolactones often give a good balance of properties intermediate between those of polyethers and adipate

Table 9.3   Major Curatives for Polyurethane Prepolymers

Curative	Applications
MBCA (methylene-bis-orthochloroaniline)	Most common for TDI
Ethacure 300 curative	Liquid MBCA alternative for TDI
Caytur 21 and 31	Heat activated for TDI and MDI
Lonzacure MCDEA	High performance for TDI
Vibracure A-157	MBCA alternative
TMP (trimethylolpropane)	Low hardness for TDI, also used with MDI
TIPA (trisopropanolamine)	Low hardness with TDI
1,4-butanediol	Most common for MDI
HQEE	Higher hardness with MDI
HER	Higher hardness with MDI
Vibracure A 120, A 122 and A 125	Low hardness with MDI

Caytur 21 and 31 (sodium chloride complex of methylene dianiline) are registered trademarks of Crompton Corp.
Lonzacure MCDEA (4,4′-methylene-bis-(3-chloro-2,6-diethylaniline) Lonza Inc.
Vibracure A 157 (trimethylene glycol di-*p*-aminobenzoate) is a registered trademark of Crompton Corp.
Vibracure A 120, A 122 and A 125 (high molecular weight polyol compositions) are registered trademarks of Crompton Corp.

polyesters. Additionally there are castor oil-based cast elastomers for use in chemically resistant and electrical applications, and polybutadiene-based polyurethanes for other special purposes. While the latter two spines have useful special properties, their physical properties, especially tear, abrasion resistance, and load bearing, are not in the same range as the ethers and esters. Putting aside these special-purpose components, there are three major types of polyols and two major types of isocyanates to make six major classes of polyurethane prepolymers.

9.6.2   Types of Curatives

The second part of the system is the curative. As mentioned earlier, a polyurethane casting involves a chemical reaction. When one mixes the prepolymer and the curative, the polymerization reaction goes forward. Curatives make up from 6 to 30% of the total weight of the final elastomer and play a major role in determining the structure of the polymer molecule and its properties. Some of the more common curatives are shown in Table 9.3.

The most commonly used diamine curative with TDI prepolymers is MBCA (methylene-bis- orthochloroaniline), although it is considered a cancer-suspect material. Another material, ETHACURE* 300 Curative, is steadily gaining wider acceptance as a MBCA alternative. ETHACURE 300 is liquid at room temperature and has recently been extensively evaluated for toxicological factors. These studies concluded that because of its low toxicity and carcinogenic potential, combined with its physical properties (liquid state

* ETHACURE 300 Curative is a registered trademark of *Albemarle Corp.*

and low vapor pressure), the material has minimal risk to humans [9]. It yields properties generally similar to those obtained with MBCA.

Caytur 21 and 31 are complexes of MDA (methylene dianiline) with sodium chloride and do not react with prepolymers until a thermal unblocking temperature is reached. They are useful for coating fabrics to make conveyor belting because of their very long working life, but cure rapidly once the unblocking temperature (130 to 140 °C; 266 to 284 °F) is reached. Lonzacure MCDEA [10] gives improved properties compared to MBCA in terms of hydrolysis resistance and dynamic heat build-up, but is considered more difficult to process and is more expensive.

MDI-based prepolymers generally use diol curatives such as 1,4-butanediol, HQEE, and HER. HQEE usually gives higher hardness compared with 1,4-butanediol and is more difficult to work with because of its high melting point. HER yields similar properties to HQEE and is lower-melting. Vibracure A 120, A 122 and A 125 are polyether- and polyester-based polyols, respectively, that are often used in conjunction with other curatives to modify hardness and other properties. TMP and other triols are sometimes used in combination with diols in MDI polymers. One of the main uses for triol curatives is where TDI ester prepolymers are used to produce low hardness (25 to 55A durometer) rolls for the printing and metal coating industries. Many of these curatives can also be used with PPDI prepolymers. With them, some outstanding performance characteristics are achieved with both diamine and diol cures, and new effects are being discovered almost monthly.

9.6.3   Processing Conditions

The other factors that can affect properties of the final product involve processing conditions. Probably the most important condition is the curative ratio. The relative amounts of prepolymer and curative have to be determined by calculation and held within close tolerances to yield the desired physical properties. The amount of curative needed to exactly match the isocyanate content of the prepolymer is called ''100% theory'' or ''100% stoichiometry.'' If 5% less or 5% more of the curative is used, it is referred to as 95% or 105% theory (or 95% or 105% stoichiometry), respectively.

There are cases where it may be desirable to modify the curative ratio from an established point to intentionally maximize some particular property of the polyurethane casting at some sacrifice of another property. For example, a cure at high theory – or 100 to 105% stoichiometry instead of the more usual 95% – improves flex life of the polymer, but means the sacrifice of some performance in compression set. What is vital is that any modification of the curative ratio must be done under close control with proper knowledge of the effects on other critical physical properties. The effect of MBCA level on several physical properties of Adiprene* L 100 (a TDI/PTMEG prepolymer) is shown in Fig. 9.5.

The other conditions important to control are the temperatures of the prepolymer, curative, and molds. These controls are critical in terms of heat history of the material before it goes into the casting operation as well as temperature during cure and post-cure

---

* Adiprene L 100 is a registered trademark of *Crompton Corp.*

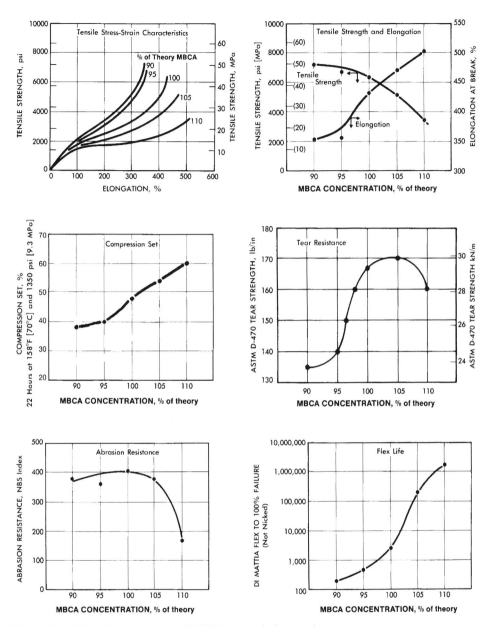

Figure 9.5   Effect of concentration of MBCA on physical properties.

of the parts. Heat history is the amount of time the prepolymer is held at elevated temperatures before mixing with the curative. When this is excessive, it causes damage to the prepolymer and affects the final properties of the elastomer. It is important to follow the manufacturer's recommendations regarding heat history and processing temperatures to obtain the best properties.

Another factor in polyurethane processing is the effect of water. Water can be a problem for polyurethanes in two ways. First, water reacts with the diisocyanate component of the prepolymer just as readily as the common diol curatives. This means that any moisture incursion into the prepolymer during storage and handling, as well as any water contamination in the curative, additives, or substrates reacts with the prepolymer. This generates carbon dioxide and causes the formation of gas bubbles in the casting. In extreme cases, this can even result in different physical properties in the final product even though the carbon dioxide is removed by strenuous vacuum degassing before casting. In humid areas, moisture exclusion can be a challenge, particularly when using MDI prepolymers in handbatch operations. With water-blown foams, water is added intentionally to the formulation to produce the necessary voids.

Although the second area where water can be a problem is not specifically a processing issue, it is important enough to be mentioned here. It occurs after the part is cured and put into service or is held in long storage. Water can react with the various linkages in the polymer chain to cause a breakdown in the structure by hydrolysis. Although all polyurethanes are subject to this process, susceptibility varies widely depending upon the particular polyurethane type and the hydrolytic environment. Some types (diol- or triol-cured MDI and PPDI polyethers) can withstand hot water and acidic or basic conditions without breaking down rapidly, while others (amine-cured TDI esters) deteriorate completely in a matter of weeks under the same conditions. Some types have been tested under a variety of exposures in tropical climates for ten years without measurable loss of properties (MBCA-cured TDI ethers) [11], while others lose most of their integrity under the same conditions in two years or less (MBCA-cured TDI esters).

9.6.4   Additives

Processors sometimes use additives to modify the physical properties of urethanes, especially when used for low durometer printing and coating rolls. With typical polyurethane systems, it is very difficult to decrease hardness to below 50 to 55A durometer without the use of a plasticizer. Benzoate esters, such as Benzoflex 9-88 SG[*] are effective and compatible with polyester-based systems, while phthalate esters, such as dioctyl phthalate (DOP), are most effective with polyethers. As with all materials used in cast polyurethane elastomers, dry grades of plasticizers are needed, and they must be protected from moisture during storage and handling.

Silica fillers are also often used in printing and coating rolls to modify processing and ink-transfer properties. Sometimes special fillers, namely, friction-reducing aids such as molybdenum disulfide, fluorocarbons (Teflon[**]) and special silicone oils, help improve wear and friction properties. On occasion, processors add protectants of several kinds, such as UV stabilizers and hydrolysis stabilizers. If possible, however, processors avoid using any additives at all.

The processor has great control over the properties of the final product by choosing the right combination of prepolymer, curative, and processing conditions. It is often

---

[*] Benzoflex 9-88 SG is a registered trademark of Velsicol Chemical Corp.
[**] Teflon is a registered trademark of DuPont Company, Inc.

possible to fine tune formulations by incorporating curative combinations and even mixtures of prepolymers.

9.7   Comparison of Polyurethanes with Other Elastomers

In comparing polyurethanes with other candidate elastomers for an application, it is useful to highlight some attributes. Table 9.4 lists the main advantages of cast polyurethane elastomers compared to the diene rubbers as a group.

As mentioned before, the main advantages of urethanes versus rubber are the higher abrasion resistance, greater cut and tear resistance, and higher load bearing ability. In addition, most cast urethanes have natural coloration ranging from completely clear to opaque white or amber. However, they accept a wide variety of pigments and dyes, so their final coloring ranges from black to brilliant fluorescent oranges, reds, or greens. This is especially useful in color-coding parts, such as in business machines, where rolls and belts are color-coded for quick identification of replacements. Even when colored black, there is so little pigment in the polyurethane, the compounds do not leave marks on concrete floors, boat hulls, or other surfaces. Because urethane compounds usually contain no additives that are not tied into the molecular structure, there can be no migration of materials out of the elastomer to deposit on other surfaces.

Diene rubbers are subject to ozone cracking, particularly around electrical equipment where ozone concentrations can be high. Polyurethanes have virtually no ozone-cracking problem. Polyurethane elastomers, particularly the polyester types, are also resistant to swelling and deterioration by oils, greases, and other non-polar solvents.

Most rubber compounds when formulated to have a 90 or 95A durometer hardness have lost many other physical properties as a result. However, polyurethane elastomers in the 85 to 95A durometer range are approaching the peak of their properties and perform extremely well at these hardnesses. Polyurethane elastomers are available up to about 75D

**Table 9.4   Advantages of Polyurethane
              Elastomers versus Diene
              Rubbers**

Abrasion resistance
Cut and tear resistance
Higher load bearing
Clarity; translucence
Non-marking, non-staining
Ozone resistance
Pourable; castable
Wider durometer range
Microorganism resistance
Oil and petroleum resistance
Low or high rebound
Versatility

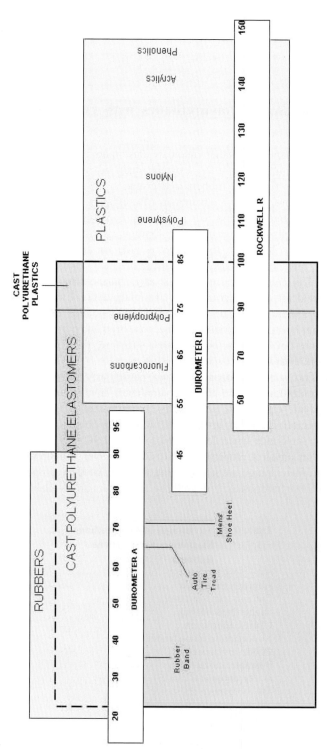

Figure 9.6   Hardness ranges of various elastomers and plastics.

hardness. Although they begin to resemble hard plastics in this range, they still have elastomeric properties. There are castable polyurethanes (Royalcast*) with a hardness as high as 85D, but their properties are more like nylon, polyacetal, epoxy, and other plastics; they can no longer be considered elastomers (Fig. 9.6).

Polyether-based polyurethane elastomers are particularly resistant to molds, fungi, and other microorganisms, making them useful in agricultural, sewage treatment, and tropical applications. While the ultraviolet resistance of polyurethanes in general is good, darkening usually occurs with outdoor exposure. Therefore, for continuous outdoor exposure, dark colors or pigments are often used, sometimes in combination with UV-stabilizer packages. Where the polyurethane must be light in color and little or no color change can be tolerated, aliphatic diisocyanate-based prepolymers with non-aromatic curatives often are the best choice.

Rebound, which relates to energy loss in mechanical cycling, is related to the hysteresis properties of the polyurethane elastomer. High hysteresis leads to development of heat in dynamic service, and it can range from very low to very high in polyurethane elastomers. This effect is discussed later in this section.

Finally, the versatility of cast polyurethane elastomers is unmatched by any other type of rubber. It is often possible to modify one property while holding others close to constant. For example, it is possible to formulate higher durometer elastomers with low rebound; lower durometer materials with high rebound; and *vice versa*. Similarly, it is possible to maximize flex life or tear resistance without sacrificing much compression set over a wide range of hardnesses. Comparison of the general properties of polyurethane elastomers with some common rubbers is shown in Table 9.5.

9.7.1   Limitations of Polyurethane Elastomers

So far, we have discussed only the advantages of polyurethanes compared to other materials. Naturally, as with anything, there are some disadvantages as well specifically related to:

- High temperature service
- Moist hot environments
- Certain chemical environments

Polyurethanes are not high temperature materials. Although cast polyurethane elastomers are thermoset polymers, there is a certain degree of thermoplasticity in their nature, and their properties tend to fall off at elevated temperatures. Generally, TDI and MDI urethanes are not useful under heavy service loads at temperatures above 105 to 110 °C (221 to 230 °F). The new PPDI-based materials promise to push that limit higher, however.

Another limitation is that all polyurethanes are subject to hydrolysis in the presence of moisture and elevated temperatures. The combination of the two factors can create a problem. While at lower temperatures, most polyurethane elastomers can withstand continual contact with water for many years, no polyurethane as yet can stand prolonged

---

* Royalcast is a registered trademark of *Crompton Corp.*

Table 9.5   General Comparison of Polyurethane Elastomers with various Rubbers

Property	Polyurethane	Nitrile	Neoprene	Natural	SBR	Butyl
Tensile strength (MPa)	20.7 to 65.5	13.8+/−	20.7+/−	20.7+/−	13.8+/−	13.8+/−
Durometer	5A to 85D	40 to 95A	40 to 95A	30 to 90A	40 to 90A	40 to 75A
Specific gravity	1.10 to 1.24	1.0	1.23	0.93	0.94	0.92
Tear resistance	Outstanding	Fair	Good	Good	Fair	Good
Abrasion resistance	Outstanding	Good	Excellent	Excellent	Good-excel.	Good
Compression set	Good	Good	Fair-good	Good	Good	Fair
Rebound	Very high to very low	Medium	High	Very high	Medium	Very low
Gas permeability	Fair-good	Fair	Low	Fair	Fair	Very low
Acid resistance	Fair-good	Good	Excellent	Fair-good	Fair-good	Excellent
Aliphatic hydrocarbons	Excellent	Excellent	Good	Poor	Poor	Poor
Aromatic hydrocarbons	Fair-good	Good	Fair	Poor	Poor	Poor
Oil and gas	Excellent	Excellent	Good	Poor	Poor	Poor
Resistance oxidation	Outstanding	Good	Excellent	Good	Good	Excellent
Resistance ozone	Outstanding	Fair	Excellent	Fair	Fair	Excellent
Low temperature resistance	Excellent	Good	Good	Excellent	Excellent	Good

contact with live steam. In between, there is a wide range of temperature and moisture conditions under which polyurethanes may or may not be suitable. Newer developments in polyurethane chemistry also show promise to push these limits further.

Lastly, there are certain chemical environments that are damaging to polyurethanes. Very strong acids and bases generally are detrimental, as are certain solvents, specifically aromatics such as toluene and ketones such as MEK and acetone; and esters such as ethyl acetate. This does not necessarily mean that polyurethanes cannot be used anywhere they might be in contact with these materials. Whether an application is acceptable requires determination of the frequency of contact and how much swell (dimensional change) can be tolerated.

9.8   Polyurethane Selection Guidelines

Following are some guidelines for selecting polyurethanes for demanding applications. Listed in Tables 9.6 and 9.7 are various physical and environmental resistance properties and the types of TDI and MDI prepolymers that have the greatest and the least likelihood of success in applications where these properties are important. This is, of course, only a general guideline. Many materials are not mentioned, such as aliphatic isocyanates for UV light resistance, or PPDI-based materials for outstanding dynamic and hydrolytic resistance properties. Additionally, there are probably many exceptions to these rules, but they can be useful as a starting point in considering materials for a given application.

Table 9.6   Guidelines for Selecting Polyurethane for Use in Demanding Applications

Property	Greatest	Least
Hardness	–	–
Elongation	–	–
Modulus	–	–
Tensile strength	Ester	Ether
Tear strength	Ester	Low cost ether
Compression set resistance	TDI	MDI
High rebound	MDI ether	TDI ester
Low temperature properties	MDI ether	TDI ester
High temperature properties	TDI	MDI
Abrasion resistance		
Sliding	Ester	Low cost ether
Impingement	MDI ether	Low cost ether
Low heat buildup	Ether	Ester
Hydrolysis resistance	MDI ether	TDI ester
Oil resistance	Ester	Ether
Heat aging	Ester	Low cost ether

Table 9.7   Guidelines for Selecting Polyurethane for Use in Demanding Applications

Property	Greatest	Least
Low duro formulation (<60A)	TDI ester	Ether
Formulation flexibility	MDI	TDI
Cost	Low cost TDI ether	MDI ether

The first property listed, and one that is of perhaps primary interest, is hardness. However, because all hardnesses can be obtained with all six main types of prepolymer systems, hardness is not a differentiator for them.

In terms of tensile strength, esters have the edge over ether compounds. However, tensile strength is rarely a key property in polyurethane applications. All types of polyurethanes can have high elongation. There is really no basis for choice there. The same is true of modulus; all types can have high or low modulus. Tear strength is an important property in many applications. Here too, the esters have the advantage over the ethers. The low cost ethers usually have the lowest tear strength.

In terms of compression set – often an important property in applications such as hydraulic seals – the amine-cured TDI compounds, both ethers and esters, tend to have better compression set resistance than diol-cured MDI compounds. So, one might choose a TDI system for the first test compound in an application requiring high compression set resistance.

Diol-cured MDI ethers have much higher rebound, or resilience, than amine-cured TDI esters. These examples illustrate the two opposite ends of the spectrum. With many other properties, differences usually are not dramatic.

MDI ethers have the lowest brittle points and best low temperature cracking or embrittlement resistance. Esters are generally not as suitable for low temperature properties. On the other hand, for high temperature properties, the TDI polymers tend to be better than the MDI polymers. A certain amount of thermoplasticity in the MDIs sometimes shows up in high temperature properties.

In terms of abrasion resistance, the type of abrasion to be encountered must be considered. For resistance to sliding abrasion, esters are generally the materials of choice. They perform better than ethers and much better than low-cost ethers. On the other hand, for impingement abrasion, where particles strike a surface at a high angle, the high resilience MDI ethers often perform much better than esters and also much better than low cost ethers. Therefore, the optimum material is determined by the abrasion mechanism occurring in the particular application being considered.

Dynamic heat buildup resulting from hysteresis is important in such applications as wheels and tires. Ethers are better than esters in terms of hysteretic heat buildup, although recent developments in low free TDI esters have narrowed the gap considerably. MDI ethers are much more resistant to hydrolysis than TDI ethers. Down at the bottom of the list in terms of this property are TDI esters. In this area, there is a significant difference between the two ends of the spectrum.

Esters have the edge in solvent resistance; ethers generally are not quite as oil- and solvent-resistant as esters. With respect to heat aging – that is, the permanent loss of properties from prolonged exposure to high temperatures – esters have the edge over ethers and as one might guess, low-cost ethers perform worst of all.

As for low durometer formulations for printing and coating rolls mentioned earlier (see Fig. 9.4), TDI esters are favored because they can tolerate a greater loading of plasticizers before significant reduction of physical properties occurs. Ethers generally do not have as high physical properties in low durometer, highly plasticized formulations.

MDI's have a greater range of formulation flexibility. To produce high durometer materials with TDI prepolymers, MBCA or another diamine type curative is necessary; however, with the MDIs, there are several diol curatives that can be used, such as 1,4-butanediol and HQEE. Many mixtures and combinations of these and other diol curatives are possible.

Throughout this list, low-cost ethers have appeared in the "least favored" column quite frequently. While their performance may not be as good as the other materials, very often there are applications where they are desirable. Cost performance is, of course, the reason to use low-cost TDI ethers. They are considerably less expensive than premium grade ethers or esters and are useful in parts that are not highly engineered, that is, in applications where the most outstanding properties of polyurethanes are not fully utilized and therefore, low-cost materials perform adequately.

MDI PTMEG ether prepolymers are listed above as the "least" desirable in terms of cost, which means that they are the most expensive. However, the true cost *in use* really depends upon the particular application because, although the MDI PTMEG ether prepolymers might be only a little more expensive than the TDI PTMEG ones, the curative cost must be factored in. Each case must be calculated individually: how much does the curative cost, what is the ratio used, and how does this contribute to total cost? Sometimes the most expensive material to buy may be the best material for the application because of its cost/performance characteristics. This is a consideration with both premium-grade and low-cost materials [12].

There are some specific applications (Table 9.8) where one type has advantages that are so outstanding as to make it dominate all other types of polyurethanes. For example, high quality roller skate wheels are practically all made from MDI ethers, mainly because of their high resilience. High resilience gives good speed performance and a smooth ride.

Table 9.8   Specific Applications

Application	Polyurethane type	Basis of choice
Roller skate wheels	MDI ether	High resilience
Printing rolls	TDI ester	Solvent resistance, good physical properties at low durometer
Oil pipeline pigs	Ester	Oil, abrasion resistance
Fork lift tires	TDI ether	Low heat buildup
Hammers	TDI ester	Tear resistance, low resilience
Sandblast curtains	MDI ether	High resilience, impingement abrasion resistance
Laundry equipment	MDI ether	Hydrolysis resistance
Paper mill rolls	TDI ether	Hydrolysis resistance, hardness stability

TDI esters are the materials of choice for printing rolls because of their high solvent resistance and good physical properties in low durometer formulations. For oil pipeline pigs (devices used for cleaning, inspecting, and maintaining pipelines), high abrasion resistance and high oil resistance are necessary to prevent wear from traveling through miles of pipe. Esters are used in this application because they combine oil resistance with high sliding abrasion resistance.

For forklift truck tires, TDI ethers long dominated the market because of their low heat buildup and high load bearing ability. More recently, high performance, low free TDI esters and cost-efficient MDI esters have taken over substantial market share because of improvements in dynamic performance. In applications where cut and tear requirements are severe because of the presence of scrap metal or similar materials, an ester might offer the best combination of load and speed performance and damage resistance and toughness. On the other hand, in areas where hydrolysis resistance is critical, such as in tropical environments or hot, humid buildings, an ether may be the best choice.

For hammers, TDI esters are favored because of the combination of high tear resistance and low resilience. Tear resistance is necessary to prevent deterioration of the face of the hammer, and low resilience causes the energy of the hammer to transfer into the object being struck, rather than allowing it to bounce back to the user.

Sandblast curtains require high resilience and abrasion resistance in the impingement mode. MDI ethers, because of their high resilience, allow the particles to bounce back without transferring much kinetic energy in the form of heat to the sandblast curtain.

MDI ethers are materials of choice for laundry equipment, such as agitators or pulsators for washing machines, because of their outstanding hydrolysis resistance. However, TDI ethers are preferable for paper mill rolls, because of their combination of good hydrolysis resistance and excellent hardness stability. In paper processing applications, it is important that the hardness and dynamic properties of the roll remain constant over a range of operating temperatures so that performance of the roll remains consistent.

9.8.1   Selecting a Polyurethane Elastomer for a New Application

To evaluate an application that may be appropriate for a polyurethane elastomer, the following basic steps are suggested:

- Decide which properties are of key importance – physical and environmental resistance
- Select polymer/curative systems that are likely candidates
- Consider engineering design principles
- Consult your suppliers for recommendations and further information
- Review your plant capabilities
- Run whatever preliminary tests are available
- Make prototype units of one or more candidate systems
- Field test in actual service, make comparisons, get approval from future customers
- Gear up for production

When one attempts to classify the uses for polyurethane elastomers, it is quickly seen that the applications for cast polyurethane elastomers are so diverse and varied that it is difficult to begin to categorize them. Some applications have been mentioned in this chapter but many others have not been discussed. Additionally, new ones seem to appear almost daily. Although some application categories are clearly defined – such as mining, rolls, oil and gas service – it turns out that the largest single category is always "miscellaneous."

For example, your motorcycle may have a cast polyurethane drive belt that has replaced the traditional chain drive. In your office machines, you may see a miniature version of the motorcycle belt driving the print head in an ink-jet printer. There are probably cast polyurethane rollers in your laser printer, chosen for long, trouble-free life. Visit a modern warehouse. The high-lift forklift trucks are riding on cast polyurethane wheels because of their toughness, load bearing ability, and non-marking behavior. Go to a hardware store and buy a cast polyurethane hammer designed to last for a generation or more. Try some of the new golf balls with tough, resilient cast polyurethane covers.

Take your family to a major amusement park and notice the polyurethane wheels on the newest, fastest, most exciting roller coasters and other rides. Put on a pair of good in-line skates and admire the smooth, effortless ride on the crystal-clear, water-white wheels. Visit a mining operation and see how many of the screens used to separate the crushed ores are made out of cast urethanes, where not too many years ago they were made from woven wire. Your favorite newspaper or magazine may have been printed using cast polyurethane rolls – and the metal used to make your new refrigerator or washing machine may have been painted with similar rolls. Ask someone involved in gas or petroleum pipeline maintenance how they manage to keep all those miles of pipe clean. Chances are that you will be told that polyurethane pigs do the job.

9.9   Millable Gums

As mentioned earlier, millable gums are polyurethanes in terms of chemical structure, but are designed to be compounded and processed in much the same way as other diene rubber compounds. They contain a degree of built-in unsaturation that allows vulcanization by sulfur or peroxides. Millable gums share many of the attributes of liquid castable polyurethane, such as abrasion resistance, cut and tear resistance, and oil and solvent resistance. Because of their similarity to other rubbers, they are readily used where various rubbers are currently compounded and molded, but where the properties of a polyurethane elastomer are needed. Millable gums find major applications in rolls, seals, gaskets, O-rings, and belts. Although their price is higher than for most other rubbers, the gums can often be blended with other stocks, such as SBR, to achieve an optimum cost/performance balance.

Millable gum polyurethanes are available in polyether and polyester types, and the same general rules regarding solvent resistance, hydrolytic stability, dynamic performance, etc., that apply to the liquid cast materials also apply here.

9.10   Thermoplastic Polyurethanes

Thermoplastic polyurethanes, or "TPUs," are similar in chemistry and properties to cast polyurethane elastomers, but are delivered to the molder as a fully reacted, complete polymer in the form of beads or pellets. Melting and injection molding or extrusion makes TPUs into finished articles. Most TPUs are based on MDI, but some are based on other isocyanates, including the new high-performance PPDI.

TPU processing lends itself well to the large volume production of small to medium sized parts, and for this reason finds wide utility in automotive applications. It is also widely used for extrusion coating of wire and cable insulation. While TPUs are convenient for many applications, they are limited in terms of property combinations and fine tuning that can be done in liquid cast prepolymer systems. Because TPUs must be melted to become a free flowing liquid, little or no crosslinking can be present. Earlier, we discussed the desirability of modifying the stoichiometry in cast systems to provide optimization of properties for specific applications. This option is also absent with TPU materials. Also, because TPU molding is a high-temperature, high-pressure process, molds must be machined from heavy-gauge steel, and are very expensive. Amortization of a mold costing $50,000 or more requires a large production run.

Chemical Terms and Tradenames

Ethacure 300 Curative: 3,5-dimethylthio-2,4(and 2,6)-toluenediamine isomers, (Albemarle Corp.)
Caytur 21 and 31: Sodium chloride complex of methylene dianiline (Crompton Corp.)
Lonzacure MCDEA: 4,4′-methylene-bis-(3-chloro-2, 6-diethylaniline) (Lonza Inc.)
HQEE: Hydroquinone di-(beta-hydroxyethyl) ether (Eastman Chemical Co.; Rheinchemie)
HER: Resorcivol di-(β-hydroxyethyl) ether (Indspec Chemical Co.)
Vibracure A 120, A 122 and A 125: High molecular weight polyol compositions (Crompton Corp.)
Benzoflex 9-88 SG: dipropylene glycol dibenzoate (Velsicol Chemical Co., Inc.)
Vibracure A 157: Trimethylene glycol di-p-aminobenzoate (Crompton Corp.)

References

1. *Chemical and Engineering News*, (August 4, 1997) p. 22
2. Smith, R. N., *An Introduction to the Chemistry of Polyurethane Elastomers*, Bayer Corporation, Pittsburgh, PA
3. Adiprene L 100 Product Bulletin, Publication ASP 1726C, (1995), Uniroyal Chemical Co., Inc., Middlebury, CT
4. Ruprecht, H. D., Recker, K., Grim, W., *Roll Covering by Rotational Casting with Fast Reacting Systems*, paper presented at SPI World Congress, September, 1991

5. Gajewski, V. G., Fuest, R. W., Singh, A., DelVecchio, J. A., *Adiprene Ribbon Flow: A New Technology for Polyurethane Roll Manufacture*, paper presented at PU China '98, Shanghai, April, 1998
6. Singh, A., Rosenberg, R. O., Chin, J., Gajewski, V. J., *Recent Developments in High Performance Castable Elastomers*, paper presented at UTECH Asia '95, Singapore, May 25, 1985
7. Clift, S. M., Clement, A. L., Quay, J. R., Dewhurst, J. E., *High Performance Polyurethane Prepolymers*, paper presented at UTECH '92, The Hague, Netherlands, March (1992)
8. Hardy, M. L., O'Malley, N. A., *A Toxicology and Exposure Comparison of DMTDA (Ethacure 300 Curative) with MOCA, MDA and TDA*, Albemarle Corp., Baton Rouge, LA, January, 1999
9. Singh, A., in *Advances in Urethane Science and Technology*, Vol. 13. Frisch, K. C and Klempner, D. (Ed.) (1996) Technomic, Lancaster, PA, pp. 112–139
10. McInnis, E. L., *Optimization of High Performance Elastomer Properties*, paper presented at Polyurethane Manufacturers Association, October, 1994
11. Adiprene/Vibrathane Bulletin, *Ten Year Weathering Study and 20 Year Hydrolytic Aging Study on Adiprene L 100* (1989) Uniroyal Chemical Co., Inc, Middlebury, CT 06749, USA
12. Palinkas, R. L., *Design of Castable PU Parts Requires Precise Materials Characterization*, Elastomerics (1991), pp. 24–27

General References

Saunders, J. H., Frisch, K. C., *Polyurethanes: Chemistry and Technology*, Vol. I and II (1962), Interscience, Division of John Wiley and Sons, New York
Ulrich, H., in *Kirk-Othmer: Encyclopedia of Chemical Technology*, Volume 23, Third Edition (1983), John Wiley & Sons, New York, pp. 576–608
Backus, J. K., in "Polyurethanes" in *Encyclopedia of Polymer Science and Engineering*, Second Edition, Vol. 13, John Wiley and Sons, New York (1989), pp 243–303
Hepburn, C., *Polyurethane Elastomers*, Elsevier Applied Science, (1992), London and New York
Schollenberger, C., in *Rubber Technology*, Morton, M. (Ed.) (1987), Van Nostrand Reinhold Co., New York

10   Thermoplastic Elastomers

Charles P. Rader

10.1   Introduction

A thermoplastic elastomer (TPE) is officially defined [1] as a member of "a diverse family of rubberlike materials that, unlike conventional vulcanized rubbers, can be processed and recycled like thermoplastic materials." All TPEs are rubbery, but not necessarily a "rubber" as defined by ASTM [1] and ISO [2]. Most softer TPEs (less than 90 Shore A hardness [3]) are true rubbers; the harder ones (greater than 90 Shore A or 38 Shore D) are generally not, but are somewhat similar to soft, impact-modified thermoplastics.

Each TPE has a melting point (T_m) above which it is fluid and suitable for fabrication, by the same processing methods and equipment as commonly used for rigid thermoplastics. Below its T_m a TPE is a soft, flexible, elastic material, often suitable for replacing a conventional thermoset rubber of comparable hardness and resistance to mechanical abuse and operating environment. The fusion/solidification process is reversible, so these unique materials are also thermoplastic and both process scrap and used articles can be recycled.

Over the past three decades, the literature on TPEs has grown progressively. A series of review articles [4,5,6], books [7,8] and symposia [9–15] have been devoted to this rapidly developing area.

10.2   Position in Spectrum of Polymeric Materials

Prior to the introduction of TPEs, a distinct division existed between rubbers and plastics. Rubbers are soft, flexible, and elastic; plastics are hard and rigid (with the exception of plasticized polyvinyl chloride, which is generally considered not to be a rubber). The number of commercial TPEs introduced over the past 30 years has blurred the distinction between rubber and plastics. A TPE is fabricated like a thermoplastic and can be recycled like many of them, but has properties and performance similar to rubbers [16].

Perhaps the best way to categorize TPEs among polymeric materials is by hardness. Figure 10.1 compares the hardness range of commercial TPEs to those of rigid thermoplastics and conventional thermoset rubbers. TPEs bridge the gap between thermoset rubbers and rigid thermoplastics. They provide the materials technologist with a continuum of materials ranging all the way from soft rubbers to hard, rigid thermoplastics.

TPEs with a hardness (Fig. 10.1) between 30 and 90 Shore A (30 to 90 International Rubber Hardness Units [IRHD]) [17] generally qualify as true rubbers [1]. Those above 38 Shore D hardness are not "rubbers," but are still considered TPEs. Furthermore, the tensile stress-strain curve of a harder (>38 Shore D) TPE has an actual yield point or at

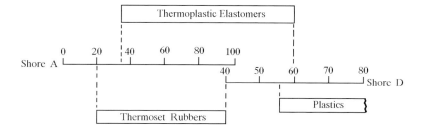

Figure 10.1   Comparison of hardness range of TPEs to those of plastics and conventional thermoset rubbers.

least a pronounced "knee." Softer TPEs exhibit a curve which progressively approaches that of a thermoset rubber as hardness decreases (Fig. 10.2).

Thus, by properties and performance, TPEs are generally considered rubbers, or at least rubberlike. Yet, by processing characteristics, they are considered to be thermoplastics, because the fabrication equipment and methods used for rigid thermoplastics can be used with TPEs. TPEs thus are used in both the rubber and the plastics industries, neither of which have communicated much with each other in the past. The person who knows how to mold and extrude a TPE works in the plastics industry, but the person who knows how to market fabricated TPE articles works in the rubber industry.

10.3   Classification of TPEs

10.3.1   Chemistry and Morphology

Commercial TPEs may be categorized on the basis of their chemistry and morphology [4]. Their chemistry includes their composition on the molecular level $(1 \times 10^{-6}$ to

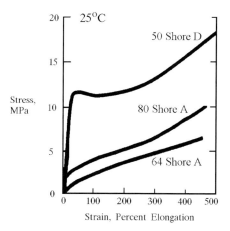

Figure 10.2   Tensile stress–strain curve of thermoplastic vulcanizate TPEs of progressively increasing hardness.

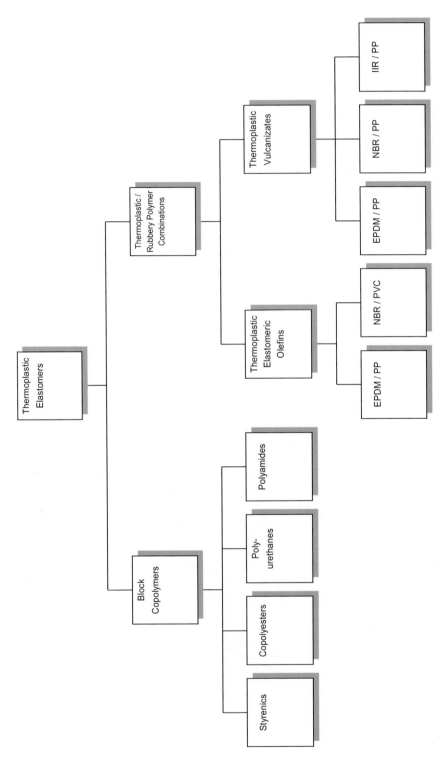

Figure 10.3   Schematic classification of commercial thermoplastic elastomers.

Solidified

Molten

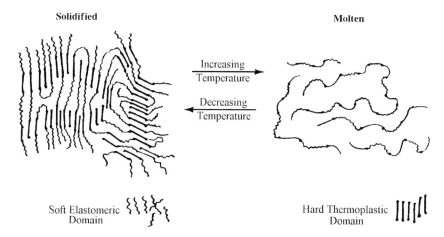

Increasing
Temperature

Decreasing
Temperature

Soft Elastomeric
Domain

Hard Thermoplastic
Domain

Figure 10.4   Melting–solidification of block copolymer TPE.

1×10^{-8} cm. scale); their morphology deals with their composition at the supramolecular level (1×10^{-4} to 1×10^{-6} cm. scale). The processability, properties, and functional performance of a TPE are critically affected by both chemistry and morphology.

Figure 10.3 summarizes schematically the classification of most TPEs of commercial importance. These materials may be categorized as either block copolymers or thermoplastic/rubbery polymer combinations. Each TPE consists of two or more polymeric phases (domains), one of which is hard and thermoplastic, and the other soft and rubbery.

Block copolymer TPEs consist of a common polymer chain with alternating hard and soft segments. In the solidified state (below T_m), the hard segments of different chains aggregate to form rigid, thermoplastic domains and the soft segments form elastomeric domains (Fig. 10.4). Chain movement within the soft domains is far greater than in the hard domains. Thus, the hard domains (thermoplastic phase) restrict the movement of the chains within the soft domains in much the same way as a sulfur crosslink or carbon black particle restricts the movement of the chains in a thermoset rubber.

When a block copolymer is heated above its T_m, bonds between the chain segments of the hard domains are disrupted, resulting in a molten, viscous material suitable for molding, extrusion, or other processing methods. When the molten TPE cools to below its T_m the hard segments reaggregate, resolidify, and assume the final molded shape (Fig. 10.4).

The morphology of a thermoplastic/rubbery polymer TPE, prepared by melt mixing (fluxing) of the two polymers, is illustrated in Fig. 10.5. These two polymers should be reasonably compatible. The thermoplastic phase must be continuous (to enable thermoplastic processing), and the rubbery phase is commonly (but not always [18]) discontinuous. Thus, the T_m of this material is essentially that of the thermoplastic, and the melting-solidification transformation is reversible above this T_m.

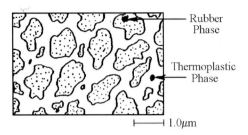

Figure 10.5   Morphology of thermoplastic/rubbery polymer TPE.

Thermoplastic vulcanizates (TPVs) [19] differ from thermoplastic elastomeric olefins (TEOs) [20] in that their rubber phase is crosslinked by vulcanization [1]. TEOs, on the other hand, have little or no vulcanization in the rubber phase.

In general, the hard (thermoplastic) phase of a TPE determines its processability, and the soft (rubbery) phase, its properties and functional performance. TPEs are selected based on their properties and projected performance in a fabricated part. Table 10.1 summarizes some of the more important properties of the different classes of TPEs.

10.3.2   Styrenic Block Copolymers

This category of TPEs [21,22] consists of block copolymers of styrene and a diene or hydrogenated diene with the structure S-D-S; S is a hard segment of polymerized styrene and D is a soft central polydiene or hydrogenated polydiene segment. Their chemical structures are given in Fig. 10.6. Both ends of the block copolymer must be styrenic for good TPE properties.

In a functioning styrenic TPE, the rigid styrenic blocks aggregate to anchor the softer diene blocks to yield a tougher, higher modulus material. These TPEs have properties superior to those of random styrene-diene copolymers, such as unvulcanized SBR rubber with the same monomer ratio, molecular weight and polydispersity.

The melting point of the styrenic TPEs is essentially that of polystyrene (PS), and the physical properties (Table 10.1) are determined more by the monomer ratio than by molecular weight, provided the PS blocks are above a certain minimum molecular weight. At low styrene levels, the materials are soft and rubbery and become progressively harder and stiffer (higher modulus) with increasing styrene content. Saturation of the soft central block by hydrogenation markedly improves resistance to oxygen and ozone, and thus improves the durability of the TPE in air.

The properties (Table 10.1) and performance of styrenic TPEs are adequate for many non-demanding rubber applications. The butadiene (B) and isoprene (I) styrenic TPEs are among the lowest cost and lowest performance TPEs (Fig. 10.7). The saturation of the center segment of the EB styrenics enables those TPEs to have a significantly higher maximum service temperature. Styrenics are more extensively used than any other generic class of TPEs [23]. Worldwide 1999 usage was approximately 660,000 MT (metric tons). This huge market is centered in applications requiring service below 70 °C, and modest resistance to hydrocarbons and moderate mechanical abuse. Such uses include shoe soles

Table 10.1   Key Properties of Different Classes of Thermoplastic Elastomers

Property	Block copolymers				Thermoplastic/rubbery polymer combinations	
	Styrenic	Copolyester	Polyurethane	Polyamide	Thermoplastic elastomeric olefins	Thermoplastic vulcanizates
Specific gravity	0.90 to 1.20	1.10 to 1.40	1.10 to 1.30	1.00 to 1.20	0.89 to 1.00	0.94 to 1.00
Shore hardness	20A to 60D	35D to 72D	60A to 55D	60A to 65D	60A to 65D	35A to 50D
Low temperature limit, °C	−70	−65	−50	−40	−60	−60
High temperature limit, °C (continuous)	100	125	120	170	100	135
Compression set resistance at 100 °C	P	F	F/G	F/G	P	G/E
Resistance to aqueous fluids	G/E	P/G	F/G	F/G	G/E	G/E
Resistance to hydrocarbon fluids	P	G/E	F/E	G/E	P	F/E

P = Poor, F = Fair, G = Good, E = Excellent

B: $\left(CH_2CH\right)_a\left(CH_2CH=CHCH_2\right)_b\left(CHCH_2\right)_c$

I: $\left(CH_2CH\right)_a\left(CH_2\underset{CH_3}{C}=CHCH_2\right)_b\left(CH_2CH\right)_c$

EB: $\left(CH_2CH\right)_a\left(CH_2CH_2CH_2CH_2CH_2CH\right)_b\left(CH_2CH\right)_c$; CH_3CH_2

Figure 10.6   Three common styrenic block copolymer TPEs; a and c = 50 to 80, b = 20 to 100. The B, I and EB segments are polybutadiene, polyisoprene, and hydrogenated poly(ethylenebutylene), respectively.

and sporting goods. In addition to their use in fabrication of rubber articles, these block copolymers are ingredients in a variety of useful formulations for sealants, asphalt, caulking, and motor vehicle lubricants.

10.3.3   Copolyesters

Copolyesters (COPs) are block copolymer TPEs with alternating hard and soft segments and an -A-B-A-B-A-B- structure where A and B are the hard and soft segments, respectively [24,25]. Between the segments (Fig. 10.8), the functional linkages are esters; within the segments, they are both ethers and esters. Their morphology is described in Section 10.3.1.

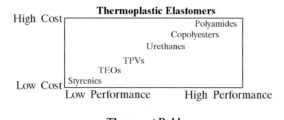

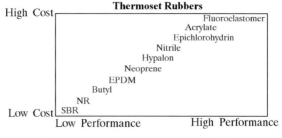

Figure 10.7   Relative cost and performance of TPEs and thermoset rubbers. A TPE at a given position on the upperchart is a rational candidate to replace a thermoset rubber at the same position on the lower chart.

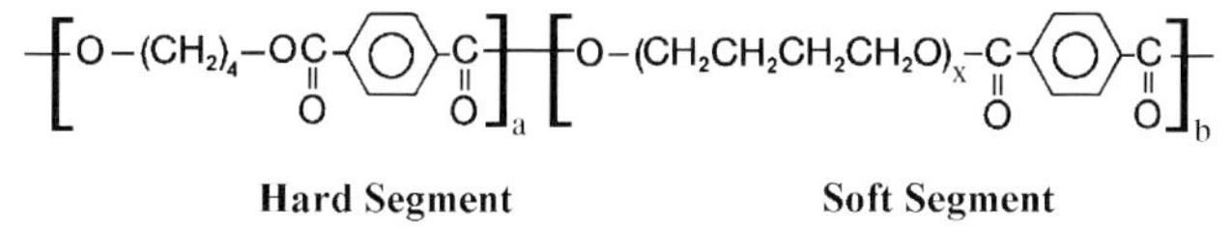

Figure 10.8   Chemical structure of a commercial copolyester; a = 16 to 40, x = 10 to 50, b = 16 to 40.

COPs have a hardness (Table 10.1) at and beyond the upper hardness range (80 to 90 Shore A) of conventional thermoset rubbers. Thus, many COPs are not true rubbers [1] but are rubbery in nature. Although generally considered a generic class of TPEs, COPs have also been described as soft, engineered thermoplastics. The high strength and modulus of a COP can more than offset their higher cost relative to that of a thermoset rubber. Thus, thinner parts and markedly lower part weights can be produced with the same performance. Further, the efficiency and economy of thermoplastic processing coupled with the lower part weight result in significant cost savings over a thermoset rubber [25].

Because many ester linkages are in the backbone of a COP chain (Fig. 10.8), these materials are susceptible to hydrolysis in both acids and bases. Therefore, they cannot be recommended for contact with concentrated acids and bases, particularly at elevated temperatures. Their resistance to other fluids is quite good, including air at temperatures up to 150 °C. As expected, their mechanical shock resistance is excellent. At low deformations (in tension, compression, shear, or torsion), COPs are highly elastic, with high resistance to flex fatigue, low heat buildup, and low creep.

10.3.4   Thermoplastic Polyurethanes

Thermoplastic polyurethanes (TPUs), the first commercial TPEs [26], have urethane linkages in the backbone of the block copolymer (Fig. 10.9). They have the same general -A-B-A-B-A-B- structure as a COP TPE, and the same soft/hard segment morphology as the styrenics and COPs [27,28]. Their soft segments are either polyester or polyether

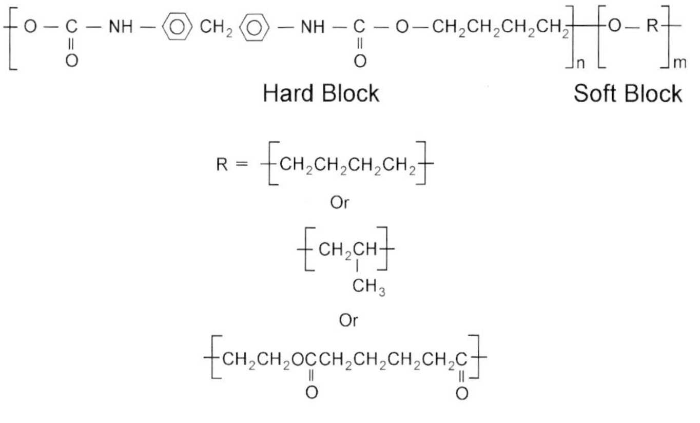

Figure 10.9   Chemical composition of commercial thermoplastic polyurethanes.

oligomeric diol blocks (800 to 3500 molecular weight), and their hard segments contain the urethane linkages.

The outstanding performance parameters of TPUs are their resistance to abrasion and low coefficient of friction against other surfaces. The hardness of most TPUs is in the high end (80 Shore A) of the thermoset rubber range. Unlike the COPs, TPUs of lower hardness (down to 50 Shore A) can be prepared. With increasing hardness, the tensile strength, modulus, and fluid resistance also increase, because of an increasing hard phase/soft phase ratio.

The melting point of the hard segments determines the theoretical upper temperature limit of a TPU (Table 10.1), whereas the glass transition point of the soft segments determines the lower temperature limit. As polar block copolymers, TPUs are very resistant to non-polar fluids (i.e., fuels, oils, greases). Conversely, they are susceptible to polar fluids, both organic and inorganic. In aqueous solutions, TPUs with polyether soft blocks are more resistant to hydrolysis than COPs; those with polyester soft blocks are comparable to COPs in hydrolytic stability.

Like the COPs, the higher cost of a TPU can be justified for uses demanding their outstanding properties, such as toughness, abrasion resistance, and low coefficient of friction. TPUs have found wide usage in shoe soles, caster wheels, and heavy duty hose, to name a few applications [29].

10.3.5   Polyamides

The polyether and polyester block polyamides (PEBAs) are the newest, highest-cost, and highest-performance class of TPEs [30,31]. They are block copolymers with the same morphology (Fig. 10.4) as the styrenics, COPs, and TPUs, but with amide linkages (Fig. 10.10) connecting the alternating hard and soft chains. These amide linkages are more resistant to chemical hydrolysis than ester or urethane linkages. Thus, the PEBAs are more resistant to chemical attack and have a higher maximum service temperature than the COPs or TPUs.

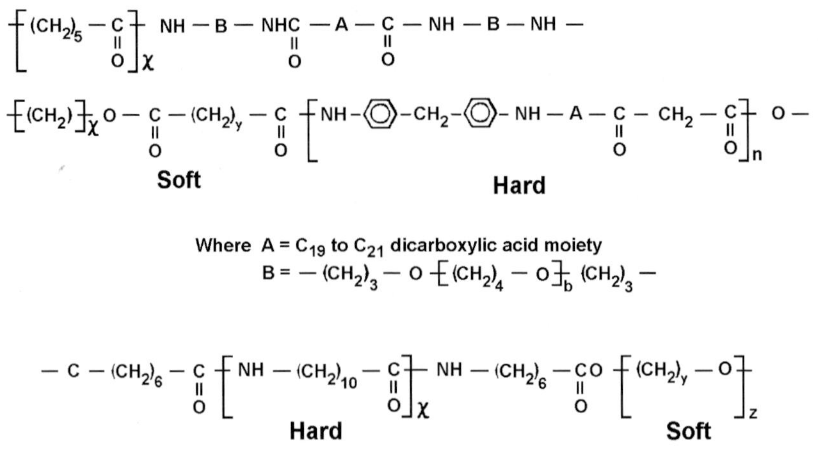

$$\left[(CH_2)_5 - \underset{\underset{O}{\|}}{C} \right]_x NH - B - NHC - A - \underset{\underset{O}{\|}}{C} - NH - B - NH -$$
$$\qquad\qquad\qquad\qquad\qquad \underset{O}{\|}$$

$$\left[(CH_2) \right]_x O - \underset{\underset{O}{\|}}{C} - (CH_2)_y - \underset{\underset{O}{\|}}{C} \left[NH - \bigcirc - CH_2 - \bigcirc - NH - A - \underset{\underset{O}{\|}}{C} - CH_2 - \underset{\underset{O}{\|}}{C} \right]_n O -$$

Soft   **Hard**

Where A = C$_{19}$ to C$_{21}$ dicarboxylic acid moiety

$$B = - (CH_2)_3 - O \left[(CH_2)_4 - O \right]_b (CH_2)_3 -$$

$$- \underset{\underset{O}{\|}}{C} - (CH_2)_6 - \underset{\underset{O}{\|}}{C} \left[NH - (CH_2)_{10} - \underset{\underset{O}{\|}}{C} \right]_x NH - (CH_2)_6 - CO \left[(CH_2)_y - O \right]_z$$

Hard   **Soft**

Figure 10.10   Chemical structures of three polyamide TPEs.

The hard segments of a PEBA generally determine its melting point and processing; the soft segments affect properties more. These soft segments may have polyether, polyester, or polyetherester linkages. PEBAs have a broad range of hardness (Table 10.1), ranging from moderately soft thermoplastic to medium hardness rubber. They have the highest maximum service temperature of any TPE, and function well at temperatures up to 170 °C.

Their polarity renders PEBAs quite resistant to non-polar, hydrocarbon-type fluids (i.e., oils, fuels, lubricants, etc.). Their resistance to attack in aqueous media is very good, but progressively decreases with increasing temperature and extremes of acidity or alkalinity. Polyester PEBAs are more sensitive to hydrolysis than polyether ones, but less sensitive to direct attack by environmental oxygen.

10.3.6   Thermoplastic Elastomeric Olefins

Simple blends of a conventional thermoplastic with a rubbery polymer [32,33,34] are called TEOs [20]. In early TPE literature, these materials were called TPOs (thermoplastic polyolefins), a term which has been used progressively less over the past decade. Each of the two polymers in a TEO has its own phase (Fig. 10.5) with the thermoplastic one continuous and the rubbery phase tending to be discontinuous, as described in Section 10.3.1. Like thermoset rubbers, TEOs can be compounded with other ingredients – carbon black, fillers, plasticizers, antidegradants, etc. – to give specific properties for a desired application. The degree of crosslinking (vulcanization) of the rubbery phase is little or none. TEOs are quite rubbery and compete with the styrenics for uses in the lower performance region of TPEs (Fig. 10.7).

By far, the two most common TEOs are EPDM/PP (EPDM rubber/polypropylene) and, to a lesser extent, NBR/PVC (nitrile rubber/polyvinyl chloride). At or near ambient temperature (0 to 40 °C), the TEOs have good rubberlike properties. With increasing temperatures, they lose the rubbery properties, with their utility severely limited above 80 °C, largely because of the lack of crosslinking in the rubber phase. The lower temperature limit of the EPDM/PP TEOs is approximately −60 °C, the glass transition temperature of the EPDM polymer. The lower limit for the NBR/PVCs can range from −40 °C up to 0 °C, depending on (1) the acrylonitrile content of the NBR polymer, (2) the amount of plasticizer present, and (3) the NBR/PVC ratio.

The chemical saturation of the polymer backbone in TEOs makes them highly resistant to attack by atmospheric oxygen (O_2) and ozone (O_3). Unfortunately, this resistance to oxidative attack is impaired by the lack of crosslinking in the rubber phase. The maximum temperature at which TEOs can be used is limited by their softening points.

The EPDM/PP TEOs have good resistance to polar fluids (water, salt solutions, acids, bases) but swell profusely in contact with non-polar fluids, to which the NBR/PVC TEOs are more resistant. TEOs are primarily used where:

1. A high level of set and creep can be tolerated
2. The needed fluid resistance is modest
3. The service temperature does not exceed 80 °C

Such use areas include external automotive, electrical insulation, and nonsealing moldings.

10.3.7   Thermoplastic Vulcanizates

This class of TPEs has been called elastomeric alloys in some earlier literature. The term thermoplastic vulcanizate (TPV) has proven to be more descriptive and understandable and is now officially sanctioned [1].

In a TPV, the rubber phase (Fig. 10.5) is highly crosslinked [16,18,35,36] by dynamic vulcanization [6,37]. This crosslinking gives rise to a pronounced improvement in a number of properties including:

1. Retention of properties at elevated temperature
2. Resistance to swelling in fluids
3. Compression and tension set
4. Creep and stress relaxation

Table 10.2 illustrates the properties improvement resulting from the dynamic vulcanization of two different EPDM rubber/PP compositions. In composition A, crosslinking greatly decreases the amount of material extractable with cyclohexane and improves the tensile properties; in composition B, the tensile strength, tension set, compression set, and oil resistance are significantly improved by crosslinking the rubber phase.

The TPVs are very similar to the TEOs in morphology (Fig. 10.5). The keys to their performance are:

1. the high degree of crosslinking (vulcanization) of the rubber phase
2. the size of the vulcanized rubber particles

Figure 10.11 shows the progressive improvement in tensile strength and tension set resulting from increasing crosslink density. Figure 10.12 shows the tensile strength improvement with decreasing particle size in the EPDM phase. At a particle size of approximately 1 μm, the properties of a TPV are quite good. TPVs represent the closest

Table 10.2   Effect of Crosslinking on Properties of Two Different EPDM/Polypropylene Compositions

Property	Composition A[a]		Composition B[b]	
	Uncrosslinked	Crosslinked	Uncrosslinked	Crosslinked
Extractable rubber, %	33.0	1.4	–	–
Crosslink density, moles/cc	0	1.6×10^{-4}	–	–
Hardness, Shore A	–	–	81.0	84.0
Ultimate tensile strength, MPa	4.9	24.3	4.0	13.1
Ultimate elongation, %	190.0	530.0	630.0	430.0
100% modulus, MPa	4.8	8.0	2.8	5.0
Compression set, %	–	–	78.0	31.0
Tension set, %	–	–	52.0	14.0
Swell in aromatic oil, %	–	–	162.0	52.0

[a] Composition A (parts by weight): EPDM rubber, 60; polypropylene, 40
[b] Composition B (parts by weight): EPDM rubber, 91.2; polypropylene, 54.4; extender oil, 36.4; carbon black, 36.4

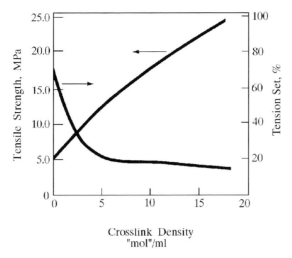

Figure 10.11   Improvement of tensile strength and tension set with crosslinking of EPDM/PP compositions.

approach of a TPE (Table 10.1) to the properties and performance of a conventional thermoset rubber.

The EPDM/PP system is by far the most common TPV, serving as a direct replacement for EPDM and neoprene rubber (Fig. 10.7). Of commercial significance are TPVs based on NBR/PP (to replace thermoset nitrile rubber), butyl/PP (to replace thermoset butyl rubber) and NR/PP (to replace thermoset natural rubber) [4].

The fatigue resistance of EPDM/PP TPEs has been found to be outstanding [38] and superior to thermoset rubbers especially compounded for that purpose. This finding is rather surprising because these TPEs also have very good compression set, even at

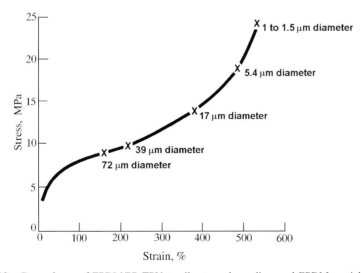

Figure 10.12   Dependence of EPDM/PP TPV tensile strength on dispersed EPDM particle size.

elevated temperatures. With thermoset rubbers, good fatigue resistance and low set tend to be mutually exclusive. Thus, a compound with good performance in one of these properties usually has poor performance in the other. TPVs provide good performance in both.

Another salient property of the TPVs is their anisotropy, or directionality. Their measured properties can vary quite significantly with the direction in which they are measured. Thus, the tensile strength and ultimate elongation of a TPV, measured parallel to the direction of the flow of molten material into the mold, commonly is 20 to 35% lower than those properties measured in a perpendicular direction. This anisotropy, also seen in the other TPE classes, is most pronounced for injection molded TPE articles. It results from the orientation of the polymer chains during the molding process.

TPVs have a broad range of hardness, from 35 Shore A to 50 Shore D. As expected, the tensile stress-strain properties (Fig. 10.2) become progressively more rubberlike with decreasing hardness and the amount of thermoplastic present. Their lower temperature limit is determined by the glass transition temperature of the rubbery polymer ($-60\,^{\circ}$C for EPDM) and their higher limit (125 to 135 $^{\circ}$C for PP) by oxidative attack of the thermoplastic phase. Different TPVs have a very broad spectrum of resistance to fluids. EPDM/PP is preferred for use in aqueous media (acids, bases, salts, etc.) and polar organic fluids. NBR/PP is preferred for contact with hydrocarbons, oils, fuels, and lubricants.

10.4   TPEs and Thermoset Rubbers

For the past four decades, the main purpose for developments in TPEs has been to replace conventional thermoset rubber. Thus, TPEs and thermoset rubber have directly competed with each other for applications and markets. It is quite appropriate, therefore, to examine the advantages and disadvantages of TPEs relative to thermoset rubbers, as follows:

1. No compounding – TPEs are used directly, with no need to add reinforcing agents, stabilizers, cure systems, etc. Batch-to-batch differences from variations in weighing and metering components are absent, leading to improved consistency in both raw materials and fabricated articles.
2. Simpler processing with fewer steps – TPE fabrication takes advantage of the ease and efficiency of thermoplastic over thermoset rubber processing (Fig. 10.13).
3. Shorter cycle times – TPE molding cycles are measured in seconds; those of thermoset rubber are measured in minutes.
4. Recycling – Scrap from processing (regrind) and articles which are past their useful lifetimes can be refabricated with no significant loss in properties [39].
5. Tighter control of material consistency and dimensions of fabricated articles – Dimensional tolerances are more precise by a factor of 2 to 3.
6. Suitability for high speed automation in both molding and assembly – This is a direct result of TPEs' improved consistency and dimensional tolerances.
7. Novel fabrication methods – Blow molding, thermoforming, heat welding, film blowing, and other thermoplastic methods may be used with TPEs. With thermoset rubber, these processes are not feasible.

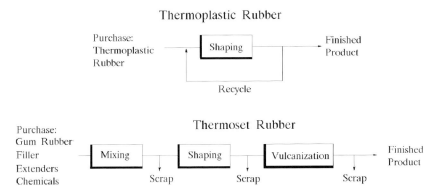

Figure 10.13   Comparison of thermoplastic rubber fabrication to that of thermoset rubber.

8. Lower specific gravity – This yields more fabricated TPE parts per unit weight of material. TPEs are purchased on a weight basis, but used on a volume basis.

Countering these advantages of TPEs are certain disadvantages which must be considered:

1. Thermoplastic processing methods and equipment, foreign to the rubber industry, are required for their fabrication. TPE fabrication by a thermoset rubber processor thus requires a capital investment in thermoplastics equipment, and operators must learn how to use this equipment.
2. The good processing economics of TPEs generally need a relatively high production volume to justify the costs of molds, extrusion dies, and development effort.
3. Drying is usually required before TPE processing. This step, familiar to thermoplastics processors, is unknown in conventional rubber processing.
4. TPEs are designed to melt at a specific temperature, above which they do not function as a rubber. Exposure of a thermoset rubber to a high temperature for a short period can be tolerated; a TPE cannot tolerate such exposure.

Each specific use of a TPE requires a comparison of these advantages and disadvantages. In most TPE applications, the TPE either directly replaces a thermoset rubber or captures a new, incipient use requiring a rubbery material. The TPE generally commands a higher price per unit weight. However, the economic advantages of thermoplastic processing can more than compensate for this higher material cost if the production volume is sufficiently large.

These relative advantages and disadvantages have fueled the phenomenal growth of TPEs, which has surpassed that of either the rubber or plastics industry. Between 1970 and 1990, TPEs grew at an annual worldwide rate of 8 to 9% [4]. Since 1990, this rate has slowly decreased to 6 to 7% [40,41], a rate likely to continue into the first decade of the 21st century. A 1995 IISRP study [23] found the worldwide consumption of TPEs to be 909,000 MT. By 2010, TPEs should capture more than 20% of the non-tire segment of worldwide rubber usage. TPEs are not expected to penetrate the massive tire market segment, which consumes slightly more than one-half of the thermoset rubber produced.

10.5   Fabrication of TPEs

10.5.1   Economy of Thermoplastics Processing

One major advantage of TPEs is the fact that they can be processed by the same methods and in the same equipment as used for rigid thermoplastics [42]. Thus, a TPE can be removed from its shipping container, dried, and then molded or extruded in the same manner as a PP or PE resin. This fact translates to a cost savings of $0.50 to $2.00 per kilogram of fabricated parts. The use of additives is normally discouraged, with the principal exception being colorants. To prevent processing problems or defective fabricated parts, most TPEs should be dried before use, with their moisture content kept below a specified level. Even non-polar, hydrocarbon-type TPEs can quickly absorb enough moisture to cause processing problems.

The rheology of molten TPEs is both complex and highly non-Newtonian (sensitive to shear). Figures 10.14(a) and 10.14(b) show the greater sensitivity of TPV viscosity to shear rate than to temperature. In TPE processing, the application of shear (pressure) is more effective than increasing temperature to induce flow in the molten material.

10.5.2   Injection Molding

Injection molding, the most common method of TPE fabrication, exploits the processing advantages of TPEs to the fullest [42]. Compression molding, used so long and so widely in thermoset rubber processing, cannot compete with TPE injection molding. With high production volumes, hot runner molding can be used to process TPEs, with no scrap resulting. The capability of TPEs to bond to compatible thermoplastics by heat welding means they can also be co- or insert-injection molded with rigid thermoplastics.

TPE melt temperatures are typically 20 to 60 °C above the melt point of the thermoplastic phase. Mold shrinkage depends on the type and grade of TPE, cross-section thickness, and flow distance. Molding machine size should be sufficient for 40 to 70 MPa clamping pressure relative to the cavity plus runner area, with a barrel capacity of six shots or less. Short, full-round runners are generally recommended, with full peripheral venting. Mold releases are not recommended, and cooling should be adequate to generate a sufficiently thick skin for extraction of the part without warping.

10.5.3   Extrusion

To extrude a TPE [42], use a thermoplastic extruder (L/D of 20/1 or greater), and not a rubber extruder, which usually has a lower L/D and heating capability. A polyolefin screw

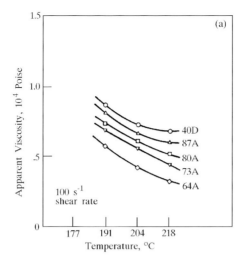

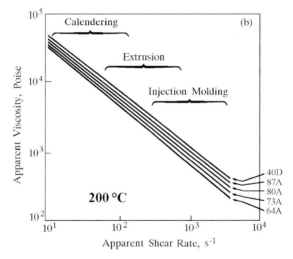

Figure 10.14   Viscosity variation of different Shore hardness TPVs with (a) temperature and with (b) shear rate.

(2:1 to 4:1 compression ratio) in any common design should be suitable. Screen packs of 20, 40, and 60 mesh provide for even flow and pressure and give a smoother extrudate. The polymer melt temperature should be 30 to 70 °C above the melting point.

Die swell increases with TPE hardness and extrusion shear rate, and decreases with increasing temperature. This swell is generally lower than that found with rigid thermoplastics or carbon black-loaded thermoset rubbers. Sheeting, tubing, and different types of profiles can be prepared by TPE extrusion. Reinforced hose and electrical wire and cable can readily be prepared by crosshead extrusion. Extrusion, rather than calendering, is the preferred method for preparing TPE sheeting.

10.5.4   Blow Molding

Massive cost savings can result from blow molding TPEs [43] to produce a variety of hollow rubber articles such as boots, bottles, and bellows. Both extrusion and injection blow molding are feasible.

Figure 10.15(a) shows schematically the extrusion blow molding of a TPE. A hollow tube of molten TPE (parison) is extruded downward into the mold. As the mold pinches the bottom shut, air in injected through a hollow pin, the mold closes around the parison, and the molten material is forced against the water-cooled mold. After cooling, the mold opens and the part is ejected.

Melt temperatures should be similar to those for TPE extrusion and injection molding, and the blow ratio (part diameter/parison diameter) should be as low as feasible. Prior to blowing, the stability and integrity of the parison are improved by the high viscosity of the molten TPE at low shear rate.

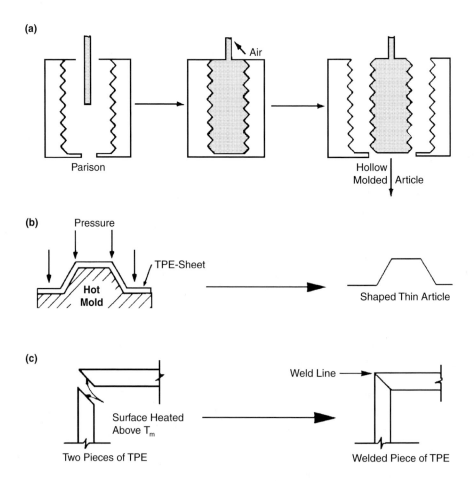

Figure 10.15   (a) Blow molding, (b) thermoforming, (c) heat welding.

10.5.5   Other Processing Methods

Sheets of extruded TPE can be fabricated into shaped articles through thermoforming. On a shaping mold (Fig. 10.15b), the TPE sheet is heated 10 to 40 °C above its softening point. Pressure, either external or vacuum, makes the softened sheet conform to the mold, giving the sheet the desired shape. This process is more practical for the harder TPEs.

Heat welding can be used because of the thermoplastic nature of TPEs. The surfaces to be bonded (Fig. 10.15c) are heated to fusion, then joined and cooled for solidification to occur. These surfaces may be heated by a variety of methods. The operation is rapid (4 to 8 sec), with a weld strength of up to 70 to 80% of the material tensile strength. Heat welding eliminates the need for an adhesive to bond a TPE to a compatible material.

10.6   Acknowledgments

The author wishes to thank Jennifer L. Digiantonio and Cari A. Wandling for their able and cheerful support in the preparation of this article. Appreciation is also due to Advanced Elastomer Systems, L.P. for permission to prepare and publish it.

References

1. ASTM D 1556, *Standard Terminology Relating to Rubber*, American Society for Testing and Materials (1997), Philadelphia, PA, Vol. 9.01.
2. ISO 1382, *Rubber - Vocabulary*, International Standards Organization (1997), Geneva, Switzerland.
3. ASTM D 2240, *Standard Test Method for Rubber Property – Durometer Hardness*, American Society for Testing and Materials (1997), Philadelphia, PA, Vol. 9.01.
4. Rader, C. P., In *Handbook of Plastics, Elastomers and Composites*, Third Edition, Harper, C. (Ed.) (1996), McGraw-Hill, New York, NY, Chapter 5.
5. Payne, M. T. and Rader, C. P., In *Elastomer Technology Handbook*, Cheremisinoff, N. P. (Ed.) (1993), CRC Press, Boca Raton, FL, Chapter 14.
6. Abdou-Sabet, S., Puydak, R. C. and Rader, C. P., *Rubber Chem. Technol.*, (1996) pp. 476-494.
7. *Handbook of Thermoplastic Elastomers*, Second Edition, Walker, B. M. and Rader, C. P. (Eds.) (1988), Van Nostrand Reinhold, New York, NY.
8. Legge, N. R., Holden, G., Quirk, R. and Schroeder, H. E., *Thermoplastic Elastomers* (1996), Hanser Publishers, Munich.
9. Rubber Division, American Chemical Society, Thermoplastic Elastomers Symposium, 138th National Meeting, Washington, D.C. (1990).
10. Rubber Division, American Chemical Society, Thermoplastic Elastomers Symposium, 142nd National Meeting, Nashville, TN (1992).
11. Rubber Division, American Chemical Society, Thermoplastic Elastomers Symposium, 146th National Meeting, Pittsburgh, PA (1994).

12. Rubber Division, American Chemical Society, Thermoplastic Elastomers Symposium, 150th National Meeting, Louisville, KY (1996).
13. Schotland Business Research, TPE '90, Third International Conference on TPE Markets and Products, Detroit, MI, March (1990).
14. Society of Plastics Engineers, TPE RETEC '93, Cincinnati, OH, October (1993).
15. Society of Plastics Engineers, TPE RETEC '95, Wilmington, DE, (September, 1995); TPE RETEC '97, Akron, OH, September (1997).
16. O'Connor, G. E. and Fath, M. A., *Rubber World*, (January, 1982).
17. Brown, R. P., *Physical Testing of Rubber*, Second Edition (1987), Elsevier, London.
18. Rader, C. P. and Abdou-Sabet, S., In *Thermoplastic Elastomers from Rubber Plastics Blends*, De, S. K. and Bhowmick, A. K. (Eds.) (1990), Ellis Horwood, New York, NY, p. 159.
19. ASTM D 6388, *Standard Classification for Highly Crosslinked Thermoplastic Vulcanizates* (*HCTPVs*), American Society for Testing and Materials (1998), Philadelphia, PA, Vol. 8.03.
20. ASTM D 5593, *Standard Classification for Thermoplastic Elastomers Olefinic* (*TEO*), American Society for Testing and Materials (1998), Philadelphia, PA, Vol. 8.03.
21. Holden, G and Legge, N. R., In *Thermoplastic Elastomers*, Legge, N. R., Holden, G., Quirk, R. P. and Schroeder, H. E. (Eds.) (1996), Hanser Publishers, Munich, Chapter 3.
22. Halper, W. M. and Holden, G., In *Handbook of Thermoplastic Elastomers*, Second Edition, Walker, B. M. and Rader, C. P. (Eds.) (1988), Van Nostrand Reinhold, New York, NY, Chapter 2.
23. *Worldwide Rubber Statistics*, 1995, International Institute of Synthetic Rubber Producers, Houston, TX.
24. Adams, R. K., Hoeschele, G. K. and Witsiepe W. K., In *Thermoplastic Elastomers*, Legge, N. R., Holden, G., Quirk, R. P. and Schroeder, H. E. (Eds.) (1996), Hanser Publishers, Munich, Chapter 8.
25. Sheridan, T. W., In *Handbook of Thermoplastic Elastomers*, Second Edition, Walker, B. M. and Rader, C. P. (Eds.) (1988), Van Nostrand Reinhold, New York, NY, Chapter 6.
26. Schollenberger, C. S., Scott, H. S. and Moore, G. R., *Rubber World* (1958) p. 549.
27. Meckel, W., Goyert, M. and Wieder, W., In *Thermoplastic Elastomers*, Legge, N. R., Holden, G., Quirk, R. P. and Schroeder, H. E. (Eds.) (1996), Hanser Publishers, Munich, Chapter 2.
28. Ma, E.C. In *Handbook of Thermoplastic Elastomers*, Second Edition, Walker, B. M. and Rader, C. P. (Eds.) (1988), Van Nostrand Reinhold, New York, NY, Chapter 7.
29. Penstrom, R., *Plastics News*, July (1996) p. 32.
30. Nelb, R. G. and Chen, A. T., In *Thermoplastic Elastomers*, Legge, N. R., Holden, G., Quirk, R. P. and Schroeder, H. E. (Eds.) (1996), Hanser Publishers, Munich, Chapter 9.
31. Farrisey, W. J. and Shah, T. M., In *Handbook of Thermoplastic Elastomers*, Second Edition, Walker, B. M. and Rader, C. P. (Eds.) (1988), Van Nostrand Reinhold, New York, NY, Chapter 8.
32. Kresge, E. N., In *Thermoplastic Elastomers*, Legge, N. R., Holden, G., Quirk, R. P. and Schroeder, H. E. (Eds.) (1996), Hanser Publishers, Munich, Chapter 5.
33. Hofmann, G. H., In *Thermoplastic Elastomers*, Legge, N. R., Holden, G., Quirk, R. P. and Schroeder, H. E. (Eds.) (1996), Hanser Publishers, Munich, Chapter 6.
34. Shedd, C. D., In *Handbook of Thermoplastic Elastomers*, Second Edition, Walker, B. M. and Rader, C. P. (Eds.) (1988), Van Nostrand Reinhold, New York, NY, Chapter 3.
35. Rader, C. P., In *Handbook of Thermoplastic Elastomers*, Second Edition, Walker, B. M. and Rader, C. P. (Eds.) (1988), Van Nostrand Reinhold, New York, NY, Chapter 4.
36. Coran, A. Y. and Patel, R. P., In *Thermoplastic Elastomers*, Legge, N. R., Holden, G., Quirk, R. P. and Schroeder, H. E. (Eds.), (1996) Hanser Publishers, Munich, Chapter 7.
37. Coran, A.Y. and Patel, R. P., *Rubber Chem. Technol.*, 53, (1980) p. 141.
38. Rader, C. P. and Kear, K. E., *Rubber and Plastics News*, May (1986).

39. Alderson, M. and Payne, M. T., *Rubber World* (1993) p. 22.
40. Reisch, M. S., *Chemical and Engineering News*, August (1996) p. 10.
41. *Chemical Week*, June (1995) p. 20.
42. Rader, C. P. and Richwine, J. R. *Rubber and Plastics News*, February (1995) p. 32.
43. D'Auteuil, J. G., Peterson, D. E. and Rader, C. P., *J. Elastomers and Plastics*, Vol. 21, October (1989) p. 265.

11   Recycled Rubber

Krishna C. Baranwal and William H. Klingensmith

11.1   Introduction

In the late 1980s and early 1990s, there was a movement in the U.S. and other nations to deal with scrap rubber problems. For example, in the U.S., what to do with the 270–280 million scrap tires and 300–500 million pounds of scrap rubber produced annually? With the passage by the U.S. Congress of the Intermodal Surface Transportation Efficiency Act (ISTEA) in the early 1990s, numerous companies, often with states' treasuries funding much of the project, were created. The act required the incorporation of rubber crumb into asphalt road paving on all federally funded road projects. After ISTEA was repealed, many businesses failed or consolidated.

The recycling businesses that have succeeded in the late 1990s found their markets in tire derived fuel and specialty areas such as indoor/outdoor surfaces and high quality ground fine particle rubber.

Despite the regulatory pressure to use scrap rubber, no standards existed until the mid-1990s and manufacturers described products in terms most favorable to their companies. In other words, 30 mesh did not necessarily mean the same thing to all manufacturers. In 1996, two ASTM standards, D 5603-96 and D 5644-96, were approved for particle size, sieve analysis, and chemical analysis of recyclables. This provided a technical base for comparison of different manufacturers' products.

The automotive industry's goal is to incorporate 25% post consumer scrap into their products with no increase in cost or decrease in performance. Currently this goal is being pursued with less vigor than previously. In fact, the level of post consumer scrap required may be lowered to a more realistic 15%. Much effort has been spent to find practical uses for recycled rubber, called SKOR (some kind of rubber) in the industry. To this date the efforts have had limited success.

Currently, those companies successfully producing recycled rubber have met market needs. Those market needs are value-driven. They cover a wide scope of technical demands from large shredded pieces for fuel to very fine particles for tires.

Much work is being done to find alternative methods to improve recyclates so as to increase their value in the future. This includes ultrasonic devulcanization, chemical devulcanization, catalytic regeneration, and various surface treatments. It is likely that one or several of these new concepts will become a commercial reality.

11.1.1   Tire Derived Fuel

It is estimated that more than 172 million tires or 64% of the total used in the U.S. were recycled into tire-derived fuel (TDF) in 1997. This represents the single largest use of

**Table 11.1   Utilization of Scrap Tires in
the US – 1997 (Millions)**

Tire derived fuel	172.0
Civil engineering	14.0
Ground up for reuse	23.0
Agricultural	2.5
Export	15.0
Pyrolysis	A few
Miscellaneous	1.5
Total	228.0

**Table 11.2   US Market Demand for Size – Reduced Rubber
From Tires (Pounds) [2]**

	1996	1998*
Pneumatic tires	48 million	140.0 million
Friction materials	8 million	8.5 million
Molded and extruded goods Rubber/plastic bound products	18 million	524.0 million
Athletic and recreation	24 million	50.0 million
Asphalt products	168 million	200.0 million
Total	400 million	582.5 million

* Estimated.

scrap tires in the U.S. The data in Table 11.1 summarizes the total utilization of scrap tires in the U.S. for 1997. TDF is expected to grow as more and more cement plants, power generation facilities, and pulp and paper plants expand their use of scrap tires for fuel [1]. Many companies and recycling firms have developed new uses for scrap tires, as shown in Table 11.2.

11.1.2   Automotive Industry's Recycling Efforts

The automotive industry is a major user of rubber products. As such, it also is a major generator of used and scrap rubber parts. Through the Vehicle Recycling Partnership, a pre-competitive, cooperative effort among Ford, GM, and Chrysler, work is being done to find the most efficient way to dismantle and reuse automotive components. Each of the automotive companies are also pursuing the use of recycled rubber in the components they purchase. Ford has been noticeably active in tire recycling efforts and is working closely with the tire companies to reach a goal of 10% recycled content. A recent announcement by Ford reports that 1.2 million Continental General tires on F-series trucks will contain recycled rubber. In addition 100,000 recycled content tires from Michelin are projected for the Ford Windstar. The recycled rubber content is reported at 5% and mostly in the tread area [3]. In the non-tire area, reported uses of recycled rubber are listed in Table 11.3.

Table 11.3   Use of Recycler Rubber in Non-Tire Automotive Parts

Ford	Lincoln Navigator	Air reflectors
Ford	Mercury Mountaineer	Fender insulators
GM	Not disclosed (report over 110 parts)	Baffles
		Sound barriers
		Water deflectors
Chrysler	Not disclosed	Splash seals

Many of the non-tire components incorporating recycled rubber are in static applications. Much work is underway looking at using recycled rubber in more dynamic applications, such as body seals, bushings, and gaskets. The goal is to utilize 25% post consumer scrap in automotive, non-tire components and 10% recycled content in tires [4].

Dr. Robert Pett of Ford reports that the automotive industry's goals are to have no more than 15% of the vehicles retired from service go to landfills by 2002. This number decreases to 5% by 2015 [5]. The automotive and transportation industry are the largest market for rubber goods, consuming an estimated 70–75% of all rubber articles produced.

Experiments incorporating ground scrap tires in road building materials have proven successful in some areas. The product is known as Crumb Rubber Modifier (CRM). California, Arizona, Texas, and Florida use crumb rubber modifier in their asphalt roads. Efforts to make this a national standard through legislation were tried in 1991. The Intermodal Surface Transportation Efficiency Act (ISTEA) required incorporating 20 pounds per ton of CRM to portions of roads being paved. Because of the increased cost involved many states resisted and in 1995 the act was repealed.

11.2   Recycling Methods

Currently, recycled rubber is reclaimed, and then processed by ambient grinding, cryogenic grinding, and wet grinding. Reclaiming silicone and butyl rubbers is common and the resulting recycled products reduce the costs and improve processing when added to virgin compounds.

11.2.1   Reclaiming

In the past, large quantities of whole tires, tread peel, tubes, and other products were reclaimed by various reclaiming agents [6]. Aryl sulfides and other, rather unpleasant smelling, chemicals were used to treat the rubber in a chemical digestion process. This reclaimed rubber was widely used in tire compounding to lower cost and improve processing and fatigue life. However, radial tires, which need higher green strength and abrasion resistance, could not use reclaim. In more recent years, environmental requirements related to the manufacture of reclaim were tightened and SBR rubber prices were lowered. The result was the almost complete elimination of reclaimed rubber manufacture

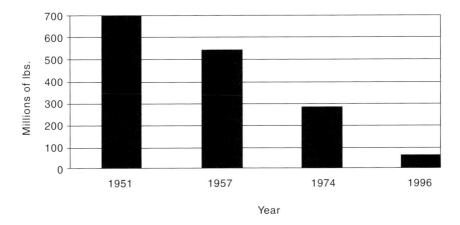

Figure 11.1   Reclaimed rubber usage.

in the U.S. Currently only two companies in the U.S., U.S. Rubber in Vicksburg, MS, and TRC in Stow, OH, are in this business. U.S. Rubber makes butyl reclaim and TRC makes silicone, EPDM, and other specialty reclaim rubber. In addition, Vredestein in the Netherlands sells low odor reclaim from whole tire, natural rubber, and butyl. There are numerous suppliers in Russia, Romania, India, Mexico, Korea, and several other countries. Figure 11.1 shows the decline in the use of reclaim. Currently it is used in mats, bumpers, chocks, low performance tires, and other, low dynamic stress, rubber articles.

11.2.2   Ambient Ground Rubber

In preparing scrap tires and rubber products for size reduction, a series of steps processes the material into chips for tire-derived fuel all the way to ultrafine 200 mesh particles for extrusions and calendered goods. The first step is to either shred tires and big rubber pieces or grind flash, pads, runners, or scrap rubber parts.

In ambient grinding, vulcanized scrap rubber is first reduced to approximately a 5 cm × 5 cm or 2.5 cm × 2.5 cm chip by shredding the tire or rubber part. The shredded rubber is then passed over magnets, separation tables, and air separators to remove metal and fiber. This material can then be further reduced by ambient grinding mills, or the shredded rubber can be frozen and then "smashed" or ground into fine particles by cryogenic grinding.

The ambient process often involves a conventional, high powered, rubber mill or a cracker mill with a close nip. This vulcanized rubber is sheared and ground into small particles. It is common to produce 10 to 40 mesh material using this relatively inexpensive method. Typical yields are 2,000 to 2,200 lbs per hour for 10 to 20 mesh and 1,200 lbs per hour for 30 to 40 mesh. The finer the desired particle, the longer the rubber must run on the mill. In addition, multiple grinds also reduce the particle size. The lower practical limit for the process is the production of 40 mesh material. Any fiber and extraneous material must be removed by an air separation or air table. Metal is removed using a magnetic separator.

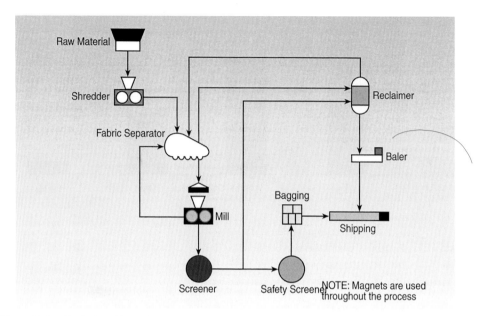

Figure 11.2   Typical ambient grinding system.

The resulting material is fairly clean. A flow chart for an ambient ground process, including a side stream for reclaiming is shown in Fig. 11.2 [7].

The process produces particles with irregular, jagged shape. In addition, a significant amount of heat is generated in the rubber during processing. Excess heat can degrade the rubber and, if the product is not cooled properly, combustion can occur upon storage.

11.2.3   Cryogenic Ground Rubber

Cryogenic grinding usually starts with chips or crumb rubber. This is cooled by a chiller, and the rubber, while frozen, is put through a mill, often a paddle-type mill. The final product has a range of particle sizes, which are sorted and either used as is or passed on for further size reduction, e.g., using a wet grind method. A typical process generates 4,000 to 6,000 lbs per hour. The size range of particles produced from this method is 60 to 80 mesh.

The cryogenic process yields fractured surfaces. Little heat is generated in the process, which decreases degradation of the rubber. The most significant feature of the process is that almost all fiber or steel is liberated from the rubber, resulting in a high yield of usable product. The price of liquid nitrogen has come down significantly recently, so cryogenically ground rubber can compete on a large scale with ambient ground products. A flow chart for a typical cryogenic process is shown in Fig. 11.3 [8].

Table 11.4 compares typical properties of ambient and cryogenically ground rubbers. Table 11.5 shows the particle size distribution for two typical 60 mesh ground rubbers using the Ro-Tap[*] method. One was prepared by ambient grinding and the other cryogenically.

---

[*] Ro-Tap type sieve shaker is available from many scientific laboratory suppliers.

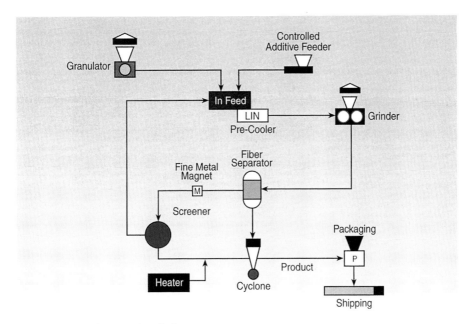

Figure 11.3   Typical cryogenic grinding system.

Table 11.6 shows the effects of using ambient ground SBR in a rubber compound. In addition, the effects of cryogenically ground butyl in a tire innerliner compound are shown in Table 11.7 and the effects of particle size and loading are shown for a cryogenically ground EPDM in Table 11.8. Table 11.9 shows comparative properties of an 80 mesh wet ground rubber (GF-80) at 10 part substitution.

Table 11.4   Characteristic of Ambient and Cryogenic Ground Rubber

Physical property	Ambient ground	Cryogenic ground
Specific gravity	Same	Same
Particle shape	Irregular	Fractured
Fiber content	0.5%	nil
Steel content	0.1%	nil
Cost	Comparable	Comparable

Table 11.5   Particle Size Distributions of Ambient and Cryogenic Ground Rubber

Amount retained	Ambient	Cryogenic
30 mesh	2%	2%
40 mesh	15%	10–12%
60 mesh	60–75%	35–40%
80 mesh	15%	35–40%
100 mesh	5%	20%
Pan	5–10%	2–10%

Table 11.6   Ambient Ground Rubber (20 Mesh) in an SBR 1502 Compound

Formulation ingredient	Level
SBR 1502	100.0
Zinc oxide	5.0
Stearic acid	1.0
TMQ	2.0
N-660 carbon black	90.0
Aromatic oil	50.0
Sulfur	2.0
MBTS	1.0
TMTD	0.5

Properties of Compound with 17, 33 and 50% Crumb Addition

	Control	17% Crumb	33% Crumb	50% Crumb
Mooney viscosity	40.0	61.0	91.0	111.0
Rheometer max torque	59.0	47.0	33.0	34.0
t_{90}, min.	2.5	2.4	1.8	2.0
Tensile strength, psi	1470.0	1150.0	870.0	560.0
Ultimate elongation, %	330.0	330.0	300.0	270.0

Table 11.7   Cryogenically Ground Butyl (80 Mesh) in the Innerliner

Formulation ingredient	Level
Butyl HT-1066	80.0
RSS 1	20.0
N-650	65.0
Mineral rubber	4.0
Durez 29095	4.0
Stearic acid	2.0
Sunthene 410	8.0
Zinc oxide	3.0
Devil A sulfur	0.5
MBTS	1.5

Cryogenically Ground Butyl at Various Levels

	Control	5%	10%	15%
Masterbatch	188.0	178.6	169.2	159.8
Cryo Ground Butyl	–	9.4	18.8	26.2
Properties				
Cure time, T_{c90}, min.	47.5	46.3	47.0	46.5
Cure rate, lbf.in/min.	0.59	0.58	0.55	0.56
Tensile strength, psi	1410.0	1350.0	1290.0	1280.0
300% modulus, psi	1120.0	1040.0	1000.0	950.0
100% modulus, psi	415.0	410.0	365.0	365.0
Air permeability, Q^*	4.71	4.70	4.47	4.16

* $Q \times 10^{-3}$ (cubic ft/0.001 inches/°F psi/day)

Table 11.8   Cryogenically Ground Rubber in an EPDM Compound

Formulation ingredient	Level
EPDM	100.0
N-650	70.0
N-774	130.0
Paraffinic oil	130.0
Zinc Oxide	5.0
Low MW PE	5.0
Stearic acid	1.0
Antioxidant	1.0
Sulfur	1.25
Sulfads	0.8
Methyltuads	0.8
Ethyl tellurac	0.8
Altax	1.0

Cryogenically Ground Butyl at Various Levels

	Control	40 mesh	60 mesh	80 mesh	100 mesh
Tensile strength, psi	1410	1290	1430	1470	1440
Ultimate elongation, %	410	330	340	400	380
300% modulus, psi	1180	1220	1230	1230	1220
100% modulus, psi	535	490	530	490	480
Hardness, psi	73	70	70	70	71
Die C tear, ppi	193	175	173	171	172

Cryogenically Ground Rubber at 20% Levels

	Control	40 mesh	60 mesh	80 mesh	100 mesh
Tensile strength, psi	1410	1230	1360	1450	1410
Ultimate elongation, %	410	320	390	390	390
300% modulus, psi	1180	1220	1300	1200	1160
100% modulus, psi	535	450	500	460	460
Hardness, psi	73	72	70	69	68
Die C tear, ppi	193	178	163	165	181

Table 11.9   Physical Properties of Radial Sidewall Compound – 10 Parts Substitution (cured at 142 °C (287 °F) for 4 min)

	Control	Sub. 10 part GF-80	Sub. 10 Parts whole tire reclaim
Modulus at 300%, MPa	3.0	3.7	3.7
Tensile strength, MPa	14.9	15.1	13.6
Elongation, %	875.0	775.0	740.0
Durometer, Shore A	58.0	58.0	59.0
Tear, MPa	2.7	2.9	2.7

11.2.4   Wet Ground Rubber

Several companies produce a wet ground rubber by using a water suspension of rubber particles and a flour grinding-type mill. The material is finely ground and mesh sizes from 60 to 120 are commonly made and used. These products are employed in many tire compounds because of their uniformity and cleanliness.

A review of scrap tire processing has been published by Astafan in *Tire Technology International '95* [9]. In addition microwave [10], ultrasonic [11], chemical devulcanization [12], microbial degradation, and mechanical shear are or have been used to produce recycled rubber. Ultrasonic and chemical devulcanization are discussed in detail in the 1996 issue of *Tire Technology International '96* [13].

In ultrasonic "devulcanization" [13], a coarse crumb rubber is exposed to high intensity ultrasonic vibrations. The resulting energy absorbed by the rubber is theorized to fracture sulfur-sulfur bonds and produce a recycled rubber that can be used as is or compounded into virgin rubber and recured. This process is being developed by National Feedscrew and Machining Company in Massillon, Ohio.

Chemical "devulcanization" involves accelerators and sulfur, which are added on a mill or in a Banbury® mixer to a coarse rubber particle. Theoretically, this combination fractures sulfur-sulfur bonds and produces a usable recycled rubber. This process is offered by STI-K of Washington, D.C. The reason for quote marks around the term "devulcanization" is that, in addition to the breaking of sulfur-sulfur bonds, a good deal of depolymerization occurs with the heat and shear used in these processes. It is difficult to quantify exactly what mechanism is causing the bond failures. In any case, the result of the process is a recycled rubber.

11.2.5   Surface Treatment and Additives for Producing Recycled Rubber

Numerous methods can be used to modify the surface or composition of recycled rubber to make it more compatible or useful. These include halogenation, liquid polymers [14], thermoplastic polymers, homogenizing agents [15], and wetting agents. These are reviewed in detail in the CWC *Best Practices Manual* [16].

11.3   Testing, Storage, and Characterization

11.3.1   Testing Standards

Recently, we summarized the quality, testing, and handling issues for dealing with scrap tires and rubber for the CWC *Best Practices Manual* [16]. In December 1996 ASTM published two documents, ASTM D 5603-96 [17] and ASTM D 5644-96 [18]. Late 1997 the Chicago Board of Trade (CBOT) also published a document which includes definitions of terms and particle size specifications of recycled rubber for buying and selling materials at the CBOT [19]. Thus, there are now specifications available for recycled rubber that vendors and customers should use to ensure material quality.

11.3.2   Material Storage

Recycling rubber from whole tires into chips and crumbs can generate heat as high as 115 °C. At these temperatures and in the presence of oxygen, spontaneous combustion can occur. Also, the presence of iron catalyzes oxidation of natural rubber. To minimize or eliminate any such problems, materials should be adequately cooled with air or water before storage or shipping.

11.3.3   Moisture Content

The currently accepted maximum moisture content is 1% in recycled rubber (ASTM D 5603-96). Typically, however, it is less than 1%. Too much moisture can cause caking and may inhibit free flow in processing. Anticaking agents such as calcium carbonate can be used to prevent this problem. Moisture buildup can lead to acidic conditions, giving slower cure rates in compounds. Therefore, recycled crumb rubber should be stored in a cool and dry place.

The moisture content is determined by heating a weighed sample at 125 °C for one hour, cooling it, and weighing it again. The difference in sample weight is the heat loss (ASTM D 1509-95).

11.3.4   Bulk Density

Because of the particulate nature of crumb rubber, it is rather difficult to measure its specific gravity. The bulk density measurement may be more appropriate. There is no bulk density specification for crumb rubber; however, ASTM D 1513 for carbon black can be used. Our recommendation is that bulk density be part of the material specifications and a range of acceptable values be determined by the vendor and customer. Another way of determining the specific gravity of crumb rubber is to make solid sheets by passing the crumb rubber through a tight mill nip and measuring the density of the compressed pieces.

11.3.5   Chemical Analysis and Material Specifications

ASTM D 5603-96 [17] lists specifications for acetone extractable, ash, moisture, carbon black, natural rubber, and rubber hydrocarbon contents for recycled rubber. As mentioned in this document, these chemical tests are done according to ASTM D 297. Typical chemical analysis results from tests in the Akron Rubber Development Laboratory on six different commercially available tire crumb samples are shown in Table 11.10 [20]. The ASTM specification [17] also lists a maximum metal content of 0.1% and fiber content of 0.5% in whole tire crumb. Fiber and metal contents in tread buffing should be zero. See Table 11.11 for specifications. In the production of recycled rubber, steel wire pieces are

Table 11.10   Analysis Results on Six Different
Crumb Rubber Samples

% acetone extractables	10–14
% ash	5.9–7.1
% carbon black	29.8–30.8
% loss on heating	0.40–0.58
% rubber hydrocarbon	45–52
% isoprene	14.1–24.2

separated by magnetic separator. Fibers are separated by use of a vibrating screen table and ''vacuuming'' off the ''fabric balls'' from the top of the screen.

To determine iron content, a pre-weighed amount of recycled rubber is spread on a non-magnetic flat surface. A small magnet is used to go over the material. The magnet should pick up steel pieces. The weight of the material thus picked up is then measured. However, for very small particles of crumb rubber (<100 mesh), a magnet may pick up recycled rubber particles as well. In that case, atomic absorption (AA) should be used to determine amounts of iron.

11.3.6   Particle Size and Distribution

The sizes and distribution of recycled cured rubber particulates are determined by the Ro-Tap method as described in ASTM D 5644-96. Six sieves are used in this mechanical shaker. The first two screens are defined in the above document for 10, 20, 30, 40, 60, 80, and 100 mesh particle size designations (see Table 11.12) [17]. The remaining four screens are to be decided between the vendor and customer. About 100 g of crumb rubber are weighed and put on the top pan with a cover and five other screens. After a fixed time of running the shaker, materials in each pan are weighed and plotted as a function of screen size, giving the particle size distributions. Vibrators and sieves are available from most

Table 11.11   Properties for Recycled Rubber (Grades 1–6) [17]

Property	%	Test method
(a) Grade 1–4		
Acetone extractables	8–22	D 297, Sec. 17, 18, 19
Ash, max	8	D 297, Sec. 34, 35, 36, 37
Carbon black	26–38	D 297, Sec. 38, 39
Loss on heating, max	1	D 1509
Natural rubber	10–35	D 297, Sec. 52, 53
Rubber hydrocarbon content (RHC), 42 min		D 297, Sec. 11
(b) Grade 1–6		
Metal content, max	0.1	See 7.3.2
Fiber content, max (grades 1, 4, 5, 6)	0.5	See 7.4
Fiber content, max (grades 2, 3)	Nil	See 7.4

Table 11.12   Recycled Rubber Product Designation [17]

Nominal product designation	Example ASTM D 5603 designation[*]	Zero screen μm	Percent retained on zero screen	Size designation screen μm	Maximum percent retained on designation screen
10 mesh	Class 10-X	2360 (8 mesh)	0	2000 (10 mesh)	5
20 mesh	Class 20-X	1180 (16 mesh)	0	850 (20 mesh)	5
30 mesh	Class 30-X	850 (20 mesh)	0	600 (30 mesh)	10
40 mesh	Class 40-X	600 (30 mesh)	0	425 (40 mesh)	10
60 mesh	Class 60-X	300 (50 mesh)	0	250 (60 mesh)	10
80 mesh	Class 80-X	250 (60 mesh)	0	180 (80 mesh)	10
100 mesh	Class 100-X	180 (80 mesh)	0	150 (100 mesh)	10

[*] When specifying materials, replace the X with the proper parent material grade designation code. For example, Class 30-2 indicates a 600 μm (30 mesh) product made from Grade 2 material: car, truck, and bus tread rubber. Class 100-6 indicates a 150 μm (100 mesh) product made from Grade 6 material: non-tire rubber.

scientific suppliers. This technique works well for coarser particles (>80 mesh). For 80 mesh and finer, small balls are formed as a result of particle agglomeration on screens, giving higher ''apparent'' particle sizes than they really are.

Several other techniques for determining sizes of finer particles are being evaluated by the authors. In our laboratory, we have developed an ultrasonic technique where a small quantity of crumb rubber is placed in a non-solvent liquid, and exposed to low levels of ultrasonic energy. The resulting dispersion is put on a glass slide and dried. An image analysis software program on a Light Optical Microscope (LOM) provides the particle size distribution. Our experience is that this technique works well even with small particles, i.e., up to 1 to 2 microns. We have compared particle size distributions of a commercial 80 mesh crumb rubber sample determined by the Ro-Tap method and LOM-ultrasonic techniques, respectively (work done at the Akron Rubber Development Laboratory, Akron, Ohio). We have found that the ultrasonic technique separates particles much better than the Ro-Tap method.

Some of the other commercially available techniques are the Coulter counter (Coulter Corporation, Miami, Florida, USA); Malvern Instruments Limited (Southborough, Massachusetts, USA); Particle Sizing Systems (PSS, Langhorn, Pennsylvania, USA); and Elcan Industries, Inc. (New Rochelle, New York, USA).

The Coulter counter, Malvern, and PSS instruments are single particle counters that use a low angle laser diffraction optical system. A small quantity of sample is dispersed in water using low concentration of water-soluble detergent or low ultrasonic energy to disperse particles. Particle sizes are measured either from scattered light (measured on photosensitive detector) or light blocked by particles. Differential volume percent, surface area percent and number percent values are plotted as a function of particle size, yielding distribution curves rather than histograms.

An air-jet screening unit from Elcan Industries, Inc., uses one screen at a time to give particle size. The sample is put on the screen and is sieved for a specified time. The material is pulled through the screen by a vacuum cleaner, and at the same time, the screen is cleaned by a rotating air-jet nozzle.

References

1. Scrap Tire Use/Disposal Study, 1996 Update, Published by the Scrap Tire Management Council, Edited by John Serumgard and Michael Blumenthal, 1996.
2. *Scrap Tire Recovery, An Analysis of Alternatives* (1998), Goodyear Tire and Rubber Company, Akron, OH.
3. *Rubber and Plastics News* (Feb. 23, 1998), Crain Communications, Cleveland, OH, pp. 22–23.
4. Personal Communication, Dr. Robert Pett, Ford Motor Company, 1997.
5. *Op. cit.* (3).
6. "Reclaim Rubber," RT Vanderbilt Handbook, 1958.
7. *Scrap Tire User's Directory*, Recycling Research Institute, Suffield, CT, 1996, p. 25.
8. *Ibid.*
9. Astafan, C., "Scrap Tire Processing in the US", in *Tire Technology International 95* (1995).
10. US Patent 4,104,205, Microwave Devulcanization of Rubber, August 1, 1978.
11. US Patent 5,258,413, Ultrasonic Devulcanization, November 1993.
12. European Patent Application EP 0690 091 A1, Application No. 95301399.2 , Filed 03.03.1995.
13. Boron, Roberson and Klingensmith "Ultrasonic Devulcanization of Tire Compounds", pp. 82–84, and Sekhar and Kromer, "De Link Concept", pp. 87–88, in *Tire Technology International 1996*, 1996.
14. US Patent 4,481,335, "Rubber Compositions and Methods", issued to Fred Stark, Nov. 6, 1984.
15. US Patent 5,510,419, "Polymer Modified Surface", issued to Burgoyne, Fisher, and Jury, April 23, 1996.
16. Klingensmith, W., Baranwal, K., *Best Practices in Scrap Tires and Rubber Recycling* (June, 1997), Clean Washington Center.
17. ASTM D 5603-96, Standard Classification for Rubber Compounding Materials–Recycled Vulcanizate Particulate Rubber.
18. ASTM D 5644-96, Standard Test Method for Rubber Compounding Materials – Determination of Particle Size Distribution of Recycled Vulcanizate Particulate Rubber.
19. "Crumb Rubber (Tire or Non-Tire)", Crumb Rubber Grades Definitions, by Chicago Board of Trade, 1997.
20. Baranwal, K., *Analysis Results on Six Different Commercial Crumb Rubber Samples* (1997), Akron Rubber Development Laboratory, Akron, OH.

12 Compounding with Carbon Black and Oil

Steve Laube, Steve Monthey, and Meng-Jiao Wang

12.1 Introduction: Carbon Black Affects Everything

Carbon black comprises about 30% of most rubber compounds. The addition of carbon black can affect virtually all phases of a rubber factory's operation, as well as all the performance characteristics of the end product, whether it is a tire component or an industrial rubber product (IRP).

Because carbon black is so important to rubber compounds, tracking the total quality cost is essential. The total quality cost of using a particular grade of carbon black, from a given source, at a specific loading, can be quite different from the cost of the carbon black alone. To obtain the maximum value from purchased carbon black, the total cost of this significant filler should be considered in terms of:

- The cost per unit weight of the black
- The carbon black's effect on compound cost, usually dependent on black and oil loading – this is especially important with expensive polymers
- Time and labor for unloading and handling of the carbon black in the factory
- The mixing time and energy consumption of the compound
- The ultimate dispersion level of the carbon black in the compound
- Factory scrap rates from mixing and subsequent processes
- Uniformity of the uncured component(s) during processing
- Effect on curing the final product/mold flow/mold removal
- Final product scrap rate
- Field performance of the final product.

12.2 Characterization of Carbon Black

At the 1988 Spring ACS Rubber Division meeting, a symposium organizer challenged participants to answer the question, "What *is* carbon black?" It was a good question. Carbon black researchers say that there are two major problems with understanding carbon black: first, we do not know how to characterize it; and second, we do not know how it reinforces rubber. Given these limitations, we outline here what we do know about the characterization of carbon black. Later, we describe what we know about how it reinforces rubber.

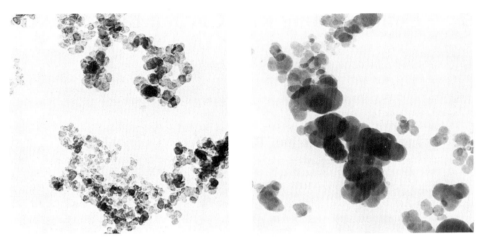

Figure 12.1   Resemblance of carbon black aggregate to clusters of grapes (electron photomicrographs of [left] N472 and [right] N774).

12.2.1   The Particle, the Aggregate, and the Agglomerate

Carbon black is a colloidal form of elemental carbon that owes its reinforcing character to the size, shape, and surface chemistry of the primary aggregates. Figure 12.1 shows that a simple analogy can be made between a cluster of grapes and the anatomy of carbon black. In this analogy, the individual grapes are particles of carbon black, and the cluster of grapes represents an aggregate. Like grapes, the carbon black aggregate is built from primary particles that are roughly spherical in shape; but the carbon black aggregate is unlike a grape cluster in that it is so strongly fused together that particles cannot be separated by the normal techniques used in rubber processing. A group of aggregates together is called an agglomerate; the grape parallel would be several clusters of grapes hanging on the vine. Because the individual particles of carbon black are fused together, the aggregate is the smallest dispersible unit of carbon black, not the particle. Dispersing carbon black, which is discussed later, is the process of breaking up the agglomerates into individual aggregates within a polymer matrix.

12.2.2   Surface Area, Structure, and Surface Activity

There are two conventional analytical measurements that characterize the ability of a particular carbon black to reinforce rubber. The first is specific surface area; the second is structure. A further parameter is "surface activity," which, from the measurement point of view, is somewhat nebulous. In addition, undesirable carbonaceous components (such as coke), non-carbonaceous materials (such as ash and moisture), and the quality of the pellets, are additional important parameters of a particular shipment of carbon black which may be measured.

Surface area refers to the amount of carbon black surface available to interact with the elastomer. For non-porous carbon black, surface area is inversely related to the size of the carbon black particles that fuse to form an aggregate. The smaller the particles (grapes), the higher the surface area that is available per unit weight.

Structure describes the number of particles fused together to form an aggregate, as well as the shape of the aggregate. The more particles in the aggregate (grapes in the cluster), the more complex is the shape and the greater the void volume (spaces) created. These void volumes can be filled with an oil for routine measurement (dibutyl phthalate, discussed later). More importantly, these voids can be filled with the polymer.

Surface activity refers to the strength of the carbon black's surface interaction with the polymer, via either physical adsorption or chemisorption.

Compound failure properties (such as tensile strength, tear strength, and abrasion resistance) are generally affected by loading and surface area; non-failure properties (such as viscosity, shrinkage, and modulus) are mostly affected by loading and structure. Surface area also affects some non-failure properties as well, most notably hysteresis. Structure can play a role, either directly or indirectly (dispersion), in the failure property of flex fatigue and, to some extent, abrasion resistance. Surface activity also plays a major role in modulus, hysteresis, and abrasion resistance.

Particle size can be directly measured only with the use of an electron microscope. Generally, the particle size of carbon black ranges from 10 to 250 nm in diameter. Practically, the particle size can be estimated by measuring the ''macro'' surface area of the carbon black using various liquids or gases to coat the surface. When nitrogen is used as an adsorbate, the relatively small molecule not only can coat the surface, but it can also penetrate micro-pores. When liquid cetyltrimethyl ammonium bromide (CTAB) is used, the relatively large CTAB molecule cannot penetrate the pores, so only ''external surface area'' (or smooth surface area) is measured. For rubber-grade blacks of less than about $120 \, m^2/g$, there is very little porosity, so CTAB adsorption is about the same as the nitrogen surface area.

A third method is to measure the amount of iodine adsorbed on the surface from solution. The amount of iodine adsorption is expressed in mg/g of carbon black, and for furnace carbon blacks, gives a number similar to the CTAB surface area. However, impurities such as oxygen and unburned feedstock can give a false, low reading to the iodine number. Actually, a depressed iodine number is one way to detect unburned feedstock on the surface of the carbon black. At the same time, iodine can penetrate pores and thus exceed CTAB values for porous blacks. CTAB is a very effective method for predicting the potential effect of surface area on rubber compounds. Although CTAB is the most relevant to rubber applications, iodine testing is still used routinely by many for quality control purposes. The following ASTM and ISO procedures can be referenced for more detail:

- ASTM D1510/ISO 1304 – iodine number
- ASTM D3765/ISO 6810 – CTAB area
- ASTM D4820/ISO 4652-2 – nitrogen surface area
- ASTM D5816/ISO 4652-2 – STSA (statistical thickness surface area)

The surface area of commercially available, rubber-grade carbon blacks ranges from about 10 to $140 \, m^2/g$. If thermal black is eliminated from the range of commercially available blacks, thereby limiting the range to blacks made from the oil-furnace process, then the range varies from about 25 to $140 \, m^2/g$. Carbon blacks with surface area less than about $45 \, m^2/g$ are referred to as semi-reinforcing for industrial products applications or "carcass" grades for tire applications. The ASTM designation (discussed below) for these carbon blacks is ASTM "500–700." Carbon blacks with surface areas of about 65 to 140 m^2/g are referred to as "reinforcing" for industrial product applications or "tread grades" for tire applications. These grades have ASTM designations of "N100–N300." It is important to note that carbon black reactors are generally not interchangeable and are designed to make either semi-reinforcing or reinforcing carbon black grades, but not both. An exception is carbon black grade N472. This conductive grade has a much higher surface area than any other conventional rubber black. Finally, the grades included in ASTM categories "N800 and N900" currently can only be made via the thermal process, not the oil-furnace process, so these grades are thermal blacks.

As with surface area, one direct way to measure structure is to count the number of particles fused together while viewing the aggregates under an electron microscope. The typical number of particles has been reported [1] to be about 136 for low structure N326; 278 for "normal structure" N330; and 331 for high structure N347.

For quality control purposes, the structure can be characterized by the void volume created by the packing of the aggregates. An oil, dibutyl phthalate (DBP), is titrated into the carbon black in a small mixing chamber, while the torque is measured. Once the void space created by the packing of the aggregates is filled with DBP, the carbon black in the chamber changes from a free-flowing powder to a coherent mass, causing a sudden sharp rise in torque. The machine shuts off automatically when this happens (ASTM D2414/ISO 4656/1). Note that the shear forces in commercial rubber mixing are much higher than the shear forces involved in mixing the carbon black with DBP in the ASTM test [2]. Thus, it is not surprising that the structure as measured by DBP is not fully indicative of the net structure after mixing in rubber. A modification to the DBP test uses pressure to crush the carbon black pellets under 165 MPa, four times before testing, in an attempt to simulate what happens to the carbon black in a mixer (ASTM D3493/ISO 6894). The resultant crushed DBP (CDBP) value is a lower number than the standard DBP. CDBP is the most appropriate way of measuring structure and its potential effect on rubber compounds.

Surface activity is the least well-defined characteristic of carbon black. In a chemical sense, it is related to the reactivity of chemical groups on the carbon black's surface. In terms of physical chemistry, it is referred to as adsorption capacity. While the adsorption or desorption of certain chemicals can be used to estimate surface activity, in-rubber properties can provide practical information.

Bound rubber often serves as a practical measure of surface activity. Bound rubber is defined as the rubber portion in an uncured compound that cannot be extracted by a good solvent. Thus, for carbon blacks with comparable surface areas, higher bound-rubber content in the compound is an indication of higher polymer-filler interaction. Surface activity is also often measured by the macro properties of cure rate and modulus. Note that the absence of surface activity can be observed by heat treating carbon black to anneal its surface. Such carbon black develops very low modulus in rubber [3].

12.2.3   Constituents Other than Carbon (Impurities)

Aside from elemental carbon, typical carbon black contains the following additional constituents:

- Ash (ASTM D1506/ISO 1125) – most of the ash comes from the quench water and the pelletizing water, but some also comes from the feedstock. Some low ash carbon blacks are being made for critical IRP applications by using low mineral content process water.
- Moisture (ASTM D1509/ISO 1126) – the referenced ASTM test measures the heat loss of the carbon black, which, for the most part, is caused by water.
- Extractables/Discoloration (ASTM D1618) – extractables come from the feedstock, especially if the black was produced with a short-quench process or otherwise has an "active surface." The extractables are measured by the percent transmittance of a toluene extract of the carbon black.
- Unwanted Carbonaceous Material (ASTM D1514/ISO 1437) – unwanted carbonaceous material or "coke" is formed from the reactor or from caking on the sides of the reactor or bulk handling system. This sieve-residue test also measures non-carbonaceous undesirables, such as material from the bricks lining the reactor and metallic particles from the processing equipment.
- Sulfur (ASTM D1619) – this impurity comes from the feedstock and is present as elemental sulfur, inorganic sulfate, and organo-sulfur compounds.

For many industrial rubber products, there is an extreme sensitivity to both carbonaceous and non-carbonaceous sieve-residue contaminants. For example, in thin-wall molded products (e.g., brake diaphragms and Class-A, show-surface extrusion products like automotive window seals), these residues can cause appearance or even performance defects. Several carbon black companies offer very-pure carbon blacks, with minimal sieve residues under various tradenames.

12.2.4   Pellets

Carbon black exits the reactor with a bulk density of about 30 to 80 g/l. Carbon black in this "fluffy" form would be very difficult to handle as well as to incorporate into rubber. A pelletizer at the manufacturing plant densifies the carbon black to about 10-times this fluffy value, making it much easier to handle. One can then test the pellets themselves. ASTM D3313 measures the hardness of the pellets. ASTM D1511 measures their size. Lastly, ASTM D1937 measures the "mass strength," which inversely relates to the tendency to pack and thus flow poorly in a bulk system.

12.2.5   ASTM Nomenclature

The ASTM nomenclature system for carbon black, ASTM D1765, is detailed in Fig. 12.2. It is composed of a letter followed by three numbers. The letter in the nomenclature system indicates the effect of the carbon black on the cure rate of a typical rubber compound. "N" is used for the vast majority of rubber-grade carbon blacks, indicating "normal" cure rate. Carbon blacks made with the "oil furnace" process fit this category. The letter "S" indicates a slow cure rate and is typically used for carbon blacks that have a high concentration of oxygen-containing groups and are therefore acidic in nature. Channel-process carbon blacks,

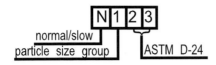

Position	Designation			
1st digit	N = normal cure rate (furnace & thermal blacks) S = slow cure rate (channel grades, modified furnace w/reduced cure rate)			
2nd digit	Particle Size Group (nanometers) now Surface area group			
	Group	Typical Na SA (m³/g)	Typical Particle Size	Former Identification
	0	> 150	1-10	-
	1	121-150	11-19	SAF
	2	100-120	20-25	ISAF
	3	10-99	26-30	HAF (EPC, MPC)
	4	50-69	31-39	FF
	5	40-49	40-48	FEF
	6	33-39	49-60	GPF
	7	21-32	61-100	SRF
	8	11-20	101-200	FT
	9	0-10	291-500	MT
3rd & 4th Digits	Assigned arbitrarily by ASTM D24, indicating differences within group (structure level, modulus level, or any physico-chemical property)			

Figure 12.2   ASTM nomenclature for rubber grade carbon blacks.

or after-treated carbon blacks, are an example. An acidic surface inhibits the amine-based accelerator systems typically used in rubber compounds [4].

The first number is inversely related to the average nitrogen surface area and directly related to the particle size; hence, a low number means a high surface area. The range of surface areas (from 0 to $150 \, m^2/g$ of carbon black) has been grouped into 10 categories, as shown in Fig. 12.2. Note that there are ASTM-recommended target values for iodine adsorption and DBP, with "typical" values given for other characteristics.

The third and fourth characters in the ASTM system are numbers arbitrarily assigned by the carbon black manufacturer that has successfully petitioned for an ASTM number. The latest ASTM list has 42 classified grades.

Each carbon black manufacturer, upon commercializing a carbon black that has non-ASTM analytical values, has the option of petitioning ASTM for a new number. This allows the manufacturer to "set the standard" for this new grade of carbon black, but it also tends to "commoditize" the grade.

12.3   Handling Carbon Black

The physical form of the pellets, as received by the rubber plant, starts the long list of items that affect the total cost/value of the carbon black. First, if the carbon black is in a bulk (railcar or hopper truck) or even a semi-bulk container (intermediate bulk container, or IBC), the carbon black must unload in a reasonable time. Once the product is unloaded from the container, it should travel through the customer's bulk system without plugging it. The carbon black must then flow into and out of the weigh hopper and next into the mixing equipment to ensure proper batch weight. Finally, the carbon black needs to disperse in the mix to the extent required in the allotted time. Poor physical form at the mixer (high dust, hard pellets) causes lengthy mixing times and/or poor dispersion. For all the above, the physical form of the carbon black is the key to ease of handling, as measured by pellet crush strength, pellet size distribution, fines (dust) level, and mass strength.

12.4   Mixing Carbon Black

12.4.1   Pellet Properties and Analyticals (also called Colloidal Properties)

The smallest dispersible carbon black unit is the aggregate, not the particle. Mixing serves to break up agglomerates (groups of aggregates) into individual aggregates. The physical form of the carbon black plays a major role, as hard pellets may not break up. Dust, on the other hand, takes a long time to wet out and get into the mix. Therefore, the carbon black manufacturer and the customer should work closely to understand pellet hardness for a particular grade of carbon black. Understanding the fine line between pellets that do not break up in the mixer (or are ''too hard'') and pellets that lead to high dust buildup at the mixer (or are ''too soft'') requires taking into account the:

- Shipping container
- Shipping distance
- Plant unloading and transport system
- Mixing equipment
- Formulation

In particular, a carbon black that is acceptable for a high shear compound may not break up and disperse in a low shear compound. Special ''soft-flavor'' pellets are ideally suited for some applications.

12.4.2   Effect of Analyticals on Dispersion

In addition to the pellet properties discussed above, the ultimate dispersion is also affected by the colloidal (analytical) properties of the carbon black itself, including surface area and

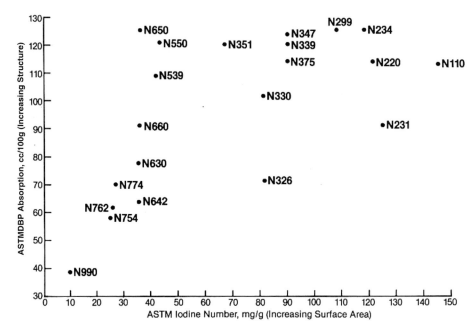

Figure 12.3   Carbon black spectrum.

structure. High-surface-area carbon blacks have a lot of cohesive force, and it takes significant energy/shear to break up the agglomerates into their constituent aggregates. Low-structure carbon blacks develop relatively low viscosity in the compound and thus relatively little shear in the mixer. Therefore, high-surface-area, low-structure carbon blacks are the hardest to disperse, and low-surface-area, high-structure blacks are the easiest. This phenomenon is of such practical importance that the spectrum of readily available carbon blacks, shown in Fig. 12.3, uses about one-half of the theoretical analytical space available. That is, as surface area increases, only carbon blacks of increasingly high structure are practical. Conversely, at high surface areas, there is no practical demand for low-structure carbon blacks, as they are too difficult to disperse. Note that it is relatively easy to wet out low-structure carbon blacks of even substantial surface area because their compact aggregates are readily coated with polymer, and they disappear into the mix early. However, they are very hard to disperse extensively in spite of the wetting factor.

12.4.3   The Mixing Process

In conventional mixing, the polymer is added first; then, after some time for polymer breakdown, carbon black is added in one or more stages. Sometimes a second, third, or subsequent stage is used to incorporate and disperse even more of the carbon black; or subsequent stages are also used to lower the viscosity of the mix. The last mixing stage is usually reserved for the addition of sulfur and curatives.

It is very important to choose the best carbon black for mixing and processing as well as final performance properties. Many times, an incorrect choice of carbon black, or an inappropriate blend of carbon blacks, can lead to excessive mixing time or passes without achieving the desired dispersion level. Therefore, compounders should choose the carbon black that both disperses best and performs as required in the end product.

The dispersion of carbon black takes place in three distinct phases within the mixing stages [5]:

1. Incorporation: First, the carbon black must be ''wetted'' by the polymer. The absence of loose carbon black can indicate the completion of wetting. More scientifically, the second power peak is used [6]. Lower structure carbon blacks wet out more quickly because they have fewer spaces in the aggregate that must be filled with polymer. Conversely, higher structure carbon blacks take longer to wet out. High levels of dust and fines, usually caused by pellets breaking down in transit and in conveyance through the customer's bulk system, cause the incorporation step to be lengthy. On the other hand, hard pellets do not break up; thus, the carbon black in them is not incorporated, much less dispersed.
2. Distribution: Distribution refers to the bulk homogenization of the mix. The incorporated or wetted carbon black must first be distributed uniformly throughout the mix and then dispersed into the mix on a more ''micro'' level. Distribution on a mill takes place via the ''cuts and end rolls'' that the mill operator performs. In an internal mixer, the rotor blades perform the ''cuts and end rolls,'' distributing the ingredients throughout the mix. There is little shear in this distribution process, so hard pellets are not broken up and fines that are not first wetted out are not distributed.
3. Dispersion: This is the ''final act'' in dispersing the carbon black agglomerates. Mixing shear must break up the carbon black agglomerates into aggregates to achieve the desired dispersion level. Agglomeration forces are stronger for carbon blacks with higher surface areas, making them harder to disperse, as noted earlier. Higher structure blacks usually disperse better because they impart a higher viscosity to the mix. With a mill, the shear that breaks up agglomerates into aggregates occurs at the nip between the two rolls. The back roll usually turns significantly faster to enhance the shearing action. In an internal mixer – either with a tangential or intermeshing rotor system – the shear is generated largely between the rotor tips and the mixing chamber shell; thus, tip clearance is very important. Hard pellets take longer to break up and can remain as undispersed lumps of carbon black when blended in a mix cycle designed for ''proper pellets.'' Any ingredient that is not wetted out – often the fines or dust – likely falls to the bottom of the mixer and is not subject to shearing forces. As noted earlier, both higher loadings of carbon black and higher structure carbon blacks develop higher viscosity and consequently, more shear in the mixer, while both lower loadings of carbon black and low-structure carbon blacks result in lower viscosity and less shear. Finally, the viscosity level of the polymer plays a major role in determining the level of viscosity and shear generated and the forces available to break up hard pellets and agglomerates.

12.5   Subsequent Processability of the Compound

Once the compound leaves the mixer, it must handle well in subsequent processes, such as calendaring, extruding, and building. The void areas in the carbon black formed by the clustering of the particles (high structure) are filled with rubber in the wetting and subsequent phases of dispersion. This removal of elastomer from the matrix results in what is known as "occluded rubber" [7], which increases the effective volume fraction of the carbon black in the compound. This causes the carbon black structure and loading to play a major role in compound viscosity and extrusion shrinkage. The polymer matrix has extremely low viscosity and high elasticity compared with the solid carbon black. The carbon black, by combining with the elastomer, inhibits shrinkage and raises viscosity. The more carbon black used or the higher its structure, the less shrinkage and higher viscosity are imparted to the compound [8,9].

12.6   Compounding Carbon Black

Although carbon black at a level of 2 to 4% makes a rubber appear to be black, it is usually added at much higher levels to reinforce the compound, not to color it.

The surface of the carbon black aggregates form physical and chemical bonds with the elastomer, dramatically improving failure properties such as tensile, elongation, tear, abrasion resistance, and flex fatigue. Figure 12.4 shows a proposed mechanism of reinforcement in which the polymer chains are attached to one or more carbon black aggregates, sometimes at multiple sites on a single aggregate, effectively helping to turn a loose bowl of "spaghetti" into a coherent mass. Failure properties are most strongly affected by the surface area of the carbon black available to interact with the elastomer. The smaller the carbon black particle size is (higher surface area per weight of carbon black), the more these failure properties are enhanced.

In addition, non-failure properties such as viscosity, extrusion shrinkage, hardness, modulus, electrical conductivity, thermal conductivity, liquid and gas permeability, and hysteresis are affected by both the loading level used and the analytical properties of the black, particularly its structure level. Non-failure properties are impacted by particle size, but loading and structure are usually the dominant parameters. Table 12.1 shows how changes in carbon black analyticals and loading levels will affect compound properties.

12.6.1   Optimum Loading

Table 12.1 also shows that carbon black loading plays a major role in compound performance. Optimum loading theory can be complex. Visualize a "pink" eraser on the tip of a pencil. This "gum" compound (no carbon black) abrades easily when rubbed

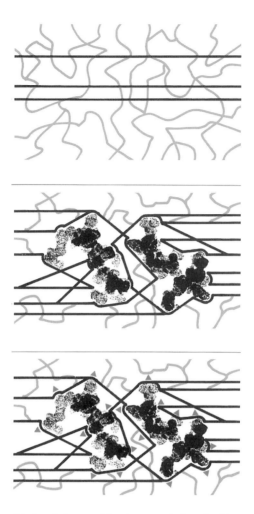

Figure 12.4   How carbon black reinforces rubber (top: unreinforced rubber consists of long and short molecules, and rubber breaks easily; middle: addition of carbon black aggregates causes rubber molecules to share the stress, and rubber does not break so easily; bottom: carbon black adds many points of friction, dissipating strain energy, and rubber does not rupture as easily).

across a sheet of paper. Now, visualize a can of carbon black with no polymer – just a can of carbon black. Imagine trying to somehow crush this carbon black onto a conveyor belt or the tread of a tire. As you can imagine, the ''pure'' carbon black offers no abrasion resistance at all. Now imagine carbon black and rubber mixed in various ratios. At lower concentrations (loadings), the rubber compound approaches the consistency of a pencil eraser and has very little abrasion resistance. At higher loadings the compound approaches the consistency of the can of carbon black and likewise has relatively little abrasion resistance.

In technical terms, carbon black can be ''underloaded'' or ''overloaded.'' The loadability is quite variable depending on the polymer. Also, the final application may

Table 12.1   The Effect of Increasing Surface Area, Structure, and Loading on Various Parameters

Parameter	Surface area	Structure	Loading
Optimum loading	Decreases*	Decreases*	Not applicable
Incorporation time	Increases	Increases*	Increases*
Oil extension potential	Increases	Increases*	"Part for part"
Dispersability	Decreases*	Increases*	Goes through optimum
Viscosity	Increases	Increases*	Increases*
Scorch time	Decreases	Decreases*	Decreases*
Extrusion shrinkage	Decreases	Decreases*	Decreases*
Extrusion smoothness	Increases	Increases*	Goes through optimum
Tensile strength	Increases*	Variable	Goes through optimum
Modulus	Increases	Increases*	Increases*
Hardness	Increases	Increases	"About 2 phr per point," a crude rule of thumb; see Table 12.3; differs from polymer base to polymer base
Elongation	Decreases	Decreases*	Goes through optimum*
Abrasion resistance	Increases*	Increases	Goes through optimum*
Tear resistance	Increases*	Variable	Goes through optimum*
Cut-growth resistance	Increases*	Decreases	Goes through optimum*
Flex resistance	Increases	Decreases	Goes through optimum*
Hysteresis	Increases*	Little	Increases* as square
Electrical conductivity	Increases	Little*	Increases*
Permeability resistance	Decreases	Little	Increases*

* Signifies a major effect.

require performance parameters that necessitate overloading or underloading in a given compound.

With the underloaded compound, the poor abrasion resistance is accompanied by high extrusion shrinkage, low viscosity, low hysteresis, and low tensile/tear values. On the other hand, the overloaded compound exhibits reduced tensile/tear values, but low shrinkage, high viscosity, and high hysteresis. Thus, overloading or underloading carbon black usually results in poor performance tradeoffs.

An example of the effects of loading on treadwear for various common tread-grade carbon blacks is shown in Fig. 12.5. Most practical tire compounds fall into the relatively narrow range of about 28 to 32% carbon black, whereas some industrial rubber product compounds contain more than 50% carbon black. The rubber industry does not talk in terms of percent of carbon black in the compound, but rather the percent of carbon black with respect to the polymer only (parts per hundred rubber, or phr). The phr of carbon black used falls generally be between 45 and 75 phr for tire and most IRP applications; however, some IRP compounds may contain more than 300 phr without being overloaded. This is especially true when dealing with highly loaded EPDM compounds. Also, optimum loadings are lower for more reinforcing carbon blacks, that is, it takes less phr to reach the "overloaded" situation with these carbon black types.

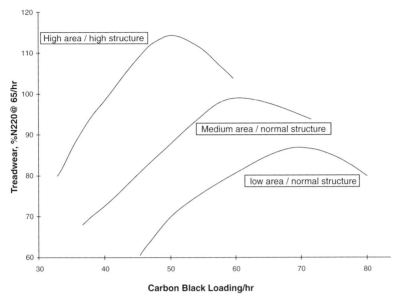

Figure 12.5   Treadwear vs. loading.

12.6.2   Importance of Dispersion

A poorly dispersed compound consists of some regions overloaded and other regions underloaded with carbon black. As discussed above, the resultant compound has relatively poor abrasion resistance, poor tensile and tear properties, and may have poor flex fatigue and variable processing properties, as well. Table 12.2 shows the dramatic drop-off in some physical properties with poor dispersion levels. To obtain the maximum reinforcement per carbon black expenditure, all compounds should first be optimized in terms of loading levels and dispersion.

Table 12.2   Effect of Banbury Mixing on Dispersion and Physical Properties in SBR Compound

Property	Banbury mixing time (min)					
	1.5	2	2.5	3	4	8
Dispersion	23.6	71.4	86.4	96.9	99.3	100.0
Mooney viscosity	133.0	122.0	114.0	97.0	83.0	68.0
Extrusion shrinkage	29.1	39.7	44.2	46.8	45.7	41.7
Tensile strength	2460.0	3110.0	3470.0	3760.0	3700.0	3750.0
300% Modulus	1850.0	2080.0	2020.0	1810.0	1820.0	1740.0
Elongation	380.0	460.0	490.0	540.0	530.0	540.0
DeMattia cut growth (Kc to 1 in.)	5.0	6.5	9.0	8.5	11.0	–
Electrical resistivity	124.0	88.0	108.0	175.0	300.0	440.0
Akron angle abrasion; loss cc/10^6 rev.	289.0	194.0	142.0	133.0	136.0	–

Material: SBR-1712 137.5; N220 69; StA 1.5; ZnO 3.0; A.O. 1.0; Santocure 1.1; Sulfur 2.0; data from Boonstra and Medalia.

12.6.3   Carbon Black Compounding Tips

12.6.3.1 Hardness

Passenger-car tire treads fall within a relatively narrow range of hardness [10]. It can be shown that compounds that are too soft wear poorly and can give poor (peak) wet traction [11]. On the other hand, compounds that are too hard can give high modulus/low elongation and a tendency toward reduced dry traction and wear. Low elongation can also cause tearing during demolding.

IRP products are often compounded to a specified hardness level and can have problems with hot mold removal, high modulus/low elongation, and poor flex fatigue, as described above.

The two major analytical properties of carbon black, surface area and structure, have a positive relationship to hardness. That is, all else being equal, both higher structure and higher surface area result in a compound with higher hardness. Finally, the level of carbon black needed to raise durometer varies greatly with the polymer. A major difference is seen between non-crystallizing polymers, such as SBR, and crystallizing polymers, such as NR. EPDM is in a class by itself, having a very high loading capability. Table 12.3 shows the phr needed to raise the durometer by about 10 points by using various carbon blacks in various polymers.

Also shown in Table 12.3, for many black/polymer types, about two parts of black, added or subtracted from a formula, raises or lowes the durometer respectively by one point. This general rule must be adjusted for carbon blacks that differ substantially in analytical properties from the "typical N330" used in Table 12.3, as well as for crystallizing vs. non-crystallizing polymers and, of course, EPDM.

The general rule is that one part of carbon black is approximately offset by one part of processing oil as far as the effect on hardness is concerned. That is, to maintain compound hardness, a good starting point is to either add or remove one part of black and one part of oil at the same time. As noted above, the hardness imparting power of carbon black is a function of both its surface area and its structure. So the general rule, which holds for an N330-type carbon black, must be modified when working with an extremely low- or high-surface area/structure carbon black. For example, if the compound contains an N660-type black, only about 0.7 phr of oil is needed to offset one part of black for hardness. At the other extreme, a black such as N134 may require about 1.3 parts of oil to offset a one -part change in black loading for hardness.

Table 12.3   Equivalent Hardness

Carbon black type	Area m^2/g	DBP cm^3/100 g	Parts to raise durometer by 10 points					
			NR	SBR	IIR	CR	NBR	EPDM
N990	8	34	42	51	38	35	48	68
N762/N774	26	63–70	28	34	25	22	32	45
N660	35	91	25	31	23	21	29	41
N330	80	103	19	23	17	15	21	30
N339	90	124	17	21	16	14	20	28

As may be deduced from the above discussion, if the addition of one part of oil is approximately equal to the addition of one part of black for a hardness change, and if a change of two parts of black are needed to change the durometer by one point, then the addition/removal of two parts of processing oil will also change the durometer of a compound by one point. Higher surface area/higher structure carbon blacks require the addition of a little more oil, and vice versa for lower surface area/lower structure carbon blacks.

12.6.3.2 Processing Oil

Oils can either be added to the polymer itself during manufacture (where they are called extender oils) or to the rubber compound (where they are called process oils). Process oils, when added to the rubber compound at 5 to 20 phr, are considered to be process aids. Extender oils are normally added by the polymer manufacturer at levels of 37.5 phr and up, as, for example, in SBR.

Oils are added either to the polymer or to rubber compounds primarily to lower viscosity and to reduce both the stress/strain resistance and the hardness of the finished product. In turn, the wetting and lubricity characteristics imparted by the oil aid in the mixing process and virtually all subsequent processes, including calendaring, sheeting, extruding, transfer molding, and mold flow. Because process oils are much less costly than either the polymer(s) used or the carbon black, adding process oil also reduces the compound's cost. However, too much process oil can lead to low hardness, low modulus, high elongation, "blooming," poor abrasion resistance, and mold fouling in injection molded applications.

Carbon black and process oil tend to affect such properties as viscosity, hardness, modulus, and elongation in opposite directions, for example, more carbon black increases viscosity, whereas more oil reduces it. As a result, oil and carbon black are frequently increased or decreased simultaneously to maintain the starting level of these properties. As noted above, some general trends emerge:

- The simultaneous addition of one part of an N300-type carbon black and one part of a processing oil approximately cancel each other with respect to hardness.
- As the carbon black becomes more reinforcing, the addition of as much as 1.3 phr of process oil is necessary to maintain hardness (N134 type); when the carbon black becomes less reinforcing, the addition of only about 0.7 phr processing oil may be necessary (N700 type).
- The durometer of a typical non-crystallizing polymer is lowered by about one point by the addition of 2 phr of process oil above the normal formula.

Commercial rubber process oils are generally obtained from the distillation of crude oil after gasoline, fuel, and other lower boiling fractions are removed. There are three principal types of rubber process oils:

1. Aromatics
2. Naphthenic
3. Paraffinic

All three classes are made up primarily of ring structures, with the classification referring to the preponderance of an aromatic or saturated ring structure. The paraffinic designation is a misnomer, as the rings are still mostly saturated, but fewer of these are unsaturated and more unsaturation occurs in side chains. In addition, there are two sub-classes:

4. Relatively napthenic
5. Relatively aromatic

ASTM D2226 recommends four classes for petroleum extender oils, with type 101 the most polar/least saturated/highest asphaltenes and type 104 the highest in saturation with the lowest asphaltenes and polar content.

The degree of classification, in turn, determines the compatibility of the oil with the polymer system according to the ''like dissolves like'' axiom of organic chemistry. A compatibility table is usually based on the viscosity-gravity constant (VGC) of the oil, which is a measure of aromaticity. This measure strongly correlates with actual molecular analyses, but can be obtained much faster and cheaper. Higher VGC means greater compatibility with SBR, BR, and neoprene, while lower aromaticity means greater compatibility with EPDM, butyl, polyisoprene, natural rubber, and thermoplastic polymers.

Other major effects of high aromaticity include staining and poor low temperature properties, but shorter mixing times because of better carbon black dispersion and shorter oil take-up times. While aromatic oils are less volatile, they are potential carcinogens and require labeling as such under the OSHA Hazard Communication Standard. Also some naphthenic oils require such labeling.

12.6.3.3 Other Vulcanizate Properties

Most carbon black suppliers have a great deal of information available to evaluate various carbon blacks over a range of loadings in a given polymer. Graphs are typically shown with measured physical properties plotted on the ''Y'' axis vs. loadings on the ''X'' axis. Typically, all major tire and IRP polymer types are covered. A second type of bulletin readily available from most carbon black suppliers features a given carbon black tested over a range of carbon black/oil loadings in a given polymer. Again, this information covers the major polymer types and most popular carbon black grades. The work illustrates the compounding principles discussed in Table 12.1 of this chapter.

12.6.3.4 Vulcanizate Hysteresis

Hysteresis, in practical terms, means energy absorption [12]. When cyclically stressed, as in most practical applications, rubber compounds are viscoelastic. That is, they store and then return part of the energy of deformation (elastic behavior) and absorb a relatively small fraction of the energy input, converting it to heat (viscous behavior). The absorbed energy causes the rubber sample to get hot. In nearly all tire applications, the absorption of energy *per se* by the compound is not wanted, as it results in higher rolling loss for the tire and thus higher fuel consumption for the vehicle. Note that there is an exception in racing tire and ultra-high performance passenger tire tread compounds, where the high energy absorption results in excellent traction. Similarly, many IRP applications also require low hysteresis compounds: motor mounts, conveyor belts, and power-transmission belts, to name a few.

Energy absorption can be measured in several ways. First, the rise in temperature can be measured directly as in various ''heat buildup'' tests. Second, a simple rebound test measures the energy input and the energy returned directly. Lastly, sophisticated dynamic test machines directly measure the forces involved under relatively precisely known and variable input conditions. Such tests measure:

- Sample temperature
- Initial sample strain or stress level
- Mode of deformation (e.g., stress or strain under conditions of extension, compression, or shear)
- Level of deformation
- Frequency of the deformation

The cyclic nature of the applied stress/strain on the sample can also be specified, e.g., pulsed strain or sinusoidal deformation. With dynamic test machines, under compression or extension, the absorbed energy lost as heat is measured directly as the loss modulus, represented by E'' (loss modulus). The recoverable or stored energy is measured as the elastic modulus, E' (storage modulus). The ratio of the energy absorbed to the energy returned – E''/E' – is widely used. This quantity is known as ''tan delta'' (tan δ) because it is the tangent of the angle formed (delta) from a vector force analysis of the loss and elastic moduli. This results in a third term, the complex modulus, E^*, which is the hypotenuse of the triangle formed from the vector analysis. As the hypotenuse, the complex modulus is related by the Pythagorean theorem to the loss and elastic moduli, as shown in Equation (12.1):

$$(E^*)^2 = (E'')^2 + (E')^2 \qquad (12.1)$$

Finally, the loss modulus is used under conditions of constant strain deformations, making this value an especially suitable parameter for tire sidewall compounds, for example. The tan δ is commonly used for constant energy deformations, or, more generally, when the exact mode of deformation of the sample is not known. For constant stress applications, a new term, loss compliance ($E''/(E^*)^2$) is used. Note that for most vulcanized rubber compounds, the storage modulus, E', is approximately equal to the complex modulus, E^* at rubbery state, so the loss compliance (D'') is approximately tan δ divided by the complex modulus. This means that high modulus lowers the loss compliance, as shown in Equation (12.2). Loss compliance = $D'' = E''/(E^*)^2$, but as $E^* \approx E'$, then

$$D'' \approx E''/(E'E^*) \approx \tan \delta / E^* \qquad (12.2)$$

Structure does not influence constant energy hysteresis such as tan delta or a simple rebound test. However, because structure affects modulus, it plays a role in hysteresis tests that are conducted with either a constant stress or constant strain mode of deformation. For example, under constant strain testing, a higher modulus compound reaches a higher stress level, that is, it has a larger area under the stress/strain curve, and more work is done on it. The compound absorbs more energy (gets hotter in a heat buildup test, for example) than a compound with the same tan δ but lower modulus. Conversely, a compound with a higher modulus deforms less when tested at constant stress conditions, thus having less work done on it. The bottom line is that in applications that tend towards constant stress (e.g., motor mounts) or constant strain (e.g. tire sidewalls) deformations, modulus and thus carbon black structure, can have a significant influence.

Surface area is the driver for constant energy tests and plays a major role even in constant stress or strain hysteresis testing. More on this is discussed in the next section on loading.

Carbon black-loaded compounds have much higher hysteresis than gum compounds. In fact, the higher the carbon black loading is, the higher the hysteresis, at least up to the point where the compound is still viable. Later, the fact that carbon black loading is the single most important factor in determining the contribution of the carbon black to the compound's hysteresis is discussed further. Because of the effect of overloading, it is important to note that an overloaded compound always has more hysteresis than an optimized compound, but may actually have lower reinforcement. Indeed, too much of a *good thing* can be bad!

After trying many "best-fit" equations to match rebound values (constant energy test) against carbon black loading, polymer type, and analytical effects, it has been found that a highly correlated relationship exists when the loading factor (% black in the compound) is squared while the surface area factor is linear, as shown in Equation (12.3) [13]:

$$\text{Hysteresis factor} = \left(\frac{\text{phr carbon black}}{\text{total phr compound}} \right)^2 \times \text{surface area} \qquad (12.3)$$

When the resultant hysteresis factor number is plotted against observed constant-energy hysteresis for a wide variety of carbon blacks over a wide range of loadings in several polymer types, the fit is excellent. This is illustrated for natural rubber in Fig. 12.6. This simple relationship can be used to predict changes in hysteresis from carbon black type and/or loading changes, or used to calculate the loading and/or surface area of a new grade of carbon black that gives the same hysteresis as the starting compound.

The mysterious parameter surface activity plays a small role in hysteresis. Higher surface activity leads to tighter bonding with the elastomer and less slippage between the polymer chains and the carbon black surface, resulting in less hysteresis. However, in the normal realm of surface activity available, this is a minor factor. To get even a 5 to 10% reduction in hysteresis via surface activity, the carbon black must be left with a relatively high extract level on the surface, which is generally undesirable. Recently, modified carbon blacks with active groups on the surface that can be chemically bonded to the elastomer with coupling agents such as bis (3-triethoxysilyl-propyl)-tetrasulfide (TESPT) have been introduced. Manufactured by co-fuming carbon blacks and silica, these products have been referred to as carbon-silica dual-phase fillers [14] and are claimed to give up to a 30% reduction in hysteresis with no loss in abrasion resistance.

12.6.4   The Tire Industry's Tradeoffs

Wear, traction, and rolling resistance are three major tire performance parameters significantly affected by the tread compound, which, in turn, is largely affected by its

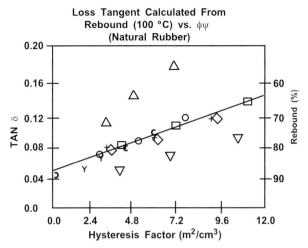

Figure 12.6   Calculated hysteresis factor vs. hysteresis for various black types; △ = graphitized; ▽ = oxidized; balance are unconventional grades.

carbon black type and loading. From a carbon black standpoint, traditional techniques available to increase wear resistance may sacrifice rolling resistance because of the concomitant increase in the hysteresis of the tread compound. For example, more reinforcing carbon black with a higher surface area can replace another carbon black "part for part," causing both tread wear resistance and hysteresis to increase. The entire range of commercially available tread blacks used varies in wear resistance about ±13% from an N220 grade,[*] as shown in the following list in which these blacks are compared to N220 (relative wear resistance 100):

- N351 = 87[*]
- N339 = 95
- N220 = 100
- N234 = 108
- N134 = 113

To obtain slightly better wear with often lower hysteresis, instead of substituting the various carbon blacks "part for part" as in the recipe above, the loading of each can be adjusted to be closer to its optimum loading. As a result, the wear level shifts higher by a few percent for all grades, and the wear differences between grades are minimized. With both of these techniques, and within the range of commercially available carbon blacks, it is extremely difficult to gain more than 1% in wear resistance without incurring a 1% penalty in constant-energy hysteresis (rebound or tan δ). Similarly, it is very difficult to remove 1% in hysteresis without sacrificing 1% in wear resistance.

Unfortunately, the tire tradeoff does not stop there. Thus far, we have talked about only two performance parameters: wear and rolling resistance. The third, traction, is also involved in a tradeoff. From a carbon black standpoint, traction is affected by the hardness and modulus imparted and by the hysteresis generated in the rubber compound. To

[*] All tested on retreaded multi-section radial tires in a 65 ESBR/35 BR; 65 black/35 oil compound.

increase the traction of a compound via carbon black, higher loadings and/or higher surface-area carbon blacks are used to increase the energy absorption of the compound. At the same time the oil content is adjusted to keep hardness and modulus in the practical range. An increase in the hysteresis of the compound results in higher rolling loss for the tire.

12.7   Hysteresis Reducing Tips

Here are three creative ways to reduce hysteresis with carbon black while minimizing losses in reinforcement.

12.7.1   "Radical Compounding"

Radical compounding is really nothing more than a good application of optimum loading theory. Ordinarily, one would not think of using a very-high-surface-area carbon black in a reduced hysteresis application, as the hysteresis varies directly with the surface area of the carbon black. However, let us combine two points from the previous discussions to explore this further:

- Point: the optimum loading of a highly reinforcing/high-surface-area carbon black is lower than that for a carbon black with lower surface area
- Point: the hysteresis varies with the square of the loading, but only linearly with surface area
- Then, a lower loading of a very-high-surface-area carbon black yields both abrasion and hysteresis benefits

This technique, illustrated in Table 12.4, is one way to gain a beneficial tradeoff in wear/ abrasion and hysteresis/rolling resistance. However, for tire tread applications, this technique has three major limitations. First, with the removal of hysteresis, an energy

Table 12.4   "Radical Compounding" in 65ESBR/35BR System

	Tire wear	$\tan \delta$ at 60 °C
N220, 65 phr black/35 phr oil	100	100
EXP 1*, 65/35	116	137**
EXP 1, 58/23	114	138
EXP 1, 50/14	120	116
EXP 1, 40/5	129	89

* $150 \, m^2/g$ product not ordinarily thought of as having the potential to use in a low hysteresis application
** Separate test reflects testing variation; rebound was 3 points lower than with the 58-phr compound

absorbing mechanism, traction may drop as noted above. In most other applications of this technique (e.g. conveyor belts) and in non-tread tire applications, this is not a factor. In addition, the wear resistance gained is large enough to possibly allow the substitution of higher T_g polymers to offset the traction loss while maintaining some wear benefit. Second, the technique can only be used when there is sufficient oil in the compound to allow a reduction in both oil and carbon black loadings simultaneously so the durometer can be maintained. Third, removal of the oil and the use of high-surface-area carbon blacks both raise compound costs.

12.7.2   Lower Loadings of High Structure Carbon Blacks

Another hysteresis reducing technique is available if the compound is produced to have a high tensile/tear value. The traditional approach is to use a low- or normal-structure carbon black (90–115 DBP) with moderately high surface area, such as N220 or N231. The low structure level results in low modulus and thus high elongation, while the moderately high surface area of the carbon black provides high tensile/tear. The resultant high ''tensile × elongation'' product is generally believed to correlate with improved cut/chip performance. The limitations of this conventional technique are quickly reached because high-surface-area/low- or normal-structure carbon blacks are very hard to disperse, as discussed previously. The resulting poor dispersion can actually cause a loss in tensile/tear values.

An interesting variation is to use a high-structure carbon black and reduce the loading to give the same low modulus/high elongation values. One advantage, based on the hysteresis factor discussion above, is that the resultant compound has lower hysteresis because of the low carbon black loading. A second advantage is that the higher structure carbon black is easy to disperse. A small potential disadvantage is the same as for ''radical compounding,'' in that the technique is limited to compounds that have sufficient processing oil content to allow the simultaneous reduction of carbon black and oil to maintain hardness. However, the reduction in carbon black loading is only on the order of 5 to 10 phr, and the higher structure of the reduced loading carbon black offsets the loss in hardness to some degree, making this technique applicable to many compounds. The technique is detailed in Table 12.5 for an all-natural rubber compound.

Table 12.5   Low Loading of ''N2OO'' High-Structure Blacks

	N220	N231	N234	N234	N234
DBP	115	92	125	125	125
Loading, phr	55	55	55	49.5	44
Tensile, relative	100	101	106	107	109
Modulus, relative	100	79	113	101	83
Elongation, relative	100	111	96	102	111
Hardness, relative	100	97	100	98	96
Tan δ, relative	100	100	103	96	89
Tensile X elongation	100	112	102	109	121
Treadwear	100	103	109	114	Not tested

All properties are relative to the N220 compound at 55 phr loading expressed as 100

Table 12.6   Carbon-Silica Dual-Phase Fillers

Type	Lab abrasion loss (21% slip, modified Lambourn abrader)	tan δ (70 °C, 10 Hz, 10% DSA)
N234	100	100
Ecoblack^{TM} *	98	67

* 100 solution SBR, 50 phr filler, Ecoblack carbon black coupled with TESPT.

12.7.3   Carbon-Silica Dual Phase Fillers

Finally, newer carbon blacks, such as the recently introduced carbon-silica dual phase fillers (tradenamed ECOBLACK fillers by Cabot Corporation), claim to reduce hysteresis while maintaining or improving abrasion (see Table 12.6).

Manufacturers of these products note that there are several advantages of these fillers over silica. They use less coupling agent and are, therefore, less expensive. They yield compounds that are semi-conductive, and the filler is not as abrasive to process equipment as silica. In addition, this dual phase system needs less heat treating. The downside is that these new fillers plus a coupling agent will add to the cost of the compound when compared to traditional carbon black.

12.8   Practical Applications: Tire Examples

12.8.1   OE Passenger-Tire Treads

The tradeoff between rolling resistance, wear resistance, and traction favors rolling resistance for North American original equipment (OE) passenger tires. This is a result of Department of Transportation (DOT) Corporate Average Fuel Economy (CAFE) regulations in the U.S. Traction requirements must also be met, but rolling resistance is the first hurdle. Wear, as expected, suffers. From a carbon black standpoint, the net result is that relatively low-surface-area reinforcing grades such as the N300 types used at relatively low loadings (40 to 60 phr, often less than optimum) are favored. In addition, the polymer system is often a blend of three or four elastomers, carefully chosen to minimize hysteresis, while enhancing traction. Natural rubber, BR, and one or more solution SBRs may be used.

12.8.2   Replacement Passenger-Tire Treads

The tradeoff with replacement tires favors wear resistance with rolling resistance and traction that is "less than optimal." In North America, these compounds are blends of ESBR and BR with moderate loadings of N300 or N200 series carbon blacks. A typical tread might be 65 ESBR/35 BR, with 65 phr of an N300 or N200 carbon black and 35 total parts of processing oil. In Europe, these treads tend to have higher SBR content, higher black loadings (70 to 90 phr), and tend to focus on N200 series blacks.

12.8.3   HP Passenger-Tire Treads

In North America, the high-performance (HP) tradeoff favors traction, and both wear and rolling resistance suffer considerably. The generally practiced technique is to use 70 to 90 (or higher) phr carbon black with high oil to get the desired durometer level. The high energy absorption from the high loading of carbon black contributes to traction. In fact, often the carbon black chosen for the HP tread is of a moderate to high surface area, $120\,\text{m}^2/\text{g}$ or higher. The high carbon black loading is often combined with high percentages of SBR in the polymer system to enhance traction, and high styrene SBRs may even be used. Wear, however, is at best on a par with a typical replacement tread, and hysteresis is very high.

12.8.4   Medium Radial Truck Treads

In smaller sizes, the tread may be a blend of natural rubber and BR, but beyond a certain size, the polymer system is 100 % natural rubber. Carbon black loadings are limited because of the inability to use much oil with natural rubber; therefore, loadings of 40 to 60 phr of an N200 or N300 series carbon black are typically used. For cut/chip applications, a lower structure N200 type, such as N231, may be used, or the lower loading technique discussed in Section 12.7.2 can be used. Silica may be used at the 10 to 15 phr level to increase cut/chip, as well.

12.8.5   Wire Coat or Skim Stocks

In this application, the polymer is nearly exclusively natural rubber and the carbon black of choice almost universally is N326, although other carbon blacks are occasionally used. N326 is preferred because it offers high green strength and reinforcement at a low compound-viscosity level for penetration into the cord. Loadings of 45 to 60 phr, sometimes with some silica as part of the adhesion system, are typical.

12.8.6   Innerliner Compounding

For many tires, the preferred polymer is a halogenated butyl, although natural rubber can be used depending on the permeability/green strength/tack/cost tradeoffs needed. The carbon black used is of the semi-reinforcing variety, usually N700 or N600 series. Recently, a novel oil-furnace carbon black, which can be used at high loadings to decrease permeability, was introduced to the market [15]. The high loading enables the rubber chemist to take advantage of the fact that even bromobutyl is about 120 times more permeable than carbon black.

12.9   Major Tradeoffs for Industrial Rubber Products

The reinforcement/hysteresis tradeoff is very similar to that of wear and rolling resistance, with two differences. In most IRP applications, there is no traction requirement, and general reinforcement such as tensile/tear is needed, rather than pure abrasion resistance. The need to maximize abrasion resistance while minimizing hysteresis is important in caster wheels, solid tires, and other "wear surfaces." Conveyor belts, in particular, need improved abrasion and cut/chip resistance with minimum energy absorption. The heat generation in many other IRP products lowers product life and, in extreme cases, can cause premature failure; tank tracks are a classic example of this phenomenon. The three techniques of "radical compounding" (Section 12.7.1), carbon-silica dual phase fillers (Section 12.7.3), and lower loadings of high-structure carbon blacks (Section 12.7.2), can all find application in IRP areas.

12.9.1   Loading/Reinforcement/Cost

Unlike tires, many IRP products use relatively more costly polymers. With polymers such as polychloroprene, chlorinated polyethylene, many butyl-based polymers, some EPDMs, fluoroelastomers, chlorosulfanated polyethylene, and polyacrylates, the pound/volume cost of the carbon black is less than that of the polymer. The IRP compounder's goal is to "load up" the compound with carbon black to lower compound costs. However, increases in viscosity, hardness, and hysteresis can cause processing and product performance problems, requiring that the loading be cut back to a more reasonable level. This problem is similar to the tire compounder's desire to "load up" the innerliner both for cost and permeability reasons; one solution may be to apply these "innerliner" carbon blacks to some IRP applications.

12.10   Compounding Tips: Industrial Rubber Products

12.10.1   Extrusion Profiles and Products

Many extruded parts are automotive-related: door and window seals, for example. The automotive industry seeks improvements in the gloss, smoothness, and blackness of these parts at lower costs. Product performance improvements, as measured by lower permanent set and increased tensile/tear, are also sought. The part producer, in turn, looks for lower compound costs, increased microwave receptivity, and reduced scrap rates. Carbon black

Table 12.7   Carbon Black's Effect on Extrudates

Extrusion property	Increasing surface area	Increasing structure
Shrinkage	Decreases	Decreases
Smoothness	Decreases	Increases
Temperature	Increases	Increases
Extrusion rate	Decreases	Variable
Green strength	Increases	Increases

plays a major role in all these requirements. Table 12.7 shows the importance of carbon black type to the extrusion market.

Improper pellet properties and impurities, as noted earlier, may cause blemishes on extruded surfaces, resulting in higher scrap rates. Excellent dispersion is therefore necessary for both consistent extrudate dimensions and surface perfection. The best carbon black for overall extrusion performance is an extremely clean, low-surface-area, high-structure carbon black. Several carbon black manufacturers have products targeted to this market.

12.10.2   Molded Products

Many molded products, such as seals and O-rings, must meet severe environmental requirements, including resistance to high temperatures and exposure to fluids. The particular processing and curing operations used to form these parts can vary widely, and a broad range of polymers can be used depending on environmental constraints. Once the application and manufacturing process have determined the polymer, then the IRP compounder can select the type and loading of carbon black. Compounds for wear surfaces may require N200- and N300-series carbon blacks. Molded compounds for dynamic applications usually use high-structure, medium-surface-area carbon blacks to limit hysteresis. Carbon black manufacturers have recently targeted this market, and can recommend specific carbon blacks. Environmental requirements often mean expensive polymer systems. Low-surface area/low-structure carbon blacks are used so the compounder can ''load up'' the expensive polymer as noted earlier. These carbon blacks also keep the compound viscosity low, which is necessary for high production rates in injection molding, for example.

12.10.3   Hose Applications

A hose is exposed to both external environmental factors, such as temperature and fluids, and internal environmental factors, i.e., whatever the hose is conveying. In the mid-1980s, a new failure mode for both automotive radiator and heater hoses was recognized: electrochemical degradation. It was discovered that compounds with lower conductivity were needed to reduce or eliminate this early failure mode. Electrical conductivity is largely influenced by the loading and type of carbon black used. Carbon black manufacturers have responded with specific products targeted at this market.

12.11   Basics of Carbon Black Manufacture

12.11.1   History

The term "carbon black" refers to materials that are essentially elemental carbon and that can be made by one of several processes. The most general statement that can be made about carbon black is that it is made by partial combustion or thermal decomposition of hydrocarbons in the vapor phase. This section discusses only the oil-furnace process of carbon black manufacture, a process developed in 1942 that "grew up" with the synthetic rubber industry. The synthetic rubber industry, in turn, grew because of concerns about the availability of natural rubber from the Far East during World War II. Today, the oil-furnace process accounts for nearly all of the carbon black used in the rubber industry. The major exception is thermal black, which is used mostly in industrial rubber product applications.

12.11.2   The Oil-Furnace Process

The oil-furnace process is a continuous process in which heated oil (feedstock) is carefully atomized and sprayed into a gas or oil flame. The key is that the amount of oxygen available is carefully limited so that the oil is "cracked" (hydrogen removed, carbon left) rather than burned to produce heat. A small amount of the feedstock is burned in the process to produce the heat necessary to crack the remaining feedstock. After the feedstock is atomized and injected into the hot primary flame, the reaction is quenched via injected water. This lowers the temperature to such a point that the end product, carbon black, is not burned off.

When precisely controlled, the non-carbonaceous components of the feedstock (primarily hydrogen) burn off and leave behind almost pure carbon, maximizing the yield of carbon black from the feedstock. Formation of carbon black in the reactor from feedstock and subsequent quenching of the reaction comprises the "front end" of the manufacturing process. The amount of oxygen permitted (combustion ratio), the amount of feedstock burned, the resultant reactor temperature, and the quench length (reaction time) are all precisely controlled and varied according to the grade of carbon black being produced.

The reactor thus produces a stream of hot combustion by-product gases, steam from the quench water, and the desired carbon black. The equipment located after the reactor – bag filter, pelletizer, dryer, and storage tanks – are collectively referred to as the "back end" of the process. The back end of the process serves primarily to separate carbon black from the by-products. To separate the carbon black, a main unit filter, or bag filter, is used. The bag filter consists of a series of temperature-resistant fiberglass bags that separate the

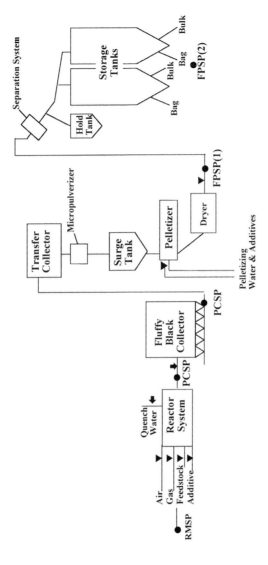

Figure 12.7   Oil furnace process; RMSP = raw material sampling point; FPSP = finished product sampling point.

carbon black from the stream of gases. At this point in the process, the carbon black has a bulk density of 30 to 80 g/l, too low for subsequent storage, shipping, and handling. Another piece of equipment, known as a pelletizer, increases the bulk density of the carbon black for subsequent handling. Because most pelletizers use both water and mechanical action to form pellets, carbon black contains a high moisture content after pelletizing, which necessitates in-line drying to decrease the moisture level. Finally, storage tanks of various configurations hold the carbon black before it is loaded into the final shipping container. An oil-furnace process schematic is shown in Fig. 12.7.

References

1. W. J. Patterson, Lecture to Akron Rubber Group, May, 1983 from Cabot internal work.
2. C. R. Herd, G. C. McDonald, and W. M. Hess, *Rubber Chemistry and Technology*, (1992) Vol. 65 (1) p. 107.
3. E.M. Dannenberg, *Rubber Age*, 98, Sept. 82; Oct 81 (1966).
4. G. R. Cotton, B. B. Boonstra, D. Rivin, and F. R. Williams, *Kauts. Gummi Kunsts.*, 22, 477 (1969).
5. J. M. Funt, *Rubber Chemistry and Technology,* Vol. 53, page 772, (1980).
6. G. R. Cotton, *Rubber Chemistry and Technology*, Vol. 57, 118 (1984) and Vol. 58, 774 (1985).
7. A. I. Medalia, *Rubber Chemistry and Technology*, Vol. 45, page 1171, (1972).
8. N. Minagawa and J. L. White, *Appl. Polymer Sci.*, 20, 501 (1976).
9. A, I. Medalia, *J. Colloid Interf.* Sci., 32, 115 (1970).
10. K. R. Dahmen and N. N. McRee, *Rubber World*, 170 (2), 66 (1974).
11. B. B. Boonstra, *Tire Science and Technology*, 2, 312, (1974).
12. S. G. Laube, Cabot Technical Bulletin TG 77 (1977).
13. A. I. Medalia, S. G. Laube, *Rubber Chemistry and Technology*, 51, 89 (1978).
14. M.-J. Wang, K. Mahmud, L. Murphy, W. J. Patterson, *Kauts. Gummi Kunsts.*, 51, 348 (1998); M.-J. Wang, W. J. Patterson, T. A. Brown, and H. G. Moneypenny, *Rubber and Plastics News*, 12, Feb. 9 (1998).
15. R. R. Juengel, D. C. Novakoski, and S. G. Laube, *Tire Technology International*, 1995.

13   Precipitated Silica and Non-Black Fillers

Walter H. Waddell and Larry R. Evans

13.1   Introduction

A wide range of non-black, particulate fillers is added to rubber compounds to improve the cured physical properties, reduce the cost, and/or impart color to the rubber product [1–3]. The chemical composition and its effect on rubber compound physical properties typically classify particulate fillers into three broad categories:

1. Non-reinforcing or degrading fillers
2. Semi-reinforcing or extending fillers
3. Reinforcing fillers.

Improvements in the physical properties are directly related to particle size, with the smaller particulate fillers imparting greater reinforcement to the rubber compound (see Fig. 13.1). Conversely, as the particle size of the filler decreases, its cost generally increases. In addition to the average particle size, the particle shape and particle-size distribution also have significant effects on the reinforcement imparted by particulate fillers. A particle with a high aspect ratio, such as kaolin clay or talc, provides greater reinforcement than a more spherical product. Particulate fillers with a broad particle-size distribution have better packing in the rubber matrix, which results in lower viscosity than that provided by an equal volume of a filler with a narrow particle-size distribution.

13.2   Mineral Fillers

Mineral fillers [4,5] are naturally occurring materials that are mined and ground to a specified particle size. Grinding may be done dry using mechanical mills. For a finer product, the ore is ground wet. Wet grinding may be autogenous, where the ore grinds by attrition with itself, or a grinding medium may be employed. Additional processing may include a combination of steps, including:

1. Separation of fine and coarse particles by screens, air or water flotation, or centrifugal filtration
2. Removal of impurities by washing, heat treatment, magnetic separation, or chemical treatment
3. Surface treatment with a variety of chemicals to improve the compatibility with the rubber matrix

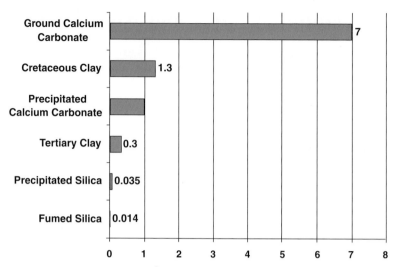

Figure 13.1   Median particle diameters (microns) of non-black fillers.

13.2.1   Calcium Carbonate

Ground calcium carbonate ($CaCO_3$), also known as whiting, limestone, marble, chalk, and calcite, is added to rubber compounds to reduce cost and impart hardness and opacity to rubber articles. Particle sizes range from 2 to 80 µm for dry-ground product and from 0.5 to 11 µm for wet-ground product. The chemical composition and crystalline nature depend on the limestone deposit mined (for example, chalk or marble) and the color of the deposit.

The most important applications of calcium carbonate by the rubber industry are in electrical wire and cable insulation where the low moisture content and natural insulating properties make it a preferred filler; in the production of articles where low cost and smooth surface appearance are desired, such as footwear; and in extruded hoses and automotive sealing parts. Typical calcium carbonate levels in rubber compounds range from 20 to 300 phr.

13.2.2   Baryte

Baryte is predominantly barium sulfate ($BaSO_4$) that is available in particle sizes from 1 to 20 µm. It is used as a filler when a high specific gravity is required in the rubber article. Typical rubber loading levels are 25 to 100 phr for articles such as stoppers and seals.

13.2.3   Ground Crystalline Silica

Crystalline silica (SiO_2) from sand or quartz can be ground and used as a degrading or extending filler for low-cost rubber articles. To minimize the health hazards of exposure to airborne crystalline silica, particle sizes are large, normally ranging from 2 to 20 µm.

13.2.4   Biogenic Silica

Naturally occurring silica (SiO_2), or dolomite, is often referred to as diatomaceous earth, because the primary deposits are the exoskeletons formed by diatoms that have extracted silicic acid from sea water and formed amorphous silica shells. Diatomaceous earth is usually very high in surface area because the shells retain the radial and/or rod-like structures of the living creature. The largest deposits are several million years old and have been partially converted to crystalline silica over time. Diatomaceous earth obtained from sedimentary rocks often contains up to 30% organic matter and inorganic impurities such as sand, clay, and soluble salts [6] requiring separation or treatment to remove crystalline content which could represent a health hazard. Biogenic silica is used as a semi-reinforcing filler or as a carrier for liquid compounding ingredients.

13.2.5   Kaolin Clay

Kaolin clay [7,8], also called kaolinite or china clay, is hydrous aluminum silicate [$Al_2Si_2O_5(OH)_4$], consisting of platelets with alternating layers of silica and alumina in the structure. The fine particles of clay are formed by the weathering of granite. Clay deposits are classified as primary, secondary, and tertiary. Primary deposits are mixtures of clay and granite that are found where the clay was originally weathered. Only 40 to 50% of the particles contained in these deposits are less than 2 μm in diameter.

Secondary deposits are formed when fine particles from primary deposits are carried by water flow and deposited in a new location. Tertiary deposits are the most important commercial deposits due to their fine particle sizes; more than 80% of the particles are less than 2 μm in diameter. Their high purity results when water carries the fine particles of a secondary deposit to a new location.

The most significant deposits of tertiary clay in the world are found in the southeastern U.S., in Cornwall, U.K., in the Amazon region of Brazil, and in Australia. Clay is broadly divided into soft and hard clay in that they produce softer and harder rubber compounds, respectively, at a given loading level. Because clay is mined as a fine particle-size material, it does not require significant grinding for use in rubber. There are five basic processes for producing clay for rubber reinforcement from the mined form:

1. Air-floated clay, in which the ore is milled to break up lumps and air classified, is the least expensive form of clay and imparts moderate reinforcement.
2. Water-washed clay involves gravity separation of impurities, bleaching, magnetic separation to improve color properties, and centrifuging to produce the desired particle size range to impart higher reinforcement by the control of pH, color, and particle size.
3. Delaminated kaolin requires chemical and/or mechanical means to break apart the platelet structure of the clay, which further increases the available surface area and reinforcement properties.
4. Metakaolin is partially calcined by heat treating to 600 °C.
5. Calcined clay is formed by heating to 1,000 °C, which produces a very white, high surface area mineral with an inert surface.

Clay is a widely used filler for rubber compounds of all types, including components of tires such as fiber adhesive compounds and the entire range of non-tire rubber applications where good reinforcement, moderate cost, and good processability are desired. Clays are commonly added to rubber compounds at levels of 20 to 150 phr.

13.2.6   Talc

Talc [$Mg_3Si_4O_{10}(OH)_2$] is a platelet form of magnesium silicate with a high aspect ratio. Because the platelets can orient in the extrusion process, it provides rubber extrudates with smooth surfaces that can be extruded at high rates. It is commonly used in compounds that have critical surface appearance requirements such as exterior automotive components or consumer goods.

Talc is used in tires in white compounds to provide a smooth appearance to the buffed sidewall. The large platelets of talc provide a barrier to gas and moisture permeability in compounds, which allows talc to be used in applications such as hydraulic and automotive hoses, barrier films, and tire innerliners. Talc is usually used in addition to other fillers with total filler content reaching 30 to 150 phr.

13.2.7   Alumina Trihydrate

Alumina trihydrate ($Al_2O_3 \cdot 3H_2O$) is not a naturally occurring mineral, but is rather an intermediate formed in the conversion of bauxite to aluminum. The major types of ATH are ground and precipitated. It is ground and classified by the same methods as other minerals, and is available in particle sizes from less than 1 μm to several micrometers in diameter. ATH provides some flame-retardant properties to rubber articles and may suppress smoke formation. The water of hydration of the ATH crystal is released beginning at 230 °C; it absorbs heat and provides water vapor to cool the rubber article and disperse smoke.

13.3   Synthetic Fillers

Synthetic fillers are generally manufactured by precipitation of soluble materials under carefully controlled conditions to provide tailored properties. They may be found as colloidal particles, which may be spherical, ellipsoid, or rod- or tube-shaped; as aggregates, which are covalently bonded groupings of individual particles; or as agglomerates that are loosely held associations of aggregates physically interacting. Reinforcement properties are a function of the colloidal particle size and shape, the aggregate dimensions and morphology, and the ability of agglomerates to break down during mixing. Additionally, the composition and surface chemistry of the filler plays a significant role in its interaction with rubber compounds.

13.3.1   Precipitated Calcium Carbonate

Precipitated calcium carbonate is formed by dissolving limestone and precipitating $CaCO_3$ as very fine particles using carbon dioxide (Aroganite Process) or sodium carbonate (Solvay Process). Typical particle sizes range from 0.02 to 2 µm. Most commercially available precipitated calcium carbonate consists of spherical, colloidal particles or aggregates consisting of a few spherical particles. Additional shapes and aggregates of varying morphology are possible and provide increased reinforcement. Precipitated calcium carbonate is used as a semi-reinforcing filler in shoe products and industrial rubber goods, particularly when resistance to alkali solutions is needed. The low moisture content and good reinforcement allow precipitated calcium carbonate to be used in wire and cable insulation applications, as well.

13.3.2   Metal Oxides

The rutile form of titanium dioxide (TiO_2) is an important filler for white and colored rubber articles. The ability of the titanium dioxide particle to scatter light provides high whiteness and opacity to rubber, which also means the filler particle can cover background colors. The anatase crystalline form of titanium dioxide is also used.

Zinc oxide, which is formed from the burning of zinc metal, was the first non-black filler used for reinforcement of rubber compounds. Although zinc oxide and magnesium oxide are still used as reinforcing fillers in some specialty compounds, particularly those that require heat resistance, their main role in rubber compounding in the last several decades has been as activators in sulfur cure systems or as curatives for chloroprene rubber compounds.

13.3.3   Precipitated Silica

Amorphous silica [9] consists of ultimate particles of the inorganic polymer $(SiO_2)_n$, where a silicon atom is covalently bonded in a tetrahedral arrangement to four oxygen atoms. Each of the four oxygen atoms is covalently bonded to at least one silicon atom to form either a siloxane (-Si-O-Si-) or a silanol (-Si-O-H) functionality. Surface silanol groups can be isolated from one another such that hydrogen bonding between the silanols cannot occur, vicinal to one another, promoting the formation of intramolecular hydrogen bonding, or geminal to one another, where two silanol groups are bonded to the same silicon atom (see Fig. 13.2).

Particulate silicas are made by precipitation from aqueous solution. The physical and chemical properties of precipitated silicas can vary according to the manufacturing process. Reinforcement and control of suspension pH, temperature, and salt content can change the ultimate particle and aggregate sizes of silicas precipitated from solution. The surface area, as determined by nitrogen [10] or CTAB [11] adsorption, is a function of the ultimate particle size. Ultimate particles can range from 5 to 50 nm in diameter.

Aggregates are three-dimensional clusters of ultimate particles (see Fig. 13.3), covalently bonded to one another via siloxane bonds, and range in size up to 500 nm in

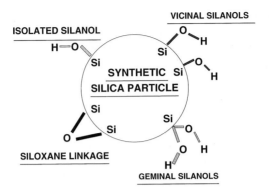

Figure 13.2   Silica surface groups: siloxane (−Si−O−Si−) and isolated, vicinal, and geminal silanols (−Si−O−H).

diameter. Aggregates can physically agglomerate through intermolecular hydrogen bonding of surface silanol groups of one aggregate to a silanol group of another aggregate, yielding structures up to approximately 100 μm in diameter. The median agglomerate particle size is generally 20 to 50 μm in diameter, but can be reduced by milling to approximately 1 μm. Precipitated silica is prepared from an alkaline metal silicate solution, such as sodium silicate in a ratio of approximately 2.5 to 3.3 SiO_2:Na, but using lower concentrations of silicate than is used in silica gel preparation [12]. In the absence of a coagulant, silica is not precipitated from solution at any pH value, but is precipitated by adding acid to sodium silicate to reduce the pH value of the hot suspension to 9 to 10, where the concentration of sodium ion exceeds approximately 0.3 N. Sulfuric acid is normally used to neutralize sodium silicate and precipitate silica.

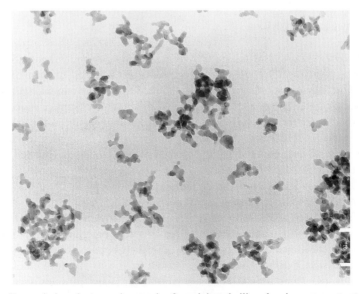

Figure 13.3   Transmission electron micrograph of precipitated silica showing aggregate structure.

Because of its small particle size and complex aggregate structure (see Fig. 13.3), precipitated silica imparts the highest degree of reinforcement to elastomer compounds among all of the non-black particulate fillers. This superior reinforcement is employed in a variety of rubber compounds for shoe soles, industrial rubber goods, and tires [13,14]. Precipitated silica is used in shoe soles for its resistance to wear and to tearing, its non-scuffing characteristics, and to obtain compounds with light color, or even transparent materials. Precipitated silica is used to improve the tear strength, resistance to flex fatigue (cracking, cut-growth), and heat aging of a wide variety of manufactured rubber goods, including conveyor and power transmission belts, hoses, motor and dock mounts, and bumper pads. Rubber rolls that are produced with precipitated silica for the abrasion resistance, stiffness, and non-marking characteristics, are used in paper processing and the dehulling of grains, particularly rice [15].

13.3.4   Silicates

Amorphous silicates [16] are precipitated from aqueous blends of soluble silicate, typically sodium silicate, and soluble salts of other metals. The most important types for reinforcement of elastomers are aluminosilicates with mixed Al_2O_3 and SiO_2 structures, magnesium aluminosilicates, and calcium silicates. The silicates have surface areas and resultant reinforcement which span the range from the highest surface area clays (ca. $30\,m^2/g$) to the lower end of the precipitated silicas (ca. $100\,m^2/g$).

13.4   Surface Treatment

Carbon black remains the particulate filler of choice for rubber articles because the inherent reinforcing effect of the non-black fillers in hydrocarbon elastomers is not comparable. This primarily results from the nonbonded interactions [17] established between the particulate filler and polymer functionalities. Surface chemistry plays an important role in the interaction of the non-black fillers and the polymer with contributions ranging from electrostatic interactions to covalent bonding to the polymer backbone. However, surface chemistry also strongly affects the interaction of the non-black filler with other chemicals in the rubber compound, particularly active metal oxides, curatives, and antidegradants.

Both surface morphology and surface chemistry play an important role in the interaction of a filler with coupling agents. For example, the dipole-induced dipole interactions between polar groups, such as siloxane and silanol, on the surface of silicas with non-polar groups (methyl, alkenyl, aryl) of hydrocarbon elastomers are weak compared to the dipole-dipole interactions resulting from hydrogen bonding between surface silanol groups in silica aggregates. In addition, the dispersive forces between a nonpolar molecule and silica are low, while those between a nonpolar molecule and carbon black are high [18]. Thus, materials that improve the compatibility between hydrocarbon elastomers and mineral fillers are of considerable interest.

HS-CH$_2$-CH$_2$-CH$_2$-Si-(-O-CH$_3$)$_3$
3-Mercaptopropyltrimethoxysilane (A-189)

H$_2$N-CH$_2$-CH$_2$-CH$_2$-Si-(-O-CH$_2$-CH$_3$)$_3$
3-Aminopropyltriethoxysilane (A-1100)

[S-S-CH$_2$-CH$_2$-CH$_2$Si-(-O-CH$_2$-CH$_3$)$_3$]$_2$
Bis(3-triethoxysilylpropyl)tetrasulfide (Si-69)

Figure 13.4   Structure of silane coupling agents.

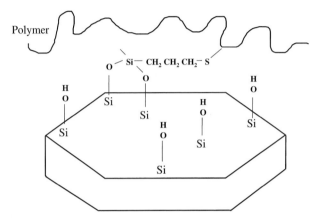

Figure 13.5   Illustration of silane coupling of polymer to silicate surface.

Organosilane coupling agents have been successfully utilized to further increase the physical properties of a number of non-black fillers including calcium silicate, clays, mica, silica, and talc. Clays pretreated with amino-functional or mercapto-functional silanes, and silicas pretreated with the mercapto-functional or tetrasulfide-containing (TESPT) silanes are commercially available (see Fig. 13.4). A manual of commercial couplants to promote adhesion between polymers and various substrates is available [19]. The reaction of a bifunctional organosilane with a silica or silicate particulate filler involves the hydrophobation of the alkoxy group of the silane with a surface silanol group of the silica or silicate, followed by reaction of the sulfur-containing function of the silane with an olefin group of the elastomer to afford a covalently bonded structure (see Fig. 13.5).

13.5   Compound Applications

Particulate fillers improve the cured physical properties of rubber compounds. All particulate fillers increase the hardness and stiffness of rubber compounds (see Fig. 13.6). All grades of carbon black, with the exceptions of N880 and N990 thermal blacks, reinforce rubber, with the finest particle size carbon blacks imparting the highest

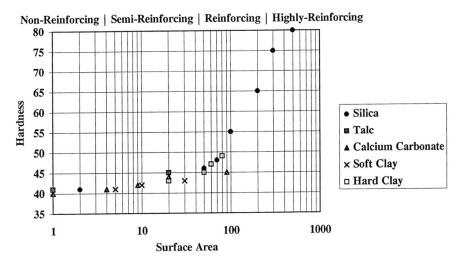

Figure 13.6   Reinforcing effect of non-black fillers with differing surface areas on hardness.

degree of reinforcement based upon improved abrasion resistance, tear strength, tensile strength, and stiffness [20]. A number of mineral and synthetic particulate fillers are used in the tire industry to extend and/or reinforce elastomers [14].

13.5.1   General Compounding Principles

The preparation of a successful rubber compound involves balancing several, often contradictory, changes in properties to provide the best performance for the application. In the broadest terms, the reinforcement potential of a filler increases as the particle size decreases, which corresponds to an increase in the surface area. Special performance characteristics of some non-black fillers, such as the ability to produce white or colored compounds, dimensional stability, vapor barrier properties, high loading levels, low cost, or electrical insulating properties, are also important reasons for their use.

In addition to the surface area, there are several important properties that affect the performance of a rubber compound. The shape of the particle is an important factor in that particles with a high aspect ratio, such as platelet or fibrous particles, have a higher surface-to-volume ratio, which results in higher reinforcement of the rubber compound. The greatest hardness is provided by rod-shaped or plate-like particles, which can line up parallel to one another during processing, compared to spherical particles of similar diameter.

Of the spherical particulate fillers, precipitated silicas, surface-treated clays, and metal silicates also produce high hardness, tensile strength, and modulus compounds. Use of clay in rubber produces compounds with high hardness. Clay is further classified as a hard clay if it reinforces rubber and also imparts high modulus, tensile strength, and resistance to abrasion. Clay is considered ''soft'' if it produces a compound with lower physical properties [8]. Fillers, such as precipitated silica, which have a three-dimensional structure of particles called aggregates (see Fig. 13.3) provide much higher mechanical interaction with the rubber molecules and result in the greatest reinforcement. No other mineral filler

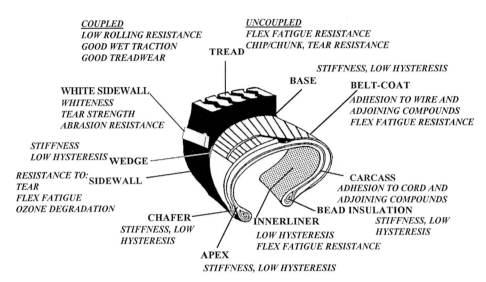

Figure 13.7   Beneficial effects of using precipitated silica in tire compounds.

approaches the precipitated silicas for imparting abrasion resistance [20]. Precipitated silica is used in a variety of tire applications (see Fig. 13.7). The reinforcement of rubber compounds by precipitated silica is directly related to the surface area of the silica [21,22], which is a function of the ultimate particle size.

13.5.2   White Sidewall

Tire white sidewall compounds are formulated for processability; cure rate compatibility with other rubber components of the uncured composite; hardness and stress-strain properties; adhesion to adjoining rubber components; resistance to tear, cut/crack propagation, abrasion, degradation by oxygen and ozone; original and UV-aged white-ness; and the lowest possible cost [23]. Typical formulations incorporate saturated elastomers such as HIIR and EPDM, which impart oxidation and ozone resistance to the compound, in combination with general-purpose diene elastomers [24]. Titanium dioxide is the whitening agent of choice; clay is used to reinforce the compound and talc as an extending filler [8]. Table 13.1 shows representative formulations of passenger tire white sidewalls for bias tires [24] (#1), radial tires for original equipment applications [25] (#2), and a low-cost compound that might be used on tires manufactured for the after-market (#3). Improvements in tear strength, flex fatigue resistance and whiteness at an equivalent compound cost are possible in Compound 2 if the clay is replaced with precipitated silica and talc [26] (see Table 13.2).

The addition of an aminosilane (see Fig. 13.4) or a mercaptosilane (see Fig. 13.4) to an EPDM/NR/SBR white sidewall formulation containing partial substitution of TiO_2 with clay increased the modulus and tear strength by 300% and decreased heat build-up and Tabor abrasion weight loss values as a function of silane level [27]. A 20 wt.% replacement of TiO_2 with mercaptosilane-treated clay in an EPDM/NR/IIR-based white

Table 13.1   Tire White Sidewall Formulations

Ingredient*	Controls #1	#2	#3
Chlorobutyl rubber, 1066	20.0	55.0	50.0 phr
EPDM, Vistalon 2504	20.0	20.0	20.0
Natural rubber, CV60	40.0	25.0	30.0
SBR 1502	20.0	0.0	0.0
Titanium dioxide, Titanox1000	30.0	35.0	20.0
Talc, Mistron Vapor	0.0	34.0	0.0
Clay, Nucap 290	30.0	32.0	70.0
Stearic acid	1.0	2.0	1.0
Sunolite 240	1.0	1.5	1.5
Sunpar 2280	0.0	4.0	0.0
Ultramarine blue	0.2	0.2	0.2
Zinc oxide	10.0	5.0	5.0
Rubbermakers' sulfur	0.5	0.8	0.5
Vultac 5 (APPS)	1.3	1.3	1.3
Benzothiazyl disulfide (MBTS)	0.8	1.0	0.8
Total	216.3	216.8	200.3

* Vistalon is a registered trademark of ExxonMobil Chemical Co.
Mistron is a registered trademark of Luzenac America, Inc.
Nucap is a registered trademark of J. M. Huber Co.
Sunolite is a registered trademark of C. K. Witco, Inc.
Sunpar is a registered trademark of Sun Refining.
Vultac is a registered trademark of Bayer, Inc.
Titanox is a registered trademark of Malvern Minerals.

Table 13.2   Optimization of White Sidewall Formulation #2

Ingredient	Control #2	Optimized #2
Clay	32.00	0 phr
TiO_2	35.00	27.50
APSS	1.30	0.80
Diphenylguanidine	0.00	0.60
Precipitated silica	0.00	17.50
Talc	34.00	42.50
Ultramarine blue	0.20	0.30
Property		
Scorch (Ts_2), min	3.40	3.40
Cure (T_{90}), min	14.60	13.30
Hardness	58.00	56.00
Abrasion Index	37.00	42.00
Hunter Whiteness Index, Original	59.60	63.90
Hunter Whiteness Index, UV aged	20.10	21.60
Tear strength, N/mm	12.80	17.60
Cut growth, mm at 100 kcycles	10.10	7.90
Cost/lb, $	1.01	1.00

sidewall formulation resulted in significantly increased 300% modulus, tensile strength, curb scuff resistance, and resistance to discoloration on aging [8].

13.5.3   Black Sidewall

The black sidewall is the outer surface that protects the casing of the tire. It is formulated for resistance to weathering, ozone, abrasion, tear, radial and circumferential cracking, and for good fatigue life [23]. Chemical antiozonants are added to the sidewall to impart effective resistance to ozone aging under static and dynamic service conditions, but these are depleted chemically by surface reactions with ozone and physically by curb scuffing and washing. N,N'-disubstituted-para-phenylenediamines are the most effective antiozonants, particularly the alkyl, aryl substituted; however, most practical antiozonants are also staining and thus can be used only in limited amounts [28]. Reinforcing precipitated silica improves the performance of conventional carbon black-reinforced sidewall compounds by extending the effectiveness of the antiozonant [29] (Table 13.3).

13.5.4   Wire Coat

The wire coat compound is composed of NR, synthetic cis-1,4-polyisoprene, carbon black, antioxidant, and a curative system with a moderate to high sulfur content. The compound is formulated to provide good adhesion to brass-coated steel wire and to adjoining rubber compounds, as well as tear, fatigue, and age resistance [23]. Silica is used with resin

Table 13.3   Tire Black Sidewall Compounds

Property	Control (0/50 phr)	Precipitated silica/black		
		(8/50 phr)	(8/46)	(8/42)
Scorch (Ts$_2$), min	4.40	4.10	4.20	4.30
Cure (T$_{90}$), min	10.70	10.40	11.00	10.40
Maximum torque, dNm	16.40	16.00	15.80	14.80
Hardness, Shore A	56.00	55.00	54.00	52.00
Tensile strength, MPa	23.00	21.90	21.10	21.90
Modulus at 100%, MPa	1.60	1.60	1.40	1.40
Elongation to break, %	658.00	682.00	681.00	714.00
Tear strength, N/mm	13.50	15.30	13.20	16.40
Cut growth, mm at 100 kcycles	25.00	25.00	16.60	16.10
G' at 0.5% strain, MPa	51.34	52.21	49.59	48.23
G'' at 2% strain, MPa	7.09	7.95	6.91	6.13
Ozone rating, days to failure	14.50	14.00	15.00	19.00

Formulation, Control-NR, CV60: 50 phr; BR 1220: 50; N330 Black: 50; Calsol 510: 10; stearic acid: 2; Sunolite 240: 1; Santoflex 13: 4; Wingstay 100: 1; zinc oxide: 3; sulfur: 1.8; Santocure MOR: 1.
Calsol is a registered trademark of R. E. Carroll, Inc.
Sunolite is a registered trademark of C. K. Witco, Inc.
Santoflex is a registered trademark of Flexsys, Inc.
Wingstay is a registered trademark of The Goodyear Tire and Rubber Co.
Santocure is a registered trademark of Flexsys, Inc.

Table 13.4   Precipitated Silica/Cobalt Containing Wire Coat Compounds

Property	phr Precipitated silica/black				
	Control	13.5/46.5[*]	20/35[†]	20/50[†]	25/42.5[**]
MDR at 150 °C					
Ts_2 Scorch, min	3.930	4.230	4.200	3.960	3.250
T_{50} Cure Time, min	6.490	7.760	8.500	7.970	7.370
T_{90} Cure Time, min	9.450	10.190	14.000	27.810	21.990
Minimum torque, dNm	2.920	3.220	3.040	4.590	4.500
Maximum torque, dNm	27.900	27.260	24.800	27.610	28.460
Hardness at 23 °C	72.000	69.000	59.000	68.000	68.000
Heat-aged	76.000	76.000	71.000	82.000	80.000
Humid-aged	76.000	77.000	72.000	81.000	79.000
Salt-aged	72.000	67.000	63.000	74.000	72.000
Hardness at 100 °C	69.000	68.000	58.000	62.000	62.000
Rebound at 23 °C, %	51.800	52.000	57.800	44.200	44.000
Rebound at 100 °C, %	68.800	69.200	69.200	57.800	58.600
Tensile strength, MPa	29.160	28.830	28.420	24.310	23.940
Elongation to break, %	494.000	523.500	641.800	544.400	540.100
Modulus at 20%, MPa	1.320	1.080	0.780	1.100	1.000
Modulus at 100%, MPa	4.250	3.430	1.950	2.820	2.800
Modulus at 300%, MPa	16.560	14.440	7.340	11.240	11.150
Tear Strength, N/mm	9.180	11.140	16.870	11.740	16.110
Cut Growth, mm at 36 kc	23.600	14.300	2.890	10.190	4.790
Rheometrics at 2% strain, 27 °C					
G', MPa	5.257	4.163	3.101	5.617	5.488
G'', MPa	0.773	0.579	0.322	1.015	0.902
tan δ	0.147	0.139	0.104	0.181	0.164
Energy of adhesion, N (% rubber coverage)					
Original	1.640	2.000	1.810	1.530	2.530
	(30)	(60)	(90)	(80)	(100)
Heat-aged	0.460	1.220	2.860	1.990	3.070
	(40)	(60)	(100)	(100)	(100)
Humid-aged	0.090	0.810	4.000	2.190	5.330
	(0)	(40)	(90)	(70)	(100)
Salt-aged	0.630	0.940	2.620	2.190	2.830
	(25)	(50)	(80)	(80)	(90)

[*] 1.5 phr cobalt neodecanoate, 4.5 phr sulfur, 0.8 phr Santocure NS (TBBS) in place of Santocure MOR accelerator.
[**] 1.2 phr cobalt neodecanoate, 4.0 phr sulfur, 0.5 phr TBBS.
[†] 1.2 phr cobalt neodecanoate, 3.8 phr sulfur, 0.8 phr TBBS.
Formulation, Control- NR, CV60: 75 phr; Natsyn 2200: 25; N326 Black: 55;
Sundex 8125: 3; Cobalt Neodecanoate: 1.5; Wingstay 100: 1; stearic acid: 2; zinc oxide: 8; Santocure MOR: 0.8; Sulfur: 4.5.
Sundex is a registered trademark of Sun Refining.
Santocure is a registered trademark of Flexsys, Inc.
Wingstay is a registered trademark of The Goodyear Tire and Rubber Co.
Natsyn is a registered trademark of The Goodyear Tire and Rubber Co.

systems in a wire coat to promote brass-coated wire-to-rubber adhesion [30], and with organocobalt complex adhesion promoters [31] to increase tear strength and composite adhesion (see Table 13.4).

Use of the silane coupling agent bis(3-triethoxysilylpropyl)tetrasulfide (see Fig. 13.4) and low surface-area silica yielded better processing compounds with reduced heat build-up in a wire coat formulation containing a cobalt-boron adhesive [32] (see Table 13.5).

Table 13.5   Silica/Cobalt-Boron Containing Wire Coat Properties

Ingredients	Control	#1	#2	#3	#4	#5
Carbon black, N326	55.000	50.000	50.000	50.000	45.000	40.000
Precipitated silica, Zeosil 85MP	0.000	10.000	10.000	10.000	15.000	20.000
Coupling agent, Si69	0.000	0.000	0.500	0.500	0.750	1.000
Accelerator, diphenylguanidine	0.000	0.000	0.000	0.200	0.000	0.000
Properties						
Maximal torque, in. lb	69.600	70.500	73.800	83.800	68.700	68.100
Minimal torque, in. lb	9.900	10.000	10.900	11.100	10.800	11.400
Delta torque, in. lb	59.700	60.500	62.900	72.700	57.900	56.700
Ts_2, min	2.300	2.800	2.900	2.200	3.200	3.300
T_{90}, min	11.300	13.500	11.900	10.500	12.300	12.300
ML (1 + 4), 100 °C, pts	44.000	41.000	43.000	46.000	43.000	47.000
Delta 5	18.700	22.500	23.100	16.800	23.200	23.100
Delta 10	23.200	27.600	28.500	21.500	27.800	28.200
Shore A 15 s, pts	58.000	57.000	58.000	60.000	56.000	56.000
Modulus at 100%, MPa	2.900	2.700	2.900	3.400	3.100	2.900
Modulus at 300%, MPa	12.400	11.500	14.200	15.300	12.800	12.100
Elongation to break, %	480.000	495.000	450.000	510.000	450.000	460.000
Tensile strength, MPa	23.000	18.500	22.800	25.700	21.500	20.700
Tear trouser at 20 °C, N/mm	17.500	20.000	24.000	15.500	22.500	24.500
Tear ASTM C at 20 °C, N/mm	47.500	51.500	50.000	51.000	52.500	55.000
Stiffness, N/mm^2	82.000	84.000	85.000	94.000	87.000	85.000
E', N/mm^2	7.180	7.330	7.520	8.240	7.700	7.400
E'', N/mm^2	0.700	0.750	0.580	0.500	0.580	0.520
E^*, N/mm^2	7.210	7.370	7.540	8.260	7.720	7.420
Tangent delta	0.097	0.102	0.077	0.061	0.075	0.070
D''	0.013	0.014	0.010	0.007	0.010	0.009
Heat build-up, °C	60.000	61.000	55.000	48.000	52.000	50.000
Permanent set, %	4.300	4.500	4.000	3.000	3.900	4.000
Static compression, %	20.000	21.000	20.000	18.000	18.600	18.500
Fatigue resistance, kcycles	116.000	123.000	130.000	135.000	119.000	127.000
Adhesion, unaged, kg	45.000	44.300	47.600	50.400	49.100	51.000
Aged adhesion						
Heat (7 d at 85 °C), kg	33.100	30.800	32.100	33.800	34.200	38.500
Humid (7 d at 70 °C, 95% rh), kg	30.900	30.100	35.000	34.600	34.600	35.200
Steam (16 h at 121 °C), kg	26.200	27.600	28.000	30.700	27.500	28.500
Salt (7 d)	23.800	28.100	28.900	29.700	31.500	33.100

Formulation, Control- NR, SMR10: 100; Renacit 7: 0.12; Black, N326: 55; ZnO: 8; stearic acid: 0.5; aromatic oil: 1.5; tackifying resin: 2; antiozonant, N-1,3-dimethylbutyl-N'-phenyl-para-phenylenediamine: 2; DCBS: 0.7; Manosperse IS 70P: 5.71; Manobond 680C: 0.67.
Renacit is a registered trademark of Elastochem.
Manosperse and Manobond are registered trademarks of Rhodia, Inc.

Adhesive and physical properties were further improved by using diphenylguanidine as the co-accelerator. High levels of silane-coupled silica yielded compounds with the highest tear strength and heat- and salt-aged adhesion values.

13.5.5   Innerliner

The innerliner is a thin layer of rubber laminated to the inside of a tubeless tire to ensure retention of compressed air. It is most commonly formulated with HIIR to provide good air retention and moisture impermeability, flex-fatigue resistance, and durability [23,25]. Clay has been the predominant non-black filler in innerliner compounds [8] (see Table 13.6). Use of calcium carbonate [33], talc, and mica [34] have been reported in multilayered

Table 13.6   Tire Innerliner Compounds

Ingredient	Control (0/50 phr)	Clay (72/0)	Clay/Black (39.7/27.5)	Clay/Black (47/32.5)
Bromobutyl rubber, 2244	100.00	100.00	0.00	0 phr
Bromobutyl rubber, MD 80-7	0.00	0.00	100.00	100.00
Carbon black, N660	50.00	0.00	27.50	32.50
Clay, Nulok 321	0.00	72.00	39.70	47.00
Flexon 621	8.00	8.00	8.00	6.00
Struktol 40 MS	7.00	7.00	7.00	6.00
Escorez 1102	4.00	4.00	6.00	5.00
Maglite K	0.25	0.25	0.25	0.15
Stearic acid	2.00	2.00	2.00	2.00
Zinc oxide	3.00	3.00	3.00	3.00
MBTS	1.50	1.50	1.50	1.50
Sulfur	0.50	0.50	0.50	0.50
Properties, cured at 150 °C				
Modulus at 100%, kg/cm^3	9.10	9.10	9.10	14.30
Modulus at 300%, kg/cm^3	30.00	36.00	30.00	38.00
Tensile, kg/cm^3	112.00	112.00	102.00	102.00
Elongation, %	870.00	870.00	900.00	900.00
Hardness, Shore A	50.00	43.00	48.00	53.00
Die B Tear, kg/cm	21.00	11.00	18.00	23.00
Tel-Tak				
to Self, kg/cm^3	125.00	101.00	87.00	91.00
to NR Carcass, kg/cm^3	30.00	30.00	47.00	37.00
Strip adhesion to natural rubber (I = Interfacial, T = Tearing)				
Room temp, kg/cm^3	24 I	22 T	27 T	25 T
100 °C, kg/cm^3	21 T	9 I	11 I	16.2 T
Cut growth, 66 °C, cm at 24 h	1.50	1.19	0.60	0.60
Days to failure	2.00	4.00	>10.00	>13.00
Air permeability, 66 °C, $\times 10^{-2}$	3.30	2.20	–	2.20
Mooney scorch at 135 °C	9.00	12.00	20.00	17.00
Mooney viscosity at 135 °C	60.00	47.00	43.00	45.00

Nulok is a registered trademark of J. M. Huber, Inc.

innerliners. Clay, silica, and talc have been found in the innerliner compounds of various passenger tires based upon direct analysis of the liner by using proton-induced X-ray emission spectroscopy [35].

13.5.6   Tread

The tread is the wear-resistant component of a tire that comes in contact with the road. It is designed for abrasion resistance, traction, speed stability, and casing protection [23]. The rubber is compounded for wear, traction, rolling-resistance, and durability [25]. Clay and silica are used in colored bicycle tire treads [36]. Silica is beneficial in NR off-the-road tire treads [36] (see Table 13.7) and in SBR agricultural tire treads [37] to reduce heat build-up and cut growth.

A NR heavy-service truck tire tread with 30 phr silica and mercaptosilane coupling agent (1% by weight of the silica weight) as a replacement for N231 black showed increased resistance to cutting and chipping [38]. Higher levels of silica could be used without a significant sacrifice in heat build-up and treadwear by using the coupling agent. Rolling resistance was reduced by 30%, wet traction was virtually unchanged, and the treadwear index was decreased only by 5% when TESPT-modified silica was used to replace all of the N220 carbon black in a NR truck tread [39].

The rolling resistance of an SBR/BR passenger tire tread was reduced 25% without substantial loss in wet or dry traction by using up to 36 phr of silica and mercaptosilane coupling agent (3% of silica) in a 72 phr filler system [40] (see Table 13.8). Use of TESPT afforded an equilibrium cure in a 20 phr silica and 40 phr N339 black tread. Lower rolling resistance with negligible changes in treadwear and wet traction was obtained [41].

Table 13.7   Off-the-Road Tire Tread Compounds

Ingredients	Control	Silicas #1	#2
Natural rubber, SMR-5	100.00	100.00	60.00 phr
Styrene-butadiene rubber, 1500	0.00	0.00	40.00
Carbon black, N330	50.00	40.00	40.00
Precipitated silica, Hi-Sil 233	0.00	10.00	10.00
Sulfur	2.50	2.50	1.50
Accelerator, MBS	0.80	1.50	1.50
Accelerator, DTDM	0.00	0.00	0.60
Properties			
Mooney Scorch at 121 °C			
min to 5 pt rise	27.60	28.00	40.50
Modulus at 300%, 2 h cure			
at 127 °C, MPa	13.62	12.96	12.41
Trouser tear, N	8.90	6.20	34.70
Goodrich Flexometer, 17.5% deflection, 1.55 MPa load, 100 °C, 30 min			
Δ T, °C	26.00	18.00	25.00
% Set	11.90	9.10	9.60
Chipping/chunking resistance, blows to fail	328.00	402.00	493.00
Pico Abrasion Index	171.00	187.00	185.00

Hi-Sil is a registered trademark of PPG Industries, Inc.

Table 13.8   Low Rolling Resistance Passenger Tire Tread Compounds

	Control	Silica #1	#2
N299 Black	72.0000	54.0000	36.0000
Precipitated silica, Hi-Sil 210	0.0000	18.0000	36.0000
Coupling agent, A189	0.0000	1.0300	2.0600
Accelerator, OBTS	1.7500	0.1000	0.1000
Accelerator, CureRite 18	0.0000	1.7500	1.7500
Stress/strain, 15 min/320 °F			
Tensile strength	2175.0000	2225.0000	2000.0000
Modulus at 300%	1100.0000	1150.0000	1000.0000
Elongation at break, %	500.0000	500.0000	500.0000
Goodrich Flexometer, 45 min/300 °F, 171/15/118			
Initial compression, %	20.3000	22.3000	23.2000
Flexural compression	0.1790	0.1670	0.1740
Permanent set	5.2000	1.3000	1.2000
Heat build-up, °F	78.0000	60.0000	52.0000
Durometer	66.0000	65.0000	65.0000
Roelig hysteresis, 45 min/300 °F			
Tan δ	0.2203	0.1673	0.1408
PS (stress cycling power load)	0.1319	0.1193	0.1072
Relative tire performance properties			
67″ Rolling resistance Index	100.0000	78.0000	73.0000
Twin roll rolling resistance index	100.0000	84.0000	80.0000
WET: avg. 20 mph + 60 mph index	100.0000	96.0000	97.0000
Dry: 40 mph index	100.0000	103.0000	105.0000

Formulation, Control – BR, 1203: 40 phr; SBR, 1204: 60; N299 Black: 72; zinc oxide: 3; antioxidant, Agerite Resin D: 2; stearic acid: 2; antiozonant, Santoflex 13: 2; paraffin oil: 37; sulfur: 1.75; accelerator, OBTS: 0.75; retarder, PVI: 0.1.
Agerite is a registered trademark of R. T. Vanderbilt Company.
Santoflex is a registered trademark of Flexsys, Inc.

TESPT-coupled (13.6% of silica) silica as a partial (40 to 60%) replacement for black improved tire ice traction (5%) and reduced rolling resistance (18%) with comparable wet traction and little or no loss in treadwear for an oil-extended NR/BR passenger tire tread [42].

A silica-filled, all-season, sSBR/*cis*-BR high-performance passenger tire tread had improved wet traction, snow traction, and rolling resistance properties, while maintaining wear life compared to that of the carbon black-filled tread [43] (see Table 13.9). The improvements in tire performance are directly related to the amount of silica present, with the best results obtained for an all-silica compound without the use of carbon black.

13.5.7   Specialty Applications

Clay, mica, and talc are used in lubricant formulations for curing tires; clay and talc in tire sealant formulations; and mica in a water-soluble paint formulation for protecting white sidewalls [14].

Table 13.9   Silica-Filled Passenger Tire Tread Formulations

Ingredient	Control #1	Silica #1	Silica #2	Control #2	Silica #3
Styrene-butadiene rubber, emulsion	100.00	100.00	100.00	0	0 phr
Styrene-butadiene rubber, solution/butadiene rubber	0.00	0.00	0.00	75/25	75/25
Carbon black, N234	80.00	0.00	0.00	80.00	0.00
Conventional precipitated silica	0.00	80.00	0.00	0.00	0.00
Reinforcing agent*	0.00	12.80	12.80	0.00	12.80
Precipitated silica†	0.00	0.00	80.00	0.00	80.00
Processing oil, Sundex 8125	37.50	37.50	37.50	32.50	32.50
Zinc oxide	2.50	2.50	2.50	2.50	2.50
Stearic acid	1.00	1.00	1.00	1.00	1.00
Antioxidant	2.00	2.00	2.00	2.00	2.00
Paraffin	1.50	1.50	1.50	1.50	1.50
Sulfur	1.35	1.40	1.40	1.35	1.40
Accelerator, sulfeneamide	1.35	1.70	1.70	1.35	1.70
Accelerator, diphenylguanidine	0.00	2.00	2.00	0.00	2.00
Properties					
Transverse adhesion on wet ground	100.00	105.00	106.00	101.00	106.00
Adhesion on wet ground	100.00	103.00	104.00	99.00	103.00
Adhesion on snow-covered ground	100.00	104.00	104.00	100.00	104.00
Rolling resistance	100.00	113.00	114.00	101.00	115.00
Rolling noise (dB)	100.00	−1.00	−1.00	100.00	−1.00
Wear service life	100.00	75.00	85.00	94.00	102.00

* X50S = 50% (w/w) bis-(3-triethoxysilylpropyl)-tetrasulfide carried on N330 carbon black.
† Precipitated silica according to Patent EP 0 157 703.

References

1. *Compounding and Processing of Elastomers*, Rubber Division, A.C.S., Intermediate Course, Akron, 1983.
2. R. O. Babbit, Editor, *Vanderbilt Rubber Handbook*, R. T. Vanderbilt Co., Norwalk, 1988.
3. M. Morton, Editor, *Introduction to Rubber Technology*, Van Nostrand Reinhold Co., New York, 1974.
4. R. A. Baker, *An Overview of Hidden Minerals of Polymer Applications*, J. M. Huber Corp., Engineered Materials Division, Atlanta, 1993.
5. H. S. Katz and J. V. Milewski, *Handbook of Fillers and Plastics*, Van Nostrand Reinhold Co., New York, 1987.
6. W. S. Stoy and F. J. Washabaugh, "Fillers" in *Encyclopedia of Polymer Science and Engineering*, Second Edition, Vol. 7, J. I. Kroschwitz (Ed.), John Wiley & Sons, New York, 1988, p. 53.
7. *Kaolin Clays and Their Industrial Uses*, J. M. Huber Corp., Engineered Minerals Division, Atlanta, 1955.
8. T. G. Florea, *Elastomerics*, 118, 22 (1986).
9. W. H. Waddell and L. R. Evans, "Amorphous Silica" in *Encyclopedia of Chemical Technology, Fourth Edition*, Volume 21, John Wiley & Sons, New York, 1997.

10. S. Brunauer, P. H. Emmett, and E. Teller, *J. Amer. Chem. Soc.,* 60, 309 (1938).
11. National French Test Method, T 45-007 (1987).
12. R. K. Iler, *The Chemistry of Silica,* John Wiley & Sons, New York, 1979.
13. M. P. Wagner, *Rubber Chem. Technol.,* 49, 703 (1976).
14. W. H. Waddell and L. R. Evans, *Rubber Chem. Technol.,* 69, 377 (1996).
15. R. F. Wolf and C. Stueber, *Rubber Age,* 87, 1001 (1960).
16. J. W. Maisel, W. E. Seeley and R. J. Woodruff, ''Unique Sodium Aluminosilicate (SSAS) Pigments as TiO2 Extenders in Rubber Compounds'', Presentation to the Rubber Division, ACS, Mexico City, 1989.
17. M. Rigby, E. B. Smith, W. A. Wakeman and G. C. Maitland, ''The Forces Between Molecules'', Clarendon Press, Oxford, 1986.
18. M.-J. Wang and S. Wolff, *Rubber Chem. Technol.,* 64, 559 (1991); M.-J. Wang, S. Wolff and J.-B. Donnet, *Rubber Chem. Technol.,* 64, 714 (1991); M.-J. Wang and S. Wolff, *Rubber Chem. Technol.,* 65, 329 and 715 (1992).
19. T. H. Ferrigno, ''The Couplant Index'', Intertech Corp., Portland, ME, 1993.
20. R. R. Barnhart, ''Rubber Compounding'' in *Encyclopedia of Chemical Technology*, Third Edition, Vol. 20, M. Grayson, Editor, John Wiley & Sons, New York, 1988, p. 365.
21. L. R. Evans and W. H. Waddell, *Rubber & Plastics* News, April 25, 1994, p. 15.
22. T. A. Okel and W. H. Waddell, *Rubber Chem. Technol.,* 67, 217 (1994).
23. R. S. Bhakuni, S. K. Mowdood, W. H. Waddell, I. S. Rai and D. L. Knight, ''Tires'' in *Encyclopedia of Polymer Science and Engineering*, Second Edition, Vol. 16, J. I. Kroschwitz, Editor, John Wiley & Sons, New York, 1989, p. 844.
24. J. C. Ambelang, ''Pneumatic Tire Compounding'', in *The Vanderbilt Rubber Handbook*, R. O. Babbit, Editor, R. T. Vanderbilt Company, Inc., Norwalk, 1978, p. 651.
25. W. H. Waddell, R. S. Bhakuni, W. W. Barbin and P. H. Sandstrom, ''Pneumatic Tire Compounding'', in *The Vanderbilt Rubber Handbook*, R. F. Ohm, Editor, R. T. Vanderbilt Company, Inc., Norwalk, 1990, p. 605.
26. L. R. Evans and W. H. Waddell, *Rubber World*, 209, 18 (1993).
27. G. M. Cameron, M. W. Ranney and K. J. Sollman, *Sver. Gummitek. Foeren. Publ.*, No. 43, VIII, 1 (1973).
28. W. H. Waddell, *Rubber Chem. Technol.,* 71, 590 (1998).
29. W. H. Waddell, J. B. Douglas, T. A. Okel and L. J. Snodgrass, *Rubber World*, 208, 21 (1993).
30. J. R. Creasey and M. P. Wagner, *Rubber Age*, 100, 72 (1968).
31. W. H. Waddell, L. R. Evans and T. A. Okel, *Tire Technol. Int. '94*, 22 (1994).
32. Ph. Cochet, D. Butcher and Y. Bomal, *Kautschuck Gummi Kunstsoffe*, 48, 353 (1995).
33. H. Nakamura to Ohtsu, Japan Patent 05017641 (1/26/93).
34. To Michelin, Japan Patent 55015396 (2/2/80).
35. W. H. Waddell and L. R. Evans, *Kautschuck Gummi Kunststoffe*, 49, 571 (1996)
36. C. Dragus, B. Mehr and S. Florea, RO 95946 B1 (9/15/88).
37. W. H. Waddell, L. R. Evans and J. R. Parker, *Rubber & Plastics News*, July 18, 1994, p 15.
38. S. N. Chakravarty, A. L. Kapur, S. N. Sau and M. Mittal, *Rubber India*, 33, 11 (1981).
39. S. Wolff, ''Silica-Based Tread Compounds: Background and Performance'', presented at TyreTech '93, Basel, Switzerland, October, 1993.
40. S. Ahmad and R. J. Schaefer to B. F. Goodrich, United States Patent 4,519,430 (5/28/85).
41. S. Wolff, *Tire Sci. Technol.,* 15, 276 (1987).
42. T. S. Mroczkowski to Pirelli Armstrong, United States Patent 5,162,409 (11/10/92).
43. R. Rauline to Michelin, United States Patent 5,227,425 (7/13/93).

14 Ester Plasticizers and Processing Additives

Synthetic ester plasticizers and processing additives are used extensively in the rubber industry to impart specific cured physical properties to a rubber compound and/or to achieve certain processing characteristics during the manufacturing process. In the first portion of this chapter, synthetic ester plasticizers are reviewed followed by processing additives in the second portion.

14.1 Ester Plasticizers for Elastomers

Wesley H. Whittington

Plasticizers used in rubber compounds play an important part in the modification of properties associated with elastomers available in today's market. A plasticizer is defined by ASTM as "a compounding material used to enhance the deformability of a polymeric compound." Both ester plasticizers and petroleum oils fit that definition. Petroleum oils are an important and significant part of elastomer plasticization. However, ester plasticizers are the topic of this chapter. Petroleum oil plasticization is discussed in Chapter 12 and elsewhere [1].

The polarity of ester plasticizers is a result of their chemical structure with the dominant feature their carbon-oxygen unit. Long carbon chains diminish the overall polarity of the ester. Petroleum oils function effectively as plasticizers for relatively nonpolar elastomers, but are essentially incompatible with polar elastomers. Conversely, polar esters are compatible with polar elastomers, but have only limited acceptance among nonpolar elastomers.

14.1.1 Derivation

Ester plasticizers result from the chemical reaction of a variety of organic acids and anhydrides with alcohols, glycols, and polyols and through transesterification of esters not generally used as rubber plasticizers. These intermediates result from sources as diverse as crude oil, natural gas, vegetable oils, and animal fats. The monoester reaction helps explain all esterifications (see Fig. 14.1).

Weak acids, which encompass most organic acids, give up an [OH⁻] radical during the esterification reaction while the alcohol gives up a [H⁺]. Monobasic carboxylic acids have one displaceable hydroxyl, dibasic have two and tribasic have three displaceable hydroxyls (see Fig. 14.2).

Two additional and important routes of acquiring the acid segment of the ester are from the acid anhydride and through transesterification of a less desirable ester (see Fig. 14.3).

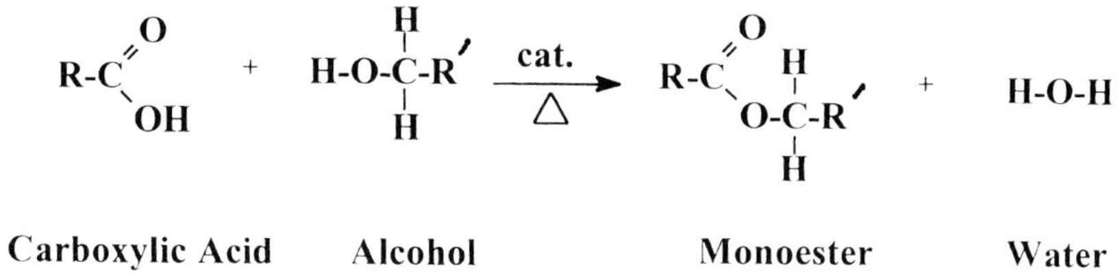

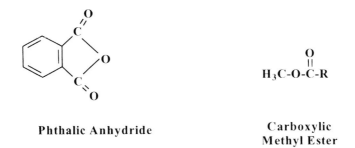

Carboxylic Acid   Alcohol   Monoester   Water

Figure 14.1   Monoester reaction.

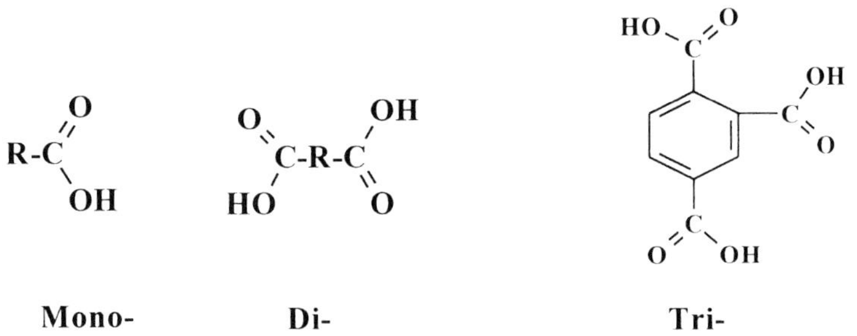

Mono-   Di-   Tri-

Figure 14.2   Mono-, di-, and tri-basic organic acids.

Correspondingly, the alcohols, glycols, and polyols used to produce ester plasticizers have one or more displaceable hydrogens per molecule. Thus, alcohols have one, glycols two, and polyols three or more displaceable hydrogens. There are other reactant groups and processes possible for ester plasticizers but, except for the phosphates, they are of relatively minor importance based on volume (pounds) potential.

The various reactant groups just mentioned can be formed into a variety of ester types. Following are examples of various types of esters along with a simple way to estimate their polarity.

1. Monoesters are the result of a reaction of monobasic acid with an alcohol.
   Example: butyl oleate (note: 2 oxygens, 22 carbons)
2. Diesters are formed by two routes.
   (a) Dibasic acids, methylesters or anhydrides are reacted with alcohols.
       Example: DOA, di(2-ethylhexyl) adipate (4 oxygens, 22 carbons)

Phthalic Anhydride

**Carboxylic
Methyl Ester**

Figure 14.3   Other organic acid segment providers.

(b) Glycols may be reacted with monobasic acids.

   Example: Triethylene glycol di(2-ethylhexoate) (6 oxygens, 22 carbons)

3. Triesters may also be formed by two routes.

   (a) Tribasic acids are reacted with alcohols.

   Example: TOTM, tri(2-ethylhexyl) trimellitate

   (b) Glycerol with monocarboxyl acids (e.g., linoleic, similar to 2(a) above), if a linoleic/oleic mixture, you could have the basis for epoxidized soyabean oil.

4. Polymeric polyesters or, more commonly, polymeric plasticizers are regularly alternating units of dibasic acids and glycols. Once such a reaction is initiated, glycol and acid units add on alternately until the supply of one or the other is exhausted, or until the reaction is terminated by some additive. The completed polymeric may have on the end of its chain:

   (a) A hydroxyl, if glycol is in excess

   (b) A carboxyl, if dibasic acid is in excess

   (c) Hydrocarbon, if terminated with a monobasic acid or an alcohol.

Tables 14.1 and 14.2 list the more common ester plasticizer building blocks. These come from unpublished work by Kuceski [2].

Table 14.1   Ester Plasticizer Building Blocks

Monobasic acid	Carbons	Nonacid groups
Heptanoic	7	–
Caprylic	8	–
Pelargonic	9	–
Capric	10	–
Lauric	12	–
Stearic	18	–
Oleic	18	one C=C
Linoleic	18	two C=C

Dibasic acid	Carbons	Nonacid groups
Maleic anhydride	4	one C=C
Succinic	4	–
Glutaric	5	–
Adipic	6	–
Azelaic	9	–
Sebacic	10	–
Dodecandioic	12	–
Phthalic anhydride	8	(ring)
Dimer	2×oleic (example)	–

Tribasic acid	Carbons	Nonacid groups
Trimellitic	9	–
Citric	6	–
Phosphoric	–	P

It can readily be seen now that carboxylic materials (organic acids, anhydrides, esters) and hydroxylic materials (alcohols, glycols, polyols), if combined in chemically acceptable groups as monoesters, diester, triesters, and polyesters, provide an almost unimaginable number of possible combinations of ester plasticizers. Fortunately, while that is theoretically possible, for practical purposes many of these combinations are not available because of cost or poor handling characteristics. Many of the esters used commonly by the rubber industry are listed in Appendix 14.1.

14.1.2   Philosophical

What do ester plasticizers do for rubber compounds? They modify the compound. Modification allows an elastomer to be customized, thereby satisfying the needs of the processor (the mixer-fabricator), the specifier (functions in his/her assembly), and the user of the assembly. Polarity, chemical configuration, and molecular size of the ester are major contributors to absorptive character, permanence, and providing the property modification required for specific applications.

The determination of whether a compound can meet certain requirements begins in the laboratory. Compounds are tested under conditions relevant to the manufacturing and service conditions the finished product will encounter. The type and quantity of ester plasticizer incorporated in a compound affect the test data obtained. Plasticizer and elastomer manufacturers often provide data showing the property modifications that can be expected with various additives, including plasticizers, in evaluation type recipes. The rubber compound manufacturer must then review and adapt that generalized information to specific application needs.

Problems may include such things as:

1. Lack of information on the elastomer chosen for the application
2. Differing property values for the recipes tested
3. Test conditions used by the manufacturer are not relevant to the new application
4. In-house plasticizers are not included in current studies and other reasons preclude increasing the number of plasticizers available

The aim of this chapter is to simplify plasticizer selection. Much of the published work for rubber concerns natural rubber, SBR, EPR, and EPDM, i.e., the high volume elastomers. Ester plasticizers are not used extensively with those elastomers. Ester plasticizers are primarily used with the polar elastomers and, of those, polychloroprene (CR) and butadiene acrylonitrile rubber (NBR) dominate the market with respect to volume consumed. Polychloroprene was frequently plasticized with large volumes of aromatic and naphthenic petroleum oils until government safety and health regulations stopped this practice. While esters were used in the past, their use in CR today has increased as a result of those regulations. NBR was and still is the largest consumer of ester plasticizers, both by total volume and variety. This is borne out by the literature evaluating ester plasticizers in NBR compared to other elastomers. Examining data for this chapter suggests that the relationship between specific ester plasticizers in a single elastomer may have similar results or follow similar trends for other elastomers.

Table 14.2   Hydroxylic Materials

	Carbons	Nonhydroxyl groups
Alcohols		
Propyl (including iso)	3	–
Allyl	3	one C=C
Butyl	4	–
Hexyl	6	–
Heptyl	7	–
n-Octyl	8	–
Octanol-2	8	$C-\overset{\mid}{C}-(C)_6$
Ethylhexyl	8	$-C-\overset{\overset{\textstyle C-C}{\mid}}{C}-(C)_4$
Isooctyl	8	$-R-\overset{\overset{\textstyle C}{\mid}}{C}-R'$
Nonyl	9	–
Iso nonyl	9	like iso octyl
Decyl	10	–
Iso decyl	10	$-R-\overset{\overset{\textstyle C}{\mid}}{C}-R'$
Undecyl	11	–
Tridecyl	13	–
Benzyl	7	⬡
Glycols		
Triethylene	6	two –COC–
Tetraethylene	8	three –COC–
Propylene	3	–OCCCO–
Dipropylene	6	one –COC–
Buylene	4	–OCCCCO–
Neopentyl	5	$-O-C\overset{\overset{\textstyle C}{\mid}}{\underset{\underset{\textstyle C}{\mid}}{C}}C-O-$
1,5 Pentane diol	5	–O–CCCCC–O–
1,6 Hexylene diol	6	(one C=C)
1,4 Cyclohexane di-methanol	8	–OC–⬡–CO–

14.1.3   Applications

What follows are summaries of data for compounds in which the variables are individual plasticizers in different elastomers. Data from polymer and plasticizer manufacturers are used to develop a range of values displayed from most negative to most positive. The ranges are then observed for trends that might work as guides for plasticizer recommendations.

Table 14.2   Continued

	Carbons	Nonhydroxyl groups
Polyols		
Glycerin	3	
Trimethyolpropane	6	
Sorbitol	6	
Pentaerythritol	5	

14.1.3.1   Low-ACN Content NBR

These data from Monsanto compare the following plasticizer types [3]:

1. Normal and branched aliphatic alcohols with dibasic acids of adipates and phthalates
2. A phthalate with one aliphatic and one benzene ring as end groups
3. Phosphates with one aliphatic and two phenyl groups as the alcohol constituents

   Aging conditions for these tests are at the relatively non-severe temperature of 100 °C. Examination of the data shows:

1. Least change as a result of air oven aging occurs with phthalates.
2. Least change after ASTM 1 oil immersion occurs with phthalates and phosphates, but many of the phosphate esters contain benzene rings and that may be the significant factor.
3. Immersion in ASTM 3 oil produces differing characteristics. Stress-strain changes appear not significantly affected by plasticizer choice. Least change in volume and, coincidentally, least change in hardness appear to be associated with straight chain alcohols or low molecular weight phthalates. Either the larger molecule or branched bulky alcohols attached to the extremities of the dibasic acid result in greater permanence.
4. Greatest compound stability after ethylene glycol immersion occurs with straight chain alcohols attached to di- and tri-basic acids.
5. The most noticeable data range difference after water agings occurs with weight change. Compounds with phosphates showed the least change.
6. Low-temperature properties for as-molded compounds were best for adipates and other dibasic acids with linear alcohol end groups (see Table 14.3).

Table 14.3   Results of Low-ACN Content NBR Testing with Various Plasticizers

Recipe: Low-ACN content NBR, Carbon Black, Sulfur-MBTS cure, Plasticizer 30 phr (15.5% of total); DOP, C7C11P, DUP, BBP, DOA, DNODA, EHdPF, IDdPF.*

Conditions and properties	Data range	Most desirable (D)** or most stable (S) in range
Original physical properties		
Ten., MPa***	8.2 to 11.9	D: IDdPF, EHdPF, C7C11P
Elong., %	380 to 460	
Hard., Duro A, pts	49 to 53	
Air oven, 70 h at 100 °C		
Ten. chg., %	−41 to +46	S: BBP, EHpDF, DOA
Elong. chg., %	−60 to −21	S: DOP, C7C11P, DOA
Hard. chg., pts.	+8 to +14	S: C7C11P, BBP, DOP
ASTM 1 oil, 70 h at 100 °C		
Ten. chg., %	−15 to +51	S: IDdPF, EHdPF, BBP
Elong. chg., %	−46 to −12	S: DNODA, DOP, DOA
Hard. chg., pts	+4 to +14	S: EHdPF, BBP, DOP
Vol. chg., %	−5 to 0	S: EHdPF, C7C11P, DOP
ASTM 3 oil, 70 h at 100 °C		
Ten. chg., %	−42 to −33	
Elong. chg., %	−26 to −9	
Hard. chg., pts	−18 to −12	S: DNODA, DUP, BBP
Vol. chg., %	+33 to +46	S: DUP, BBP, IDdPF
Ethylene glycol, 70 h at 100 °C		
Ten. chg., %	−17 to +31	S: C7C11P, IDdPF, DUP
Elong. chg., %	−27 to −12	S: DNODA, DUP, BBP
Hard. chg., pts	+2 to +11	S: DOP, C7C11P, DUP
Water, 70 h at 100 °C		
Ten. chg., %	−18 to 0	
Elong. chg., %	−26 to −24	
Hard. chg., pts	−4 to +7	
Wt. chg., %	+4 to +10	S: IDdPF, EHdPF, BBP
Low-temperature, D 746, °F		
As molded	−65 to −50	D: DOA, DNODA, C7C11P

* A table of plasticizer acronyms appears in Appendix 14.1.
** For this and other tables that follow: most desirable is defined as highest tensile, highest elongation (not the result of undercure), and lowest hardness; most stable is defined as least percent change or, if for hardness, least points Durometer change.
*** An abbreviation list appears as Appendix 14.2.

14.1.3.2   *Neoprene Blend GN 88/WHV 12*

These compounds [3] undergo a ZnO cure and exhibit few data ranges where the span from lowest to highest values is significant. Test conditions are the same as those used for the low-ACN NBR compound, 70 h at 100 °C.

Table 14.4   Test Results of Neoprene Blends GN 88/WHV 12 with Various Plasticizers
Recipe: Neoprene GN-88/WHV-12, Carbon Black, ZnO Cure, Plasticizer 30 phr (12.3% of total); Plasticizers: DOP, C7C11P, DUP, BBP, DOA, DNODA, EHdPF, IDdPF.

Conditions and properties	Data range	Most desirable (D) or most stable (S) in range
Original physical properties		
Ten., MPa	12.0 to 14.1	D: BBP, IDdPF, DUP
Elong., %	215 to 240	
Hard., Duro A, pts.	78 to 80	
Air oven, 70 h at 100 °C		
Ten. chg., %	−18 to −1	
Elong. chg., %	−31 to −16	
Hard. chg., pts.	0 to +12	S: EHdPF, IDdPF,BBP
Wt. chg., %	−5 to −.55	S: DUP, C7C11P
ASTM 1 oil, 70 h at 100 °C		
Ten. chg., %	−8 to −11	
Elong. chg., %	−36 to −20	
Hard. chg., pts.	+2 to +8	S: IDdPF, BBP, DNODA
Vol. chg., %	−4 to 0	
ASTM 3 oil, 70 h at 100 °C		
Ten. chg., %	−35 to −3	S: DUP, DNODA, C7C11P
Elong. chg., %	+3 to +33	S: DUP, BBP, EHdPF
Hard. chg., %	−32 to −22	
Vol. chg., %	+48 to +62	S: DNODA, DOA, C7C11P
Ethylene glycol, 70 h at 100 °C		
Ten. chg., %	−16 to −4	
Elong. chg., %	−27 to −7	S: DOP, C7C11P, DUP
Hard. chg., %	−9 to −3	S: IDdPF, EHdPF, BBP
Water, 70 h at 100 °C		
Ten. chg., %	−27 to 0	
Elong. chg., %	−27 to −7	
Hard. chg., pts	−14 to−2	S: DOP, DUP, C7C11P
Wt. chg., %	−11 to −23	S; DOP, DUP, C7C11P
Low-temp., ASTM D746, °F		
As molded	−50 to −30	D: DOA, DNODA

Hardness changes after air oven aging and ASTM 1 oil, and ethylene glycol immersion appear most stable with phosphates and butyl benzyl phthalate. The common ester characteristic is the benzene rings external to the di- and tri-basic acids.

After ASTM 3 oil immersion, the least changes in weight, tensile, and elongation occur with multibasic acid esters having linear alcohol end groups. Least weight change is with adipate, while tensile and elongation stability occur with phthalates.

14.1.3.3   Different Elastomers with the Same Plasticizer

Would two or more plasticizers mixed into compounds of different elastomers hold relatively similar positions to each other through a series of test conditions? The likelihood is that because of the differing polarities in the elastomers, which results in differing attractions between a given plasticizer and the two elastomers, their aging data will not follow the same trends. Data for two series of compounds, a Neoprene and a Low-ACN NBR, were surveyed. Each elastomer was mixed in three recipes, each containing either IDdPF, DOA, or DOP. These compounds, when surveyed for data of lowest and highest values, showed somewhat similar trends for original properties. For example, highest original tensile occurred with IDdPF; lowest original tensile occurred with DOA. However, when specimens of the compounds were subjected to hot air or fluids such as water or ASTM 2 oil, there is no correlation of significance [3].

14.1.3.4 Medium Acrylonitrile-Content NBR

A study [4] with monomeric plasticizers for air oven and oil agings at 125 °C by The C.P. Hall Company contains:

1. Branched aliphatic alcohol (2-ethylhexyl) end groups for the dibasic aliphatic, phthalic and, trimellitic acids.
2. Long chain alkyl and ether alcohols of the dibasic acids: adipic, glutaric and sebacate substitute.
3. Eight and ten carbon aliphatic monobasic acids on multigroup ethylene glycols.
4. An aromatic phosphate.

Interesting findings from the study include:

1. Compound scorch times were longest with 2-ethylhexyl end groups on the dibasic acid; they were shortest with ether end groups on dibasic acids and when glycols were the internal segments for monobasic acid end groups.
2. Original physical property differences were nearly insignificant.
3. Oil aging property changes were least with monomeric diesters with constructions of ether and diether groups on dibasic acids and with phosphate.
4. Water immersions show little of significance except that the plasticizers with ether end groups appear either to be extracted from their compounds or aid in resisting absorption of water. The former is more likely because of the hydrophilic character of the ether groups.
5. Low-temperature results for as-molded specimens was best for adipates, but after air oven and Fuel C immersion, low-temperature results are better for ether esters attached to dibasic acids (Table 14.5).

14.1.3.5 Medium ACN NBR

Uniroyal has published a study [5] covering 21 esters and three petroleum oils as plasticizers. The data from that study deals only with adipate and phthalate mono-merics; however, there are other interesting findings. These common diester configura-tions have dibasic acids as the center portion of the molecule. Diesters with glycols as the middle portion of the plasticizer were not included in this work. Findings include:

1. Petroleum oils are not acceptable plasticizers for NBR. Uniroyal mixed and tested three compounds with petroleum oils as plasticizers, one each with aromatic,

Table 14.5   Test Results for Medium-High Acrylonitrile-Content NBR and Various Plasticizers
Recipe: Medium-high ACN NBR, Carbon black, low-sulfur, high accelerator, Plasticizer 20 phr (10.2% of total): DOA, DBEA, DBEEA, 7006, 7050, 83SS, TOTM, TrEGCC, TrGDEH, TeGDEH, TrAF.

Conditions and properties	Data range	Most desirable* (D) or most stable (S) in range
Mooney viscosity at 121 °C		
Min visc.	25 to 33	
Scorch, min.	5.5 to 13.3	D: Long–DOA, DOP, TOTM
		S: Short-7050, TeGDEH, TrGDEH
Rheometer scorch at 170 °C	1.8 to 2.8	
Original physical properties		
Mod. at 300%, Mpa	4.9 to 7.7	–
Ten., ultimate, Mpa	11.7 to 13.8	–
Elong. at break, %	550 to 600	–
Hard., Duro A, pts	57 to 60	–
Air oven, 70 h at 125 °C		
Ten. chg., %	+8 to + 32	S: 83SS, TOTM
Elong. chg., %	−49 to −30	S: DBEA, DBEEA, TOTM
Hard. chg., pts	+8 to +25	S: TOTM, 83SS, 7050
Wt. chg., %	−11 to −1	S: TOTM, 83SS 7050
ASTM 1 oil, 70 h at 125 °C		
Ten. chg., %	+10 to +33	S: TOTM, 83SS
Elong. chg., %	−42 to −32	–
Hard. chg., pts	+8 to +19	S: 7050, TrAF, DBEA=DBEEA
Vol. chg., %	−14 to −9	S: 7050, TrAF
ASTM 3 oil, 70 h at 125 °C		
Ten. chg., %	0 to +22	S: TrAF, DOA, 7006
Elong. chg., %	−29 to −18	S: DBEA, 83SS, 7006
Hard. chg., pts	−2 to −7	S: TrAF, 83SS, DOP
Vol. chg., %	+0.76 to +3.7	S: TrAF, 7006, DBEEA
Fuel C, 70 h at 23 °C		
Ten chg., %	−66 to −47	–
Elong. chg., %	−55 to −47	–
Hard. chg., %	−25 to −18	–
Vol. chg., %	+38 to +43	–
Dry out, 24 h at 70 °C after fuel C		
Hard chg., pts	+11 to +16	–
Vol. chg., %	−14 to −11	–
Dist. water, 70 h at 100 °C		
Ten. chg., %	−5 to +10	S: TrAF, DOP
Elong. chg., %	−25 to −17	S: DOA, TrAF, DBEA, 7006
Hard. chg., pts	−2 to +4	S: DBEEA
Vol. chg.,	−1.2 to +7.4	S: 7006, 83SS, DBEEA
Low-temperature, D2137		
As molded	−43 to −24	D: DBEA, DOA, 7006=TrEGCC
After air oven	−34 to −20	D: 83SS, DBEEA, TrEGCC
After ASTM 1 oil	−27 to −20	–
After ASTM 3 oil	−27 to −21	–
After fuel C/dry out	−37 to −25	D: DBEA, DBEEA
After dist. water	−27 to −22	–

* Most desirable = longest scorch time.

paraffinic, and naphthenic oils. All compounds exuded. The C. P. Hall Company Laboratory also recently attempted a similar study. Batches were mixed and test slabs molded that represented each of the three types at 20 phr. Test slabs were also prepared for compounds that contained 10 phr DOP/10 phr petroleum oil, using each of the three oils. All six compounds showed exudation (bleeding) within four days, from unstressed test slabs.

2. Some plasticizers provide excellent low-temperature properties for compounds in the as-molded condition. But these compounds approach a common value after aging. Plasticizers that give good as-molded low-temperature properties usually do not exhibit the best values after exposure to ordinary aging conditions.

The adipate and phthalate data that follow show:

1. Insignificant differences in the property changes for air and oil agings associated with the acid portion of the various diesters.
2. Adipates show slightly better values for all low-temperature tests except brittle point of a compound containing C7C11P. Normal alcohol end groups help a diester provide good low-temperature properties.
3. After air and ASTM 1 oil aging with the adipates, the ether alcohol end groups of a diester show a combination of better tensile and elongation retention coupled with least weight change. This did not hold with the higher swelling associated with ASTM 3 oil.
4. The ether alcohol also provided best low-temperature properties for adipate esters.
5. A different kind of finding occurred with the phthalates. The alcohol end groups contain no ethers, but only hydrocarbons with some linearity and some branching. In this part of the study:
    (a) Following air aging, compounds with the highest molecular weight show the least change in stress-strain and weight.
    (b) Following ASTM 1 oil immersion, the least changes are associated with lowest molecular weight materials.
    (c) Following ASTM 1 oil immersion, the least stress-strain change is associated with lowest molecular weight materials and the order nearly reverses itself with weight change.

Plasticizers are generally extracted when rubber compounds are immersed in hydrocarbon fluids. Depending upon the polar characteristics of the aging fluid, it may be absorbed by the rubber compound; hence, a change in the balance of swell, i.e., a shifting, widening, or narrowing the range of swell for high and low aniline point fluids may occur. ASTM 1 oil is not absorbed readily into the NBR compound; it causes extraction and results in less stress-strain change than would occur with a fluid that is absorbed.

Immersion in ASTM 3 oil shows absorption of that test fluid. Some extraction also occurs. With this situation, plasticizer polarity is important. Elastomers exert some affinity upon esters, but this immersion fluid shows an affinity for the ester as well. As the compound is swelled by the absorbed ASTM 3 oil, plasticizer leaves the compound. The saponification values (and polarity) increase from DIDP to DOP to DBP; the volume change values in ASTM 3 oil likewise increase in that order. The less polar the monomeric plasticizer, the more easily it is extracted by the high swelling oil. A compound with a plasticizer that is the least extractable also appears to experience the

Table 14.6 Comparisons of Adipate/Phthalate, Branched/Linear Alcohols and Ether/Hydrocarbon Alcohols

Recipe: Med. ACN NBR, Carbon Black, Sulfur - Sulfur donor cure, Plasticizer at 20 phr (9.5% of compound)

Condition and property	Range of data	Most stable (S) or most desirable (D)
Adipates: DOA, DBEEA, 97A		
Air oven, 70 h at 125 °C		
TC + EC *	+57 to +75	S: DBEEA, DOA, 97A
Wt. chg., %	−10 to −4	S: DBEEA, 97A, DOA
ASTM 1 oil, 70 h at 125 °C		
TC + EC	+53 to +62	S: DBEEA, DOA, 97A
Wt. chg., %	−13 to −11	S: DBEEA, DOA=97A
ASTM 3 oil, 70 h at 125 °C		
TC + EC	+33 to +42	S: DBEEA, DOA, 97A
Wt. chg., %	+1.3 to +2.9	S: DOA, 97A, DBEEA
Low-temperature		
Brittle point, °C	−42 to −38	D: DBEEA, DOA, 97A
TR-10, °C	−37 to −34	D: DBEEA, DOA=97A
Gehman, T100, °C	−40 to −38	D: DBEEA, 97A, DOA
Phthalates: DBP, DOP, DN711P, DIDP	Air oven, 70 h at 125 °C	
	TC + EC 61 to 70	
S: DIDP, DN711P, DBP, DOP	Wt. chg., % −10 to −3.7	
S: DIDP, DN711P, DOP, DOP	ASTM 1 oil, 70 h at 125 °C	
	TC + EC 46 to 57	
S: DBP, DOP=711P, DIDP	Wt. chg., % −13 to −10	
S: DBP, DOP, DIDP, 711P	ASTM 3 oil, 70 h at 125 °C	
	TC + EC 34 to 46	
S: DBP, DIDP, DOP, 711P	Wt. chg., % +1.7 to +3.5	
S: 711P=DIDP, DOP, DBP	Low-temperature	
	Brittle point, °C	

least change in stress-strain. Air aging data suggests that least change is associated with highest molecular weight. Low-temperature data shows normal alcohol end groups provide better low-temperature flexibility and brittleness than branched alcohols attached to dibasic acids (see Table 14.6).

14.1.3.6 Medium-High ACN-Content NBR

A study by Goodyear [6] deals with five polymeric plasticizers and 16 monomerics. The study also includes an unplasticized compound. Interesting findings from this study involved Fuel A immersions at 23 °C and ASTM 1 and 3 oil immersions at 149 °C. Goodrich [7,8,9], Uniroyal [5], and other NBR manufacturers have done similar studies. This data looks specifically for differences in data ranges for polymerics and monomerics and at the relationship with an unplasticized compound. Observations included:

1. The results for changes with monomerics were usually greater than for polymerics. Instances where monomerics did not span the entire range and show positive advantage were:
   (a) Less elongation change after Fuel B immersion
   (b) Less volume change after ASTM 3 oil immersion
   (c) Colder low-temperature flexibility by Gehman
   (d) Colder low-temperature nonbrittleness by impact

All those noted advantages are expected for NBR plasticized with monomerics esters.

2. With fluid aging, as volume changes become increasingly larger, elongation losses generally become greater.
3. Compounds containing polymeric plasticizers appear not to suffer as great a change upon aging as do those with monomerics.
4. Volume change and hardness change usually proceed in opposite directions: as aging fluid is absorbed, hardness decreases; or as plasticizer is lost through extraction or volatilization, hardness increases.
5. Plasticizers are compound property modifiers. If the polymer and cure system are proper for an application, plasticizers can cause noticible differences in compound properties.

Some caution is recommended when using polymeric plasticizers with NBR. Testing shows that polymeric plasticizers, when used at high concentrations, can be too polar and exude from cured unstressed test sheets. The exudation is more noticeable or severe the higher the acrylonitrile content, the lower the plasticizer viscosity, and the more tightly the compound is crosslinked (e.g., peroxide cures). Most polymerics do not exude from medium-high ACN NBR if used at 15 phr or less. At 20 phr or more of polymeric plasticizer, better permanence is achieved by using either lower saponification value materials or higher viscosity (in excess of 20,000 cps), or both. If more than 15 phr polymeric plasticizer must be included in a compound, it is recommended that for every phr in excess of 15, add 1 phr of a diester or other more solvating plasticizer. Other ingredients in the compound also influence permanence, such as absorption properties of a reinforcing filler or vulcanized vegetable oil (see Table 14.7).

14.1.3.7   NBR/PVC Polyblends

NBR/PVC polyblends are of two types:

1. Blends in which the NBR is to be cured. These blends may be of two types. If the blend is a rubber-type application, NBR is the continuous phase material. If the blend is a TPE-type material, NBR is the discontinuous phase.
2. Blends in which the NBR is not cured, PVC is the continuous phase. The NBR functions as a nonextractible plasticizer for the PVC.

In both applications, the PVC is likely to be softened significantly with plasticizer. Most of the same incorporated plasticizers are useful in both NBR and PVC. NBR/PVC blends are very highly plasticized for applications requiring 25 to 30 Duro A hardness. During service, those compounds are frequently exposed to high concentrations of hydrocarbon

Table 14.7   Test Results of Medium-High ACN-Content NBR and Various Plasticizers

Recipe: Med-High ACN Content NBR, FEF Black, Low-Sulfur/High accelerator cure, Plasticizer
Plasticizer: Twenty-one in the study, each at 30 phr (15.1% of total). Five were polymerics. One compound, no plasticizer. Polymerics: Glutarates at 11K, 12K, 2K cps Visc., Adipate at 4K and Sebacate at 100K (K = 1000). Monomerics: 7050, TrEGCC, DOTP, TXIB, TNONDTM, TIDM, uTM, DOP, DBP, DUP, 600, 707A, DBCFm.

Test conditions	Ave. all plasticizers	Monomeric range	Polymeric range	"0" Plast. value
Air oven, 70 h at 121 °C				
Ten. chg., %	+8	−9 to +25	−2 to +12	+11
Elong. chg., %	−34	−41 to −26	−41 to −28	−27
Hard. chg., %	+10	−3 to +24	+2 to +5	−1
Wt. Chg. %	−8	−1 to −15	−1 to −2	−2
Fuel A, 70 h at 23 °C				
Ten. chg., %	−4	−13 to +5	−9 to 0	−7
Elong chg., %	−11	−20 to −2	−9 to −2	−9
Hard. chg., %	−4	−6 to +6	−9 to −7	−3
Vol. chg., %	−3	−7.0 to +5.3	+1.0 to +4.0	+1
Fuel B, 70 h at 23 °C				
Ten. chg., %	−42	−60 to −23	−43 to −39	−33
Elong. chg., %	−30	−50 to −11	−48 to −37	−36
Hard. chg., %	−18	−24 to −12	−23 to −20	−21
Vol. chg., %	+27	+8.9 to +45	+25 to + 31	+26
Fuel C, 70 h at 23 °C				
Ten. chg., %	−40	−53 to −22	−58 to −48	−51
Elong. chg., %	−46	−45 to −19	−73 to −47	−42
Hard. chg., %	−18	−20 to −12	−25 to −22	−25
Vol. chg., %	+35	+22 to +43	+33 to +48	+45
ASTM 1 oil, 70 h at 149 °C				
Ten. chg., %	−15	−49 to +19	−13 to −6	−11
Elong. chg., %	−48	−64 to −32	−41 to −33	−30
Hard. chg., %	+11	+6 to +20	+2 to +11	+4
Vol. chg., %	−12	−20 to −5.9	−9.7 to −4.1	−2.4
ASTM 3 oil, 70 h at 149 °C				
Ten. chg.,%	−18	−44 to −12	−21 to −10	−21
Elong. chg.,	−32	−52 to −11	−33 to −14	−24
Hard. chg.,	−2	−5 to +6	−9 to −3	−18
Vol. chg., %	+8	−4.4 to +12	+6.3 to +21	+13
Water, 70 h at 100 °C				
Vol. chg., %	−2	−12 to +4.4	−3.4 to +8.9	+7.1

solvents. For these applications, one can take advantage of polymeric plasticizer extraction resistance and improved longevity of the vulcanized part results. The problem of exudation by polymeric plasticizers that occurs with NBR alone may not occur with NBR/PVC blends, because of high holding capacity of PVC.

14.1.3.8 Ethylene Acrylic and Polyacrylate Elastomers

Ethylene acrylic (EA) and polyacrylate (PA) [10] are very polar elastomers; they both can be used for the same types of applications and plasticized effectively by polar plasticizers.

Even though they are compatible, many of the plasticizers used for the nitrile elastomers are not used with EA and PA. The major reason is the high temperature requirements of EA and PA are greater than for nitrile. Many plasticizers acceptable for NBR are too volatile for EA and PA elastomer applications. However, the high temperature service plasticizers used for EA and PA can be used for nitriles.

The data presented for EA and PA elastomers was generated by The C. P. Hall Company, but plasticizer evaluations are also available from elastomer producers. Plasticizers examined are:

1. Monomerics of dibasic acid: dibutoxyethoxyethoxyethylglutarate (DBEEEG) from two sources
2. Monomerics same as above, except adipate rather than glutarate
3. Monomerics of polyethylene glycol with monobasic acid end groups (PGd2EH) from two sources.
4. Two adipic acid, low viscosity polymerics of similar structure

The data for the similar plasticizers indicates that slight differences are expected if retesting is done. It is also likely that best overall compound properties are achieved by using a mixture of polymeric and monomeric esters (see Table 14.8).

14.1.3.9 Chlorosulfonated Polyethylene (CSM)

The C. P. Hall Company tested this material [11] for low-temperature properties, air oven, ASTM Fuel B, and ASTM 1 oil with compounds containing:

1. Aliphatic dibasic acid esters
2. Polymerics: Low molecular weight adipate and phthalate
3. Epoxide esters, a monomeric and an oligomeric

Compound data of significance includes:

1. Highest original tensiles occurred with polymerics but, for other originals, the sebacate and substitute sebacate gave the most desirable properties.
2. Following air oven aging, the epoxides show greatest stability.
3. ASTM Fuel B immersion followed by air oven dry out resulted in monomerics showing least change after Fuel B (they are extractable), but compounds with polymerics offer least change after the combination of Fuel B/Air Oven (they show greater permanence).
4. Polymerics offer the most property stability after compound aging in ASTM 1 oil (see Table 14.9).

14.1.3.10 Chlorinated Polyethylene (CPE)

After NBR, the largest amount of data found for a single elastomer is for a CPE [12]. The number of esters tested necessitated employment of a screening system to limit the number of esters under consideration. From among five individual studies, two rubber formulations, each representing a different ester were chosen; their data was combined by retabulation. The two were chosen based upon best overall property retention (least

Table 14.8   Test Results of EA and PA with Various Plasticizers

Recipe: Ethylene acrylic, 1-octadecanamine, stearic acid, alcohol phosphate, carbon black, hexamethylenediamine carbamate, *N,N'*-diphenylguanidine, plasticizer at 12.9% of compound: 7050, DBEEEG, DBEEEA, PGd2EH, P-675, P-670.

Conditions properties	Data range	Most desirable* (D) or most stable in (S) range
Press cure 10 min. at 165 °C,		
Oven post cure 1 hr at 177 °C.		
Original physical properties		
Tensile, ultimate, MPa	9.0 to 10.2	D: P-675, 7050, P-670
Elong. at break, %	370 to 410	D: DBEEEG,7050, DBEEEA
Hard. Duro A, pts.	53 to 56	D: DBEEEG=P-670=DBEEEA
Low-temp., D-2137, °C	−42 to −47	D: DBEEEG, DBEEEA=PGd2EH
Air oven aging, 168 h at 177 °C		
Ten. chg., %	+1 to +23	S: P-675, DBEEEA, 7050
Elong. chg.,%	−33 to −54	S: P-675, PGd2EH, DBEEEG
Hard. chg., pts	+10 to +24	S: P-675, PGd2EH, P-670
Wt. chg., %	−6.3 to −14	S: PGd2EH, P-670, P-675
Air oven aging, 1000 h at 150 °C		
Ten. chg., %	−5.4 to +1.6	S: P-670, P-675, PGd2EH
Elong. chg., %	−34 to −8.1	S: DBEEEA, DBEEEG=PGd2EH
Hard. chg., pts	+6 to +20	S: PGd2EH, P-670, P-675
Wt. chg., %	−4.4 to −13	S: PGd2EH, P-670, P-675
ASTM 1 oil, 168 h at 149 °C		
Ten. chg., %	+15 to +35	S: P-675, P-670, PGd2EH
Elong. chg., %	−20 to −30	S: DBEEEA=P-670, PGd2EH
Hard. chg., pts	+5 to +13	S: P-675, P-670=7050
Vol. chg., %	−1.3 to −7.0	S: P-670, P-675, 7050
Trans. fluid, 168 h at 150 °C		
Ten chg., %	−18 to −1.8	S: PGd2EH=DBEEEG, DBEEEA
Elong. chg., %	−38 to −23	S: PGd2EH, P-675=DBEEEG
Hard. chg., %	−12 to −6	S: DBEEEG, DBEEEA, PGd2EH
Vol. chg., %	+16 to +21	S: PGd2EH=DBEEEG, 7050
Trans fluid, 1000 h at 150 °C		
Hard. chg., %	−3 to +5	S: 7050=DBEEEG=PGd2EH
Vol. chg., %	+16 to +21	S: DBEEEG=PGd2EH, DBEEEA

* D = greatest tensile, greatest elongation, lowest hardness, coldest low-temp.

change in property values), after all stated aging conditions. The following plasticizers were considered in this review for CPE: IOES, OET, ESO (G-62, G-60, S-75), 325P, DOP, TOTM, TIOTM. Mooney Scorch was judged best by longest time to 5 point rise and ODR by longest time to 2.5 point rise. Original Stress/Strain was judged by the largest values, while hardness was by most efficient or lowest value (see Table 14.10).

This table shows if the specification requires greatest property stability after air aging, epoxidized esters are best, but if it is for stability after oil immersion, the choice is likely to be trimelliatates or diesters with saturated, nonether end groups or monomeric esters. Diesters with ethers in the end groups gave good original low-temperature properties; otherwise, disadvantages were more numerous than advantages.

Table 14.9   Test Results of CSM with Various Plasticizers

Recipe: Chlorosulfonated PE, Carbon Black, Maglite D, Wax, Peroxide/pentaerythritol/TAC, Plasticizer 30 phr (14% of total) 83SS, 7050, DOS, DBES, EpGdO, 62ESO, P-7068.

Conditions properties	Data range	Most desirable[*] (D) or most stable in (S) range
Mooney viscosity at 135°C		
Min. visc.	6.2 to 13.7	–
t_5, min.	19.3 to 37.5	D: P-670, P-7068=62ESO
ODR at 182 °C		
ML	6.0 to 27.0	–
MH	56.1 to 118	–
t_s2, min.	2.4 to 3.5	D: DBES=DOS=EpGdO
t_c' (90)	8.3 to 9.7	–
Original physical properties		
Ten., ultimate, Mpa	11.0 to 15.9	D: P-7068, P-670
Elong. at break, %	290 to 480	D: 83SS, DBES, G-62
Hard., Duro A, pts.	61 to 69	D: DBES=83SS, DOS
Tear, die C, ppi	139 to 178	D: 83SS, DBES, 62ESO
Low-temp., D2137	−36 to −52	D: DOS, DBES, 83SS
Air oven aging, 70 h at 121 °C		
Ten. chg., %	−10 to +6	S: 62ESO, 83SS, DOS
Elong. chg., %	−13 to −4	S: 62ESO, P-7068, EpGdO
Hard. chg., pts	+2 to +9	S: EpGdO, P-670 = P-7068
Wt. chg., %	−1.2 to −5.7	S: 62ESO, EpGdO, P-670
ASTM fuel B, 70 h at 40 °C		
Hard. chg., %	−31 to −20	S: DOS, P-670, P-7068
Vol. chg., %	+39 to+54	S: DOS, DBES, P-7068
Followed by air oven, 22 h at 70 °C		
Hard. chg., pts	+5 to +18	S: P-670, P-7068=62ESO
Wt. chg., %	−15 to −7.1	S: EpGdO, P-670, P-7068
ASTM 1 oil, 70 h at 121 °C		
Ten. chg., %	−9 to +15	S: 62ESO, P-7068, DOS
Elong. chg., %	−18 to −7	S: P-7068, EpGdO, P-670
Vol. chg., %	−9.8 to +2.8	S: 62ESO, P-670,P-7068

[*] D = longest t_s2, longest t_5

Thus, they are not included as considerations for chlorinated polyethylene polymers (see Table 14.11).

14.1.4   Application Trends

We have considered the plasticizing characteristics of several esters in five elastomer types. The contribution of ester plasticizers depends upon their molecular configurations and size, the attraction between the ester and elastomer, and the conditions of testing. Considerations for configurations include linear vs. branched alcohols, aliphatic vs. aromatic structures, diesters of dibasic acid and of polyethylene glycol, and lastly, overall polarity. Polar elastomers accept a wide range of plasticizers, depending upon the relative

Table 14.10   CPE: Which Plasticizer Type Is Best? (1 = first choice, 2 = second choice)

	Epoxy	Monomeric	Polymeric	Trimellitate
Mooney scorch	1			
ODR		1	1	
Original physical properties				
Tensile	2			1
Elongation	1	2		
Hardness		1		
Low-temp. tor. stiff.	2	1		
Air oven at 150 °C				
Ten. chg., %		1		
Elong. chg., %	2			1
Hard. chg., pts.	1		2	
Wt. loss, %	1			
ASTM 1 oil				
Hard. chg., %		2	1	
Vol. chg., %		2	1	
ASTM 3 oil				
Hard. chg., %		1	1	1
Vol. chg., %		1	1	1
Vol. chg. range				
ASTM 3-ASTM 1		1	1	1

Table 14.11   Test Results of CPE with Various Plisticizers
Recipe: CPE 85/EPR 15 blend, Carbon black, Peroxide, Plasticizers Plasticizers: ESO (G-60, G-62, 7170), 759, TOTM, TIOTM, IOES, OET, 325P, DOP

Conditions and properties	Data range	Most desirable (D) or most stable (S) in range
Mooney viscosity at 121 °C		
Scorch, t_5, min.	15 to >25	D: G-60=G-62=7170=759
ODR at 160 °C		
Scorch, ts 2.5 min.	16 to >34	D: DOP, 325P, TOTM=TIOTM
Original physical properties		
Ten., ultimate, MPa	7.8 to 27.4	D: TOTM, TIOTM, G-62
Elong. at break, %	265 to 600	D: 759, OET, G-60, IOES
Hard. Duro A, pts.	65 to 73	D: OET, IOES=DOP=759
Low temp. D-1043,T10000, °C	−34 to −23	D: IOES, OET=G-60=DOP
Air oven, 168 h at 150 °C		
Ten. chg., %	−38 to −26	S: 759, DOP, IOES
Elong. chg., %	−79 to −38	S: TOTM, G-60, TIOTM
Hard. chg., pts.	+13 to +26	S: G-62=7170=325P
Wt. chg., %	−11 to −1.3	S: G-60, 7170, G-62
ASTM 1 oil, 70 h at 100 °C		
Hard. chg., pts.	−5 to +4	S: TIOTM, IOES, TOTM
Vol. chg., %	+3.6 to +15	S: OET, TOTM, TIOTM
ASTM 3 oil, 70 h at 100 °C		
Hard. chg., pts.	−29 to −16	S: DOP, TOTM, 325P
Vol. chg., %	+54 to +71	S: TOTM, IOES, OET=325P
Vol. chg. range		
(ASTM 3–ASTM 1)	44 to 63	S: 325P, TOTM, IOES

Table 14.12   Differences resulting from the Use of Low vs. High Viscosity Plasticizers

	Low viscosity	High viscosity
Permanence	Low	High
Low-temperature properties	Good	Poor
Handling	Easy	Difficult
Saponification value	Low	High

characteristics of the two materials. Laboratory testing conditions that can be correlated to service conditions, affect plasticizer permanence.

Ester plasticizers may be classified in several ways. The simplest way may be by viscosity. Low viscosity plasticizers are monoesters, diesters, triesters, and epoxides. High viscosity plasticizers are polymeric polyesters. The basic differences in compound properties as a result of using these two types are shown in Table 14.12.

It should be recognized that there may be wide variations in each comparison listed above and even some overlapping, e.g., the low-temperature properties associated with the poorest of the low viscosity plasticizers may not be as good as that for the best of the high viscosity polymerics.

The following are some general properties associated with ester plasticizer configurations:

1. For ease of incorporation and permanence of unstressed specimens, as the polarity for one elastomer relative to another increases, so should the polarity of the plasticizer. A plasticizer may be too polar or nonpolar for a stated elastomer.
2. Polymeric plasticizers exhibit more permanence than monesters, diesters, and triesters.
3. Esters made with dibasic and tribasic acids generally impart more change in rubber compound properties with changes in the configuration of the alcohol end group than with their respective acid groups.
4. Diesters with linear end groups generally provide better low-temperature properties than diesters with branched end groups. But branching generally results in easier incorporation during mixing.
5. Ether esters of dibasic acids show better resistance to volatility and extraction by nonswelling hydrocarbon fluids than those with end groups of strictly aliphatic hydrocarbons.
6. Common dicarboxylic aliphatic acids show the trends listed in Table 14.13 if the alcohol is held constant, e.g., 2-ethyl-hexyl, but the carbon content of the acid increases.
7. A phthalic segment has eight carbons and offers high compatibility with polar elastomers. However, the benzene ring appears more inflexible than the aliphatic chains. The results are high hardness, poor low-temperature properties, and high volatility. Additionally, it is highly extractible by hydrocarbon fluids.
8. Increasing the number of carbons in the linear alcohol diesters, as the dibasic acid is held constant, results in the tendencies shown in Table 14.14.
9. The structures of polymeric plasticizers are treated as proprietary information by manufacturers. However, the application of polymerics is somewhat related to

Table 14.13   Property Change Trends Among Dicarboxylic Aliphatic Acids

Acids	Carbons	Trends with increasing carbons
Glutaric	5	Improved resistance to volatility and hydrocarbon
Adipic	6	extraction. Improved low-temperature. Decreasing
Azelaic	9	polarity.
Sebacic	10	

Table 14.14   Trends Resulting from Increasing Carbon Atoms in Linear Alcohol Esters

Alcohols	Carbons	Trends with increasing carbons
Butyl	4	More tendency to exude. Less easily
Octyl	8	absorbed. Decreasing polarity.
Nonyl	9	
Decyl	10	
Tridecyl	13	
Butoxyethyl	6	These two alcohols by having additional
Butoxyethoxyethyl	8	oxygens available will have increased polarity compared to 6-carbon or 8-carbon alcohol as shown above.

viscosity and saponification value. The higher the viscosity is, the greater the permanence. The higher the saponification value is, the higher the polarity. It is suggested that particular needs for polymerics be discussed with a manufacturer of those materials.

The range of ester plasticizers includes a wide variety of materials acceptable for elastomers classified as polar. A number of elastomers, even though not mentioned herein, can be combined with plasticizers based on the information in this chapter. Polymeric plasticizers generally are not used for their contribution to low-temperature properties; however, some detract less than others. Plasticizers acceptable at high test temperatures (150 to 200 °C) may not be needed to satisfy lower test temperatures.

14.2   Process Additives

Christopher R. Stone

Process additives are chemicals that, when added to a rubber compound at a relatively low loading, improve the processability of the compound. Ideally, this improved processability is achieved without adversely affecting the physical properties of the component produced.

These products cover a wide range of chemicals, from organic chemical peptizers, such as dibenzamido-diphenyl disulphide, through resinous homogenizers and tackifiers to metal soaps of fatty acids that improve filler dispersion and reduce compound viscosity by lubrication at the molecular level. These and many more types of chemicals influence many different aspects of rubber processing. They may, like the peptizers, affect the very start of the production process by reducing the polymer molecular weight during a mastication stage or during the early part of the compound mixing. They may be used to influence the quality of compound mixing or they may be added to influence the way the compound behaves during subsequent down-line processing, such as extrusion, calendering or one of the many types of molding.

It is essential that any process additive be carefully selected to maintain the balance between process efficiency and product performance. In this respect, it is better to consider the processing requirements during the initial development of a compound rather than discover processing deficiencies after a compound has been developed to meet certain finished property specifications and put into production. If the process additive is included in the formulation at the compound development stage, both final product properties and processing can be optimized together.

14.2.1   Control of Viscosity

Natural rubber, as produced by the tree, has a molecular weight far too high for current equipment to handle as solid rubber. Its molecular weight must be reduced early in the process. Synthetic rubbers, however, are made with their molecular weight distributions designed for processing as well as for finished product properties. For this reason, control of viscosity for natural rubber is discussed separately from control of viscosity for synthetic rubbers.

14.2.1.1   Viscosity Control of Natural Rubber

The very high viscosity of natural rubber is reduced by decreasing the molecular weight of the polymer, by physically breaking the polymer molecules by mechanical work in a process known as mastication. This process may be carried out on an open mill but is usually carried out in an internal mixer. The shear forces, developed in the polymer matrix, cleave the polymer molecules. This purely mechanical process is most efficient at low temperatures (below 90 °C), but above 120 °C oxidative scission of the polymer chains becomes the dominant mechanism [13]. On an open mill, it is difficult to achieve temperatures above 90 °C; but in an internal mixer, it is difficult to prevent the temperature from exceeding 90 °C.

Mastication of rubber is energy intensive and when carried out as a separate step in the mixing procedure, it reduces the output capacity of the mixer [14]. Chemical peptizers [15] have been widely used for many years to increase the efficiency of oxidative polymer chain scission. They also reduce the temperature of the onset of oxidative chain scission, improving the efficiency of the mechanical mastication because effective oxidative chain scission occurs [16] when the temperature is between 90 and 120 °C.

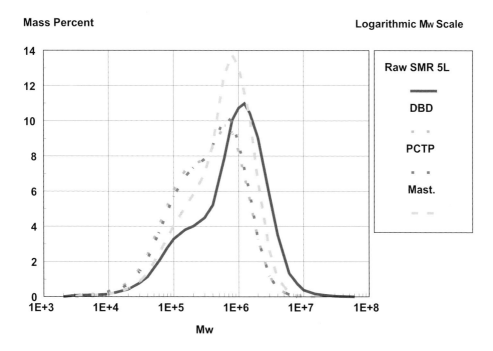

Mass Percent

Logarithmic Mw Scale

Raw SMR 5L

DBD

PCTP

Mast.

Mw

Figure 14.4 Natural rubber mastication, comparison of peptizers and mastication.

Fatty acid soaps have also been shown to improve the processability of natural rubber compounds [17]. These materials have shown good results when added during a separate mastication of the rubber and when added during masterbatch mixing using rubber that has not undergone pre-mastication.

Molecular weight distribution curves, produced using gel permeation chromatography for raw SMR 5L and the same polymer masticated with 0.2 phr of metal-activated dibenzamido-diphenyl disulphide (DBD) chemical peptizer, with 0.2 phr of metal-activated pentachlorothiophenol (PCTP) chemical peptizer and without any additive, are shown in Fig. 14.4. The mastication was carried out in a 1.5 liter, tangential rotor, internal mixer. It was run for three minutes with a rotor speed of 60 rpm and a start and water circulation temperature of 90 °C.

It must be remembered that gel permeation chromatography can only measure polymer in solution and about 25% of the raw natural rubber is insoluble gel. Thus, the distribution curve for the raw SMR 5L only characterizes 75% of the material. However, it can be clearly seen that both chemical peptizers have significantly reduced the polymer molecular weight over the entire molecular weight range. They decreased the high molecular weight fraction but they also caused a considerable increase in the low molecular weight tail. This can cause stickiness in the masticated rubber leading to factory processing problems. The curve of the rubber masticated without additive shows a smaller reduction in molecular weight compared with those with the chemical peptizers but the molecular weight reduction is largely confined to the higher molecular weight portion of the rubber. The low molecular weight tail is not increased.

Mass Percent   **Logarithmic Mw Scale**

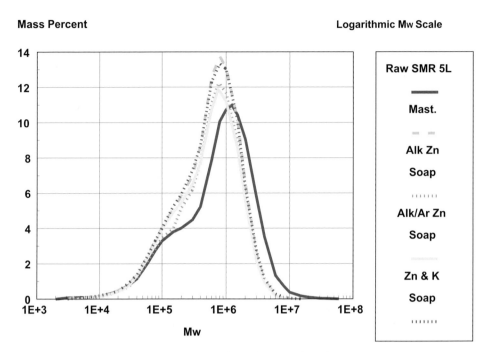

Figure 14.5   Natural rubber mastication, comparison of soaps and mastication.

The molecular weight distribution curves for rubber masticated without additives and rubber masticated with 3 phr of three different fatty acid soap products are shown in Fig. 14.5. The mastication conditions were the same as before. The additives are an alkyl zinc soap, an alkyl/aryl zinc soap, and a zinc/potassium soap. The sample masticated with the zinc/potassium soap is almost identical, with respect to molecular weight distribution, as the rubber masticated without additive. The rubbers masticated with the alkyl zinc soap and the alkyl/aryl zinc soap show very similar curves but with even less molecular weight reduction in the molecular weight range between one hundred thousand and one million.

The increase in the low molecular weight tail associated with the chemical peptizers may also result in poor physical properties. The greater number of polymer free ends may have an adverse effect on the fatigue performance or the visco-elastic dynamic properties.

There are no discernible differences between the energies of mastication of the batches containing chemical peptizers and the batch with no additives. All of the three batches containing the process additives used about 11% less energy than the other batches.

The Mooney viscosities of these masticated samples, shown in Fig. 14.6, clearly show that the chemical peptizers yielded the largest reductions. A reduction of 60% is given on mastication with the DBD peptizer and 53% on mastication with the PCTP peptizer. Straight mastication only achieved a reduction of 21% whilst mastication with the process additives gave reductions around 40%. Thus, the unsaturated fatty acid soap process additives result in an 11% reduction in the energy of mastication but double the reduction in Mooney viscosity, compared with mastication without additive.

ML (1 + 4) @ 100°C

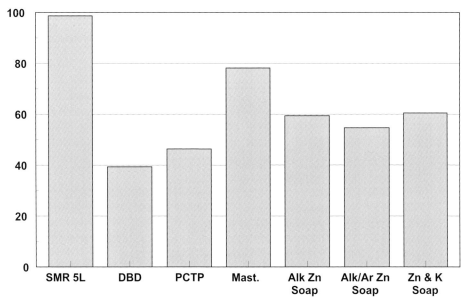

Figure 14.6   Mooney viscosity, SMR 5L raw and masticated.

These masticated rubbers were mixed into a high quality, truck tire, tread formulation. The same formulation was mixed using the unmasticated SMR 5L with the same three unsaturated fatty acid process additives at the same loadings but this time added during masterbatch mixing. The same formulation was also mixed using the unmasticated SMR 5L without any process additives. This last compound would not be practical in actual production but was added as a theoretical control.

The Mooney viscosities of both the masterbatches and final mixes of these compounds are shown in Fig. 14.7. The compounds mixed using rubber pre-masticated with chemical peptizers had the lowest masterbatch and final mix viscosities. The two-stage mix without additive shows its impracticality with a masterbatch viscosity over 200. However, after final mixing, its viscosity came in line with the other compounds.

There are no major differences between the Mooney scorch times or the Rheometer cure data for all ten compounds. All the static physical properties tested on the cured compounds gave results with either no differences or no clear trends. Goodrich heat build up and blow-out test results on cured samples, shown in Fig. 14.8, clearly show some of the disadvantages of chemical peptizers [18].

14.2.1.2   Viscosity Control of Synthetic Rubber

In general, synthetic rubber polymers are manufactured with viscosities suitable for processing. However, to achieve specific compound properties, it is often necessary to use high levels of reinforcing fillers, making the compounds difficult to process. It is also

MS (1 + 4) @ 100°C

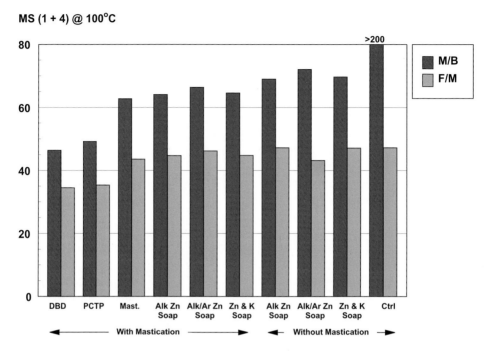

Figure 14.7   Mooney viscosity, mixed compound.

HBU, °C   **Blow Out Time, min**

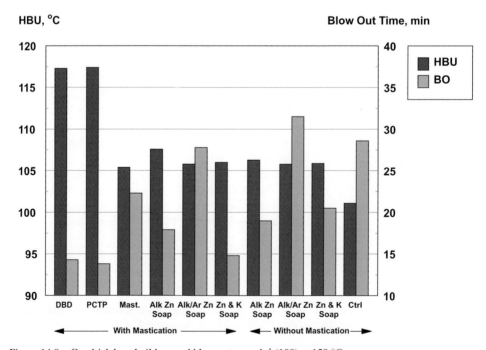

Figure 14.8   Goodrich heat build-up and blow-out, cured t'_c (100) at 150 °C.

not always easy to disperse high levels of such surface active fillers and still maintain acceptable levels of mill room productivity. Process additives can assist with both these problems.

Many of the specialty synthetic polymers are not compatible with all process additives and care must be taken in their selection in these cases. It is usually best to consult with the technical service departments of leading suppliers of process additives for advice on the correct additive for each application.

14.2.2   Mode of Action of Process Additives

Most process additives provide some form of lubrication, either within the bulk of the compound or at its surface. Lubrication within the bulk of the compound reduces the apparent viscosity; this effect may vary with the shear rate applied to the compound. Lubrication of this type improves the compound flow under shear, for example, by increasing extrusion or calendering speed. Lubrication at the surface can assist with mill or calender roll release and even mould release and also improve die definition on extrusion.

We have no direct evidence as to the exact mechanism of action of surfactant-type process additives but by drawing parallels with known mechanisms from similar systems, we can propose a reasonable hypothesis. From surfactant technology [19–26], it is known that concentrated solutions of soap in water have high viscosities and that a multi-phase system of a soap, water and a mineral oil form a relatively stiff product that we know as lubricating grease. Both of these systems exhibit excellent lubrication at high shear but are stiff at rest or very low shear. Thus, surfactants behave differently, depending upon the media in which they are dispersed.

Surfactant molecules align themselves into aggregates known as micelles when dispersed in either a polar or a non-polar medium. However, the orientation of the surfactant molecule is reversed by the change of medium. In a polar medium, the polar functional group of the surfactant is attracted to and even soluble in the polar medium and the non-polar aliphatic chain is oriented away from the polar medium, causing the formation of the micelle structures. In a non-polar medium, illustrated by Fig. 14.9, the aliphatic, non-polar chain is attracted to or even soluble in the non-polar medium and the polar functional group is oriented away from the non-polar medium, causing the formation of the micelle structures. When significant shear forces are applied, the weak intermolecular forces between the multi-layer laminar micelles are easily broken and the viscosity of the medium is dramatically reduced.

The actual behavior of the surfactant varies depending upon its structure. The factors affecting it the most are the nature, and specifically, the strength of the polarity of the functional group; the length of the aliphatic hydrocarbon chain; the presence or absence of unsaturation within the hydrocarbon chain and the degree of branching of the hydrocarbon chain.

To date, there is no direct evidence that such surfactant micelles form and behave, as described above, in polymer systems. However, it seems reasonable to argue that rubber, as a high molecular weight hydrocarbon, is not significantly different from mineral oil. On this basis, it also seems reasonable that certain surfactants should also form laminar

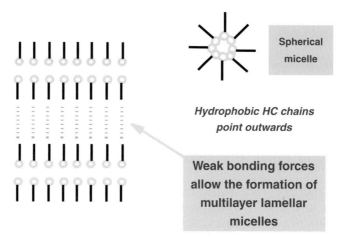

Figure 14.9   Micelle formation in a non-polar medium, mineral oil.

micelles within the polymer matrix as shown in Fig. 14.10. These micelles would be expected to behave similarly to those in oil systems, in that that they would not reduce viscosity at very low shear but would give large reductions at medium to high shear. The viscosity reduction would result from effective inter-chain lubrication between the polymer molecules themselves. This effect would produce the lubrication within the bulk of the rubber compound and be very useful in processes such as extrusion and calendering.

There is evidence to support this hypothesis in that differences between surfactants used as rubber process additives cause changes to the viscosity versus shear behavior of the compound consistent with known oil surfactant technology. The general structures of the most common forms of surfactant process additives are shown in Fig. 14.11. Zinc

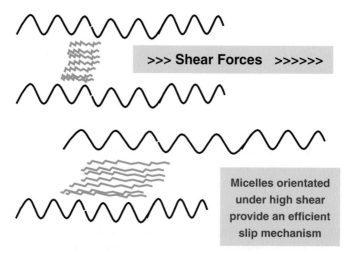

Figure 14.10   Proposed micelle behavior in polymer under high shear.

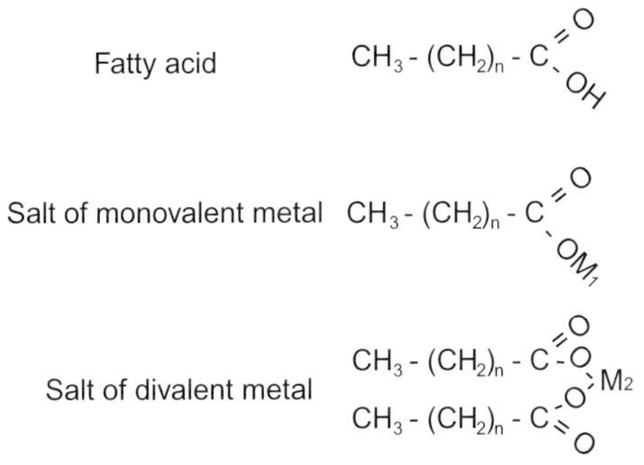

Figure 14.11   Typical surfactant process additives.

soaps form the largest group of surfactants used as rubber process additives. They are intrinsically soluble in rubber in that they do not bloom when added at quite high levels. However, they cannot be totally soluble or they would not be able to form micelles. Only the hydrocarbon portion is truly soluble and the degree of solubility of this portion is one factor influencing the performance of the product.

With chain lengths of less than ten carbon atoms, micelles cannot form. Above this threshold, micelles can form and the lubrication effect with shear is seen. The surfactant process additive tends to be crystalline if its chain length distribution is narrow and the product solubility may be too low and bloom may result. If this narrow distribution is also at a high molecular weight, the melting point is likely to be too high for easy dispersion. Wider molecular weight distributions of the hydrocarbon chain of the surfactant process additive, results in a lower melting point, easy dispersion in the rubber and less tendency to bloom.

Branching of the hydrocarbon chain portion of the surfactant process additive increases the solubility and reduces the melting point. If sufficient branching is present, the solubility can be so high that the total product becomes truly soluble in the rubber. No micelle formation is possible in this case and no viscosity reduction with shear is observed. This is the case with zinc-2-ethyl hexanoate.

The polarity of the functional group of the surfactant process additive also strongly influences its performance. High polarity functional groups tend to reduce the solubility in rubber somewhat and give the product more activity as a surface lubricant. They also provide an attraction for polar fillers, improving the dispersion of such fillers in the rubber compound. Surfactant process additives with functional groups with lower polarity arc largely confined to acting as bulk lubricants.

14.2.2.1 Surface Lubricants

Very few process additives only cause one type of process modification. As already discussed, many surfactants can influence the following: the bulk viscosity, the surface lubricity, and the filler dispersion of rubber compounds. However, many have one predominant mode of action. Fatty acid esters, whilst belonging to the surfactant group of process additives, can be made to have greater effects on the surface lubrication than on the bulk viscosity.

14.2.2.2 Process Additives for Homogenizing and Improving Filler Dispersion

Many modern rubber compounds are based on blends of two or more different polymers to achieve specific compound property requirements. However, these requirements will only be met if the different polymers are blended very well.

No two polymers are truly soluble in each other. In polymer mixtures, one polymer forms the continuous phase and the other polymers form domains dispersed throughout the continuous phase. The continuous phase is not always formed by the polymer present in the greatest amount. In the case of natural rubber blended with polybutadiene, the natural rubber forms the continuous phase until the polybutadiene level rises above 80% of the blend.

Fillers are not necessarily at equal concentration in all of the polymers present in the blend. Fillers have a natural partitioning distribution between the various polymers used in blends and although this distribution may be affected by mixing techniques, there is always a tendency for the polymer domains to change after mixing so as to restore the natural partitioning of the filler.

Ideally, a polymer blend compound has very small domains of the polymers forming the discontinuous phase, with these being uniformly distributed throughout the continuous phase polymer. The filler materials are also well dispersed and partitioned correctly between the various polymer phases present.

Most process additives intended for improving the homogeneity of polymer blends are based upon polymer-modified, hydrocarbon resin blends. The exact way these additives affect mixing to achieve the improvement is not fully understood. One proposed mechanism is that they must have some solubility in both polymers and in this way, the solubility parameters of the two polymers are brought closer together. Another proposal is that they improve the adhesive forces between the different polymers during the early stages of mixing, thus increasing the shear forces and improving the mixing efficiency. Work is underway to study the mode of action of these homogenizing resins.

Whilst it may not be known how they work, it is clear that these process additives do work. Physical properties, both static and dynamic, of blend compounds mixed using these products are superior to those mixed without them.

These same homogenizing resins also improve filler dispersion in both polymer blend compounds and compounds based on single polymers. They also often impart some bulk viscosity reduction as well, but this is usually a much smaller effect and less shear rate-dependent than that produced by surfactant process additives. Energy savings and/or time reductions in mixing are often possible when using these materials.

There are potential problems associated with the use of processing additives intended to improve homogenization of polymer blends. These materials are usually based upon resin blends and such materials also behave as extenders, increasing the viscous modulus component of the visco-elastic properties. Because of this, compounds containing these process additives may exhibit increased dynamic heat build-up, increased compression set and higher rates of stress relaxation.

The basic physical properties of compounds, such as hardness, modulus and tensile strength, are not usually adversely affected by the use of these process additives because the improved filler dispersion compensates for the extension effect. Many surfactant process additives also improve filler dispersion. The choice of the functional group of the surfactant determines which type of filler responds the most in this respect.

14.2.3   Application of Process Additives

Covering all the potential processing problems and their solutions is beyond the scope of this chapter. However, it is possible to show how process additives can influence real compounds in real processes by giving a few examples.

Figure 14.12 shows the improvement in injection molding performance given by the addition of 3 phr of a process additive to a polychloroprene compound. The process additive is of the surfactant type and is a blend of calcium soaps and amides of fatty acid esters. The compound formulation is: Neoprene WHV 34; Neoprene WRT 66; MgO 4; ZnO 5; N 550 Carbon Black 40; N 774 Carbon Black 30; Wax 2.2; Antioxidant 3; Polyethylene glycol ester plasticizer 15; Rape seed oil 8; ETU 1 and TMTD 1.5. The Mooney scorch time and the physical properties are virtually unchanged by the addition of the process additive.

Figure 14.13 shows the effects of adding 4 phr of an ester of saturated fatty acids to an NBR compound. This material provides lubrication within the bulk of the compound but also gives some effects associated with plasticizer extension. The viscosity at low shear, as measured by the Mooney viscometer, is reduced considerably but the viscosities at higher shears, as measured by the capillary rheometer, are also greatly reduced. The extension

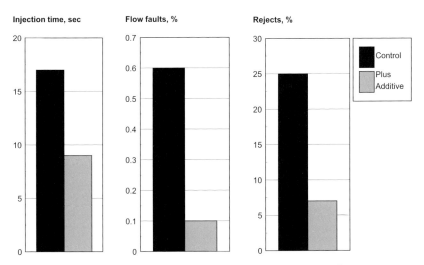

Press: Werner & Pfleiderer GSP 400/63 S. Mould volume: 6x70 cm³.
Cure: 1.5 min @ 200°C. Injection pressure: 126 bar. Clamping pressure: 25 MPa

Figure 14.12   Polychloroprene compound injection molding.

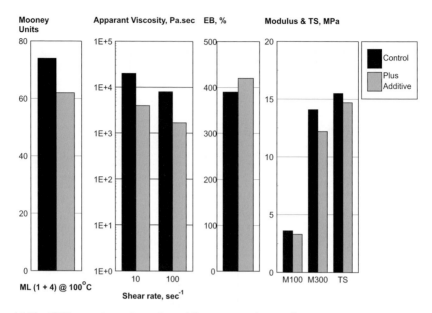

Figure 14.13   NBR extrusion and transfer molding compound properties.

effects show up as an increase in elongation at break and a reduction in the moduli and tensile strength values. Hardness and compression set, not shown, are virtually unchanged.

Figure 14.14 shows the increase in the weight of compound transferred in thirty seconds into a spiral mold, which demonstrates the improvements in transfer molding

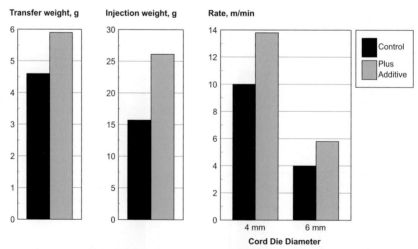

Transfer spiral mould at 170°C, pressure: 25 MPa for 30 seconds. Injection Press: Boy 15 SE
Mould 85 x 85 x 3 mm with central injection at 200°C. Preplasticer screw at 70°C. 60 sec cycle.
Extruder: Troester GS30/K 10D 30mm Diameter cold feed extruder, L/D ratio 10:1
Screw Speed: 60 rpm, Temperatures: Die 100°C, Screw 80°C, Barrel 80°C

Figure 14.14   NBR compound, extrusion, injection and transfer molding.

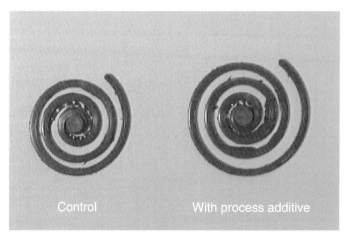

Figure 14.15   Transferred spiral moldings of control (left) and compound with process additive (right).

performance. The injection molding improvement was shown in a factory trial where the control compound would not produce a complete molding. The extrusion performance is also considerably improved. The compound formulation is: pre-blend of a 36% ACN NBR 100 and N772 carbon black 100; plus ZnO 5; stearic acid 1; TMQ 1; DOP 10; DTDM 1.5; TMTD 1.7.

The transferred spiral moldings of both the control and the compound with process additive are shown in Fig. 14.15. The additional flow achieved by the compound containing the process additive (right side) is evident.

Figure 14.16 shows the improvement in extrudate quality given by the addition of 3 phr of a process additive to a silica reinforced passenger "Green Tire" tread compound. They were extruded through a Garvey die using a Troester GS 30/K 10D 30 mm diameter cold feed extruder, having an L/D ratio of 10:1. The extruder was run at 70 rpm with the following temperatures: screw 90 °C, barrel 90 °C, head 100 °C and die 110 °C. The

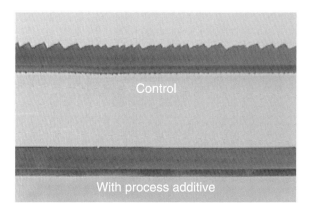

Figure 14.16   Difference in extrudate quality between compound without process additive (upper) and with process additive (lower).

process additive was of the surfactant type, a blend of zinc and potassium soaps of fatty acids. This has a relatively high polarity for a better response with a filler with a polar surface. Stearic acid was left out of the compound containing the process additive to avoid bloom. Many fatty acid soaps, when added to compounds, cause stearic acid to bloom. The compound formulation was as follows: Medium Vinyl Solution SBR 75; High CIS BR 25; High Dispersion Silica 80; 50:50 TESPT Silane and N330 Black 12.5; Aromatic oil 32.5; ZnO 2.5; Stearic acid 1; 6PPD 2; Wax 1.5; Sulphur 1.4; CBS 1.7 and DPG 2.

References

1. Dimeler, G. R., "Plasticizers for Rubber and Related Polymers," Educational Symposium No. 9, October 1982 Meeting, Rubber Division of ACS.
2. Kuceski, V. P., "Plasticizers," unpublished work prepared for Rubber Division of ACS course, "Intermediate Rubber Technology."
3. "Plasticizer Performance in Synthetic Rubber," *Plasticizers and Resin Modifiers*, Monsanto Company, pp. 3–8.
4. Laboratory Report, "Monomeric Plasticizers in NBR," Laboratory Project 0172-112, The C. P. Hall Company, 4/5/83.
5. "Plasticizers for Paracril", Uniroyal Chemical, PR-91.
6. Shell, R. L., *Chemigum N615 – Evaluation of Plasticizers*, Goodyear Chemical, 4/4/77.
7. VanderMass, W. B., "New Plasticizer Evaluation," B. F. Goodrich Chemical, Application Development Report 3378-895, 6/21/74.
8. Morsek, R. J., "Plasticizer Evaluation," B. F. Goodrich Chemical, Application Development Report 3378-454, 12/8/70.
9. Marguglio, L. A., "Plasticizer Evaluation," B. F. Goodrich Chemical, Application Development Report 3378-2110, 7/28/75.
10. Laboratory Report, "Plasticizers for Acrylic Elastomers", The C. P. Hall Co. (presented at a meeting of the Southern Rubber Group, Inc., 6/23/98).
11. Svoboda, R. D. "Plasticizers for Hypalon", The C. P. Hall Company, Laboratory Project Report 0580-K478
12. Publication Numbers 12003-1, 12003-2, 12002-5, 12002-3, DuPont Dow Elastomers LLC (née Dow Chemicals USA), Recipe: Polymer blend CPE (CMO136) – 85/EPDM Epsyn 70-A-15, black filled, peroxide cured, plasticizer 25 phr (11.9% if total).
13. F. H. Cotton, (1931) *Trans. IRI*, 6, 487.
14. Chakravarty, S. N., and Pandit, R. R. (1976), *Kauts. Gumi Kunst.* 29(11).
15. Fairclough, P. J., and Swift, P. (1973) *NR Technology*, 4, 46.
16. Crowther, B. G. (1981) *NR Technology*, 12(2).
17. Crowther, B. G. (1989) *Zinc soaps: new examination of their properties and applications in the rubber industry*, paper presented at Rubbercon andcq;89.
18. Engels, H.-W., and Abele, M. (1991), *Rubber World*, 205(5).
19. Reiss-Husson, R., and Luzzatti, V. (1964) *J. Phys. Chem.*, 68, 3504.
20. Reiss-Husson, R., and Luzzatti, V. (1966) *J. Colloid Interface Sci.*, 21, 534.
21. Brown, J. C., Pusey, P. N., Goodwin, J. W., and Ottewill, R. H. (1975) *J. Phys. A: Math. Gen.*, 8, 664.
22. Celuba, D., Thomas, R. K., Harris, N. M., Tabony, J., and White, J. W. (1978) *Faraday Discussion*, 65/6.

23. Eckwall, P., Danielsson, I., and Stenius, p. (1972) *Surface Chemistry and Colloids*, S, 108ff, Butterworths, London.
24. Luzzatti, V., and Husson, F. (1962) *J. Cell. Biol.*, 12, 207.
25. Lindman, B., and Wennerström, H. (1980) *Micelles: Amphiphile Aggregation in Aqueous Solutions*, Springer, Berlin, Heidelberg, Topics in Current Chemistry series.
26. Brown, G. H., and Wolken, J. J. (1979) *Liquid Crystals and Biological Structure*, Academic, New York.

Appendix 14.1   Common Esters for Rubber

Adipates

DBEA	dibutoxyethyl adipate
DBEEA	dibutoxyethoxyethyl adipate
DNODA	di(normal C8C10) adipate
DOA	dioctyl (2-ethylhexyl) adipate
7006	alkyl alkylether diester adipate
DIDA	diisodecyl adipate
DINA	diisononyl adipate
97A	C9C7 adipate

Epoxies

ESO	epoxidized soyabean oil
G-60	epoxidized soyabean oil
G-62	epoxidized soyabean oil
7170	epoxidized soyabean oil
IOES	isooctyl epoxy stearate
OET	octyl (2-ethylhexyl) epoxy tallate
6.8	epoxidized soyabean oil
10.4	epoxidized linseed oil
EpGdO	epoxidized glycerol dioleate

Glutarates

7050	dialkyl glutarate
DIDG	diisodecyl glutarate
DBEEEG	dibutoxyethoxyethyl glutarate
759	dibutoxyethoxyethoxyethyl glutarate

Glycol Diesters

TeEGEH	tetraethylene glycol di-2-ethylhexoate
TrEGCC	triethylene glycol caprate caprylate
TrEGEH	triethylene glycol di-2-ethylhexoate
PGd2EH	polyethylene glycol di-2-ethylhexoate

Miscellaneous

TXIB	texanol isobutyrate
DBCFm	di-butylcarbitol formal
600	monomeric pentaerythriol ester
707A	polymeric pentaerythritol ester

Phosphates

EHdPF	2-ethyl hexyl diphenyl phosphate
IDdPF	isodecyl diphenyl phosphate
TrAF	triaryl phosphate

Phthalates

BBP	butylbenzyl phthalate
C7C11P	dinormal 7,9,11 carbon phthalate
DBP	dibutyl phthalate
DIDP	diisodecyl phthalate
DOP	dioctyl (2-ethylhexyl) phthalate
DUP	diundecyl phthalate
DINP	diisononyl phthalate
DIOP	diisooctyl phthalate
DN711P	dinormal 7,9,11 phthalate
DOTP	dioctyl terephthalate
DTDP	ditridecyl phthalate

Appendix 14.1   Continued

Polymerics

25P	polymeric sebacate, 200,000 cps viscosity
300P	polymeric, 3,300 cps viscosity
315P	polymerics, 7.5K cps viscosity
325P	polymerics, 5.3K cps viscosity
330P	polymeric, 5.8K cps viscosity
7035P	polymeric glutarate, 11K cps viscosity
7046P	polymeric glutarate, 12K cps viscosity
7092P	polymeric glutarate, 24K cps viscosity
P-670	adipate polymeric, low viscosity
P-675	adipate polymeric, low viscosity
P-7068	phthalate polymeric, low viscosity

Sebacates

83SS	dibutoxyethoxyethyl substitute sebacate
DBES	dibutoxy ethyl sebacate
DBS	dibutyl sebacate
DOS	dioctyl (2-ethylhexyl) sebacate

Trimellitates

TIDTM	tri-isodecyl trimellitate
TNODTM	tri-n-octyl n-decyl trimellitate
TOTM	trioctyl (2-ethyhexyl) trimellitate
uTM	unidentified trimellitate
LTM	linear (mixed normal alcohol) trimellitate
TINTM	tri-isononyl trimellitate
TIOTM	tri-isooctyl trimellitate

Appendix 14.2   Abbreviations and Definitions

ACN	Acrylonitrile
Chg.	Change
Elong.	Elongation
Min.	Minimum
Mod.	Modulus
pts	Points
Ten.	Tensile
Visc.	Viscosity
Ave.	Average
Dist.	Distilled
Hard.	Hardness
min.	minute(s)
ODR	Oscillating Disc Rheometer
Temp.	Temperature
Vol.	Volume
Wt.	Weight

15 Sulfur Cure Systems

Byron H. To

15.1 Introduction and Historical Background

Vulcanization (or curing) is a chemical process designed to reduce the effects of heat, cold, or solvents on the properties of a rubber compound and to create useful mechanical properties. This is most often accomplished by heating with vulcanizing agents, such as elemental sulfur, organic peroxides, organic resins, metal oxides, or urethanes. This process converts a viscous entanglement of long chain molecules into a three dimensional elastic network, as shown in Fig. 15.1.

This chapter covers only the use of sulfur along with activators, accelerators, and retarders to provide the complete cure system for compounds based on general purpose elastomers.

Before the discovery of vulcanization with sulfur by Charles Goodyear [1] in 1839, John Hancock [2] manufactured waterproof garments, gloves, inflatable cushions, shoes, and boots with just raw natural rubber. These rubber products suffered severely from poor physical properties, poor resistance to light and swelling in liquids and, above all, sensitivity to the extremes of temperatures, i.e., the products became stiff when cold and sticky when hot. Goodyear's, and subsequently Hancock's, method of heating rubber and sulfur together resulted in the following rubber compound with improved physical properties.

Natural rubber   100 parts
Sulfur   8 parts

This compound was vulcanized for five hours at 140 °C with good physical properties. In addition, the product remained stable over a wide range of temperatures and was more resistant to swelling in liquids. However, the aging properties were poor.

The shortcomings of using sulfur alone as a vulcanizing agent were recognized early by Goodyear and others, and the search started for additional materials to enhance the properties of the vulcanizate and decrease the time required when using sulfur alone. A number of inorganic oxides such as zinc oxide were tested in the years that followed and did result in shorter cure times, but provided little improvement in the physical properties. One compound was:

Natural rubber   100 parts
Sulfur   8 parts
Zinc oxide   5 parts

The maximum physical properties of this compound were developed in only three hours at 140 °C. This activating effect of zinc oxide remains an essential part of present day compounding.

Raw rubber   Vulcanized rubber

Figure 15.1   Cross linking.

The next important discovery in the history of compounding was made by Onslager who, in 1906, found that aniline accelerated the vulcanization process. Aniline is now regarded as the original organic accelerator. The use of aniline was, however, objectionable because of toxicity effects. Then it was found that the reaction of aniline with carbon disulfide yielded an even more powerful accelerator, thiocarbanilide, which was much less toxic and therefore more acceptable than aniline. A typical compound might be composed of:

Natural rubber   100 parts
Sulfur   6 parts
Zinc oxide   5 parts
Thiocarbanilide   2 parts

This combination of zinc oxide and thiocarbanilide reduced the sulfur level, improved significantly the aging properties, and further shortened the vulcanization time. As a result of this success, many trials with derivatives of thiocarbanilide were conducted, and in 1921 mercaptobenzothiazole (MBT) was introduced to become the first really scorch safe accelerator offering many advantages in compounding.

Before the foundations for modern compounding were final, one more important discovery was made concerning the part played by fatty acids in the efficiency of the vulcanizing system. It was found that the fatty acid constituent in the non-rubber hydrocarbon portions of natural rubber further activated the vulcanization process with organic accelerators. The addition of stearic acid or similar products therefore became standard practice as a precaution against possible deficiencies in the raw rubber, such as:

Natural rubber   100 parts
Sulfur   3 parts
Zinc oxide   5 parts
Stearic acid   1 parts
MBT   1 parts

This compound cures in 20 minutes at 140 °C and provides vulcanizates with a good balance of processing, curing, and physical properties.

15.2   Vulcanizing Agents

Sulfur is by far the most widely used vulcanizing agent in conjunction with activators (metal oxides and fatty acids) and organic accelerators. These are used primarily with general purpose rubbers such as natural rubber (NR), styrene butadiene rubber (SBR), and polybutadiene rubber (BR) – all of which contain unsaturation (double bonds) as opposed to basically saturated rubbers (e.g., butyl, EPDM, etc.). Sulfur donors are used to replace part or all of the elemental sulfur to improve thermal and oxidative aging resistance. They may also be used to reduce the possibility of sulfur bloom and to modify curing and processing characteristics. Two chemicals have been developed over the years to function as sulfur donors – alone or in combination with sulfur – tetramethylthiuram disulfide (TMTD) and dithiodimorpholine (DTDM). They are used to provide significantly improved heat and aging resistance plus reduced heat build-up, in many cases.

15.3   Activators

Activators are both inorganic and organic chemicals used to activate or extract the full potential from the organic accelerators. Zinc oxide is the most widely used inorganic activator (lead and magnesium oxides are also used, but less often), while stearic acid is normally the organic activator of choice. Many rubber compounds today incorporate a combination of zinc oxide and stearic acid in sulfur-cured compounds.

Only a few studies [3,4] have been published on the effects of variations in the concentrations of activators. While many textbooks and publications list the reasons for using accelerator activators, almost no data are available on their specific effects. For these reasons, the presence or absence of the common activators in a simple natural rubber compound has been examined, and rheographs obtained for these compounds are shown in Fig. 15.2.

In the absence of an accelerator, the activators zinc oxide and stearic acid, are ineffective in increasing the number of crosslinks produced (Compound #2 in Fig. 15.2).

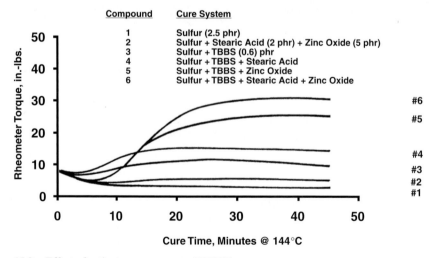

Figure 15.2   Effect of activators on cure rate (100 NR).

Without activators, a sulfenamide accelerator with sulfur produces a significant increase in rheometer torque (crosslinks) in a reasonable period of time (Compound #3). The addition of stearic acid alone to this compound produces a less dramatic effect (Compound #4) than adding zinc oxide alone (Compound #5). For zinc oxide to be fully effective, it must be present in a form that can react with the accelerator system. This means that its particle size must be very fine or the zinc must be in a soluble form. Most natural rubber and some synthetics contain enough fatty acids to form soluble zinc salts that can react with the accelerators. To ensure that enough fatty acids are available, it is common to add 1 to 4 phr of stearic acid or a similar fatty acid. The fatty acid serves as a plasticizer or lubricant to reduce the viscosity of the compound and to solubilize the available zinc. This permits the development of crosslinks by the organic accelerator as is shown in Compound #6.

15.4   Accelerators

Functionally, accelerators are typically classified as primary or secondary. Primary accelerators usually provide considerable scorch delay, medium to fast cure, and good modulus development. Secondary accelerators usually produce scorchy, very fast curing stocks.

Although the development of organic accelerators goes back to the early 1900s, today's primary accelerators are thiazoles and sulfenamides. The sulfenamides are reaction products from MBT or MBTS (mercaptobenzothiazole or the disulfide of MBT) and amines. Examples of these amines are cyclohexyl (CBS), tertiary butyl (TBBS), morpholine (MBS), and dicyclohexylamine (DCBS). The effects of differences in amines are reflected in differences in scorch safety and cure rates.

Typical secondary accelerators are DPG, DOTG, TMTD, TMTM, ZMDC, and ZBPD. Table 15.1 provides identification and comparisons of the different classes of accelerators based on their rates of vulcanization. The secondary accelerators are seldom used alone, but generally are found in combination with primary accelerators to gain faster cures. However, this practice usually results in shorter scorch safety. The comparisons between accelerator classes shown in Fig. 15.3 is typical. Figures 15.4 and 15.5 show comparisons of the primary accelerators CBS, TBBS, MBS, and MBTS in natural rubber and SBR

Table 15.1

Class	Vulcanization rate	Acronyms
Aldehyde-amine	Slow	
Guanidines	Medium	DPG, DOTG
Thiazoles	Semi-fast	MBT, MBTS
Sulfenamides	Fast-delayed action	CBS, TBBS, MBS, DCBS
Sulfenimides	Fast-delayed action	TBSI
Dithiophosphates	Fast	ZBPD
Thiurams	Very fast	TMTD, TMTM, TETD, TBZTD
Dithiocarbamates	Very Fast	ZDMC, ZDBC

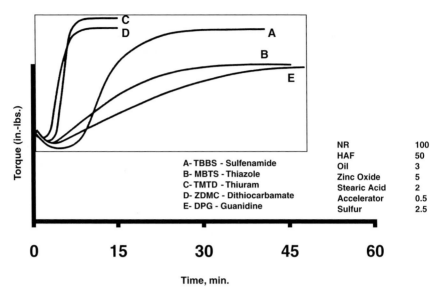

Figure 15.3   Comparison of accelerator classes in natural rubber.

respectively. Major differences in scorch safety, cure rate, and state of cure are observed. MBS gives the greatest scorch safety, while TBBS provides a faster cure rate and higher state of cure (modulus). Similar comparative results are seen when secondary accelerators are used to speed up cure times. In these cases, TMTD and TMTM give higher modulus and longer scorch safety than DPG or ZDMC, but are still more scorchy than TBBS alone.

A large number of secondary accelerators can be used with primary accelerators, thereby providing a great deal of flexibility in processing and curing properties. Tables 15.2

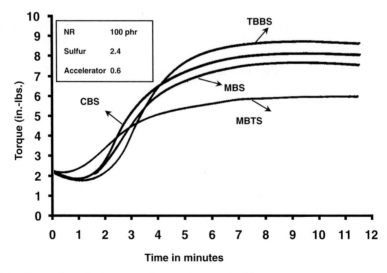

Figure 15.4   Comparison of primary accelerators in natural rubber.

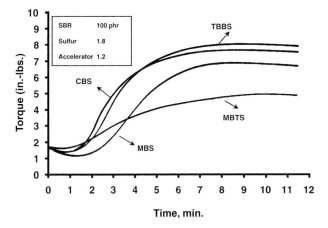

Figure 15.5   Comparison of primary accelerators in SBR.

and 15.3 examine some of the more common secondary accelerators and their effect with sulfenamide accelerator, TBBS. The effects of these secondary accelerators with the other sulfenamide accelerators are similar. Seven different secondary accelerators are evaluated with TBBS in both NR and SBR compounds with a MBTS/DPG as the control.

In natural rubber, all of the activated sulfenamide compounds provide more scorch delay than the activated thiazole (control). Of the secondary accelerators tested, ZDMC is the scorchiest while TETD provides the longest scorch delay. Compounds containing thiurams or dithiocarbamates show cure times at least equal to that of the control but with

Table 15.2   Comparison of Secondary Accelerators in Natural Rubber

Stock	1	2	3	4	5	6	7	8
Sulfur	2.5							
MBTS	1.2	–	–	–	–	–	–	–
DPG	0.4	–	–	–	–	–	–	–
TBBS	–	0.6						
TMTD	–	0.4	–	–	–	–	–	–
TMTM	–	–	0.4	–	–	–	–	–
TETD	–	–	–	0.4	–	–	–	–
ZDMC	–	–	–	–	0.4	–	–	–
ZDEC	–	–	–	–	–	0.4	–	–
ZDBC	–	–	–	–	–	–	0.4	–
DOTG	–	–	–	–	–	–	–	0.4
Mooney Scorch at 120 °C								
t_5, minutes	7.2	16.8	21.5	23.5	13.7	16.7	20.2	21.7
Rheometer at 145 °C								
t_{90}, minutes	9.2	7.5	9.5	9.8	7.0	7.8	9.3	16.5
Stress/strain data								
(Cure to t_{90} mins. at 145 °C)								
100% Modulus (mpa)	2.62	3.38	3.52	3.03	3.14	2.86	2.83	2.83

NR 100, FEF Black 40, Aromatic Oil 10.0, Zinc Oxide 5.0, Stearic Acid 1.5, 6PPD 2.0.

Table 15.3   Comparison of Secondary Accelerators in SBR

Stock	1	2	3	4	5	6	7
Sulfur	1.8	⟶					
MBTS	1.2	–	–	–	–	–	–
DPG	0.4	–	–	–	–	–	–
TBBS	–	0.5	⟶				
TMTD	–	0.3	–	–	–	–	–
TMTM	–	–	0.3	–	–	–	–
TETD	–	–	–	0.3	–	–	–
ZDMC	–	–	–	–	0.3	–	–
ZDBC	–	–	–	–	–	0.3	–
ZBPD	–	–	–	–	–	–	0.3
Mooney scorch at 135 °C							
t_5, minutes	10.4	12.3	22.0	14.5	13.2	18.7	24.4
Rheometer at 160 °C							
t_{90}, minutes	9.2	7.4	9.3	8.6	8.7	12.0	21.3
Stress/strain data							
(Cure to t_{90} mins. at 160 °C)							
100% Modulus (mpa)	2.00	2.14	2.07	1.97	1.89	1.86	1.59

SBR 1500 100, N-330 Black 50, Aromatic Oil 10, Zinc Oxide 4, Stearic Acid 2, 6PPD 2.

much longer scorch delays. Only DOTG as a secondary accelerator shows a longer cure time than the control. Therefore, significant improvement in scorch protection with no increase in cure time can be obtained through the use of activated sulfenamide cure systems.

At the level of accelerators used, all the activated sulfenamide systems produced a higher modulus than the activated thiazole system. The results obtained for SBR (Table 15.3) are similar to those obtained with natural rubber. From a knowledge of these relationships, compounders are able to select the class of accelerators that best fits their needs. Further refinements are possible, particularly with the sulfenamide accelerators, by selecting a particular member of the accelerator class. The development of activated sulfenamide cure systems to meet specific processing requirements and curing properties means both a selection and a refinement process. The selection of the primary and secondary accelerators to be used is a quantum jump based upon the experience of the compounder. After a decision has been made, a systematic study is required to fit these accelerators to the specific process conditions encountered. Response surface experimentation is an effective way to determine the effects of concentrations and ratios of accelerators and sulfur.

15.5   Conventional, Semi-Efficient, and Efficient Cures

Chemical antidegradants are widely used to confer aging resistance to rubber products. A second approach involves modifying cure systems to generate vulcanizates with more di-sulfidic and mono-sulfidic crosslinks which have greater chemical and thermal stability than the polysulfidic crosslinks, and main chain modifications common to conventional vulcanizates. Such cure system modifications are accomplished via sulfur donors or high ratios of accelerator to sulfur. Sulfurless or low sulfur cure systems, as these cure systems are sometimes called, are also known as efficient vulcanization (EV) and semi-efficient (semi-EV) systems [5].

In natural rubber, EV and semi-EV systems can provide remarkable resistance to reversion and aging, but often with some fatigue compromise. In SBR, excellent aging and marching modulus control can be realized generally without fatigue compromises. The choice of cure system depends in part on processing conditions required. Sulfur donors normally give longer processing safety and better green stock storage than the use of high accelerator/low sulfur systems; however, the latter may be better in long overcure situations.

Vulcanizate modification offers a second approach for protection as compared to the use of chemical antidegradants. The former approach involves more development and testing time, because basic changes must be made in the curative system with additional risks than would normally be encountered by the simple addition of an antioxidant. However, the results from such a move may result in dramatic improvement of aged properties. The ultimate in protection of sulfur curable elastomers may be a combination of vulcanizate modification plus a properly chosen, high quality antidegradant system.

15.6   Retarders and Inhibitors

Retarders are materials which, when used with sulfur vulcanized polymers, provide longer scorch safety with relatively little change in other properties. Acidic materials such as salicylic acid, benzoic acid, and phthalic anhydride are used as retarders, primarily with thiazole-based cure systems. These materials are extremely efficient in some applications to extend scorch delay, but they may reduce the rate of cure. Acidic retarders are generally not effective with sulfenamide accelerators and may function as activators instead.

The development and commercialization of the pre-vulcanization inhibitors, N-cyclohexylthio-phthalimide (CTP) [6] revolutionized compounding for scorch control. CTP provides scorch safety without significantly affecting the other compound properties. Figures 15.6 and 15.7 clearly show the effects of CTP on scorch and curing properties. One limitation of CTP is that some bloom may occur when excessive levels (0.5 phr or higher) are used. Such high levels are seldom required in practical areas. In addition to being an effective scorch controller, CTP has found use in a number of other applications:

- Single stage mixing, i.e., mixing all ingredients – including curatives – in one step in an internal mixer, thus reducing mixing time.
- With higher mixing temperatures resulting in reduced mixing times.
- Providing greater flexibility in choice of accelerators, sulfur or sulfur donor levels, and antidegradants (e.g., amine-based antiozonants, which may impart scorchiness).

Table 15.4 provides some typical cure systems (combinations of activators, sulfur, sulfur donor, accelerators, and CTP) for use in carbon black-filled NR and SBR compounds.

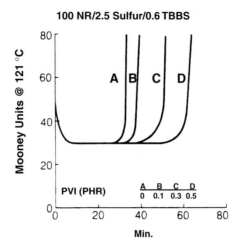

Figure 15.6   Effect of CTP on Mooney scorch.

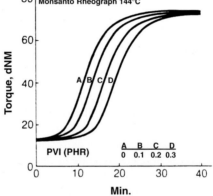

Figure 15.7   Effect of CTP on cure profiles.

Table 15.4   Cure Systems for NR and SBR

Cure system	Natural rubber			SBR		
	Conv.	Semi-EV #1	Semi-EV #2	Conv.	Semi-EV #1	Semi-EV #2
Zinc oxide	5.0	5.0	5.0	3.0	3.0	3.0
Stearic acid	2.5	2.5	2.5	2.0	2.0	2.0
Sulfur	2.5	1.5	1.5	2.0	1.0	1.0
DTDM	–	1.0	–	–	1.0	–
TBBS	0.6	0.6	1.6	1.2	1.2	2.0
CTP	0.1	0.3	0.3	0.2	0.2	0.2

15.7   Recent Developments

N-tert-butyl-2-bezothiazole sulfenimide (TBSI) is a primary amine-based accelerator that exhibits a moderate to long scorch safety, a slow cure rate, and good modulus development similar to the properties found with the secondary amine-based sulfenamide accelerators such as MBS. TBSI is a derivative of TBBS with a second MBT moiety inserted in the TBBS molecule [7]. Because of its unique molecular structure, TBSI exhibits improved reversion resistance and storage stability under hot and humid conditions. Figures 15.8 and 15.9 compare the processing and curing characteristics of TBSI to some of the common sulfenamide accelerators in both NR (30 min. at 145 °C) and SBR (30 min. at 165 °C). TBSI also provides similar performance to either MBS or DCBS when used as the accelerator in a steel skim compound, as illustrated in Fig. 15.10.

The need to replace secondary amine-based accelerators, such as the thiurams and dithiocarbamates, "due to the formation of nitrosamine" continues to be a significant challenge to compounders. *N*-nitrosamines are formed when secondary amines such as dimethylamines are reacted with various nitrosating agents such as nitrogen oxides that are emitted from gas powered vehicles. Exposure limits have been established in Germany on

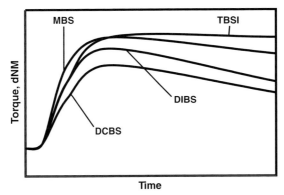

Figure 15.8   TBSI vs. secondary amine sulfenamides: natural rubber (time in min.).

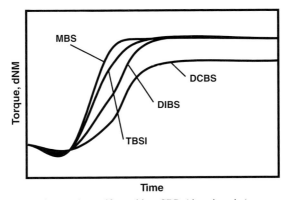

Figure 15.9   TBSI vs. secondary amine sulfenamides: SBR (time in min.).

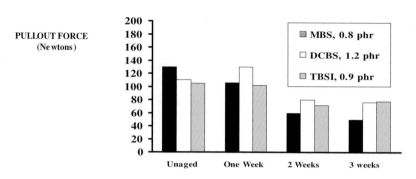

PULLOUT FORCE
(Newtons)

0.5 Embedment
62.5% Cu Wire: 3X.2 + 6X.35

Figure 15.10   Comparison of accelerator performance in tire skim compound.

Table 15.5   Dithiophosphate Acceleration in Natural Rubber

Cure system	Control	Semi-EV	EV	Dithiophos.
Sulfur	2.5	1.0	–	0.80
TBBS	0.6	0.8	1.1	0.80
DTDM	–	1.0	1.0	–
TMTD	–	–	1.1	–
ZBPD	–	–	–	0.80
CTP	–	–	–	0.35
Mooney scorch at 120 °C				
t_5, min.	23.0	39.4	10.9	17.5
Rheometer at 145 °C				
t_{90}, min.	20.3	19.3	16.1	11.4
Rheometer at 180 °C				
Reversion (torque unit reduction at 60 min.)	9.3	8.3	1.0	4.5
Cure to $t_{90} \times 5$ at 181 °C				
(Compared to cure at t_{90} at 145 °C)				
% retained modulus	50	68	89	75
% retained tensile	46	76	92	80
Stress/strain data				
(Cure t_{90} at 145 °C)				
300% modulus (mpa)	17.0	17.5	14.8	15.2
Ultimate tensile (mpa)	29.2	30.7	29.1	29.5
Ultimate elongation (%)	490	490	530	530
After aging 48 h at 100 °C				
% retained modulus	69	82	92	86
% retained tensile	65	79	83	79
High temperature cure				
(Cure t_{90} at 180 °C)				
(Compared to cure at t_{90} at 145 °C)				
% retained modulus	77	88	92	93
% retained tensile	72	94	98	94
Fatigue-to-failure (100% strain)				
Kilocycles to fail (unaged)	130	115	58	109
After aging 48 h at 100 °C				
Kilocycles to fail	8	39	36	34

12 *N*-nitrosamines, five of which are derived from secondary amines used extensively in the rubber industry including di-n-butylamine, diethylamine, diisopropylamine, dimethylamine, and morpholine. The use of dithiophosphate accelerators has been demonstrated somewhat successful in providing a non-secondary amine-based cure system for natural rubber compounds. Specifically, Zinc-*O,O*-di-n-butylphosphorodithioate (ZBPD) [8] imparts good aging resistance and excellent reversion resistance in NR. Table 15.5 compares ZBPD to the EV and semi-EV cure systems in natural rubber.

Accelerator ZBPD gives excellent reversion resistant natural rubber compounds with good retention of physical properties on thermal and thermal oxidative aging as well as high temperature curing. The compound cure rate is significantly increased with reduction in scorch safety and initial fatigue property when compared to both the conventional sulfenamide acceleration and the semi-EV cure systems. Processing safety can be improved either by adjusting the levels of sulfur and accelerator, or by adding a prevulcanization inhibitor. Initial fatigue can also be improved by adjusting the curatives level.

Maintaining properties and performance throughout a rubber product's service life is directly related to maintaining the integrity of the vulcanizate structure under both thermal and thermal oxidative conditions. Historically, this has been achieved by reducing the sulfur content in the crosslinks by using efficient or semi-efficient cure systems. However, as with many changes in rubber compounding, there is a trade-off which, in this case, is a reduction in performance in dynamic fatigue and tear resistance. Two recently developed materials have allowed compounders to forget this compromise, namely hexamethylene-1,6-bisthiosulfate (HTS), a post vulcanization stabilizer and 1,3-bis(citraconimidomethyl)benzene (BCI-MX), an anti-reversion agent.

HTS promotes the formation of flexible hybrid crosslinks [9] as shown in Fig. 15.11. By reducing the average length of sulfur chains at their points of attachment to the polymer backbone, thermal stability is improved. On the other hand, a long chain within the crosslink structure enhances flexibility under dynamic conditions. HTS eliminates the

Sulfur
Accelerator
ZnO
Stearic Acid

$$-O_3SS(CH_2)_6SSO_3-$$

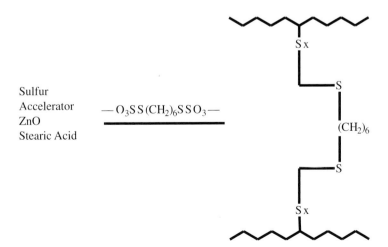

Figure 15.11   Formation of hybrid crosslinks.

Table 15.6   Effect of HTS on Thermal Aging and Fatigue of Natural
              Rubber

Stock	1	2	3
Sulfur	2.5	2.5	1.20
TBBS	0.6	0.6	1.75
HTS	–	2.0	–
% retained 300% modulus			
$10 \times t_{90}$ at 145 °C	75	91	100
$5 \times t_{90}$ at 180 °C	56	68	69
$\dfrac{t_{90} \text{ at } 180\,°C}{t_{90} \text{ at } 145\,°C}$	73	89	91
Fatigue-to-failure (100% strain)			
Kilocycles to fail			
t_{90} at 145 °C	197	241	123
$10 \times t_{90}$ at 145 °C	127	169	90

compromise between thermal aging and dynamic properties [10], as illustrated in Table 15.6.

BCI-MX can substantially reduce deterioration of a rubber compound's physical properties caused by reversion. It reacts to form heat stable crosslinks in sulfur-cured rubber. These new crosslinks produce a vulcanized network that is resistant to overcure and provides good high temperature performance. BCI-MX is unique in that it only becomes active after crosslinks begin to revert through the crosslink compensation mechanism [11]. A schematic representation showing the incorporation of BCI-MX crosslinks during the curing and reversion process is shown in Fig. 15.12. Compensation for the loss of polysulfidic crosslinks during reversion is attained by incorporating the thermally stable BCI-MX crosslinks. Figure 15.13 compares rheographs of a BCI-MX cure system to that of a conventional system. Quantitatively, the conventional system has 18% reversion and the BCI-MX system has none.

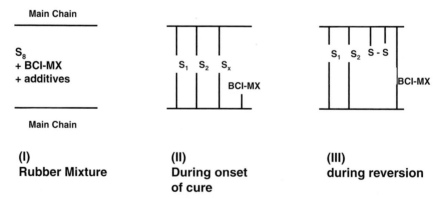

Figure 15.12   Schematic representation showing incorporation of BCI-MX crosslinks during curing and reversion.

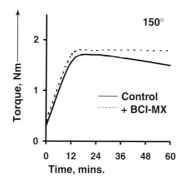

Figure 15.13   Comparison of BCI-MX cure system vs. conventional system.

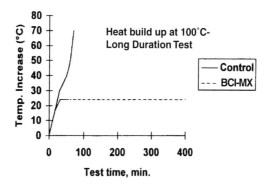

Figure 15.14   Comparison of BCI-MX vs. conventional cure with Goodrich flexometer heat build-up data.

An extended heat build-up test with a Goodrich Flexometer (Fig. 15.14) shows that the BCI-MX-containing compound attained its equilibrium temperature in less than an hour and maintained that temperature for six hours without failure, while the control compound did not survive one hour.

Nomenclature

Acronyms	Chemical name
BCI-MX	1,3-Bis(citraconimidomethyl)benzene
CBS	N-cyclohexyl-2-benzothiazoles sulfenamide
CTP	N-cyclohexylthio-phthalimide
DCBS	N-dicyclohexyl-2-benzothiazole sulfenamide
DOTG	N,N'-di-ortho-tolylguanidine
DPG	N,N'-diphenylguanidine
DTDM	4,4-dithiodimorpholine
HTS	Hexamethylene-1,6-bis thiosulfate
MBS	N-Oxydiethylene-2-benzothiazole sulfenamide
MBT	Mercaptobenzothiazole

Nomenclature *continued*

Acronyms	Chemical name
MBTS	Mercaptobenzothiazole disulfide
TBBS	*N*-tert-butyl-2-benzothiazole sulfenamide
TBSI	*N*-tert-butyl-2-benzothiazole sulfenimide
TBZTD	Tetrabenzylthiuram disulfide
TMTD	Tetramethylthiuram disulfide
TMTM	Tetramethylthiuram monosulfide
ZBPD	Zinc *O,O*-di-n-butylphosphorodithioate
ZDBC	Zinc dibutyldithiocarbamate
ZDMC	Zinc dimethyldithiocarbamate

References

1. C. Goodyear, U.S. Patent 3,633 (1844).
2. T. Hancock, British Patent 9,952 (1843).
3. *Vulcanization: Encyclopedia of Polymer Science and Engineering*, Volume 17, Second Edition.
4. Trivette, Jr., C. D., Morita, E,. and Young, E. J. *Rubber Chemical Technology*, (1962) 35, 1370.
5. Cooper, W., *J. Polymer Science*, (1958) 28, 195.
6. Coran, A.Y, .and Kerwood, J. E., U.S. Patent 3,427,319 (Feb. 11, 1969).
7. Tisler, A. L., and To, B. H., "Improved Rubber Properties Created with Sulfenimide Accelerator," presented at a Meeting of the Rubber Division, American Chemical Society, Philadelphia, October (1993).
8. Dunn, J. R., and Scanlan, J., *J. Polymer Science*, (1959), 35, 267.
9. Anthione, G., Lynch, R. A., Mauer, D. E., and Moniotte, P.G., "A New Concept To Stabilize Cured NR Properties During Thermal Aging", ACS Rubber Division Meeting, 1985.
10. To, B. H., *Rubber World*, (1998) 218, 5.
11. Hogt, A. H., Talma, A. G., deBlock, R. F., and Datta, R. N., U.S. Patent 5,426,155, 1955.

16   Cures for Specialty Elastomers

Byron H. To

16.1   Introduction

Vulcanization systems for specialty elastomers such as ethylene-propylene-diene terpolymer (EPDM), nitrile rubber (NBR), polychloroprene (CR), and isobutylene-isoprene (butyl) elastomer (IIR) are generally different from those for natural rubber (NR), styrene butadiene rubber (SBR), and polybutadiene rubber (BR) and their blends. This difference results from the higher saturations in the polymer backbones in the specialty polymers, which means higher ratios of accelerator to sulfur are required. This chapter deals with selected cure systems for vulcanizing these four specialty polymers.

16.2   Cure Systems for EPDM

Properly compounded EPDM [1] exhibits many desirable vulcanizate properties including resistance to ozone, heat, ultra-violet light, weathering and chemicals. Because of this attractive combination of properties, EPDM has taken over a wide variety of applications. However, the relatively low unsaturation of EPDM requires complex cure systems to achieve the desired properties. Several cure systems are discussed in this chapter, highlighting the advantages and disadvantages of each. Simpler accelerator systems with applicability across a wide range of EPDM types are also reviewed.

Nearly every conceivable combination of curing ingredients has been evaluated in various EPDM polymers over the years. A number of these systems have shown particular merit, including the following:

Cure System 1	"Low Cost"	Sulfur	1.5
		TMTD	1.5
		MBT	0.5

This is one of the first cure systems developed for EPDM. It exhibits a medium cure rate and develops satisfactory vulcanizate properties. The primary advantage of this system is its low cost, while a major drawback is its tendency to bloom.

Cure System 2	"Triple 8"	Sulfur	2.0
		MBT	1.5
		TEDC	0.8
		DPTT	0.8
		TMTD	0.8

This common, non-blooming, cure system has been labelled the "Triple 8" for obvious reasons. It provides excellent physical properties and very fast cures but tends to be scorchy and is relatively expensive.

Cure System 3	"Low Set"	Sulfur	0.5
		ZBDC	3.0
		ZMDC	3.0
		DTDM	2.0
		TMTD	3.0

Excellent compression set and good heat aging properties are the chief characteristics of this system. Its major drawbacks are the tendency to bloom and very high cost.

Cure System 4	"General Purpose"	Sulfur	2.0
		MBTS	1.5
		ZBDC	2.5
		TMTD	0.8

This general purpose, non-blooming system offers good performance and is widely used.

Cure System 5	"2121" System	ZBPD	2.0
		TMTD	1.0
		TBBS	2.0
		Sulfur	1.0

An attractive balance of fast cure, good physical properties and good resistance to compression set and heat aging are the features of this cure system, which was derived from a complex, statistically designed experiment to optimize the level of each ingredient [2]. Tables 16.1, 16.2, and 16.3 summarize the properties obtained with these cure systems when evaluated in three EPDM polymers varying in type and amount of unsaturation. The data confirm the features of each cure system described above. The polymers used are:

Polymer	Third monomer type	% Unsaturation
Nordel[1] 1070	1,4 hexadiene	2.5
Vistalon[2] 5600	ENB	4.5
Vistalon 6505	ENB	9.5

[1] Nordel is a registered trademark of E.I.DuPont de Nemours and Company.
[2] Vistalon is a registered trademark of Exxon Chemical Company.

The development of faster curing, more unsaturated EPDM polymers has made it possible to utilize simpler cure systems, such as activated thiazoles and activated sulfenamides systems, similar to those used in NR, SBR and other highly unsaturated polymers. The use of these cure systems in EPDM polymers is illustrated in Table 16.4. The data in this table compare one of the faster cure systems described earlier, the "Triple 8" system, with these simpler activated systems. The simpler systems offer lower cost, bloom-free compounds, are safer to process, and yield satisfactory physical properties, but they are slower curing. The addition of a second activating accelerator, such as zinc dialkyldithio-carbamate, speeds up the cure with no significant change in physical properties.

Table 16.1   Low Unsaturation EPDM

Cure system	1	2	3	4	5
Low cost	x				
Triple 8		x			
Low set			x		
General purpose				x	
"2121"					x
Mooney scorch at 135 °C					
Minimum viscosity	41.00	49.00	43.00	46.00	41.00
t_5, min.	11.40	6.00	17.50	9.50	15.20
t_{35}, min.	14.40	8.30	24.80	12.40	19.70
Rheometer at 160 °C					
Maximum torque unit	23.50	29.60	24.50	27.50	22.50
t_2, min.	3.50	2.50	4.50	3.00	5.80
t_{90}, min.	17.50	17.30	14.50	15.50	18.00
Stress-strain data					
(Cure t_{90} at 160 °C)					
Shore 'A' hardness	67.00	71.00	69.00	71.00	66.00
100% Modulus (mpa)	3.31	4.86	3.59	4.14	2.66
Ultimate tensile (mpa)	11.66	12.83	11.04	11.83	11.14
Ultimate elongation (%)	320.00	280.00	325.00	295.00	430.00
(After aging 70 h at 120 °C)					
Shore 'A' hardness	72.00	77.00	73.00	77.00	73.00
100% Modulus (mpa)	5.87	9.45	5.55	9.18	4.86
Ultimate tensile (mpa)	13.46	13.90	11.56	13.11	12.21
Ultimate elongation (%)	235.00	160.00	225.00	155.00	280.00
Compression set					
(22 h at 120 °C)					
Cure t_{90} + 5 min. at 160 °C					
% Set	68.00	67.00	40.00	67.00	68.00

Masterbatch: EPDM (100), N-550 (100), N-774 (100), paraffinic oil (110), antioxidant TMQ (2.0), zinc oxide (5.0), stearic acid (2.0).

A common problem with the widely known EPDM cure systems is that, while they produce low compression set, they also exhibit severe bloom. This adverse combination of properties can be overcome with the "2828" system, which consist of 2.0 DTDM/0.8 TMTD/2.0 ZDBC/0.8 DPTT, as shown in Table 16.5. This cure system provides compression set comparable to the low set cure system discussed earlier, with no bloom observed.

16.3   Cure Systems for Nitrile

Nitrile rubber [3] is a general term describing a family of elastomers obtained by copolymerizing acrylonitrile and butadiene. Although each resulting polymer's specific

Table 16.2   Medium Unsaturation EPDM

Cure system	1	2	3	4	5
Low cost	x				
Triple 8		x			
Low set			x		
General purpose				x	
"2121"					x
Mooney scorch at 135 °C					
Minimum viscosity	41.00	46.00	3.08	39.00	38.00
t_5, min.	7.30	4.20	11.00	7.00	10.50
t_{35}, min.	9.80	6.20	17.20	10.00	14.50
Rheometer at 160 °C					
Maximum torque unit	28.00	31.00	25.00	29.00	28.00
t_2, min.	3.20	1.50	3.40	2.50	4.20
t_{90}, min.	12.80	9.30	8.00	13.80	12.00
Stress-strain data					
(Cure t_{90} at 160 °C)					
Shore 'A' hardness	74.00	76.00	74.00	76.00	74.00
100% Modulus (mpa)	4.21	4.28	3.07	4.04	4.14
Ultimate tensile (mpa)	10.45	11.04	9.69	11.07	11.35
Ultimate elongation (%)	305.00	275.00	375.00	310.00	400.00
After aging 70 h at 120 °C					
Shore 'A' hardness	78.00	80.00	77.00	79.00	78.00
100% Modulus (mpa)	1.43	1.23	1.62	1.21	1.93
Ultimate tensile (mpa)	12.66	12.35	11.07	11.42	12.90
Ultimate elongation (%)	207.00	175.00	235.00	175.00	280.00
Compression set					
(22 h at 120 °C)					
Cure t_{90} + 5min. at 160 °C					
% set	67.00	67.00	50.00	65.00	63.00

Masterbatch: EPDM (100), N-550 (100), N-774 (100), paraffinic oil (110), antioxidant TMQ (2.0), zinc oxide (5.0), stearic acid (2.0).

properties depend primarily upon its acrylonitrile content, all of them generally exhibit excellent abrasion resistance, heat resistance, low compression set, and high tensile properties when properly compounded. Probably the predominant feature dictating their use is their excellent resistance to petroleum oils. Cure systems for nitrile rubber are somewhat analogous to those for SBR except that magnesium carbonate-treated sulfur is usually used to aid its dispersion into the polymer. Typical cure systems employ approximately 1.5 phr of the treated sulfur with the appropriate accelerators to obtain the desired rate and state of cure. Common accelerator systems include thiiazoles, thiurams, thiazole/thiuram, or sulfenamide/thiuram types. Examples of these systems are shown in Table 16.6.

As operating requirements for nitrile rubber become more stringent, improved aging and set resistance become important. These improvements can be achieved by reducing the amount of sulfur and by using a sulfur donor such as TMTD or DTDM as a partial or total replacement for the elemental sulfur. Examples of these sulfur donor cure systems are

Table 16.3   High Unsaturation EPDM

Cure system	1	2	3	4	5
Low cost	x				
Triple 8		x			
Low set			x		
General purpose				x	
''2121''					x
Mooney scorch at 135 °C					
Minimum viscosity	41.00	48.00	44.00	37.00	37.00
t_5, min.	9.10	5.30	14.00	8.00	10.00
t_{35}, min.	13.50	7.70	29.50	12.80	14.50
Rheometer at 160 °C					
Maximum torque unit	30.00	35.00	29.00	33.00	28.00
t_2, min.	2.80	1.50	3.40	2.50	3.40
t_{90}, min.	11.10	8.00	9.00	11.20	9.50
Stress-strain data					
(Cure t_{90} at 160 °C)					
Shore 'A' hardness	75.00	77.00	75.00	76.00	76.00
100% Modulus (mpa)	6.69	8.07	6.14	7.59	5.62
Ultimate tensile (mpa) 10.28	10.69	9.59	11.25	9.59	
Ultimate elongation (%)	160.00	135.00	165.00	155.00	175.00
After aging 70 h at 120 °C					
Shore 'A' hardness	79.00	83.00	78.00	81.00	80.00
100% Modulus (mpa) 10.35	–	8.76	–	7.99	
Ultimate tensile (mpa 11.59	11.83	10.21	11.66	10.63	
Ultimate elongation (%)	115.00	85.00	120.00	8.05	140.00
Compression set					
(22 h at 120 °C)					
Cure t_{90} + 5min. at 160 °C					
% Set	66.00	62.00	43.00	65.00	67.00

Masterbatch: EPDM (100), N-550 (100), N-774 (100), paraffinic oil (110), antioxidant TMQ (2.0), zinc oxide (5.0), stearic acid (2.0).

shown in Tables 16.7 and 16.8. The advantages of these systems are improved set resistance and aging resistance while adequate processing safety and fast cures are maintained. Note also in Table 16.8 that when equal levels of DTDM/TBBS/TMTD are used and only the accelerator concentration is adjusted, a wide modulus range results and processing safety and fast cure rate maintained. This method controls crosslink density in sulfur donor systems. Table 16.9 illustrates the relationship between these cure systems in terms of modulus and processing safety.

It has been shown that reducing TMTD and replacing it with a sulfur donor or sulfur donor/TBBS results in longer processing safety and the required state of cure. Where processing safety is of major importance, all the TMTD can be replaced by a sulfur donor/TBBS combination. Table 16.10 outlines an example of this system.

Changing the ratio of TBBS to the sulfur donor in favor of the sulfur donor increases processing safety at the same state of cure. Increasing the sulfur donor level and

Table 16.4   Activated Cure Systems in EPDM

Stock	1	2	3	4	5
CBS	1.20	–	1.20	–	Triple 8
MBTS	–	1.50	–	1.50	
TMTD	0.70	0.80	0.70	0.80	
ZDEC	–	–	0.70	0.80	
Sulfur	1.50	1.50	1.50	1.50	
Mooney scorch at 135 °C					
t_5, min.	9.20	7.40	7.40	6.30	2.40
Rheometer at 160 °C					
Maximum torque unit	60.00	60.00	60.00	60.00	70.00
t_2, min.	3.00	2.70	3.10	2.30	1.10
t_{90}, min.	17.70	22.00	13.60	17.00	10.00
Stress-strain data					
(Cure t_{90} at 160 °C)					
100% Modulus (mpa)	5.69	5.89	5.97	6.38	6.97
Ultimate tensile (mpa)	10.10	10.79	10.48	10.89	11.17
Ultimate elongation (%)	220.00	220.00	210.00	200.00	150.00
Compression set					
(22 h at 100 °C)					
Cure t_{90} at 160 °C					
% Set	54.00	51.00	43.00	48.00	47.00
After overcure 1 h at 160 °C					
(22 h at 100 °C)					
% Set	27.00	24.00	23.00	22.00	24.00

Masterbatch: Medium Unsaturation EPDM (100), N-550 (200), paraffinic oil (120), zinc oxide (5.0), stearic acid (1.0).

maintaining TBBS constant increases the state of cure with the same processing safety. In some cases, low sulfur systems based on sulfur donors have been adopted for the express purpose of increasing processing safety. Using low sulfur systems yields processing safety longer than that obtainable by any other means.

16.4   Cure Systems for Polychloroprene

Polychloroprene [4] rubbers are vulcanized with metallic oxides. Generally, 5 phr of zinc oxide and 4 phr of magnesium oxide are the rule. Sulfur is also used as a curative, and it raises the state of cure. Zinc oxide is the vulcanizing agent and the magnesium oxide controls scorch. Table 16.11 compares three cure systems used in sulfur-cured polychoroprene compounds.

Ethylene thiourea (ETU) has traditionally been the accelerator of choice for attaining maximum physical properties in polychloroprene compounds. A second system is based on

Table 16.5   Cure Systems for Low Set EPDM Compounds

Stock	1	2	3	4
Sulfur	0.50	0.50	–	0.50
DTDM	2.00	1.70	2.00	–
TMTD	3.00	2.50	0.80	1.00
ZDMC	3.00	2.50	–	–
ZDBC	3.00	2.50	2.00	–
DPTT	–	–	0.80	–
CBS	–	–	–	2.00
ZBPD	–	–	–	3.20
Mooney scorch at 135 °C				
t_5, min.	14.20	13.70	16.40	14.20
Rheometer at 160 °C				
Maximum torque unit	23.00	22.10	18.50	15.80
t_2, min.	5.00	4.80	6.70	5.20
t_{90}, min.	11.20	11.20	15.20	11.50
Stress-strain data				
(Cure t_{90} at 160 °C)				
Shore 'A' hardness	70.00	69.00	70.00	66.00
100% Modulus (mpa)	2.97	2.69	2.28	1.89
Ultimate tensile (mpa)	8.83	8.34	8.00	6.48
Ultimate elongation (%)	550.00	545.00	670.00	640.00
After aging 70 h at 120 °C				
Shore 'A' hardness	72.00	73.00	70.00	71.00
100% Modulus (mpa)	3.79	3.17	2.55	2.79
Ultimate tensile (mpa)	9.17	8.89	7.93	8.00
Ultimate elongation (%)	435.00	425.00	560.00	450.00
Compression set				
(70 h at 120 °C)				
Cure t_{90} + 5 min. at 160 °C				
% Set	60.00	58.00	57.00	76.00

Masterbatch: Medium Unsaturation EPDM (100), N-550 (50), N-762 (150), paraffinic oil (120), antidegradant 6PPD (2.0), zinc oxide (5.0), stearic acid (1.0).

TMTM/DOTG/sulfur. This system provides greater scorch safety, but a lower state of cure. In some cases, small quantities of ETU may be added to speed up the cure and raise the state of cure. A third system sometimes used especially in the W-type polymer to speed up cure is based on thiocarbanilide (A-1) acceleration. Thiocarbanilide-accelerated compounds exhibit good processing safety and fast and level cure with excellent tensile properties and compression set resistance.

An unexpected advantage of the A-1 system is its response to a prevulcanization inhibitor such as N-(cyclohexylthio)phthalimide (CTP) added to provide longer scorch delay. However, a sacrifice in compression set and modulus is observed, as shown in Table 16.12. Also included in this table is an A-1/ZBPD combination, which offers very fast cure, but scorchy compounds with excellent compression set resistance. This cure

Table 16.6   High Sulfur Cure Systems for Nitrile Rubber

Stock	1	2	3
MC treated sulfur[*]	1.50	1.50	1.50
TMTM	0.40	–	–
MBTS	–	1.50	–
TBBS	–	–	1.20
TMTD	–	–	0.10
Mooney scorch at 120 °C			
t_5, min.	6.80	8.10	5.70
Rheometer at 160 °C			
t_{90}, min.	8.70	15.20	4.70
Stress-strain data			
(Cure t_{90} at 160 °C)			
Shore 'A' hardness	73.00	71.00	75.00
100% Modulus (mpa)	4.21	3.59	5.04
Ultimate tensile (mpa)	16.35	16.25	17.32
Ultimate elongation (%)	380.00	475.00	355.00
After aging 70 h at 100 °C			
Shore 'A' hardness	80.00	78.00	82.00
100% Modulus (mpa)	3.58	2.44	2.87
Ultimate tensile (mpa)	13.89	11.05	9.87
Ultimate elongation (%)	323.00	323.00	202.00
Compression set			
(22 h at 100 °C)			
% Set	31.00	50.00	55.00

[*] MC = magnesium carbonate treated sulfur for improved dispersion.
Masterbatch: Medium acrylonitrile NBR (100), N-550 (40), N-770 (40),
plastiziser DOP (15), zinc oxide (5), stearic acid (1), antioxidant TMQ (1)
antiozonant 6ppd (2).

system may be particularly applicable in continuous vulcanization processes, where rapid onset of cure is an important requirement.

16.5   Cure Systems for Butyl and Halobutyl Rubber

Because of its low unsaturation, butyl rubber (IIR) [5] possesses excellent resistance to weathering, heat, and ozone, as well as exhibiting excellent fatigue resistance. Of course, its most predominant attribute is low gas permeability which makes it the preferred polymer for innertubes, innerliners, bladders, and other air containment parts.

Compared to NR, butyl rubber requires more active accelerator systems and lower sulfur loadings to achieve a suitable balance of cure rate and physical properties. A

Table 16.7   High Sulfur vs. Low Sulfur Cure Systems for Nitrile Rubber

Stock	1	2
MC treated sulfur	1.50	0.30
MBTS	1.50	–
TBBS	–	1.00
TMTD	–	1.00
Mooney scorch at 120 °C		
t_5, min.	8.10	8.10
Rheometer at 160 °C		
t_{90}, min.	15.20	10.50
Stress-strain data		
(Cure t_{90} at 160 °C)		
Shore 'A' hardness	71.00	69.00
100% Modulus (mpa)	3.59	3.11
Ultimate tensile (mpa)	16.25	15.11
Ultimate elongation (%)	475.00	485.00
After aging 70 h at 100 °C		
Shore 'A' hardness	78.00	74.00
100% Modulus (mpa)	2.44	2.77
Ultimate tensile (mpa)	11.05	13.44
Ultimate elongation (%)	323.00	432.00
Compression set		
(22 h at 100 °C)		
% Set	50.00	24.00

Masterbatch: Medium acrylonitrile NBR (100), N-550 (40), N-770 (40), plastiziser DOP (15), zinc oxide (5), stearic acid (1), antioxidant TMQ (1), antiozonant 6PPD (2).

combination of thiazole and thiuram accelerators are commonly used in innertube compounds. The requirements for butyl tubes include good heat resistance and low compression set upon stretching or maintaining dimensions after inflation. A major problem in most tube compounds is the weakness of the splice, which results in premature failures resulting from separation. One method to improve the splice is to develop longer scorch times, resulting in better flow and better knitting at the spice prior to cure. This must be accomplished with no loss in other key properties. Table 16.13 compares a semi-EV cure system to a conventional sulfur cure system.

The semi-EV system, based on the sulfur donor DTDM, significantly improves heat and compression set resistance as well as processing safety. All these factors contribute to better product performance.

Concerns about the potential hazards of exposure to *N*-nitrosamines in the rubber industry had prompted development of alternatives to secondary amine-based accelerators and curatives. Recent work [6] shows that zinc-*O,O*-di-n-butylphosphorodithioate (ZBPD), a non-secondary amine-derived accelerator, provides a good balance of cost and

Table 16.8   Sulfurless Cure Systems for Nitrile Rubber

Stock	1	2	3	4
MC Treated Sulfur	1.50	–	–	–
MBTS	1.50	–	–	–
TBBS	–	1.00	3.00	2.00
TMTD	–	1.00	3.00	2.00
DTDM	–	1.00	1.00	2.00
Mooney scorch at 120 °C				
t_5, min.	8.10	10.70	7.00	7.90
Rheometer at 160 °C				
t_{90}, min.	15.20	14.70	12.50	13.30
Stress-strain data				
(Cure t_{90} at 160 °C)				
Shore 'A' hardness	71.00	68.00	71.00	73.00
100% Modulus (mpa)	3.59	3.07	4.28	5.73
Ultimate tensile (mpa)	16.25	15.42	16.35	16.77
Ultimate elongation (%)	475.00	485.00	360.00	290.00
After aging 70 h at 100 °C				
Shore 'A' hardness	78.00	73.00	75.00	78.00
100% Modulus (mpa)	2.44	2.67	3.81	4.76
Ultimate tensile (mpa)	11.05	13.42	14.55	13.86
Ultimate elongation (%)	323.00	422.00	320.00	240.00
Compression set				
(22 h. at 100 °C)				
% Set	50.00	22.00	13.00	12.00

Masterbatch: Medium acrylonitrile NBR (100), N-550 (40), N-770 (40), plastiziser DOP (15), zinc oxide (5), stearic acid (1), antioxidant TMQ (1), antiozonant 6PPD (2).

Table 16.9   Effect of Cure Systems on Scorch and Modulus

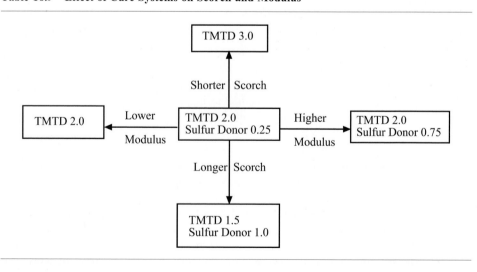

Table 16.10   Effect of Cure Systems on Processing Safety and State of Cure

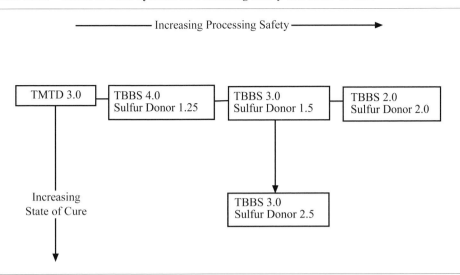

performance requirements in butyl compounds. Table 16.14 lists the seven cure systems in this comparative study. Figures 16.1, 16.2, and 16.3 graphically summarize some of the properties evaluated.

Table 16.15 compares the processing and curing properties of these cure systems. Stock 7 (S/ZBPD/TBBS 1.0/5.0/1.0) provides longer scorch safety and shorter cure time than Stock 1 (conventional sulfur cure). Stock 3 (S/MBT/ZBPD/TBBS 1.25/1.0/1.25/1.25) matches the closest to Stock 2 (semi-EV cure), with a slightly longer cure time. All ZBPD cure systems yield similar physical properties to those of the conventional sulfur cure as shown in Fig. 16.2. Upon aging at 125 °C, the ZBPD systems provide equal or better retention of tensile and elongation than the conventional sulfur cure. Stock 4 (S/MBT/ZBPD/TBBS 2.0/1.0/1.25/1.25) provides the best results, as shown in Fig. 16.3.

Conventional butyl rubber cannot co-vulcanize with highly unsaturated elastomers, such as natural rubber or SBR, because its low unsaturation precludes successful competition for the vulcanizing agents. As a result, blends of butyl rubber with NR, SBR, or BR do not have good vulcanizate properties. The introduction of chlorine or bromine to the butyl molecule, which created a new class of butyl rubber known as halobutyl rubbers, broadened the vulcanization rate and enhanced co-vulcanization with these general purpose elastomers, while maintaining the unique properties of butyl rubber.

Halobutyl can be vulcanized with a variety of curing systems [7]; some general sulfur-based cure systems are:

(a) Benzothiazoles/sulfenamides
     With the exception of MBT, which is scorchy, benzothiazoles and sulfenamides function initially as retarders but produce ultimately higher state of cure. The vulcanizates give a good balance of processing and curing properties. Small amounts of thiurams or guanidines can be used as secondary accelerators for cure rate control.

Table 16.11   Polychloroprene Cure Systems

Stock	1	2	3
ETU	0.50	–	–
TMTM	–	1.00	–
DOTG	–	1.00	–
Sulfur	–	0.50	–
A-1	–	–	0.70
Mooney scorch at 135 °C			
t_5, min.	7.70	34.50	9.50
Rheometer at 160 °C			
Maximum torque unit	31.20	29.30	25.60
t_2, min.	2.20	5.20	2.30
t_{90}, min.	20.80	25.00	5.80
Stress-strain data			
(Cure t_{90} at 160 °C)			
Shore 'A' hardness	61.00	58.00	57.00
100% Modulus (mpa)	2.38	1.93	1.86
Ultimate tensile (mpa)	18.32	16.84	17.39
Ultimate elongation (%)	470.00	550.00	540.00
Compression set			
(22 h at 100 °C)			
% Set	12.00	23.00	10.00

Masterbatch: Neoprene W (100), N-990 (20), N-774 (40), aromatic oil (15), antioxidant TMQ (1.0), zinc oxide (5.0), magnesium oxide (4.0), stearic acid (1.0).

Table 16.12   Improved Scorch Safety in Polychloroprene

Stock	1	2	3
A-1	0.70	0.70	0.70
ZBPD	–	0.50	–
CTP	–	–	0.20
Mooney scorch at 120 °C			
t_5, min.	18.00	7.70	30.00
Rheometer at 160 °C			
t_{90}, min.	7.40	4.80	10.70
Stress-strain data			
(Cure t_{90} at 160 °C)			
Shore 'A' hardness	62.00	63.00	60.00
300% Modulus (mpa)	13.46	11.76	8.56
Ultimate tensile (mpa)	20.04	20.01	18.63
Ultimate elongation (%)	435.00	475.00	530.00
Compression set			
(22 h at 100 °C)			
% Set	17.00	19.00	24.00

Masterbatch: Neoprene W (100), N-990 (20), N-774 (40), aromatic oil (15), antioxidant TMQ (1.0), zinc oxide (5.0), magnesium oxide (4.0), stearic acid (1.0)

Table 16.13   Cure Systems (I) for Butyl Rubber

Stock	1 Conventional	2 Semi-EV
Sulfur	2.00	0.50
TMTD	1.00	1.00
MBT	0.50	–
DTDM	–	1.20
TBBS	–	0.50
Mooney scorch at 121 °C		
t_5, min.	18.50	36.20
Rheometer at 160 °C		
t_{90}, min.	21.80	21.00
Stress-strain data		
(Cure t_{90} at 160 °C)		
Shore 'A' hardness	68.00	68.00
300% Modulus (mpa)	5.52	7.11
Ultimate tensile (mpa)	10.97	11.32
Ultimate elongation (%)	510.00	600.00
After aging 70 h at 120 °C		
% Tensile retention	57.00	76.00
Compression set		
(70 h at 120 °C)		
% Set	81.00	56.00

Masterbatch: Butyl 218 (100), GPF (70), paraffinic oil (25), zinc oxide (5).

Table 16.14   Cure Systems (II) for Butyl Rubber

Stock	1	2	3	4	5	6	7
Sulfur	2.00	0.50	1.25	2.00	1.00	0.50	1.00
TMTD	1.00	1.00	–	–	–	–	–
DTDM	–	1.20	–	–	–	–	–
MBT	0.50	0.50	1.00	1.00	3.00	3.00	–
ZBPD	–	–	1.25	1.25	2.00	3.50	5.00
TBBS	–	–	1.25	1.25	–	–	1.00
Mooney scorch at 135 °C							
t_5, min.	10.00	17.50	19.00	16.00	6.00	6.00	13.00
Rheometer at 172 °C							
t_{90}, min.	12.50	11.00	14.00	17.00	15.00	9.00	10.00
Stress-strain data							
(Cure t_{90} at 172 °C)							
300% Modulus (mpa)	5.00	4.00	4.00	4.50	3.50	2.50	4.00
Ultimate tensile (mpa)	12.00	12.00	11.50	11.00	12.00	11.00	12.00
(Aged 72 h at 125 °C)							
% tensile retained	71.00	77.00	75.00	84.00	72.00	85.00	70.00
% elongation retained	81.00	78.00	84.00	93.00	76.00	78.00	80.00

Masterbatch: Butyl 218 (100), GPF (70), paraffinic oil (25) zinc oxide (5).

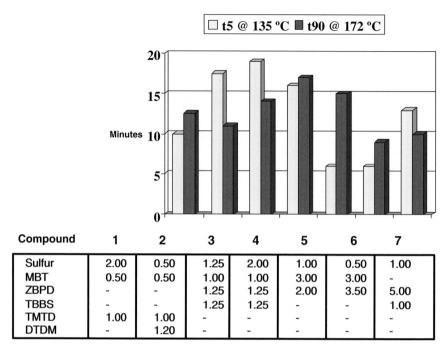

Figure 16.1   Processing and curing comparisons which include some non-secondary amine accelerators.

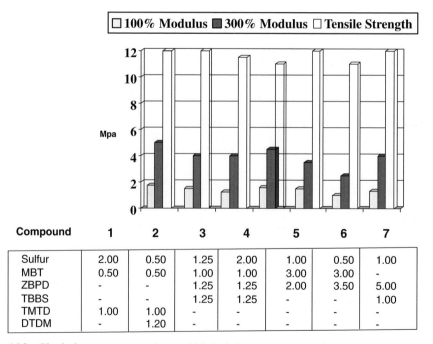

Figure 16.2   Physical property comparisons which include some non-secondary amine accelerators.

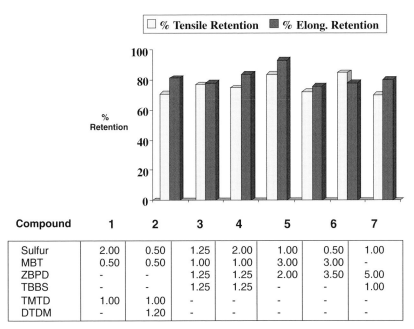

| % Tensile Retention | % Elong. Retention |

Compound	1	2	3	4	5	6	7
Sulfur	2.00	0.50	1.25	2.00	1.00	0.50	1.00
MBT	0.50	0.50	1.00	1.00	3.00	3.00	-
ZBPD	-	-	1.25	1.25	2.00	3.50	5.00
TBBS	-	-	1.25	1.25	-	-	1.00
TMTD	1.00	1.00	-	-	-	-	-
DTDM	-	1.20	-	-	-	-	-

Figure 16.3   Heat aging (72 h at 125 °C) comparisons which include some non-secondary amine accelerators.

(b) Thiurams

Thiurams polysufides are used extensively because they yield compounds with good physical properties and are relatively inexpensive. At low levels, especially in bromobutyl/natural rubber blends, these materials are scorch retarders. But as their proportion is increased, the retarding effect is less pronounced. They are, therefore, used in conjunction with other curatives, such as elemental sulfur or a sulfur donor, e.g., dithiodimorpholine, if higher state of cure and good scorch safety are required.

(c) Dithiocarbamates

Zinc dithiocarbamates are used in applications requiring high states of cure and low compression set. Such compounds are susceptible to scorch, so accelerator levels should be kept to a minimum.

(d) Alkyl phenol disulfides

Alkyl phenol disulfides are particularly effective in co-vulcanizing halobutyls with high unsaturation rubber. These curatives give very good physical properties, heat-aging, and ozone resistance in these polymer blends. They can be used in open steam cures as well as press cures. However, the relatively high reactivity of these curatives can result in reduced scorch times. Scorch can be retarded with MBTS or magnesium oxide.

(e) Amines and thioureas

Diamines and thioureas react with the halogen in the halobutyl to crosslink the polymer. The main advantage of amine cures is their outstanding ozone resistance. Cure is improved when acid acceptors such as magnesium oxide are present. Ethylene thiourea cures halobutyls similarly to the primary diamines, but gives better physical properties, is less scorchy, and is easier to handle.

Nomenclature

Abbreviations	Chemical names
A-1	Thiocarbanilide
6 PPD	N-(1,3-dimethylbutyl)-N-phenyl-paraphenylenediamine
TMQ	Polymerized 1,2-dihydro-2,2,4-trimethylquinoline
CBS	N-cyclohexyl-2-benzothiazole sulfenamide
CTP	N-(cyclohexylthio)phthalimide
DOTG	N,N-di-ortho-tolylguanidine
DPTT	Dipentamethylene-thiuram tetrasulfide
DTDM	4,4′-dithiodimorpholine
MBT	Mercaptobenzothiazole
MBTS	Mercaptobenzothiasole disulfide
TBBS	N-tert-butyl-2-benzothiazole sulfenamide
TEDC	Tetraethyldithiocarbamate
TMTD	Tetramethylthiuram disulfide
TMTM	Tetramethylthiuram monosulfide
ZBPD	Zinc-O,O-di-n-butylphosphorodithioate
ZDBC	Zinc dibutyldithiocarbamate
ZDEC	Zinc diethyldithiocarbamate
ZDMC	Zinc dimethyldithiocarbamate

References

1. Morton, M., *Rubber Technology*, Third Edition, Chapter 9 (pp. 271–272).
2. Flexsys America L.P. Technical Publication, 1985 "Development of Acclerator Systems Through A Systematic Design Procedure"
3. Morton, M., *Rubber Technology*, Third Edition, Chapter 11 (pp. 333–335).
4. Morton, M., *Rubber Technology*, Third Edition, Chapter 12 (pp. 346–347).
5. Morton, M., *Rubber Technology*, Third Edition, Chapter 10 (pp. 294–295, 304–308, 311–316).
6. Flexsys America L.P. Technical Publication, 1994 "Alternative Non-Secondary Amine Compounding for Butyl".
7. Exxon Chlorobutyl Rubber Compounding and Applications, 1976 (SYN-76-1290), Exxon Chemical.

17 Peroxide Cure Systems

Leonard H. Palys

17.1 Introduction

Guidelines, thought processes, and examples of selecting and using organic peroxides in crosslinking are reviewed in this chapter. There are several different classes of organic peroxides used in crosslinking and polymer modification. Peroxide half-life and the specific types of free radicals produced beyond simple homolytic cleavage can play an important role in this selection. Other considerations, such as the type of elastomer chosen, effect of various additives, and the use of coagents can influence crosslinking efficiency. A comparison between peroxide and sulfur vulcanization is also provided here.

17.1.1 What is an Organic Peroxide?

Organic peroxides [1] are compounds that possess one or more oxygen-oxygen (R-OO-R) bonds. These chemical compounds are used commercially to produce free radicals. These free radicals create the desired crosslinked networks in elastomers by hydrogen abstraction and/or addition to double bonds. The structure of the peroxide affects its thermal and chemical stability, as well as the energy level of the free radicals generated.

17.1.2 Classes of Organic Peroxides

There are seven major classes of organic peroxides as shown in Table 17.1: diacyl peroxides, peroxydicarbonates, peroxyesters, peroxyketals, dialkyl peroxides, hydroperoxides and ketone peroxides [2]. All of these major classes are manufactured commercially by the largest organic peroxide producer in the U.S., ATOFINA Chemicals, Inc., Organic Peroxides Division, under the Luperox tradename.

The most commonly used peroxides for crosslinking elastomers are the Dialkyl peroxides followed by t-butyl Peroxyketals and select Diacyl and Peroxyester type peroxides. These crosslinking peroxides are available from the largest U.S. distributor to the rubber industry, R. T. Vanderbilt Company Inc., under the Varox tradename.

The other classes of peroxides are used in polymerization and other non-elastomer curing applications which are outside the scope of this book. For example, ketone peroxide and hydroperoxide [3] type peroxides are decomposed into free radicals via redox (oxidation-reduction) using select transition metal salts or amine compounds [4]. Applications for these peroxides include the room temperature cure of styrene monomer containing unsaturated polyester resins for the manufacture of bathtub/showers, vanities, boats, etc. Most of the peroxyesters and all of the peroxydicarbonates require refrigerated

Table 17.1   The Seven Major Commercial Organic Peroxide Classes[*]
(some shown are not applicable for rubber use)

Organic peroxide class	Structure	10 h half-life
Diacyl peroxides	$\underset{R-C-OO-C-R}{\overset{\displaystyle O \qquad\ O}{\overset{\displaystyle \| \qquad \|}{}}}$	21–75 °C
Peroxydicarbonates	$\underset{RO-C-OO-C-OR}{\overset{\displaystyle O \qquad\ O}{\overset{\displaystyle \| \qquad \|}{}}}$	49–51 °C
t-Alkyl peroxyesters	$\underset{R-C-OO-(\text{tert R})}{\overset{\displaystyle O}{\overset{\displaystyle \|}{}}}$	38–107 °C
Peroxyketals	$R'' \diagdown \ {}_{\diagup} OO-(\text{tert R})$ C $R' \diagup \ {}_{\diagdown} OO-(\text{tert R})$	92–112 °C
Dialkyl peroxides	$(\text{tert R}')-OO-(\text{tert R}'')$	115–131 °C
tert-Alkyl hydroperoxides	$(\text{tert R})-OOH$	Not applicable
Ketone peroxides	$\underset{\overset{\displaystyle \| \qquad\quad \|}{R \qquad\quad R}}{\overset{R' \qquad\quad R'}{\overset{\displaystyle \| \qquad\quad \|}{HOO-C-OO-C-OOH}}}$	Not applicable

[*] Certain peroxides such as the peroxydicarbonates, hydroperoxides, ketone peroxides, and some peroxyesters, as well as some individual members of the other classes, are not suitable for general elastomer crosslinking applications because of half-life and/or room temperature stability considerations.

storage and shipment, and are unsuitable for crosslinking elastomers. These unique peroxides produce free radicals for use in various commercial polymerization processes that transform vinyl monomers into polymers and copolymers.

17.1.3   General Peroxide Selection Guidelines

Important considerations when selecting an organic peroxide for crosslinking and polymer modification are: half-life activity, minimum cure time, SADT (Self Accelerating Decomposition Temperature), maximum storage temperature, free radical energies, peroxide masterbatches, and high performance "HP" peroxides for controlling scorch and improving productivity.

Safety is always an issue in industry. Peroxides normally used by the rubber industry are room temperature-stable compounds that are generally quite safe to handle. However, as with any chemical, strict attention must be paid to the MSDS for personal protective equipment (PPE), health, safety, storage, and specific handling issues. Current MSDS information can be obtained from the peroxide manufacturer or distributor. It is important to avoid all possible contamination (amines, strong mineral acids, organometallics, etc.)

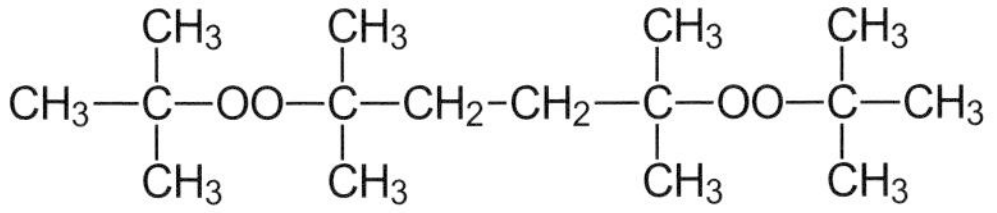

Figure 17.1   2,5-dimethyl-2,5-di(t-butylperoxy)hexane (DMBPHa), a dialkyl type organic peroxide used in polymer modification and crosslinking applications.

when working with any organic peroxide. For example, benzoyl peroxide is extremely sensitive and reactive to amine and/or metal salt contamination. Again, it is important to review the current MSDS and literature for the specific peroxide to be used. The peroxides used by the rubber industry are primarily extended products, i.e., peroxides dispersed on an inert filler like calcium carbonate, clay, wax, or even in a polymer matrix. In general this dilution further increases product safety and simplifies handling.

17.1.3.1   Half-Life

Half-life is defined as the temperature at which one half of the peroxide decomposes in a given time, e.g., 1 hour, 10 hours. As an example, the structure and half-life data for 2,5-dimethyl-2,5-di(t-butylperoxy)hexane (DMBPHa), a Dialkyl type organic peroxide [5,6] is provided in Fig. 17.1. If the concentration of DMBPHa in a rubber is 3% and this composition is reacted for one minute at 181 °C, the theoretical DMBPHa concentration would then be 1.5%; i.e., one-half of the original 3% peroxide concentration, because the 1 minute half-life temperature for DMBPHa is estimated to be 181 °C.

Peroxide half-life or its rate of thermal decomposition is an important first step in determining which peroxide to consider for a polymer modification or crosslinking application. To simplify this process, decomposition rate data for about 50 commercial organic peroxides are summarized in an ATOFINA Chemicals, Inc., *Halflife* bulletin supplied with software files on a computer disk [7,8]. Process temperatures can be entered into a spreadsheet to easily calculate percent of peroxide remaining with time. In a crosslinking application, the loss of peroxide coincides with gains in carbon-carbon linkages. Half-life times at various temperatures can also be calculated.

17.1.3.2   Minimum Cure Time

When crosslinking elastomers or modifying polymers, it is important to adjust the time/temperature profile to ensure ten half-lives of peroxide decomposition. One half-life of peroxide decomposition results in 50% of the original peroxide decomposing. This would relate to a $T_{c50}(min)$ on a rheometer cure curve or the time to attain 50% of the final cure. T_{c90} (min) relates to approximately four half-lives of peroxide decomposition, wherein approximately 90% of the peroxide is decomposed (Table 17.2).

The minimum cure time for crosslinking operations should be the half-life time, times ten. In Fig. 17.1, the peroxide 2,5-dimethyl-2,5-di(t-butylperoxy)hexane (DMBPHa) has a 1 minute half-life of 181 °C. Thus minimum cure time for this peroxide is ten minutes, i.e., (1 minute half-life times 10 half-lives of decomposition).

Table 17.2   Half-life versus Percent of Peroxide Decomposed

Number of half-lives	% of original peroxide decomposed
1	50
2	75
3	87.5
4	93.75
5	96.9
6	98.4
7	99.2
8	99.6
9	99.8
10	99.9

17.1.3.3   SADT (Self Accelerating Decomposition Temperature)

SADT is the lowest temperature at which a peroxide in its largest shipping container undergoes self accelerating decomposition [9]. One should know this temperature and respect it. This test is carried out in a large, temperature-controlled oven over seven days, and looks for any activity in the peroxide. If nothing happens in seven days, then the test is repeated at temperatures five degrees higher. If the peroxide undergoes a noticeable decomposition within seven days, the temperature is dropped five degrees. SADT is determined when the decomposition is repeatable, and occurs after about three days in the oven.

17.1.3.4   Maximum Storage Temperature (MST)

This maximum temperature at which a peroxide should be stored ensures that no significant loss in peroxide assay occurs in six months. The MST is related to the quality of the product and the process. Room temperature storage is the primary requirement for peroxides used in the elastomer crosslinking industry.

One of the lesser known situations which can pose potential MST problems is crosslinking with benzoyl peroxide. Benzoyl peroxide is a room temperature-stable peroxide because of its crystalline structure, but it has a rather low half-life (1 hr. at 92 °C) identical to a refrigerated (50 °F max.) storage peroxide like the liquid t-amylperoxy-2-ethylhexanoate. When making solutions of benzoyl peroxide, the crystal structure is lost and the thermal stability reverts to a refrigerated storage requirement. Thus, when making solutions of benzoyl peroxide, toluene, and silicone rubber for coating or fabric impregnating, it is important to only prepare what can be used that day.

17.1.3.5   Energy of Peroxide Free Radicals

Peroxides generate free radicals. The types of free radicals, their energy level, and the half-life of the peroxide itself are very important considerations when choosing a peroxide for a crosslinking, polymer modification, or grafting operation. Free radicals produced by peroxides

Table 17.3   Bond Dissociation Energies

Precursor	BDE, kJ/mol[*]	kcal/mol	
$(R)_3C-H$	381	91.1	More stable weaker radicals
$(R)_2CH-H$	406	97.0	Poor hydrogen abstractors
RCH_2-H	418	99.9	More discriminate
CH_3-H	439	104.9	Good hydrogen abstractors
$RO-H$	439	104.9	Less discriminate
$R-\overset{\overset{O}{\|\|}}{C}-OO-H$	444	106.1	Stronger radicals
C_6H_5-H	469	112.1	Less stable

Peroxide radical fragments with energies less than 100 kcal/mol are more stable, weaker radicals but are more discriminating in their reactions. Peroxides that generate these types of radicals provide good activity/selectivity with unsaturation. Radicals with energies greater than 100 kcal/mol are stronger, less discriminating, but provide good crosslinking efficiency because of good hydrogen abstraction capabilities.

[*] To convert kJ/mol to kcal/mol, divide by 4.184.
[**] This line represents a polyethylene (PE) backbone. Free radicals above this line are too weak to abstract hydrogens off PE. Peroxides which generate free radicals ≥ 100 kcal/mol possess high crosslinking efficiency due to their ability to abstract secondary hydrogens commonly present in most crosslinkable polymers, including PE.

can have different energy levels, which affects peroxide performance when crosslinking by hydrogen abstraction versus addition to a double bond. A list of hydrogen bond dissociation energies for various radical fragment precursors is provided in Table 17.3.

Polymers such as polyethylene (PE) have mostly secondary hydrogens attached to the carbon chain. In Fig. 17.2, it would take a free radical with at least 100 kcal/mole of energy to successfully abstract this hydrogen atom off the backbone and initiate a crosslinking reaction. We have found that peroxides which generate radical fragments associated with bond dissociation energies greater than 100 kcal/mole (418 kJ/mole) generally provide good crosslinking efficiency when the primary mechanism for crosslinking is hydrogen abstraction.

Knowing which peroxides produce what type of radicals and decomposition by-products, coupled with their half-life information, makes it easier to replace or add different peroxides in a crosslinking formulation. If a crosslinking system contains no unsaturation (no double bonds or monomer coagents), then peroxides which generate high energy radicals are more efficient. For crosslinking formulations with ~ 15+% unsaturation and higher, peroxide efficiency differences are still important, but become less noticeable.

17.1.3.5.1 ASTM Abbreviations for Peroxides

ASTM D-3853 abbreviations for peroxides used in the crosslinking industry are provided in parentheses below. For brevity, these peroxide abbreviations are used in the remainder of this chapter.

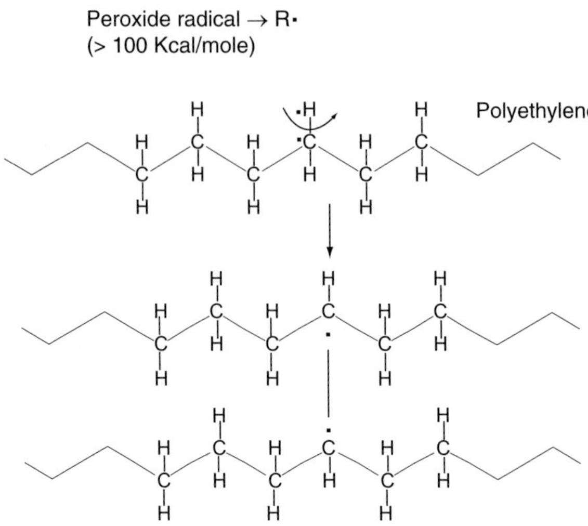

Peroxide radical → R·
(> 100 Kcal/mole)

Figure 17.2   Peroxide attack (abstraction of secondary hydrogens) on two different PE chains.

The peroxides used in the crosslinking industry that only produce high energy radicals are:

2,4-dichlorobenzoyl peroxide (DCBP)
benzoyl peroxide (BPO)
t-butyl perbenzoate (TBPB)
dicumyl peroxide (DCP)
t-butylperoxyacetate (TBPA)
OO-t-butyl-O-isopropyl monoperoxycarbonate (TBIC)
OO-t-butyl-O-(2-ethylhexyl) monoperoxycarbonate (TBEC)
1,3- and 1,4-di(t-butylperoxy)diisopropyl benzene (BBPIB)
2,5-dimethyl-2,5-di(t-butylperoxy)hexyne-3 (DMBPHy)

The peroxides used in the crosslinking industry that produce a mixture of high and lower energy radicals are:

1,1-di(t-butylperoxy)-3,3,5-trimethylcyclohexane (DBPC)
1,1-di(t-butylperoxy)cyclohexane (DBC)
n-butyl-4,4-di(t-butylperoxy)valerate (BPV)
ethyl-3,3-di(t-butylperoxy)butyrate (EBPB)
ethyl-3,3-di(t-amylperoxy) butyrate (EAPB)
1,3-and-1,4-di(t-amylperoxy)diisopropyl benzene (BAPIB)
2,5-dimethyl-2,5-di(t-butylperoxy)hexane (DMBPHa)

Peroxides that produce solid decomposition products that can migrate or "bloom" to the surface are DCBP, BPO, and BBPIB.

17.1.3.6 Peroxide Polymer Masterbatches

Peroxides are sold in pure form and as 40% to 90% assay, free flowing compositions, extended on inert fillers and waxes. Recently, peroxides have been commercially available as a uniform 40% masterbatch in various polymers and as a 90% assay powder dispersed on inert filler [6]. An organic peroxide producer [10] has commercialized ultra-clean, filler-free, 40% peroxide compositions in several polymers including EPM, EPDM, and EVA. The peroxide types available as polymeric masterbatches are DBPC, DBC, DCP, BBPIB, and DMBPHa. Table 17.4 lists the advantages of using a peroxide masterbatch to reduce mixing time while greatly improving peroxide dispersion.

For example, an EPM elastomer composition was divided into two portions. One was mixed with 3 phr of a 40% BBPIB peroxide on calcium carbonate, the other with 3 phr of a 40% BBPIB peroxide-EPM masterbatch. Both compositions were compounded on an open mill for ten minutes. Then each was added to a Banbury for mixing, and samples were taken out at 5 min., 10 min., and 15 min. intervals.

Twelve individual ODR rheometer evaluations were conducted at $180\,^{\circ}C$ for each mixing time interval. The standard deviation for the 12 individual (M_H-M_L) crosslinking results were determined for each peroxide formulation. The highest and lowest ODR rheometer values were provided immediately below the reported standard deviations in Table 17.4. The M_H-M_L values correspond to the relative degree of crosslinking. A lower standard deviation value for the 12 separate ODR crosslinking evaluations indicates that the peroxide is more uniformly distributed. Understandably, a uniform peroxide dispersion is essential for consistent physical property performance.

Surprisingly, the data in Table 17.4 proves that the 40% BBPIB peroxide-polymer masterbatch (BBPIB 40 MG) leads to a uniform distribution of peroxide in a very short time, compared to the use of a BBPIB peroxide extended on ordinary $CaCO_3$ filler. The standard deviation for a crosslinkable system made with 40% BBPIB peroxide on calcium carbonate with 15 min. of mixing on a Banbury is equivalent to only five minutes of

Table 17.4   Comparison of Peroxide Masterbatch versus Peroxide Extended on an Inert Filler Crosslinking EPM with m/p Di (t-butylperoxy) diisopropylbenzene

Standard Deviation of (M_H-M_L) for 12 ODR rheometer runs at $180\,^{\circ}C$
(Actual M_H-M_L low-high data points in parentheses)

Banbury mixing time	5 min.	10 min.	15 min.
BBPIP 40 P	4.20	3.39	2.46
(60% $CaCO_3$ filler)	(26 to 38 dN-m)	(28 to 38.3 dN-m)	(28 to 36.5 dN-m)
BBPIB 40MG	2.54	1.54	0.63
(60% polymer)	(30.1 to 36.7 dN-m)	(30.6 to 33.8 dN-m)	(33.6 to 35.5 dN-m)
polymeric masterbatch			

BBPIB 40 P = 40% m/p di(t-butylperoxy)diisopropylbenzene on calcium carbonate.
BBPIB 40 MG = 40% m/p di(t-butylperoxy)diisopropylbenzene in EPM copolymer masterbatch (R. T. Vanderbilt Company).
10 minutes on an open roll mill to create a 120 kg stock compound.
1 kg samples taken out after 5 min., 10 min., and 15 min. of mixing in a Banbury. Twelve ODR Rheometer runs were conducted using 12 separate samples per batch. Standard deviation was determined.

mixing time when using the 40% BBPIB peroxide EPM polymer masterbatch is used. Peroxide-polymer masterbatches shorten required mixing time, leading to improved peroxide uniformity, less risk of premature crosslinking or scorch during compounding, improved final part quality, and more consistent physical properties from batch to batch and part to part.

17.1.3.7   High Performance (HP) Peroxide Formulations for Improved Productivity

A new family of proprietary Dialkyl peroxide formulations has been developed to improve productivity of injection molding, transfer molding, lay-up compression molding, and extrusion operations in a wide variety of elastomers [11]. These high performance "HP" peroxides are proprietary, filler-extended formulations based on DCP peroxide (Varox DCP-40KE-HP), BBPIB peroxide (Varox 802-40KE-HP) and DMBPHa peroxide (Varox DBPH-50-HP). These HP peroxides are all commercially available, but only from R. T. Vanderbilt Company [12].

The HP peroxide formulations result in longer elastomer processing times without scorch or premature crosslinking, plus elastomers have lower viscosities at process temperatures for easier flow through narrow channels. In addition, HP peroxides can provide improved productivity, i.e, reduced cure times at higher cure temperatures with equivalent scorch time protection. The final product also exhibits equivalent to better final physical properties because of more uniform cure behavior. For example, in commercial operations, improved flow through narrow mold runners (for production of various gaskets) and easier, cleaner, tear trim flash removal have resulted. These HP peroxides are useful for the crosslinking of various elastomers including EPDM, EPM, FKM, FVMQ, VMQ, CM, CSM, and the new metallocene ethylene octene copolymers.

With elastomer compounds that are difficult to process or highly filled, a common practice to avoid scorch or premature crosslinking is to use a more thermally stable, higher half-life peroxide. If using DCP peroxide results in scorch, the common practice is to switch to BBPIB peroxide, as shown in Table 17.5. The BBPIB peroxide provides longer scorch times or safer processing times, but the cure time is now twice as long, based on the T_{c90} values. According to Table 17.5, a better approach to providing superior scorch time protection and a desirable cure time is to use Varox DCP 40KE-HP, a proprietary high performance peroxide formulation based on DCP.

Additional data on the HP peroxides Varox DCP-40KE-HP and Varox 802-40KE-HP are provided in Table 17.6. These unique peroxide formulations provide both improved cure performance and final physical properties. Despite the lower concentration of only 3.8 parts Varox 802-40KE-HP versus 4.8 parts BBPIB peroxide at 40% assay, similar M_H-M_L values are obtained, along with a significantly lower compression set of 28 versus 36, determined at 150 °C for 70 hours. Thus, the HP peroxide is more efficient in crosslinking this poly(ethylene 24% octene) type elastomer.

Additionally, productivity improves with the HP peroxide Varox DBPH-50-HP compared to the standard DMBPHa peroxide as shown in Fig. 17.3. Further examples of these high performance HP peroxides are provided in ACS publications [13, 14]. The HP type peroxide provides a longer scorch time at any corresponding temperature,

Table 17.5   Disadvantages of Higher Half-Life Peroxides to Avoid Scorch

Crosslinking EPDM Peroxide		PHR	
DCP-40KE	7.0	–	–
BBPIB-40KE	–	4.3	–
Varox DCP-40KE-HP	–	–	7.0
Maximum torque M_H(dN-m) moving die rheometer, $1°$ arc			
185 °C	14.3	15.1	14.8
Scorch time $T_{S0.4}$ (seconds)			
185 °C	17.4	20.4	22.8
150 °C	101.0	139.0	208.0
Cure time T_{C90} (seconds)			
185 °C	117.0	208.0	104.0
Mooney viscosity (MV) at 135 °C (Mooney units)			
	44.1	43.1	42.2
Mooney scorch time t_5 (minutes) at 135 °C			
	7.8	16.2	27.0

DCP-40KE = 40% dicumyl peroxide on burgess clay.
BBPIB-40KE = 40% m/p di(t-butylperoxy)diisopropylbenzene on clay.
Varox DCP-40KE-HP = Proprietary, scorch resistant version of dicumyl peroxide (38% assay) on burgess clay, commercially available only from R. T. Vanderbilt Company.

compared to the standard peroxide. Increased productivity is obtained by increasing the crosslinking temperature by 10 °C when using the Varox DBPH-50-HP peroxide. This provides a significant reduction in the overall cure time, while maintaining the scorch time protection required for mold filling operations.

Table 17.6   Crosslinking Poly(ethylene 24%octene) Engage 8100 from DuPont Dow Elastomers, Adjusting Peroxide Level to Obtain Equivalent Crosslinking

Peroxide		PHR		
DCP-40KE	8.30	–	–	–
Varox DCP-40KE-HP	–	5.40	–	–
BBPIB-40KE	–	–	4.80	–
Varox 802-40KE-HP	–	–	–	3.80
Crosslinking poly(ethylene octene) MDR @ 185 °C, $1°$ arc				
M_H (dN-m)	20.00	19.10	19.50	19.60
M_L (dN-m)	1.92	1.66	1.74	1.65
T_{C90}(seconds)	125.00	108.00	219.00	200.00
$T_{S0.4}$(seconds)	19.00	26.00	22.00	25.00
Molded sheet cured @ 185 °C for 7 to 8 min.				
Compression set (%)	35.00	34.00	36.00	28.00

DCP-40KE = 40% dicumyl peroxide on clay filler.
Varox DCP-40KE-HP = Proprietary scorch resistant 38% dicumyl peroxide on clay from R. T. Vanderbilt.
BBPIB-40KE = 40% m/p-di(t-butylperoxy)diisopropylbenzene on clay.
Varox 802-40KE-HP = Proprietary scorch resistant 38% m/p-di(t-butylperoxy)diisopropylbenzene on clay from R. T. Vanderbilt.

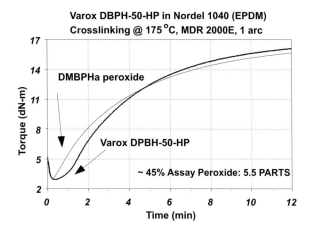

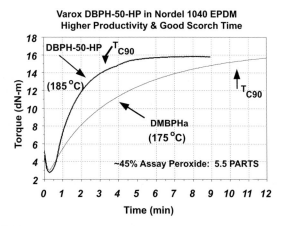

Figure 17.3   Crosslinking EPDM with Varox DBPH-50-HP a proprietary, scorch resistant version.

17.2   Peroxides Used in Crosslinking

Four classes of organic peroxides [15] are used commercially for crosslinking [16] and polymer modification [17] of polyethylene and elastomers: Diacyl, Peroxyester, Peroxyketal, and the Dialkyl type peroxides. Each class of peroxide has a distinctive half-life (thermal stability) and utility in the crosslinking industry. The most common peroxide used in the crosslinking industry has traditionally been the Dialkyl class. However there are several other peroxide types which provide distinct advantages, such as no aromatic odor, no bloom, and liquid form, in many manufacturing operations.

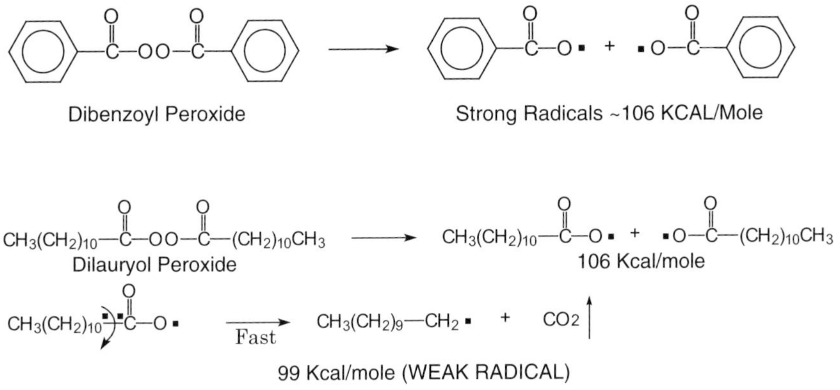

Figure 17.4   Diacyl type peroxide, (dibenzoyl peroxide versus dilauroyl peroxide). Dibenzoyl peroxide decomposes to produce benzoyloxy radicals which have a strong energy level of ~106 kcal/mol. Dilauroyl peroxide undergoes multiple bond decomposition to ultimately produce relatively low energy radicals.

17.2.1   Diacyl Peroxides

The best Diacyl peroxide used for crosslinking without troublesome chlorinated decomposition by-products is BPO (dibenzoyl peroxide). The class of Diacyl peroxides is relatively low in half-life range as shown in Table 17.1. BPO is a solid and is commonly used for crosslinking silicone rubber because shear heating is low during compounding.

Although the half-life of BPO is one of the lowest for the crosslinking peroxides (10 hr. 73 °C, 1 hr. 92 °C), it yields one of the highest energy free radicals, even higher than the most commonly used Dialkyl type peroxide, dicumyl peroxide. As shown in Fig. 17.4, BPO decomposes thermally to produce high energy benzoyloxy radicals, ~106 kcal/mole. Further decomposition to phenyl radicals and formation of CO_2 is not favored, as phenyl radicals have an estimated energy level of ~112 kcal/mole (see Table 17.3). Thus, BPO undergoes a single bond decomposition reaction.

The very high energy level of the radicals generated by BPO crosslink both non-vinyl and vinyl-containing silicone rubber (VMQ). Thus, BPO is often referred to in the silicone rubber industry as "non-vinyl specific" [18]. The peroxide is available in solid granular form (Luperox A98*), but the powder form (Luperox ACP35*) and the paste form (Luperox ANS55*) provide better dispersion in silicone rubber. The suggested use range of this peroxide is 106 to 138 °C, for completed cures in 2 hours to about 5 minutes, respectively.

Increasing the concentration of "non-vinyl specific" BPO peroxide in silicone rubber provides higher degrees of cure than possible with the commonly used "vinyl specific" Dialkyl peroxides, BBPIB and DMBPHa. Once all the vinyl groups of the silicone rubber are used up, adding more of the "vinyl specific" peroxides provides no further increases in cure, but does decrease cure time, e.g., T_{c90}.

Dilauryl peroxide (DLP) is also a "Diacyl" type peroxide. However, DLP peroxide undergoes multiple bond decomposition where the high energy radicals (106 kcal/mole)

---

* Luperox is a registered trademark of ATOFINA Chemicals, Inc.

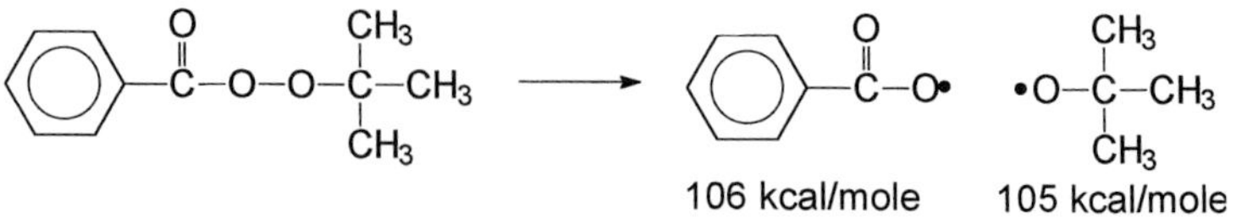

106 kcal/mole   105 kcal/mole

Figure 17.5   Decomposition of t-butyl perbenzoate (TBPB), an efficient, crosslinking peroxide of the Peroxyester class.

quickly rearrange via beta-scission to produce a weak, low energy (99.9 Kcal/mole) alkyl radical. As a result, DLP is used exclusively for polymerization reactions and exhibits very poor crosslinking performance where hydrogen abstraction is required.

17.2.2   Peroxyester and Monoperoxycarbonate Peroxides

As shown in Fig. 17.5, t-butylperbenzoate (TBPB, Luperox P) is a pure liquid peroxide of the peroxyester type, useful in crosslinking a wide variety of elastomers as well as various crosslinkable plastisols in the 130 to 170 °C range. TBPB peroxide is effective at lower temperatures because of its somewhat low half-life temperature (1 hr half-life at 125 °C). This peroxide is quite efficient because one half of this molecule generates the benzoyloxy radical, the same higher energy radical obtained from benzoyl peroxide (BPO). However, because of its low molecular weight of 194 g/mole, the expected higher volatility of this peroxide must be considered in each individual process. TBPB has been used in silicone rubber and curing EVA and EPDM. The suggested use range of this peroxide is 142 to 180 °C, for completed cures in 2 hours to about 5 minutes, respectively.

In Fig. 17.6 is shown a pure liquid peroxide (Luperox TBEC) with a chemical name of OO-t-butyl-O-(2-ethylhexyl)monoperoxycarbonate (1 hr half-life at 122 °C). This peroxide provides good crosslinking efficiency, and has a high molecular weight of 246 g/mole. It is used in silicone, EVA, EPM, and EPDM crosslinking. The suggested use range of this peroxide is 142 to 180 °C, for complete cures in 2 hours to about 5 minutes, respectively. It is non-aromatic, non-discoloring, and does not produce decomposition by-products which would lead to bloom.

17.2.3   Peroxyketal and Dialkyl Type Peroxides

These are the two most common classes of peroxides used in the rubber industry. Traditionally, the Dialkyl class, in particular DCP, BBPIB, and DMBPHa, are the most commonly used peroxides for elastomer crosslinking applications.

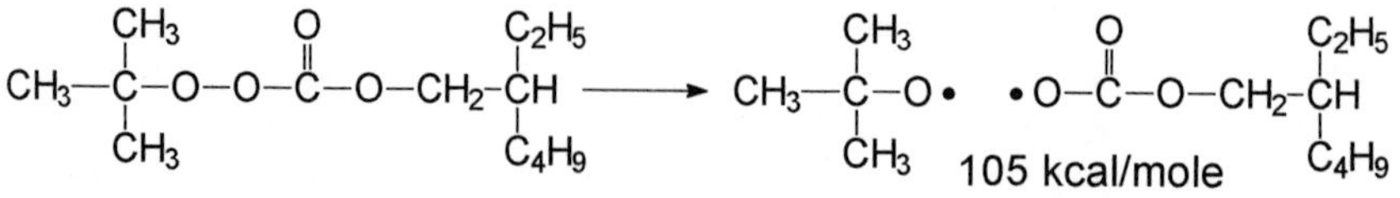

105 kcal/mole

Figure 17.6   Decomposition of OO–t-butyl–O–(2-ethylhexyl)monoperoxycarbonate, a peroxide that generates high energy radicals, good for crosslinking. It does not have a tendency to decarboxylate, as the resulting radical would also be high in energy.

Table 17.7   Peroxyketal Peroxides

Structure	Name	Half-life Data ($^{\circ}$C)	
		10 hr.	1 hr.
$CH_3-\overset{\overset{\displaystyle CH_3}{\vert}}{\underset{\underset{\displaystyle CH_3}{\vert}}{C}}-OO\diagdown\diagup OO-\overset{\overset{\displaystyle CH_3}{\vert}}{\underset{\underset{\displaystyle CH_3}{\vert}}{C}}-CH_3$ (3,3,5-trimethyl-cyclohexane ring)	1,1-Di(t-butylperoxy)-3,3,5-trimethyl-cyclohexane Luperox 231, Varox 231 ASTM term = (DBPC)	96	115
$CH_3-\overset{\overset{\displaystyle CH_3}{\vert}}{\underset{\underset{\displaystyle CH_3}{\vert}}{C}}-OO\diagdown\diagup OO-\overset{\overset{\displaystyle CH_3}{\vert}}{\underset{\underset{\displaystyle CH_3}{\vert}}{C}}-CH_3$ (cyclohexane ring)	1,1-Di(t-butylperoxy)-cyclohexane Luperox 331, Varox 331 non-ASTM term = (BPC)	97	116
$CH_3-CH_2-\overset{\overset{\displaystyle CH_3}{\vert}}{\underset{\underset{\displaystyle CH_3}{\vert}}{C}}-OO\diagdown\diagup OO-\overset{\overset{\displaystyle CH_3}{\vert}}{\underset{\underset{\displaystyle CH_3}{\vert}}{C}}-CH_2-CH_3$ (cyclohexane ring)	1,1-Di(t-amylperoxy)-cyclohexane Luperox 531 non-ASTM term = (APC)	93	112
$CH_3-CH_2-O-\overset{\overset{\displaystyle O}{\Vert}}{C}-CH_2-\overset{\overset{\displaystyle OO-\overset{\overset{\displaystyle CH_3}{\vert}}{\underset{\underset{\displaystyle CH_3}{\vert}}{C}}-CH_3}{}}{\underset{\underset{\displaystyle OO-\overset{\overset{\displaystyle CH_3}{\vert}}{\underset{\underset{\displaystyle CH_3}{\vert}}{C}}-CH_3}{}}{C}}-CH_3$	Ethyl-3,3-di(t-butylperoxy) butyrate Luperox 233, Varox 233 ASTM term = (EBPB)	114	134
$CH_3-CH_2-O-\overset{\overset{\displaystyle O}{\Vert}}{C}-CH_2-\overset{\overset{\displaystyle OO-\overset{\overset{\displaystyle CH_3}{\vert}}{\underset{\underset{\displaystyle CH_3}{\vert}}{C}}-CH_2-CH_3}{}}{\underset{\underset{\displaystyle OO-\overset{\overset{\displaystyle CH_3}{\vert}}{\underset{\underset{\displaystyle CH_3}{\vert}}{C}}-CH_2-CH_3}{}}{C}}-CH_3$	Ethyl-3,3-di(t-amylperoxy) butyrate Luperox 533 non-ASTM term = (EAPB)	112	132

Source: Product Bulletin, *Peroxyketals*, ATOFINA Chemicals, Inc., Philadelphia, PA.

A listing of the commercial Peroxyketal and Dialkyl type peroxides for crosslinking is provided in Tables 17.7 and 17.8, respectively. Thermal decomposition mechanisms for the DBPC (Peroxyketal), DCP, and BBPIB (Dialkyl) peroxides are provided in Figs 17.7 and 17.8.

Peroxyketals are lower in half-life than Dialkyl peroxides, thus providing faster cure rates at a given temperature. The Peroxyketal peroxides are non-aromatic liquids that are

Table 17.8   Dialkyl Peroxides (di-tertiary alkyl peroxides)

Structure	Name	Half-life Data (°C)	
		10 hr.	1 hr.
	Dicumyl Peroxide Luperox DCP, Varox DCP; ASTM term = (DCP)	117	137
	Di(t-butylperoxy)-m/p-diisopropylbenzene Luperox F, Varox 802; ASTM term = (BBPIB)	119	139
	2,5-Dimethyl-2,5-di(t-butylperoxy)hexane Luperox 101, Varox DBPH; ASTM term = (DMBPHa)	120	140
	t-Butyl Cumyl Peroxide Luperox 801; ASTM term = (TBCP)	123	143
	2,5-Dimethyl-2,5-di(t-butylperoxy)hexyne-3 Luperox 130, Varox 130; ASTM term = (DMBPHy)	131	152

Source: Product Bulletin, Dialkyl Peroxides, ATOFINA Chemicals, Inc., Philadelphia, PA.

often dispersed on inert fillers for easier handling. They do not produce "bloom" i.e, solid decomposition by-products.

Peroxyketal peroxides generate a mixture of weak and high energy free radicals. The weaker radicals shown in Fig. 17.7 are less likely to abstract a hydrogen from a polyethylene type backbone as per the hydrogen bond dissociation energy information in Table 17.3. Thus, Peroxyketals are less efficient than the high energy radical producing Dialkyl type peroxides when crosslinking saturated polymers (no double bonds). In highly unsaturated elastomer formulations (e.g., golf ball formulations containing polybutadiene and acrylic coagents), the Peroxyketals are equivalent in crosslinking efficiency to the Dialkyl peroxides.

To illustrate this, in Table 17.9 a Peroxyketal is compared to a Dialkyl peroxide in a fully saturated polymer that contains no double bonds. Crosslinking the fully saturated poly(ethylene octene) copolymer requires efficient hydrogen abstraction by strong free radicals. The DBPC Peroxyketal peroxide crosslinks the polymer about seven times faster at 165 °C, compared to the DCP Dialkyl peroxide at 170 °C because of the Peroxyketal's lower half-life. However, the DBPC Peroxyketal produces a mixture of weak and strong free radicals, compared to strictly high energy radicals from the DCP Dialkyl peroxide (compare Fig. 17.7 to Fig. 17.8). For this reason, DCP is approximately twice as efficient as DBPC on an equal weight basis.

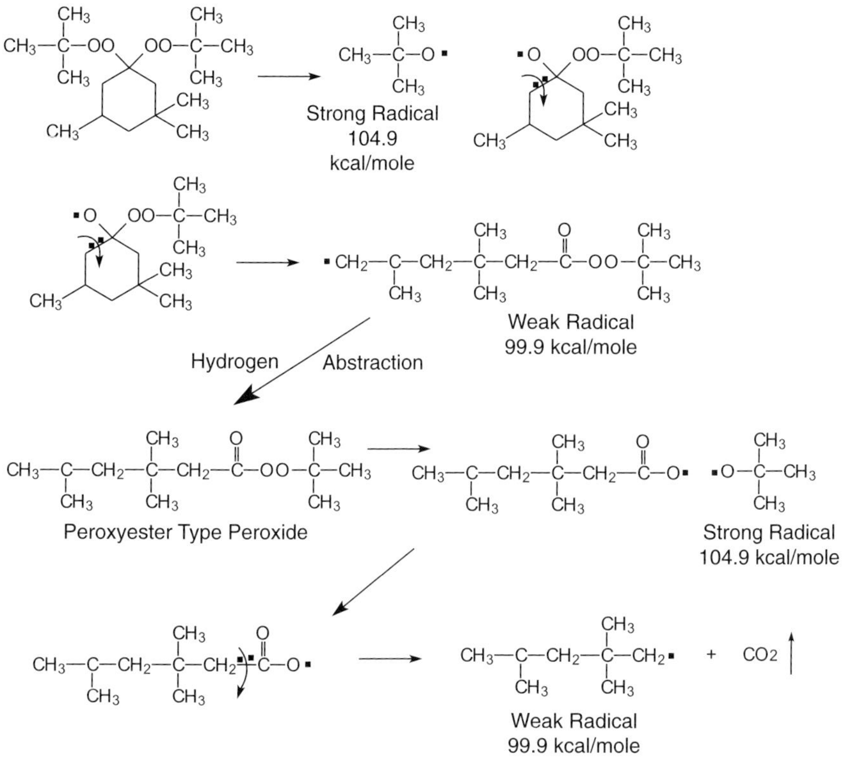

Figure 17.7   Peroxyketal peroxides: thermal decomposition of 1,1-Di(t-butylperoxy)-3,3,5-trimethyl-cyclohexane (DBPC).

Peroxide crosslinking in an elastomer which contains a fair amount of unsaturation (i.e., double bonds) is examined in Table 17.10, which compares the cure performance of two Peroxyketals, DBPC and EBPB to DCP, a Dialkyl peroxide, in an EPDM compound containing 9% diene unsaturation. The difference in efficiency between the Peroxyketals and the Dialkyl peroxide is less pronounced because of the unsaturation present in the terpolymer. The DBPC efficiently and readily crosslinks unsaturated elastomers at a much lower process temperatures (149 °C vs. 167 °C), while providing faster cure times (T_{c90}), compared to DCP. Note that DCP has the lowest half-life of all the commercially available Dialkyl type peroxides.

In Table 17.11, the Peroxyketals are compared to a Dialkyl peroxide in a highly unsaturated polybutadiene cure formulation. The Peroxyketal type peroxides DBPC and particularly EBPB exhibit equivalent crosslinking efficiency to DCP. The lower energy radicals produced by the DBPC and EBPB Peroxyketals easily add to the double bonds of the diene and/or abstract weak allylic hydrogens to the same degree as the higher energy radicals generated by DCP, a Dialkyl type peroxide. Peroxyketals and Dialkyl peroxides are equally effective in highly unsaturated elastomers. However, peroxyketal type peroxides can provide faster cure times at a given cure temperature.

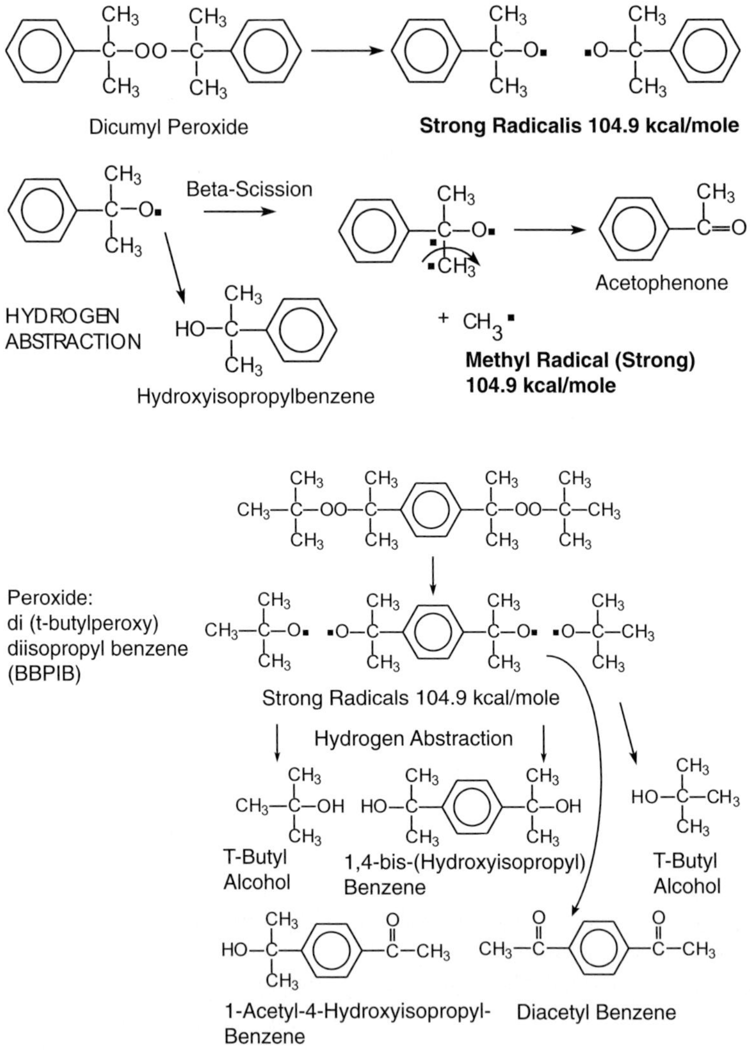

Figure 17.8   Dialkyl peroxides: thermal decomposition of dicumyl peroxide (DCP) and of di(t-butylperoxy)diisopropylbenzene (BBPIB).

17.2.4   Performance Characteristics of Dialkyl Type Peroxides

The peroxide DMPBHa (Table 17.8) is the peroxide used in controlling the molecular weight (known as vis-breaking or controlled rheology) of polypropylene. It is the preferred peroxide for fluoroelastomers [19,20] and silicone rubber and in high performance applications requiring bloom-free and odor-free characteristics.

Based on decomposition mechanisms, DMBPHa generates a mixture of lower energy and high energy free radicals. Compared to DCP or BBPIB, the DMBPHa is less efficient on a moles of peroxide moiety basis. Comparative data for these three peroxides in a

**Table 17.9   Cure Performance of a Peroxyketal versus a Dialkyl Peroxide
1,1-Di(t-butylperoxy)-3,3,5-trimethylcyclohexane versus Dicumyl Peroxide in
Engage Saturated Poly(ethylene octene) Copolymer**

Component	Parts
Poly(ethylene-24% octene) copolymer	100
N774 Carbon black	75
2280 process oil	10
Zinc oxide	5
AgeRite MA antioxidant	2

Parts of 40% peroxide extended on inert filler				Peroxide
40% Peroxyketal	9.3	18.6	–	–	(DBPC)
40% Dialkyl	–	–	4	8	(DCP)

Crosslinking Engage 8100 (Engage is a registered trademark of DuPont Dow Elastomers) using
Alpha Technologies MDR 2000E with a 1° arc

Cure temp (°C)	165 °C	165 °C	170 °C	170 °C
M_H (dN-m)	8.90	19.10	9.50	20.30
M_L (dN-m)	2.48	2.17	1.74	1.68
M_H-M_L (dN-m)	6.45	16.90	7.76	18.60
T_{C90} (minutes)	1.31	1.26	8.70	8.30
$T_{S0.4}$ (seconds)	21.00	25.20	39.00	28.00

(DBPC) = 40% peroxyketal = 40% 1,1-di(t-butylperoxy)-3,3,5-trimethylcyclohexane on calcium
carbonate filler.
(DCP) = 40% dialkyl peroxide = 40% dicumyl peroxide on Burgess KE clay filler.

Nordel[*] 1040 EPDM [21,22] and in Engage[*] poly(ethylene octene) [23,24] have been
reported. For example, in EPDM, 0.017 moles of DMPBHa was needed to provide the
same level of crosslinking as 0.01 moles of either DCP or BBPIB. Thus, DCP and BBPIB
are similar in molar efficiency when crosslinking EPDM or poly(ethylene octene).

The DCP and BBPIB peroxides are the most widely used for crosslinking, offering
excellent cost-performance. However as shown in Fig. 17.8, these peroxides generate
aromatic   decomposition   by-products   such   as   acetophenone,   or   the   solid
bis(hydroxyisopropyl)benzene, 1-acetyl-4-(hydroxyisopropyl) benzene and 1,4 or 1,3
diacetylbenzene decomposition by-products which create "bloom." On a weight basis,
less BBPIB or DMBPHa is needed compared to DCP to achieve the same degree of
crosslinking. The BBPIB and the DMBPHa each have two peroxide groups per molecule
and a higher active oxygen content than DCP, which contains only a single peroxide
group.

DMBPHy (Luperox 130) with a chemical name of 2,5-dimethyl-2,5-di(t-butylperoxy)hexyne-3, is another important, less commonly known, Dialkyl peroxide. It is a very
efficient peroxide for crosslinking reactions because of its ability to generate strong, high

[*] Nordel and Engage are registered trademarks of DuPont Dow Elastomers.

Table 17.10   Cure Performance of Peroxyketals versus Dialkyl Peroxides
1,1-di(t-butylperoxy)-3,3,5-trimethylcyclohexane (DBPC), ethyl-3,3-di(t-butylperoxy)butyrate
(EBPB) vs dicumyl peroxide (DCP) in poly(ethylene propylene diene) EPDM using an ODR, 3° arc

Formulation	Parts
EPDM*	100
N-990	40
N-550	25
Sunpar 2280	10
ZnO	5
40% Peroxide	5

* EPDM fast cure; 52% ethylene.

Peroxide	DBPC	EBPB	DCP
Cure temperature °C	149.0	167.0	167.0
Net torque M_H-M_L (dN-m)	28.0	32.0	49.0
Cure time, T_{C90} (mins.)	10.0	12.2	13.3

DBPC = Peroxyketal type peroxide: 1,1-di(t-butylperoxy)-3,3,5-
trimethylcyclohexane (40% on calcium carbonate).
EBPB = Peroxyketal type peroxide: ethyl-3,3-di(t-butylperoxy)butyrate
(40% on calcium carbonate).
DCP = Dialkyl type peroxide: dicumyl peroxide (40% on clay).

energy, free radicals. This peroxide has one of the highest thermal stabilities of the Dialkyl family and is used to crosslink polymers which require elevated processing temperatures such as polyethylene, particularly, HDPE (high density polyethylene). Table 17.12 lists ODR cure performance data comparing DCP and DMBPHy in LLDPE (linear low density polyethylene). Note that the DMBPHy peroxide is evaluated at a 30 °C higher crosslinking temperature than DCP because of the vast differences in half-life activity, yet it provides a longer T_{c90}. The higher thermal stability of the DMBPHy peroxide permits (150–120 °C) higher compounding temperatures for various polymers than DCP or BBPIB does, without decomposing the peroxide. The final polymer composition containing peroxide can then be used in other processes such as injection molding, or extrusion of wire and cable products that are cured via continuous vulcanization steam tubes or autoclaves.

17.2.5   t-Amyl and t-Butyl Type Peroxides

Within certain classes of peroxides (e.g., Peroxyesters, Peroxyketals, and Dialkyls) there are two major types of peroxides: t-butyl and t-amyl. As shown in Fig. 17.9, t-butyl type peroxides produce high energy t-butyloxy radicals (104.9 kcal/mole) and methyl radicals (104.9 kcal/mole), which are excellent for all crosslinking applications, including polymers such as PE, that cannot be crosslinked via sulfur vulcanization. The t-amyl type peroxides, e.g., 1,1-di(t-amylperoxy)cyclohexane, generate lower energy (99.9 kcal/mole) ethyl radicals which are poor crosslinkers of PE. However, they are particularly suited for crosslinking systems that contain high unsaturation or formulations which contain monomers, coagents,

Table 17.11   Cure Performance of Peroxyketals vs Dialkyl Peroxides in 1,4-Polybutadiene, a System Containing High Unsaturation

Formulation	Parts
1,4-Polybutadiene	100
Peroxide (40% assay)	4

Peroxide type	Peroxyketal	Peroxyketal	Dialkyl peroxide
Abbreviation	DBPC	EBPB	DCP

Crosslinking at 163 °C with an ODR, 1° arc

M_H-M_L (dN-m)	39.0	47.0	44.0
T_{C90} (min)	3.8	10.8	13.9
T_{S2} (min)	0.6	1.0	1.3

DBPC = 40% peroxyketal = 40% 1,1-di(t-butylperoxy)-3,3,5-trimethylcyclohexane on calcium carbonate filler.
EBPB = 40% peroxyketal = 40% Ethyl-3,3-di(t-butylperoxy)butyrate on calcium carbonate filler.
DCP = 40% dialkyl peroxide = 40% dicumyl peroxide on calcium carbonate filler. Tacktene XC-575 polybutadiene.

and unsaturated polymers together with polymers which are susceptible to radical attack. The t-amyl type peroxides produce "selective" free radicals which target double bonds and significantly reduce residual monomer or coagent levels. Thus, the "t-amyl" type peroxides are useful in the manufacture of various thermoplastic elastomers (TPE, TPV) via dynamic vulcanization. In TPE and TPV, the more fragile PP phase is less likely to be degraded [25] because unsaturated coagents and elastomers would be the primary target of the lower energy ethyl radicals produced by t-amyl type peroxides [26,27,28].

The t-amyl type peroxides are also suitable for PE melt flow modification [29] and have been found to minimize the formation of gels. Commercially, 0.05% to 0.1%

Table 17.12   Cure Performance of Various Dialkyl Type Peroxides in LLDPE

LLDPE		100	100
AgeRiteD quinoline type antioxidant		0.50	0.50
Peroxide:	DCP	2.00	–
Peroxide:	DMBPHy	–	2.00
ODR rheometer data temp (°C)		168.00	185.00
Maximum torque	M_H (dN-m)	33.56	31.64
Minimum torque	M_L (dN-m)	3.11	2.40
Degree of crosslinking	$M_H - M_L$	30.50	29.20
Scorch time	T_{S2} (min)	2.50	3.50
Cure time	T_{C90} (min)	14.70	19.80

DCP = dicumyl peroxide.
DMBPHy = 2,5-dimethyl-2,5-di(t-butylperoxy)hexyne-3.

$$CH_3-\overset{\overset{\displaystyle CH_3}{|}}{\underset{\underset{\displaystyle CH_3}{|}}{C}}-O\cdot \quad \xrightarrow{\text{Slow Rxn}} \quad CH_3-\overset{\overset{\displaystyle O}{\|}}{C}-CH_3 \;+\; CH_3\cdot$$

t-Butoxy radical — Acetone — Methyl Radical
104.9 kcal/mole — 104.9 kcal/mole

$$CH_3CH_2-\overset{\overset{\displaystyle CH_3}{|}}{\underset{\underset{\displaystyle CH_3}{|}}{C}}-O\cdot \quad \xrightarrow{\text{Fast Rxn}} \quad CH_3-\overset{\overset{\displaystyle O}{\|}}{C}-CH_3 \;+\; CH_3CH_2\cdot$$

t-Amyloxy Radical — Acetone — Ethyl Radical
104.9 kcal/mole — 99.9 kcal/mole

peroxide, dispersed on PE powder or pellets is reacted at 200 °C in an extruder. This yields a PE polymer with improved melt strength to prevent sagging during processing, increased dimensional stability, and higher impact and tensile strength for production of extruded pipe or blown film.

17.2.6   Effect of Additives When Crosslinking with Peroxides

Table 17.13 illustrates the effect of different process oils and plasticizers on cured compound modulus, when crosslinking with organic peroxides. Note that all the oils and plasticizers reduce the relative degree of crosslinking as implied by differences in $(M_H\text{-}M_L)$. The aliphatic type additives display higher cure states than their aromatic counterparts. Peroxide free radicals abstract labile hydrogens from aromatic species, creating non-reactive free radicals and thus, reduce the number of efficient radicals available for hydrogen abstraction or addition across double bonds.

   Other additives can have a dramatic effect on peroxide cures. In Table 17.14, all of the antioxidants show reduced $(M_H\text{-}M_L)$ values; thus, the concentration of peroxide should be increased to compensate for the loss in crosslinking EPDM. When using amine-type antioxidants, 3 phr of DCP are needed to obtain the same cure state as 1 phr of DCP with no antioxidant. The quinoline-type antioxidant had the least effect on the cure state, followed by the amine-type, with the hindered phenol antioxidant resulting in the largest reduction in final cure state. An excellent "synergistic" antioxidant which can simultaneously improve the modulus and heat aging of peroxide cured EPDM is Vanox ZMTI [33].

17.3   Role of Monomeric Coagents in Peroxide Crosslinking

When monomeric coagents [34] are added to an elastomer or polymer, unsaturation is added to the system. This provides higher crosslink densities for a given peroxide

Table 17.13   Effect of Process Oil and Plasticizer when Crosslinking with Organic Peroxides. Crosslinking EPM with 2,5-dimethyl-2,5-di(t-butylperoxy)hexyne-3 (DMBPHy)

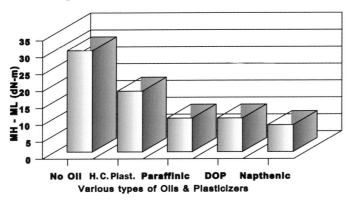

Crosslinking formulation	
Ingredient	phr
EPM	100.0
Oil	25.0
130 (peroxide)	1.1

(DMBPHy) = 2,5-dimethyl-2,5-di(t-butylperoxy)hexyne-3 cure performance data using an ODR R-100 rheometer.

Table 17.14   Effect of Antioxidants on the Efficiency of Organic Peroxides. Crosslinking EPDM with Dicumyl Peroxide (DCP)

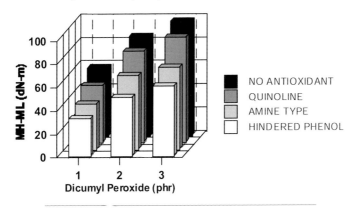

Crosslinking formulation	
Ingredient	phr
EPDM	100
Dicumyl peroxide	as shown in bar graph
Antioxidant	0.5

Cure performance data using an ODR R-100 rheometer.

concentration. Higher crosslinking is possible because it is easier and more efficient to crosslink by free radical addition to unsaturation (double bonds) than hydrogen abstraction. Peroxides generate rigid carbon-carbon bonds and the incorporation of coagents can lead to different types of crosslinked networks that may improve tensile strength, hot tear strength, heat aging, and compression set. Furthermore, low molecular weight liquid or meltable solid coagents have been known to lower the crosslinkable composition's viscosity, thus improving processability. Just as each peroxide type has its tradeoffs, so do the various coagents. One must consider the polymer to be crosslinked and the possible consequences of decreased scorch safety time.

There are numerous monomeric coagents, including the following commercially notable ones:

- Maleimide type: N,N′-phenylenebismaleimide (MBM)
- Allylic type: triallyl cyanurate (TAC), triallyl isocyanurate (TAIC), diallyldiglycolcarbonate (ADC), allyl methacrylate (AMA)
- Methacrylate type: ethyleneglycol dimethacrylate, trimethylolpropanetrimethacrylate, zinc dimethacrylate
- Acrylic type: ethyleneglycol diacrylate, trimethylolpropane triacrylate, zinc diacrylate
- Polymeric coagents: liquid 1,2-polybutadiene resins

17.3.1   Crosslinking PE with Coagents and Peroxides

When crosslinking PE, the most suitable coagents are the allylic types, e.g., TAC or TAIC. The allylic coagents do not readily homopolymerize compared to the acrylic and bismaleimide types and participate well with strong hydrogen abstracting peroxides, such as DCP, BBPIB, and DMBPHy. Thus, bismaleimide and acrylic type coagents are generally less efficient in polymers which contain no unsaturation like PE. In crosslinking LDPE, as shown in Table 17.15, TAC provides the highest torque, high insolubles, and a low swell ratio, which indicates a tighter crosslinked network (lower uptake of xylene).

Table 17.15   Use of Peroxide-Coagent Blends for Crosslinking LDPE @ 185 °C using an ODR-R100, 3° arc

Co-Curing system	Torque (in-lbs)	Cure time TC90 (min.)	Scorch time TS2 (min.)	Percent Insolubles	Swell ratio
none	37	4.5	1.3	88	8.2
TAC 2 phr	58	3.8	1.1	93	5.5
TAC/EGDM 1/1 phr	49	4.2	1.2	92	6.1
AMA 2 phr	47	4.0	1.5	92	6.0
AMA/EGDM 1/1 phr	46	4.3	1.4	91	6.4

TAC = triallyl cyanurate.
AMA = allyl methacrylate.
EGDM = ethylene glycol dimethacrylate (SR-206, Sartomer Company).
Peroxide: BBPIB = di(t-butylperoxy)diisopropylbenzene.
Peroxide conc. 2 phr.

Table 17.16   Crosslinking EPDM* with Dicumyl Peroxide and Bismaleimide or Acrylic Type Coagents

Ingredient	Parts		
EPDM	100	100	100
N774	75	75	75
P. Oil	10	10	10
TMQ	2	2	2
DCP	7	7	7
Coagents:			
MBM	–	1	–
TMPTM	–	–	1
Moving die rheometer temperature 180 °C, 1° arc			
M_H dN-m (0.1 N-m)	14.6	22.9	20.4
M_L dN-m (0.1 N-m)	3.06	3.20	2.97
T_{SI} (seconds)	24.6	16.8	24.6
T_{C90} (seconds)	176	113	139
Mooney viscometer data 135 °C			
t_5 (min)	7.8	4.2	3.3

* EPDM = Nordel 1040 (DuPont Dow).
** Minimum Viscosity (MV), in Mooney Units (MU).
MBM = N,N′ phenylene bismaleimide (Vanax MBM, RT Vanderbilt).
TMPTM = trimethylolpropane trimethacrylate (SR-350, Sartomer Company).
TMQ = polymerized 1,2-dihydro-2,2,4-trimethylquinoline (antioxidant).
P. Oil = Paraffinic oil.
DCP = 40% dicumyl peroxide on clay filler.

17.3.2   Crosslinking EPDM with Coagents and Peroxides

As discussed previously, the effectiveness of various monomeric coagents for PE is quite limited. However, when crosslinking EPDM, a wider variety of coagents can be used. In Table 17.16 the coagent MBM (N,N′ phenylene bismaleimide) is compared to TMPTM (trimethylolpropane trimethacrylate) on an equal weight basis using DCP peroxide. One part of either coagent produces an approximate 40% increase in rheometer torque. The solid bismaleimide is typically more efficient and provides faster cures than the liquid trimethacrylate coagent. However, the liquid trimethacrylate coagent generally provides improved processing of the final compound because of the lower M_L rheometer minimum torques and lower MV (Minimum Mooney Viscosity).

Using acrylic coagents for lower compound viscosity and improved EPDM physical properties has been verified by Walter Nagel [30] of Sartomer Company, who compared MBM, TAC, and CHDMDA (cyclohexane dimethanol diacrylate – scorch retarded) on an equal weight basis. The novel CHDMDA diacrylate type coagent is unexpectedly more efficient than MBM, and provides better hot tear strength and compression set values in a radiator hose EPDM formulation. In EPDM, allylic type coagents are generally slower reacting than either the bismaleimide or diacrylate coagents; thus, as expected, TAC provides a longer T_{c90} even when compared to the peroxide control with no coagent.

17.3.3   Crosslinking HNBR with Coagents and Peroxides

Another reason to use coagents when crosslinking elastomers is to reduce the concentration of peroxide curative. Adding coagents increases the level of unsaturation, making the crosslinking more efficient, thus the peroxide concentration can be reduced. Fully saturated HNBR cannot be cured by sulfur vulcanization, but only with organic peroxides. Peroxides, as well as sulfur vulcanization, create low molecular weight decomposition by-products, which may result in off-gassing and/or porosity during processing. Increasing the coagent levels and reducing peroxide levels will help to minimize porosity and off-gas.

When crosslinking HNBR, several coagents have been successfully employed: TAIC, TAC and MBM; however, scorch (unwanted, pre-mature crosslinking) is often a concern. To solve this problem, MBM used in combination with less peroxide, specifically Varox DCP-40KE-HP [11], a proprietary scorch resistant DCP, provided the desired balance of cure and scorch time performance. Recently, Bender [31] found that MBM provided good cure while not reducing tear resistance, compared to the allylic coagent TAIC in HNBR. In addition, Bender found that increasing peroxide level, followed by increasing coagent, led to increased crosslinking and lower compression set, with no appreciable difference between MBM and TAIC coagents.

17.4   Advantages and Disadvantages of Peroxide Crosslinking versus Sulfur Vulcanization

Peroxides generate free radicals that are capable of crosslinking a wide variety of saturated and unsaturated polymers, e.g., PE and EPDM, respectively. Peroxides create strong carbon-carbon type crosslink bonds which allow the finished product manufacturer to better utilize the full engineering potential of the elastomer. In contrast, the carbon-sulfur and the sulfur-sulfur type bonds of sulfur vulcanization are much weaker, as shown in Table 17.17. Peroxides can offer the benefits of low odor (e.g., with DMBPHa peroxide), production of high quality colored elastomers (no staining like sulfur vulcanization), and better heat stability of the final crosslinked part. The stronger carbon-carbon bonds created by organic peroxides are the reason for the better (lower) compression set crosslinked articles. Thus, crosslinking with peroxides is the preferred method for producing high value/and better performing technical articles.

Higher temperature compression set testing for peroxide cure systems at 150 °C versus ~100 °C for sulfur, plus the lack of reversion, is evidence for better heat aging properties

Table 17.17   Sulfur versus Peroxide Bond Strength

Bond type	Cure system	Bond strength kJ/mol
C−C	Peroxide	350
C−S	Low sulfur	285
S−S	Traditional sulfur cure	115 to 270

Table 17.18   Peroxide Crosslinking versus Sulfur Vulcanization

	Sulfur vulcanization	Peroxide crosslinking
Keltan 4906 EPDM (DSM Copolymer)	100.0	100.0
N660	75.0	75.0
Parafinic oil	20.0	20.0
Zinc oxide	3.0	–
Stearic acid	1.0	–
AgeRite D	1.0	1.0
Dicumyl peroxide (40% on clay)	–	6.0
DPTT	2.8	–
MBTS	1.6	–
TMTM	1.6	–
Sulfur	0.4	–
Moving die rheometer at 170 °C, 1° arc		
M_H (dN-m)	30.0	28.0
Percent compression set	71.0	14.0
150 °C, 70 hrs.		

DPTT = dipentamethylenethiuram tetrasulfide
MBTS = benzothiazyl disulfide
TMTM = tetramethylthiuram monosulfide

associated with peroxide cures. A comparison of peroxide versus sulfur cures is provided in Table 17.18. Weak monosulfidic bonds produced by the efficient EV vulcanization system cannot provide such performance. The need for higher performance parts, especially in automotive applications, make peroxides the curative of choice. Note also that sulfur vulcanization can cure only unsaturated (double bond-containing) polymers and cannot be used with the new metallocene-based poly(ethylene octene) copolymers, such as Engage by DuPont Dow, which have been engineered to provide improved heat aging.

Unlike peroxides, sulfur vulcanization requires zinc oxide and/or fatty acid metal salts like zinc stearate, which must be used in combination with organic accelerators (mercaptobenzothiazoles, sulfenamides, thiurams, and dithiocarbamates). The possible formation of potentially harmful nitrosamine type compounds are of concern to the rubber manufacturer. In addition, the presence of zinc compounds leads to questions about the stability of crosslinked elastomers, e.g. electrochemical corrosion of automotive hose, which has recently been documented by G. Vroomen et. al. [32].

The breaking and reforming of weaker monosulfidic and polysulfidic bridges which produce poor compression set and heat aging of sulfur vulcanized products also provides certain advantages. Greater elongation, higher tear strength, and better flex fatigue are typically attained with sulfur cure. Furthermore, sulfur vulcanization is less affected by acidic clays, highly aromatic oils, high concentrations of oil and fillers, and aromatic and amine antioxidants, e.g., ODPA (octylated diphenyl amine) or BHT (butylated hydroxytoluene). When elastomers are cured with organic peroxides, standard, non-acidic Burgess type clay fillers and paraffinic oils should be used (Tables 17.13 and 17.14). However, increasing the oil and filler concentrations in a peroxide formulation results in a decrease in cure because of absorption of the peroxide on the filler and reaction of the free radicals with the oil. Increasing the peroxide concentration overcomes this problem. In regard to antioxidant systems for peroxide cure, polymerized quinoline or Vanox ZMTI – zinc 2-mercapto-toluimidazole type antioxidants [33] or a combination of these should be used.

References

1. Sanchez J. and Myers, T. N. (ATOFINA) "Organic Peroxides", *Kirk-Othmer Encyclopedia of Chemical Technology*, Fourth Edition, Volume 18, (1996), pp. 203–310.
2. Sanchez J. and Myers, T. N. In Polymeric Materials Encyclopedia, *Peroxide Initiators (Overview)*, 4927–4938, (1996).
3. *Tertiary Alkyl Hydroperoxides*, Product Bulletin; Elf Atochem North America.
4. Porter, N. A., Organic Peroxides, Ando, W. (Ed.), (1992), John Wiley & Sons, Inc., NY, 101–156.
5. The Vanderbilt Rubber Handbook, Ohm, R. F. (Editor), R. T. Vanderbilt Co., 30 Winfield St., Norwalk, CT., phone (203)-853-1400; *VaroxDBPH or VaroxDBPH-50* p. 331, 13th Edition, (1990).
6. Product Bulletin, *Dialkyl Peroxides (Luperox101)*, ATOFINA, 2000 Market Street, Philadelphia, PA. 19103, phone: 215-419-7000.
7. Technical Publication, *HALFLIFE-Peroxide Selection Based on Half-Life*, 2nd ed., and Copyrighted Software; ATOFINA; 2000 Market Street, Philadelphia, PA., 19103, phone (215) 419-7000, (1992).
8. Callais, P. A.; Kamath, Sanchez, J.; Gravelle, J. M. *Plast. Compound.*, (1992) 15(1), 49–52.
9. Cassoni, J. P., Novits, M. F., and Palys. L. H., "Safe Handling and Processing in the Elastomer Industry using Organic Peroxides" Paper No. 52, Presented at a meeting of the Rubber Division, ACS, Las Vegas, NV May 29–June 1, (1990).
10. Luperox Masterbatches, ATOFINA 4, cours Michelet – La Defense 10 – Cedex 42, Paris; Luperox Masterbatches, ATOFINA; 2000 Market Street, Philadelphia, PA., 19103, phone (215) 419-7000.
11. Palys, L. H., Callais, P.A., Novits, M. F. and Moskal, M. G. (ATOFINA), "New Peroxide Formulations For Crosslinking Chlorinated Polyethylene, Silicone and Fluoroelastomers" Paper No. 92, Presented at a meeting of the Rubber Division, ACS, Anaheim, California, May 6–9, (1997).
12. Varox "HP" type peroxides; R. T. Vanderbilt Co., Norwalk, CT, USA; phone (203)-853-1400; Toll free in USA: (800)-243-6064; website: www.rtvanderbilt.com.
13. Palys, L. H., Callais, P. A. and Novits, M. F. (ATOFINA), "Crosslinking Elastomers with Improved Productivity Using Novel Scorch Resistant Peroxide Formulations", Paper No. 119, Presented at a meeting of the Rubber Division, ACS, Orlando, Florida, October 26–29, (1993).
14. L. H. Palys, P. A. Callais, M. F. Novits and M. G. Moskal (ATOFINA), "New Peroxide Crosslinking Formulations For Metallocene Based Poly(ethylene octene) Copolymers" Paper No. 88, Presented at a meeting of the Rubber Division, ACS, Louisville, Kentucky, October 8–11, (1996).
15. Kamath, V. R., Palys, L. H., *Polyethylene Melt Flow Modification*, SPE Regional Technical Conference (RETEC), Plastics Waste Management, Paper No. 6, October (1990).
16. Harpell, G. A., Walrod, D. H., *Organic Peroxides for Cure of Ethylene-Propylene Rubbers*, Rubber Chemistry & Technology, Vol 46., No. 4, 1007–1018, (1973).
17. Bremner, Tim., Rudin, Alfred, *Modification of High Density Polyethylene by Reaction with Dicumyl Peroxide*, Guelph-Waterloo Centre for Graduate Work in Chemistry and Department of Chemistry, University of Waterloo, Waterloo, Ontario Canada, (1990).

18. Bobear, W. J., Rubber Chem. Technology 40, 1560 (1967).
19. R. D. Stevens, T. L. Pugh and D. L. Tabb, (DuPont Dow Elastomers) "New Peroxide-Curable Fluoroelastomer Developments"; ACS Rubber Division Meeting, LA, California, April 23–26 (1985), Paper No. 21.
20. Denise A. Kotz, (DuPont Dow Elastomers) "Improved Processing and Other Advances in Peroxide-Curable Viton Fluoroelastomers", Paper No. 84, Presented at a meeting of the Rubber Division, ACS, Louisville, Kentucky October 8–11, (1996).
21. F. K. Jones, J. L. Laird, and B. W. Smith; "Characterization of EPDMs Produced by Constrained Geometry Catalysts"; Paper No. 37, Presented at the Rubber Division, ACS, Montreal, Quebec, Canada, May 5–8, (1996).
22. J. L. Laird and J. Riedel; "Evaluation of EPDM Materials as Produced by Constrained Geometry Catalyst Chemistry Against Current Commercial EPDM Products and Performance Requirements"; Paper No. 9, Presented at the Rubber Division, ACS, Cleveland, Ohio, October 17–20, (1995).
23. L. T. Kale, J. J. Hemphill, D. R. Parikh, K. Sehanobish & L. G. Hazlitt Dow Chemical, Freeport, TX (now DuPont Dow Elastomers), "Performance of Ethylene/1-Octene, Ethylene/1-Pentene and Ethylene/1-Butene Elastomers Made Using INSITE Technology in Peroxide-Cured Durable Formulations" Paper No. 10, Presented at the Rubber Division, ACS, Cleveland Ohio, October 17–20, (1995).
24. M. S. Edmondson, T. P. Karjala, R. M. Patel & D. R. Parikh; "Ethylene-Propylene-Diene Elastomers via Constrained Geometry Catalyst Technology" Paper No. 36, Presented at the Rubber Division, ACS, Montreal, Quebec, Canada, May 5–8, (1996).
25. Callais, P.A.; Kazmierczak, *The Maleic Anhydride Grafting of Polypropylene with Organic Peroxides*, Dallas, TX ANTEC (Annual Technical Conference of the Society of Plastic Engineers) (1990).
26. Palys, L. H., Kamath, V. R., Chaser Catalysts Improve Graft Polyols Processing, Modern Plastics, No. 7, Vol 65, July pp 70–74, (1988).
27. Kamath, V. R., Palys, L. H., US Pat. 4,804,775, Process for Reducing Residual Monomers in Low Viscosity Polymer-Polyols (1986).
28. Kamath, V. R. Solution Polymerization of Acrylic Monomers, EP 273090, (1988).
29. Palys, L. H., *Modification of Polyolefins*, ATOFINA and Werner & Pfleiderer International Seminar Reactive Modification of Polyolefins, Hotel Nikko, Dusseldorf, Germany, October 13, (1995).
30. Nagel, Walter, "New Coagent for Improved EPDM Properties" Paper No. 79, Presented at a meeting of the Rubber Division, ACS, Louisville, Kentucky October 8–11 (1996).
31. H. Bender and E. Campomizzi, "Modelling of Peroxide Cured HNBR Compounds for Seal Applications" Paper No. 14, presented at a meeting of the Rubber Division, ACS, Indianapolis, Indiana May 5–8 (1998).
32. G. Vroomen, J. Noordermeer and M. Wilms, "Automotive Coolant Hose Technology – Keltan EPDM, Peroxide Curing" Paper No. 42, presented at a meeting of the Rubber Division, ACS, Nashville, Tennessee September 29–October 2, (1998).
33. R. F. Ohm, P. A. Callais and L. H. Palys, "Cure Systems and Antidegradant Packages for Hose and Belt Polymers" Paper No. 52, presented at a meeting of the Rubber Division, ACS, Chicago, Illinois, April (1999).
34. Costin, Richard, and Nagel, Walter, "Coagent Selection for Peroxide-Cured Elastomers", Paper No. 4, presented at a meeting of the Rubber Division, American Chemical Society, Indianapolis, Indiana, May 5–8, 1998.

18 Tackifying, Curing, and Reinforcing Resins

Bonnie Stuck

18.1 Introduction

Resins are used in the rubber industry as tackifiers, cure agents, and reinforcing agents. Tackifying resins are probably the largest and most diversified group. Phenol-formaldehyde, thermoreactive, resol-curing resins generally are used to cure polymers such as butyl rubber for high temperature resistance. Reinforcing resins include phenol-formaldehyde, novolak resins and styrene resins. Table 18.1 lists the three major functional groups of resins. Chemically, there are many types of resins and variations, but because of limited space, they are not all discussed in this chapter. Phenol-formaldehyde based resins are by far the largest group of resins used in the rubber industry, so this chapter focuses on them.

18.2 Phenol-Formaldehyde Resins

Phenol-formaldehyde resins have long been used in the rubber industry as tackifiers, reinforcers, or curing agents. Phenolic tackifiers can be found in many compounds that require good tack for component building purposes. Phenolic reinforcing resins are found in the compounds where high stiffness and hardness are needed. Phenolic curing resins are used to cure butyl tire-curing bladders, air bags, TPE elastomers, and products that require high heat resistance.

Phenol-formaldehyde resins can be chemically classified into two major groups: novolaks and resols. The novolak resins are made from the reaction of phenol with formaldehyde under acidic conditions (see Fig. 18.1). The phenol can be an alkylphenol or a combination of phenol and alkylphenol. The molar ratio of formaldehyde/phenol is less than one for novolaks. Novolak resins are thermoplastic resins that soften at elevated temperatures unless crosslinked with a methylene donor. Tackifying and reinforcing resins are variations of novolak resins.

The resol resins are made from the reaction of phenol with formaldehyde under basic conditions (see Fig. 18.2). The phenol is usually an alkylphenol. The molar ratio of formaldehyde/phenol is greater than one for resols. Resol resins are thermoreactive. Adhesive and curing resins are produced from different variations on resol resins.

Table 18.1 Resin Types

Curing resins	Reinforcing resins	Tackifying resins
a. Phenol-formaldehyde resol type	a. Phenol-formaldehyde novolak type	a. Phenol-formaldehyde novolak type
	b. High styrene resins	b. Hydrocarbon
	c. Methylene donors	c. Rosin derivatives
	d. Resorcinol	d. Phenol-acetylene condensation
	e. Resorcinol-formaldehyde	e. Terpene derivatives
	f. Poly-butadiene resins	f. Cumarone Indene
	g. Styrene-acrylonitrile	g. Tall oil derivatives
	h. Poly-vinyl-chloride resins	

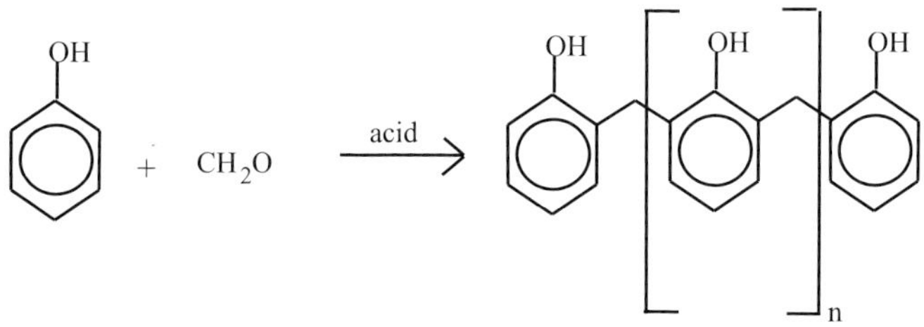

Phenol Formaldehyde Novolak phenol formaldehyde

Figure 18.1 Formation of a novolak resin.

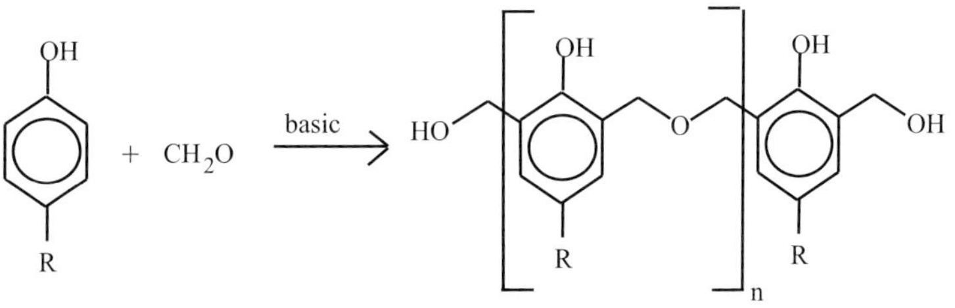

Alkylphenol Formaldehyde Resol phenol formaldehyde

Figure 18.2 Formation of a resol resin.

18.2.1 Types of Phenol-Formaldehyde Resins

18.2.1.1 Reinforcing Resins

Reinforcing resins are novolak resins that typically have a branched structure. These resins are made from the reaction of a phenol and formaldehyde under acidic conditions to produce a branched novolak resin as shown in Fig. 18.3.

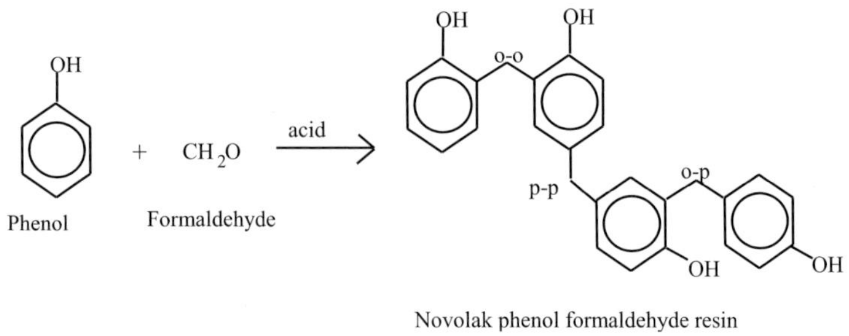

Novolak phenol formaldehyde resin

Figure 18.3   Formation of a reinforcing resin.

The ratio of ortho-ortho (o-o), para-para (p-p), and ortho-para (o-p) bonds is dependent on the type of catalyst.

Novolak reinforcing resins can be modified by the following:

- Part of the phenol in the reaction can be replaced by other phenolic or non-phenolic compounds
- Part of the formaldehyde can be replaced by other aldehydes
- Varying the type of catalyst during the polycondensation can form linear versus branched structures
- Common chemical modifications to reinforcing resins are made with cashew nut oil, cresol, or tall oil

Reinforcing novolaks are thermoplastic and must be crosslinked with a methylene donor such as HMT (hexamethylenetetramine) or HMMM (hexamethoxymethylo-melamine) to make them thermosetting and reinforcing. The amount of methylene donor needed to crosslink novolak resins depends on the hardness level needed, the amount of novolak resin in the formulation, and whether the methylene donor is HMT or HMMM. Generally, the amount of methylene donor used is 8 to 15% of the total novolak resin in the formulation. Higher amounts of HMMM are used to get the same hardness level as HMT. Formulations with less than 10 phr of novolak resin may require larger amounts of methylene donor because of dilution by the overall compound formulation.

Reinforcing phenol-formaldehyde resins can be used to make a high hardness (90+ Shore A durometer) bead filler that has good processing properties. A model formulation is shown in Table 18.2.

A high hardness tread compound with good flex properties, improved tear resistance and good processing properties is possible with phenolic reinforcing resins. Table 18.3 shows a model formulation.

A hard subtread compound with low rolling resistance can be made with a phenol-formaldehyde reinforcing resin that gives good tire handling properties and low hysteresis in low rolling resistance tires, high performance tires, and truck tires, as shown in Table 18.4.

In the automotive industry, EPDM rubber profiles are used to weatherproof windows, doors, and car hoods. Phenolic resins have been used to make weather-stripping profiles

Table 18.2 High Hardness Tire Bead Apex Compound

Ingredient	PHR
First stage (Banbury):	
Natural rubber SMR20	60.00
Butadiene rubber 1220	40.00
Carbon black HAF LS N326	80.00
Aromatic oil	5.00
Stearic acid	1.50
6PPD Antidegradant	1.00
Zinc oxide	7.00
Phenolic resin	15.00
Second stage (2-Roll Mill):	
HMT or HMMM methylene donor	1.50
MBS accelerator	0.80
Insoluble sulfur (95% active)	2.20
PVI scorch inhibitor	0.25

that are made of co-extruded dual hardness rubber compounds. These co-extruded profiles have replaced the traditional profile containing a soft rubber compound and a metallic insert. Table 18.5 lists the components of a high hardness EPDM model compound (50+ Shore D durometer) made with phenol-formaldehyde reinforcing resins.

Phenolic reinforcing resins are compatible with the most common elastomers in the rubber industry; however, the degree of compatibility varies somewhat depending on the polarity of the elastomer. For example:

Table 18.3 High Hardness Tread Compound

Ingredient	PHR
1st step	
SBR 1500	100.00
N339 carbon black	65.00
Silica	8.00
Phenolic resin	8.00
Wax	1.00
6PPD antidegradant	2.25
Zinc oxide	2.50
Stearic acid	2.00
Aromatic oil	21.00
2nd step	
HMMM methylene donor	1.20
Sulfur	1.60
TBBS accelerator	1.40

Table 18.4   Low Rolling Resistance Hard Subtread

Ingredient	PHR
First stage	
SMR- or SIR-20 NR	50.00
1203 High Cis BR	50.00
N650 carbon black	35.00
Aromatic oil	0
Naphthenic oil	0
Phenolic resin	10.00
Zinc oxide	3.50
Stearic acid	2.00
TMQ antioxidant	1.50
Wax	1.50
6PPD antioxidant	1.50
Final stage	
Sulfur	1.80
TBBS accelerator	2.00
DPG accelerator	0.25
HMMM methylene donor	1.20

Table 18.5   High Hardness EPDM Compound

Ingredient	PHR
Masterbatch	
EPDM	100.0
Carbon black N660	150.0
Calcium carbonate	20.0
Process oil	40.0
Poly ethylene glycol	3.0
Process aid	3.0
Stearic acid	2.0
Zinc oxide	8.0
High styrene resin	5.0
Phenolic resin	15.0
Calcium oxide	10.0
Final	
HMT methylene donor	1.5
Sulfur	1.5
MBT accelerator	1.5
TMTD accelerator	0.8
ZDBC accelerator	1.4
CBS accelerator	1.5

- NBR: Phenolic resins are very compatible with NBR polymers, and large amounts (25 to 100+ phr) of resin can be incorporated to form very hard, ebony type compounds. Lesser amounts (10 to 20 phr) of resin in NBR form softer and more flexible vulcanizates.
- EPDM: Phenolic reinforcing resins have good compatibility with EPDM polymers. Levels of up to 30 phr can be used. Reinforcing novolak resins can increase the hardness and abrasion resistance of EPDM compounds.
- SBR, BR, and NR: Phenolic resins are not as compatible with SBR, BR, and NR polymers. However, levels of 10 to 20 phr can increase hardness and abrasion resistance. NBR rubber at levels of 15 to 25 phr is often combined with SBR, BR, or NR to increase the compatibility of the resin, enabling resin levels greater than 20 phr to be incorporated.
- CR: Phenolic resins have the least compatibility with CR polymers. Levels of 5 to 10 phr are recommended. NBR rubber (15 to 25 phr) is often used in combination with CR rubber to increase the compatibility of the resin allowing for resin phr levels above 15.

18.2.1.2   Tackifying Resins

Tackifying resins are produced from different variations on novolak resins. Generally, they have a linear structure as shown in Fig. 18.4. The alkyl group (R) is usually an octyl or *t*-butyl group.

Table 18.6 shows a model formulation with a phenol-formaldehyde tackifying resin. Generally, phenolic-formaldehyde resins with higher molecular weights and higher softening points have superior tack, especially after aging under conditions of high humidity and temperature.

18.2.1.3   Curing Resins

Phenol-formaldehyde curing resins are classified as resols and are thermoreactive resins. They are used in curing butyl rubber and other elastomers with low unsaturation. These curing resins are used in butyl tire curing bladders, air bags, seals, masking tapes, TPE elastomers, and other products subjected to high heat and steam. The chemistry of a

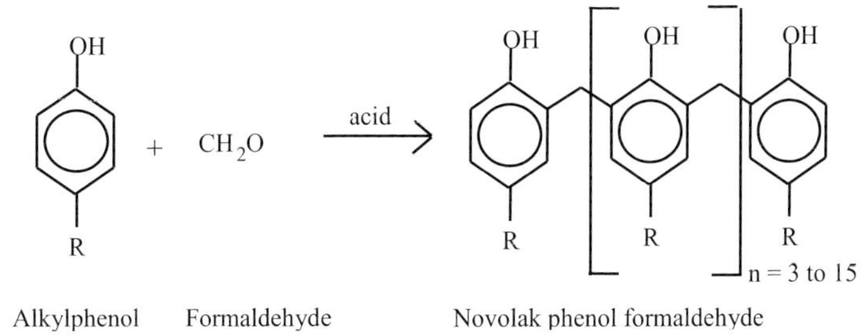

Figure 18.4   Formation of a tackifying resin.

Table 18.6   Tackifying Resin Compound

Ingredient	phr
SMR 20 natural rubber	45.00
BR 1220 BR	30.00
SBR 1500	25.00
N650 carbon black	50.00
Naphthenic oil	15.00
Stearic acid	2.00
Zinc oxide	2.50
6PPD antidegradant	3.50
Wax	3.00
Tackifying resin	5.00
Sulfur	1.75
MBTS accelerator	0.90

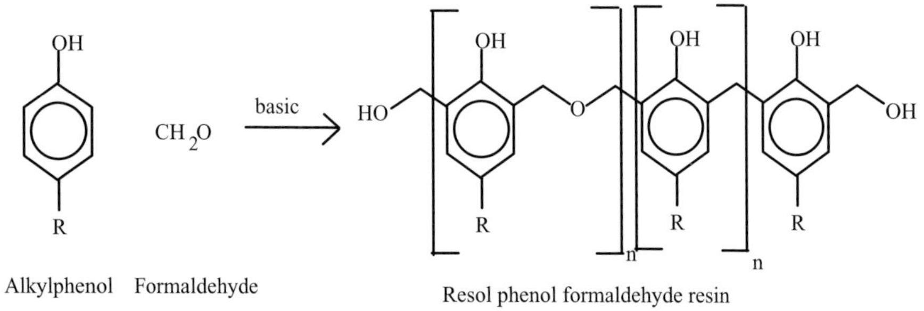

Alkylphenol   Formaldehyde

Resol phenol formaldehyde resin

Figure 18.5   Formation of a curing resin.

phenol-formaldehyde curing resin is shown in Fig. 18.5. The amount of curing resin needed depends on the properties desired and the level and type of halogen donor used. In most cases, the amount of curing resin is between 7 and 10 phr.

Although curing resins alone can cure the rubber, to obtain a reasonable curing time the addition of a halogen donor is recommended. A polychloroprene rubber (CR) at 3 to 5 phr gives the best overall results. A typical butyl tire curing bladder compound formulation is shown in Table 18.7.

Table 18.7   Butyl Bladder Compound

Ingredient	phr
Butyl rubber	100
CR	5
N330 carbon black	50
Castor oil	5
Zinc oxide	5
Phenol-formaldehyde cure resin	10

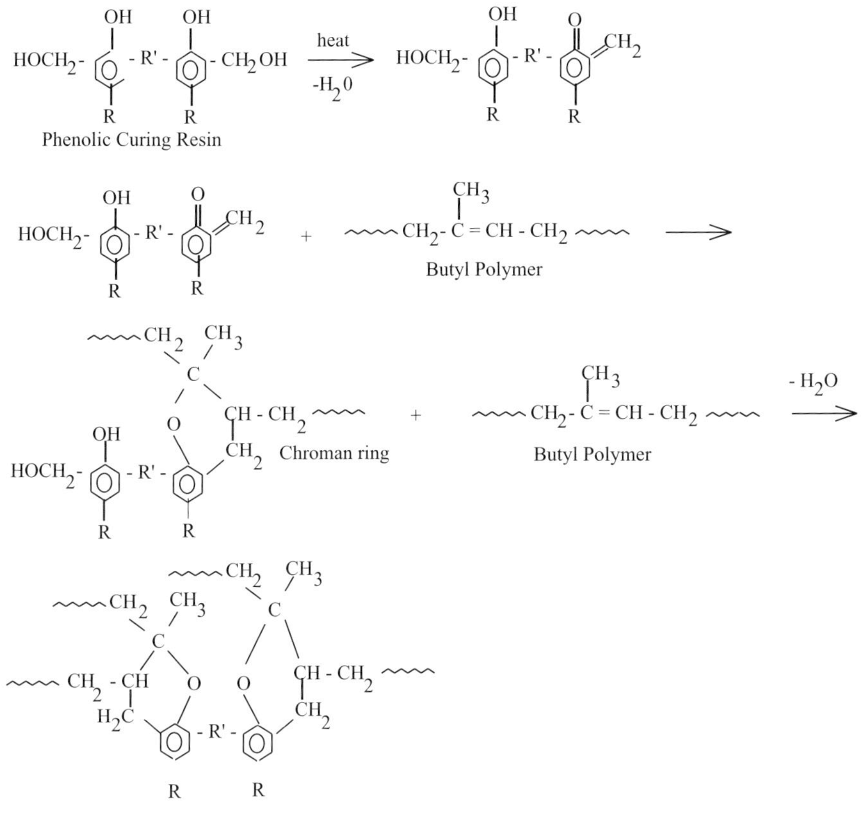

Phenolic Curing Resin

Butyl Polymer

Chroman ring   Butyl Polymer

Crosslinked Butyl Polymer

Figure 18.6   Reaction of curing resin with butyl rubber.

The ratio of methylol groups/ether bridges/methylene bridges depends on the molar ratio of alkylphenol, the type and amount of catalyst, the temperature, and the pH. The curing resin reacts with the butyl rubber as shown in Fig. 18.6.

18.3   Methylene Donor Resins

As mentioned previously, methylene donor resins, such as HMT (hexamethylenetetramine) or HMMM (hexamethoxymethylomelamine), are used in combination with phenol-formaldehyde, novolak reinforcing resins to make them thermosetting and reinforcing. HMT and HMMM are also used to crosslink resorcinol-formaldehyde resins, pure resorcinol, other resorcinol-based resins. The resorcinol-based resins are primarily used to promote the adhesion of rubber-coat compounds to brass-coated wire or resorcinol-formaldehyde-latex (RFL) dipped textiles, such as polyester, nylon, and aramid. There are other methylene donors used in the rubber industry, but HMT and HMMM are the most commonly used.

18.4   Resorcinol-Based Resins

As mentioned above, the resorcinol-based resins are commonly combined with an HMT or HMMM methylene donor resin to promote adhesion of rubber coat-compounds to brass-coated wire or RFL dipped textiles.

18.5   High Styrene Resins

Styrene-butadiene resins with high levels of styrene (50% and above) are often used as reinforcing resins in rubber compounding. They are relatively low in cost and do not require a crosslinking agent like phenol-formaldehyde novolak resins do. However, high styrene resins are thermoplastic and soften at temperatures above their glass transition temperature. The glass transition temperature of pure styrene is 100 °C, but most commercially available high styrene resins have lower glass transition temperatures and tend to loose their hardness at about 70 °C. In contrast, the methylene donor, crosslinked, phenol-formaldehyde resin network is very stable at temperatures in excess of 100 °C.

18.6   Petroleum-Derived Resins

Hydrocarbon-based resins are used extensively as economical tackifiers in uncured rubber compounds. The hydrocarbon resins are primarily derived from petroleum and include aliphatic, aromatic, and hydrogenated resins and combinations of hydrocarbon resins with rosin-based resins. Generally, hydrocarbon resins provide good initial tack, but are not as good as phenol-formaldehyde novolak resins for high humidity conditions and long-term tack. Cumarone-Indene resins are also petroleum-based resins.

18.7   Wood-Derived Resins

This group includes a large variety of resins derived primarily from pine trees. These include the rosin, terpene, terpene-phenolic, tall oil derivative, pine tars, gum rosin, and wood rosin resins. These resins are economical resins that supply good initial but sometimes inconsistent tack because they vary, depending on the source of pine trees. However, recently more refined rosin-based tackifiers have become available that provide consistent and economical tack. These highly refined rosin tackifiers are often used in combination with phenol-formaldehyde novolak resins to provide good initial and long-term tack.

References

1. Chemical Variations on Novolak Phenol-Formaldehyde Reinforcing Resins in a NR/BR Based Compound – Part 2 B. L. Stuck, J-C. Souchet, C. Morel-Fourrier (CE117)
2. CECA Phenol-Formaldehyde Reinforcing Resins in a Farm Tread Compound – Part 2 by B. L. Stuck, J-C. Souchet, C. Morel-Fourrier (CE119)
3. CECA Phenol-Formaldehyde Reinforcing Resins in a Hard Low Rolling Resistance Passenger Tire Subtread Compound by B. L. Stuck, J-C. Souchet, C. Morel-Fourrier (CE127)
4. CECA Resins - Manufacturing, Chemistry and Laboratory Analysis by B. L. Stuck, J-C. Souchet, C. Morel-Fourrier (CE104)
5. CECA R7510 and R7578 Phenolic Tackifying Resins (CE124)
6. Chemical Variations on Novolak Phenol Formaldehyde Reinforcing Resins for EPDM Profiles -Part 3 by B. L. Stuck, J-C. Souchet, C. Morel-Fourrier (CE126)
7. Hofmann, W. *Rubber Technology Handbook* (1989) Hanser Publishers, Munich

Abbreviations

6PPD: N-(1,3-dimethylbutyl)-N′-phenyl-p-phenylenediamine
BR: butadiene rubber
CBS: N-cyclohexyl-2-benzothiazolesulfenamide
CR: chloroprene rubber
DPG: diphenylguanidine
EPDM: ethylene propylene diene rubber
HMMM: hexamethoxymethylomelamine
HMT: hexamethylenetetramine
IR: isoprene rubber
MBS: 2-(4-morpholinyl) mercaptobenzothiazole
MBT: 2-mercaptobenzothiazole
MBTS: mercaptobenzothiazole disulfide
NBR: butadiene acrylonitrile rubber
NR: natural rubber
PVI: N-cyclohexylthiophthalimide
RFL: resorcinol formaldehyde latex
SBR: styrene butadiene rubber
TMTD: tetramethylthiuram disulfide
ZDBC: zinc dibutyldithiocarbamate

19   Antidegradants

Fred Ignatz-Hoover

19.1   Introduction

The term "antidegradants" is a relatively recent descriptive term covering materials that protect rubber products against degradation by a variety of forces. The major need for antidegradants is in the unsaturated elastomers such as NR, SBR, BR, nitrile rubber, and neoprene. While the saturated or mostly saturated elastomers and plastics such as PP, EP, IIR, and EPDM also require protection, the unsaturated elastomers containing unsaturated bonds with allylic or tertiary benzylic hydrogen atoms are somewhat reactive and more prone to degradation. Most prominent degradants are oxygen, ozone, dynamic fatigue (or flexing), heat, and light (UV) [1]. The degradative results of these forces fall into one of the following categories:

1. Chain scission: The loss of physical properties such as tensile strength, elongation, and hardness primarily resulting from oxidation, which is quite often accelerated by heat, light, or metal ions. Chain scission reactions accompany the decomposition of hydroperoxides in certain elastomers. Elastomers containing segments of PP, poly-isoprene (e.g. NR, synthetic IR), and PE exhibit this chemistry.
2. Crosslinking: An increased stiffness and hardness most often seen in SBR or BR. In the case of butadiene-containing elastomers, peroxide decomposition leads to crosslinking.
3. Fatigue cracking during dynamic flexing: Catastrophic failure resulting from rupture and tearing of the rubber during flexing.
4. Surface cracking and crack growth: Resulting from ozone attack when rubber is under stress, particularly when stretched either under static or dynamic conditions.

Several factors are important to consider when selecting an antidegradant, including the type of elastomer, the end use application, and the service conditions. Among the important factors to consider are:

- End use of the rubber (static or dynamic application)
- Color of the product: is it black or white (or light-colored)?
- Environment (oxygen, ozone, heat, light, acid rain, exposure to solvents, and detergents, etc.)
- Type(s) of rubber
- Cure system
- Cost

Antidegradants are generally classified into one of two classes, either as antioxidants or antiozonants. Antioxidants provide protection from oxygen degradation by slowing the reaction of oxygen with rubber. Antiozonants prevent deterioration of the rubber from attack by ozone.

19.2   Properties of Antidegradants

Cost performance, or which materials perform satisfactorily at the lowest cost, is the primary criterion used to select the proper antidegradant package. Other criteria involved include staining and discoloration, volatility, solubility, stability, physical state, and concentration or usage level.

19.2.1   Discoloration and Staining

The first perfomance property considered is discoloration and staining. In general, phenolic antioxidants are non-discoloring and amine antidegradants are discoloring. Staining is caused by antioxidant migration and physical transference to adjacent material. The extent and intensity of staining is determined by the chemical nature of the antidegradant and its mobility.

Thus, for light-colored stocks, amine antidegradants should not be used. Phenolic types are the antidegradant of choice in this case. For rubbers that contain carbon black, discoloration is less problematic and the more potent amine antioxidants are used. For applications where migration staining is a problem, the high potency phenolics or a low-volatility amine antioxidant, with only slight potential for contact staining, may be used.

19.2.2   Volatility

Both the molecular weight and the chemical structure of a molecule determine its volatility [2]. Generally, the greater the molecular weight is, the lower the volatility. The type of molecule, however, has a greater effect than its molecular weight. For example, hindered phenols have little potential for hydrogen bonding and are highly volatile compared to amine antidegrandants of the same molecular weight [3].

Throughout the service life of a rubber article, volatility is important from the standpoint of loss of antidegradant. The amount of antidegradant lost depends on amount of surface exposed (the surface to volume ratio), temperature of service, and circulation of air over the surface. The volatility of antidegradants affects rubber testing and results may vary widely depending upon the volatility of the antidegradant and use of open or closed aging chambers. Use separate heated chambers for each sample set to prevent cross contamination.

19.2.3   Solubility and Migration

A third important property of antidegradants is their relative solubility in rubber and in the water, detergents, or solvents to which the rubber comes in contact [4]. Ideally, it would be desirable to have high solubility in rubber but poor solubility in water and organic solvents. Solubility in the rubber and the contacting media results in a partitioning or transfer of a portion the antidegradant into the contact media, causing a net depletion of antidegradant in the rubber article.

The solubility of antiozonants in rubber is especially important because their effectiveness depends on the use of high levels of this antidegradant, its solubility, and mobility. Ozone is so active that it reacts rapidly at the surface of the rubber before significant diffusion into the part can occur [5]. Thus, the antiozonant must be soluble in the rubber and must migrate to the surface to be effective [6]. The migration rate depends not only upon molecular weight but also upon solubility in the rubber. Mobility and the efficiency of antiozonant migration to the surface allows for effective protection against ozone attack.

The solubility of an antidegradant in rubber depends upon its chemical structure, the type of rubber, and temperature. Poor solubility results in blooming when quantities larger than the solubility limit are used. "Bloom" is an accumulation of a compounding ingredient in crystalline form on the surface of a cured or uncured compound. Blooming occurs when a chemical is readily crystallizable, has a limited solubility in the rubber, and is present in the rubber compound at levels exceeding the solubility limit. The diaryl-paraphenylenediamines (DPPD) are examples of this. Thus, the bloom of crystalline and semi-crystalline wax on the surface of a rubber article creates a passive barrier to ozone [7]. Under static conditions, this waxy surface barrier protects the rubber from degradation by ozone [8]. However, in dynamic applications, flexing the rubber cracks and flakes the waxy protective surface. In these cases, failure by ozone degradation can be catastrophic because of the concentration of ozone attack at a flaw of the waxy surface barrier. The use of chemical antiozonants in combination with waxes overcomes this problem. In fact, this combination provides improved protection compared to either means by itself. Bloom in uncured compounds can cause interply adhesion problems in layered rubber products both during building and service. Phenolic and phosphite antioxidants generally have a reasonable solubility in rubber and blooming does not occur at normal use levels.

The solubility of antidegradants in solvents or water is also important [9]. Their extraction in service, such as in hydraulic hose or in articles subject to detergent cleaning, laundering, or dry cleaning, can be a serious problem [10]. Recently, acid rain has also been implicated in the depletion of amine antidegradants from rubber compounds.

19.2.4   Chemical Stability

The stability of an antidegradant toward heat, oxygen, etc. is important if it is to have maximum effectiveness for long periods. Hindered phenolic antidegradants should not be heated in presence of acidic materials, because dealkylation and loss of effectiveness occurs. Many antioxidants are subject to oxidative reactions with the development of colored species; the amine antioxidants are particularly sensitive. In this sense, these chemicals are sacrificial because reaction of the oxygen with them is dominant, thereby preserving the integrity of the rubber article.

Reactivity is also a consideration. Antioxidants fall into two general catagories, primary and secondary [11]. Primary antioxidants are hydrogen atom donors and act as "chain stoppers", stopping the free radical chain reaction of oxidation. Secondary antidegradants are considered to be peroxide decomposers. These chemicals convert peroxides to more stable products without the detrimental free radical chain reactions.

19.2.5   Physical Form

The physical form of an antidegradant is one more characteristic that governs its selection. Although polymer manufacturers prefer materials that are liquid and easily emulsified, many rubber compounders use solid, free flowing, non-dusty materials or liquid materials that can be pumped and directly injected into mixing equipment. In fact, liquids are becoming more popular as rubber processing and tire building become more automated. Materials to be avoided are semi-crystalline substances which stratify during storage, highly viscous liquids, and low melting resins. In the antiozonant field, a trend appears to be toward materials that are inherently liquid or are liquid blends, which are combinations of dialkyl and alkyl- alkyl-paraphenylenediamines. 6PPD, a low melting solid, is often conveyed through heated plumbing systems liquifying it and simplifying material handling.

19.2.6   Antidegradant Concentration

Determining the proper level of antidegradant to use is a complex question whose answer depends upon cost, polymer type, end-use application, staining requirements, solubility limits, etc. Most materials have an optimum level, based on laboratory aging studies; amounts greater than this optimum are not required.

The compounder is encouraged to use enough material so that this optimum concentration remains even after extended use in which some amount of the material may be destroyed or rendered inactive.

This optimum level is difficult to specify, but generally, the paraphenylenediamines offer best oxidation resistance at about 0.5 phr and best antiozonant protection in the 2 to 4 phr range. Here, solubility becomes a limiting factor, particularly with diaryl paraphenylenediamines. The phenolic antioxidants are effective at the 0.5 to 1.0 phr range; performance is not usually improved by going to higher levels.

19.3   Antidegradant Types

Antidegradants are generally classified into two classes with two characteristic performance features. Antidegradants are considered either staining and discoloring, or non-staining and non-discoloring. In general, performance features favor either antioxidant performance or antiozonant performance. Several examples of antidegradants and the characteristic performance features are discussed below.

It is appropriate to note here that an effective, non-staining, non-discoloring antiozonant has never been developed, i.e., one with performance equivalent to the staining antiozonants. Discoloration results from the action of light on the antioxidant itself; hence, the more effective and sensitive amine antioxidants show the most discoloration. In addition to direct discoloration, the tendency of some chemicals to migrate causes a problem. Staining can easily result from the migration of a discoloring antioxidant from a black compound to a light-colored surface.

19.3.1   Non-Staining, Non-Discoloring Antioxidants

These fall into six categories:

1. Hindered phenols
2. Hindered "Bis" phenols
3. Hydroquinones
4. Phosphites
5. Organic sulfur compounds
6. Hindered amine and nitroxyl compounds

There are many products in the first and second classes, but only a few in the other three classes. All are used to some degree to protect against oxygen, heat, light, and discoloration.

19.3.1.1   Hindered Phenols

The examples above describe only two of the many chemicals of this class produced and used worldwide. Products in this class are more volatile than other higher molecular weight antioxidants and thus, are not as persistent. Nevertheless, they do provide moderate protection at a low-to-moderate cost. It is also of interest that phenol itself is not an antioxidant, i.e., substitution in one or both ortho positions (to the OH) is necessary. These products are used in latex, general mechanical goods, and shoe products.

19.3.1.2   Hindered "Bis" Phenols

These are two examples of the "bis" phenols where two hindered phenol groups are linked together with an alkylene or alkylidene bridge in the ortho or para position. Other "bis" phenols are bridged by sulfur:

The "bis" phenols are generally the most potent and persistent of the non-staining antioxidants, but are also the most expensive. The alkylene- or alkylidine-bridged antioxidants tend to give the best non-discoloring characteristics, while the sulfur-linked products may have a slight edge in heat resistance at higher temperatures, e.g., less loss in tensile strength and elongation. Both classes are also quite effective in protecting rubber foam products. 4,4'-Butylidene-bis-(6-t-butyl-m-cresol) is particularly effective in natural rubber latex for many uses. Although not strictly a "bis" phenol, a polyphenol linked with dicyclopentadiene provides good persistence and reasonable light protection. Phenolic antidegradants are prone to losses through volatilization; the bis compounds, because of their higher molecular weight, are less prone to these losses.

19.3.1.3   Substituted Hydroquinones

This is somewhat of a specialty class, used primarily for tack retention in uncured rubber films and adhesives, and as stabilizers for some synthetic rubbers (NBR, BR). They are only active in uncured rubber, not in sulfur-cured rubber products, where they not only do not work as antioxidants, but impart severe scorch problems.

19.3.1.4 Phosphites

Tris (nonylphenyl) phosphite is one class of the phosphites, which are derived from PCl$_3$ and various phenols. These materials, similar to hydroquinones, are used only for stabilization of synthetic rubber during manufacturing and storage, i.e., they are not active antioxidants in cured rubber products. They prevent gel build-up and viscosity increases in raw polymers, particularly in emulsion SBR.

There are trends today for using combinations of phosphites with phenolic antioxidants to gain unique protective characteristics in solution elastomers (e.g., BR, SBR) and also in a number of plastics.

19.3.1.5 Organic Sulfur Compounds

These compounds are, for the most part, no longer generally used as antioxidants. One compound still with market significance is zinc dibutyldithiocarbamate. This product stabilizes butyl rubber after production and serves as an antioxidant for rubber-based adhesives. Because these materials are also very active accelerators, and scorchy ones at that, their use as antioxidants must be very selective and limited.

Another dithiocarbamate is nickel dibutyldithiocarbamate (NBC) used as an antiozonant for static ozone protection only in NBR, CR, and SBR to a limited extent. It also provides heat resistance for hypalon and acts as a peroxide decomposer. It should never be used with natural rubber because it is a pro-oxidant for NR.

Mercaptobenzimidazole (MBI) and its zinc salt (ZMBI) both function as synergists with some conventional antioxidants. More recently, the tolylimidazoles have come into use because of some questionable toxicological properties of the benzimidazoles. Dilauryl dithiodipropionate has been substituted for some ''bis'' phenols for similar reasons, and is particularly effective in some plastics.

19.3.1.6 Hindered Amine and Nitroxyl Compounds

Hindered amine and nitroxyl compounds are generally more expensive but highly active materials. The high cost and performance features of these chemicals have led to their usage in color- and light-sensitive materials, especially saturated and elastomeric materials, such as PP, EP, and EPDM.

The choice of a non-staining, non-discoloring antioxidant in preference to a discoloring one compromises cost/performance benefits in terms of protection. In general, the discoloring antioxidants are less expensive per pound and more active than non-discoloring ones. Selection of a non-discoloring antioxidant should therefore be restricted to applications where control of discoloration and/or staining is critical.

19.3.2   Staining/Discoloring Antioxidants

This class of materials consists of various types of mono- and di-amines, which are very effective antioxidants, but impart mild to severe discoloration and staining characteristics. In addition, the paraphenylene diamines are potent antiozonants and antifatigue agents. Typical of some of the most widely used products are:

1. Phenylnaphthylamines (PANA, PBNA)
2. Dihydroquinolines
3. Diphenylamine derivatives
4. Substituted paraphenylenediamines
5. Amine-based "Bound-in" or "Polymer bound" antioxidants

19.3.2.1   Phenylnaphthylamines

The phenylnaphthylamines are among the oldest of the antioxidants and were used extensively until the early 1990s. Their decline is the result of suspected toxicity problems rather than displacement by a technically superior product. This class of chemicals offers good heat, oxidation, and flex fatigue protection.

19.3.2.2   Dihydroquinolines

19.3.2.2.1 Polymerized Dihydroquinolines

This group includes very potent and widely used general purpose antioxidants, including polymerized 1,2-dihydro-2,2,4-tri methylquinoline. Degrees of polymerization and differences in molecular weight are the major variations among these products. Examples are Flectol* H and Flectol Pastilles; Flectol H has the higher molecular weight and exhibits a higher softening point. Consequently, the lower migratory properties of the high molecular weight material provide superior performance in staining and antioxidant persistance. The high molecular weight material is also preferred in latex applications because it can be more easily ball-milled and dispersed. Both Flectol products are useful in protecting rubber articles against degradation by heavy metal pro-oxidants such as copper, manganese, nickel, and cobalt. Scorch deficiencies make their use in neoprene rubbers less than desirable.

19.3.2.2.2 Monomeric Dihydroquinolines

Functioning only as an antioxidant and not an antiozonant, the only remaining commercially significant antidegradant of this class is the 6-Ethoxy derivative. This product (ETMQ) is very staining and discoloring and thus, cannot be used with white or light-colored stocks.

---

* Flectol is a registered trademark of Flexsys America L.P.

19.3.2.3 Diphenylamine Derivatives

One group of this class often provides antioxidant protection similar to or slightly less than the polymerized dihydroquinolines, depending upon aging temperatures. They are very mildly staining and discoloring, and are very effective in neoprene and nitrile rubbers.

A second group of this class is derived from the reaction of ketones with diphenylamine, e.g., high temperature reaction product of acetone and diphenylamine (BLE-25, AgeRite Superflex)*. These are quite effective antioxidants and better for flex cracking and fatigue resistance than the alkylated diphenylamines, but they are also much more staining and discoloring.

19.3.2.4 Substituted Paraphenylenediamines (PPDs)

These are very effective antioxidants, but they are discussed in depth under the section on antiozonants. Some PPDs are very efficient stabilizers in protecting synthetic rubbers during production and storage. Both the diaryl PPDs and alkyl-aryl PPDs in particular are used for this purpose.

19.3.2.5 Amine-based, ''Bound-in'' or ''Polymer Bound'' Antioxidants

A relatively recent development in antidegradant technology is the incorporation of hindered phenols as well as amines within the diene elastomer structure itself. These ''bound-in'' antioxidants are more stable or non-migratory, and thus, more durable. An example of a phenolic antioxidant of this type is one prepared by the reaction of 2,6-di-t-butylphenol with allyl chloride. In the resultant material, the allyl group provides the active site for inclusion with rubber molecules. An excellent review of the structure and patent references of polymer bound antioxidants is provided by Kuczkowski and Gillick [12].

Amine-based, ''polymer-bound'' antioxidants are similar to bound-in phenolics in that vinyl or acrylic unsaturation is built into the amine antidegradant material.
This then copolymerizes with the conjugated diene monomer.

A second method of incorporating amines is through generation of nitroso anilines or dinitrosodiphenylamines to form products by addition to the polymers. In all cases, the objective is to retain the antioxidants under severe environmental conditions.

Characteristics of quinone diimine antidegradants have recently been reviewed [13]. N-phenyl-N$'$-1,3-dimethylbutyl-p-quinone diimine (6-QDI) has unique properties in sulfur-vulcanized diene rubber. After vulcanization, a percentage of the 6-QDI is no longer extractable from the polymer. This portion of the 6QDI has chemically bonded either to the polymer or to the carbon black or both. In this fashion, it serves as a bound antioxidant, showing significant improvements over combinations of 6PPD and TMQ (a very effective antioxidant combination) [14]. In addition, another portion of the 6QDI is converted to 6-PPD (N-phenyl-N$'$-1,3-dimethybutyl-p-phenylenediamine), probably the most widely used

---

* BLE-25 is a registered trademark of Uniroyal Chemical Co.; AgeRite Superflex is a registered trademark of B. F. Goodrich Co.

and effective antiozonant used today. This 6PPD is free to migrate to the surface, functioning as an antiozonant. These observed effects of 6QDI or other similar products made available to the rubber compounder, provide significant improvements over currently used antidegradants systems.

19.3.3   Antiozonants

The selection of antiozonants depends very much on the application, and there are very different requirements for static ozone protection compared to dynamic ozone protection. Most applications involve both static and dynamic conditions of strain, e.g., a tire experiences both conditions when it is parked and driven, respectively. Although there are many conditions encountered and many levels of ozone concentration in various environments, there are only four classes of materials classified as antiozonants, and some of these are limited in performance. These are:

1. Petroleum waxes, which provide a protective layer at the surface
2. Nickel dibutyldithiocarbamate
3. 6-Ethoxy-2,2,4-trimethyl-1,2-dihydroquinoline (ETMQ)
4. Substituted paraphenylenediamines (PPDs)

There are also a few non-staining, non-discoloring products available and referred to as antiozonants, but these are not well identified and function only in isolated special areas.

19.3.3.1 Petroleum Waxes

Petroleum waxes are supplied as two major types: paraffin and microcrystalline. Each type is characterized by an optimum migration temperature where the mobility and solubility of the wax in rubber are balanced so sufficient bloom can form at the surface to provide optimum protection. Microcrystalline waxes, for example, have higher molecular weights compared to paraffin waxes and exhibit optimum bloom levels at higher ambient temperatures. Thus, there are environmental conditions where one type of wax offers more effective ozone protection compared to the other. Because these situations cannot be reasonably predicted or controlled, a blend of waxes with different molecular weights is commonly used.

One disadvantage with waxes is that they do not offer protection against ozone attack under dynamic conditions (cyclic stress-relaxation). Because of their inextensibility, the wax film ruptures, leaving the elastomer unprotected. To compensate for this deficiency, a PPD antiozonant is added. In contrast to waxes, PPDs provide a chemical barrier against ozone attack effective under both static and dynamic conditions.

In general, wax levels greater than 2.5 phr should be used only after careful testing to assure that dynamic protection is not seriously affected. Examples of some commonly used waxes are Sunproof Improved, Vanwax H, Hallco Wax C-1018.*

---

* Sunproof is a registered trademark of Uniroyal Chemical Co.; Vanwax is a registered trademark of R. T. Vanderbilt Co.; Hallco is a registered trademark of C. P. Hall.

19.3.3.2   Nickel Dibutyldithiocarbamate (NBC)

This product has already been discussed in the section on sulfur-containing antioxidants, but deserves consideration as an antiozonant, even though it is limited in scope. It is primarily used in neoprene and nitrile.

19.3.3.3   6-Ethoxy-2,2,4-trimethyl-1,2-dihydroquinoline (ETMQ)

ETMQ was the first antiozonant to be produced commercially (~1950) and it is still used because of its relatively low cost and because it is specified in many military tires. Even now it is usually used in combination with one of the PPDs. It has about 60% as much activity as an antiozonant as the best of the PPDs. It works by permitting the formation of many, very small cracks, thereby relieving strain and slowing crack growth. However, severe staining and discoloration is a limiting factor with this product (although this was less of a concern in the early 1950s because ETMQ was the only antiozonant available).

19.3.3.4   Substituted Paraphyenylenediamines (PPDs)

By far, the most common protectants used by the rubber industry are the paraphenylene-diamines. These materials are total antidegradants, protecting the rubber article against ozone and oxygen, as well as fatigue, heat, and metal ions. They are relatively expensive but their ability to protect against all the common degradative forces overcomes their high price. There are three classes of commercially available PPDs, differing in the nature of the substituent R groups.

19.3.3.4.1 Dialkyl PPDs

In this class, both substituent groups are alkyl, most frequently a branched C7 such as 77PD. They tend to be the scorchiest of the PPDs but often provide superior ozone protection in a static environment, particularly in the absence of wax. They also tend to be more sensitive to oxidation and may not persist in rubber products as long as other materials. Furthermore, they are not particularly effective in dynamic environments.

Combinations of dialkyl PPDs with alkyl-aryl-PPDs provide better protection against ozone attack under both static and dynamic conditions than can be obtained by any other means. In fact, the dialkyl PPDs are nearly always used in combination with other PPD types.

19.3.3.4.2 Alkyl-Aryl PPDs

In this class, one substituent is a phenyl ring; the other is an alky (e.g., branched C6 alkyl as in 6-PPD). The alkyl-aryl PPDs represent the best all-around class available. They offer superior dynamic protection, and when combined with wax, superior static protection. They are normally free of bloom problems.

6-PPD [N-phenyl-N′(1,3-dimethylbutyl)-p-phenylene-diamine] is often preferred in modern compounding because of its specific benefits, including:

- Less effect on processing safety than other alky-aryl and dialkyl-p-phenylenedia-mines, where scorch is a problem
- Less volatility than other widely used alkyl-aryl and dialkyl-p-phenylenediamines, leading to its use in long-life applications in hot climates
- Excellent SBR (styrene butadiene rubber) stabilization and carry-through antiozonant properties, reducing dosage of antiozonant in final rubber mix

19.3.3.4.3 Diaryl PPDs

In this class, the substituents are both aryl, normally phenyl or mixed tolyl. Diaryls generally cost less but have lower activity than the alkyl-aryls. Diaryl PPDs may cause bloom problems at effective use levels, i.e., 2 phr or more.

19.4   Examples of Antidegradant Activity

19.4.1   Oxidation Resistance

The relative effectiveness of several antioxidants in natural rubber is shown in Fig. 19.1 and Table 19.1. Here, the overall superiority of TMQ is evident. Also, note that this effectiveness is proportional to concentration, and no plateau is reached as with other antioxidants.

TMQ is also effective in SBR, and Fig. 19.2 shows how the usual modulus increase found in SBR/BR upon aging is reduced by addition of TMQ. For the unprotected SBR compound, the 100% modulus doubles in 2-1/2 days; with 1 phr Flectol Pastilles, this same increase in modulus does not occur until five days have elapsed [14].

19.4.2   Effect of Antidegradants on Fatigue Life

A number of antidegradants are compared in Fig. 19.3 as to their effect on the fatigue life of a NR stock. All the additives produce a rapid increase in fatigue life even at low concentrations and a limiting value is approached at about 3 phr additive. The longest fatigue lives are obtained with alkyl-aryl PPDs; 6PPD is an example. The fatigue life of stocks decreases on aging. Some results obtained with NR stock oven-aged for four days at 90 °C (194 °F) are shown in Table 19.2. These compounds have been compared at the same strain energy value of 140 psi, or approximately 100% strain [14].

Table 19.2 illustrates the superior activity of the p-phenylenediamines in both unaged stock and after aging. The PBN samples, for example, while providing acceptable protection for some requirements before aging, show a marked decrease in activity after aging. The sample containing 6PPD however, after the same aging period, is still showing properties superior to those of the original stock containing PBN.

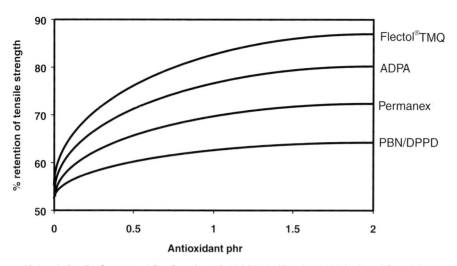

Figure 19.1   Aging Performance After four days @ 90 °C (194 °F) – NR. 90/10 Phenyl β-naphthylamine/ Diphenyl-p-phenylenediamine. Acetone Diphenylamine condensation product [14]. * Flectol is a registered trademark of Flexsys America L.P. Permanex is a registered trademark of Flexsys America L.P.

Table 19.1   Antioxidant Comparison-in NR

1.0 Flectol	TMQ	PBN/DPG	ADPA	Permanex ODPA
Aged @ 90 °C (194 °F)				
% Retention of tensile strength after:				
4 days	81.0	52.0	72.0	69.0
6 days	54.0	28.0	48.0	43.0

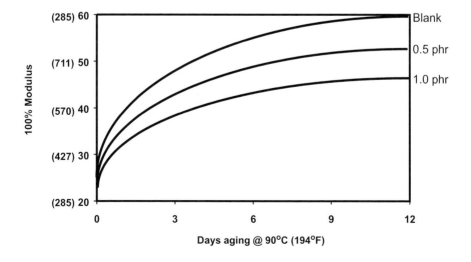

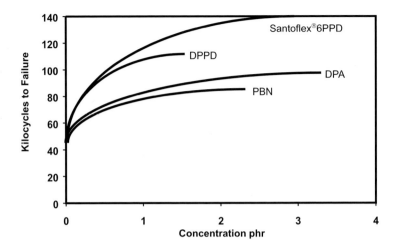

Table 19.2   Influence of Aging on Fatigue Life of NR Stocks

Additive	phr	Unaged	Aged 4 days 194 °F (90 °C)	% retention tensile strength
Nil	–	57,000 cycles	8,000 cycles	50.0
PBN	1	140,000 cycles	36,000 cycles	60.0
Santoflex 6PPD	2	185,000 cycles	160,000 cycles	78.0

For dynamic ozone protection, 6PPD antidegradant is generally best under both unaged and aged conditions, followed by the dialkyl class. 77PD may exhibit superior static ozone performance, particularly in the absence of wax. Ordinarily, 77PD is not recommended as the sole antidegradant because of its inferior dynamic performance, especially after aging. Many formulators use a blend of 6PPD and 77PD (e.g., 2:1) to combine the advantageous effects of both systems: 6PPD for dynamic protection and 77PD for static ozone protection (see Table 19.3).

19.4.3   Combinations of Antiozonants and Antioxidants

Because tires operate in an oxidative environment, as well as in ozone, it should not be surprising that an effective antioxidant contributes to overall performance. Of course, PPDs are very effective antioxidants, but it seems a waste to sacrifice a PPD to oxidative loss if a lower cost antioxidant suffices.

The data in Table 19.4 indicates that the addition of TMQ antioxidant along with 6PPD can extend ozone and fatigue protection even though TMQ offers no ozone or fatigue protection by itself at conventional levels. The extension of ozone and fatigue life is even more dramatic after the oxidative aging of this sidewall compound. Laboratory results were confirmed in long-term tire tests in a very convincing way [15].

Table 19.3   Comparison of Antiozonants in Natural Rubber Stock

	1	2	3
Santoflex 6PPD	–	3.0	1.0
Santoflex 77PD	–	–	2.0
Test method			
intermittent belt test			
Ozone chamber	0.0	10.0	9.0
Total 216 hours			
Continuous Flexing			
Outdoors 20 days	0.0	10.0	10.0
	(Sample broke)		
Static ozone chamber			
324 hours			
10% extension	3.0	7.0	10.0
20% extension	1.0	1.0	10.0

Crack Rating: 10 ⟶ 0 increased cracking

Table 19.4   Combinations of Santoflex 6PPD and Flectol TMQ

	NR/BR			
	1	2	3	4
phr Santoflex 6PPD	–	–	2.0	2.0
phr Flectol TMQ	–	1.0	–	1.0
Oxidative aging				
% Retention of tensile strength				
Aged 5 days @ 90 °C	40.0	64.0	72.0	76.0
Aged 5 days @ 110 °C	10.0	15.0	18.0	23.0
Ozone resistance				
Hours to 70% retention of				
100% modulus; 25 pphm ozone				
Unaged	14.0	13.0	57.0	67.0
Aged 2 days @ 90 °C	8.0	8.0	41.0	53.0
Fatigue life				
Kilocycles to fail at 100% strain				
Unaged	60.0	55.0	129.0	141.0
Aged 2 days @ 90 °C	50.0	50.0	106.0	112.0
Aged 7 days @ 50 °C in oxygen bomb	40.0	45.0	95.0	121.0

19.4.4   Resistance to Metal Poisoning

Earlier, it was noted that the alkyl-aryl PPDs and dihydroquinoline-type antioxidants provide protection against metal poisoning, particularly resulting from copper, manganese, and cobalt, among others. However, most of the problems encountered are in

Table 19.5   Copper Inhibition-White-Filled NR*

	Control	DN-PPD	Flectol TMQ
phr antioxidant	0.0	1.0	1.0
% Retention tensile strength			
Aged @ 90 °C (194 °F) for 5 days	8.5	34.0	34.0
Strain	1.0	5.0	2.0
Discoloration	1.0	3.0	2.0

Rating: 1 ————————► 6 color increasing.
* Pale Crepe-100, Fixe-50m, TiO₂-5, ZnO-5, Stearic Acid-1, Sulfur-2.5, MBTS-1.0, Copper Oleate-0.25.

non-black compounds. To demonstrate the effectiveness of antidegradants here, a white-filled NR compound containing 0.25 phr of copper oleate was aged without and with TMQ and dinaphthyl-PPD (long used in this application). Results are shown in Table 19.5 [15].

References

1. Lenz, R.W., *Organic Chemistry of Synthetic High Polymer*, Interscience Publishers, New York, 1967, pp.733–767.
2. Spracht, R. B., Hollingshead, W. S., Bullard, H. L., and Wills, D. C., Rubb. Chem. Technol. 37, 210 (1964).
3. Juve, A. E., and Shearer, R., India Rubber World, 128, 623 (1953).
4. Dickenson, P. B., J. Rubber Res. Inst. Malaysia, 22, 165 (1969).
5. Lake, G. J., Rubb. Chem. Technol. 43, 1230 (1970).
6. Lattimer, R. P., Hooser, E. R., Diem, H. E., Layer, R. W., and Rhee, C. K., Rubb. Chem. Technol., 53, 1170 (1980).
7. Roberts, A. D. *Natural Rubber Science and Technology*, Oxford University Press, Oxford, 1988 pp. 215–216.
8. Roberts, A. D. *Natural Rubber Science and Technology*, Oxford University Press, Oxford, 1988 pp. 213–215.
9. Grinberg, A. A., Zototarevskoya, L. K. and Tavadia, E. P., Soviet Rubber Technol., 28(6) 25 (1969).
10. Cain, M. E., Grazeley, K. F., Gelling, I. R., and Lewis, P. M., Rubb. Chem. Technol., 45, 204 (1972).
11. Pospisil, J., Developments in Polymer Stabilization Volume 7, Elsevier Applied Science Publisher, London, 18984, pp. 1–63.
12. Kuczkowski, J. A. and Gillick, J. G. Rubb. Chem. Technol., 57, pp. 621–651 (1984).
13. Ignatz-Hoover, F., Maender, O. W., Lohr, R., Rubber World, 218(2), 38 (1998).
14. Cain, M. E., Gelling, I. R., Knight, G. T., Lewis, P. M., Rubber Industry, 216–226 (1975).
15. Monsanto Chemicals Ltd. Technical Bulletin, O/RC-14 (1969).
16. Monsanto Chemicals Ltd. Technical Bulletin, IC/RC-25 (1972).

Abbreviations

6-QDI	N-Phenyl-N′-1,3-dimethylbutyl-p-quinonediimine
6-PPD	N-(1,3 dimethylbutyl)-N″-phenyl-p-phenylene-diamine
BCI-MX	1,3-Bis(citraconimidomethyl)benzene
BHT	Butylated Hydroxy Toluene
BR	Butadiene Rubber
CTP	N-(Cyclohexylthio)phthalimide
CR	Chloroprene Rubber
DCBS	N,N′-dicyclohexyl-2-benzothiazolesulfenamide
DPPD	Di Phenyl Paraphenylenediamine
DTDM	4,4′-Dithiodimorpholine
EP	Ethylene Propylene Rubber
EPDM	Ethyl Propylene Diene Rubber
ETMQ	Ethoxy Trimethyl Quinoline
HMMM	Hexamethoxymethylmelamine
HTS	Hexamethylene-1,6-bisthiosulfate disodium salt dihydrate
IIR	Butyl Rubber
MBI	Mercaptobenzimdazole
NBC	Nickel dibutyl carbamate
NBR	Nitrile Rubber
NR	Natural Rubber
PANA	Phenyl-α-naphthyl amine
PBNA	Pheynyl-β-naphthyl amine
PP	Polypropylene
SBR	Styrene Butadiene Rubber
TBBS	N-tert-butyl-2-benzothiazolesulfenamide
TMQ	Polmerized 1,2-dihydro-2,2,4-trimethylquinoline
ZMBI	Zinc Mercapto Benzimidazole

20 Compounding for Brass Wire Adhesion

Alex Peterson

20.1 Introduction

Steel-cord reinforced ply compounds in tire, hose, and belting products have performance characteristics dependent on the quality and durability of the rubber compound, wire reinforcement, and adhesive interface connecting these materials. The basic concepts for bonding wire to rubber are similar to those for other bonding applications. The rubber compound should have sufficient scorch delay and viscosity for satisfactory wetting and contact with the wire surface prior to vulcanization. The rubber compound should then develop robust wire-to-rubber adhesion through primary and secondary bond formation, diffusion, and physical anchoring during vulcanization.

Brass-plated carbon steel wires are commonly used to reinforce conventional, sulfur-cured, natural rubber-based ply compounds. The chemical interaction of the brass and rubber during vulcanization generates a copper sulfide adhesion layer that bonds the wire to the rubber compound. The brass and rubber compound, therefore, have an interdependent role in the development of the adhesive interface. The relationship of brass wire adhesion to the chemical composition of the brass wire surface and the rubber compound ingredients has been extensively reviewed in the literature [1–7].

Compounding for brass wire adhesion is complex because all the ingredients affect processing, cure, and mechanical properties as well as adhesion at the wire-rubber interface. Optimal formulations meeting target performance properties can be developed through designed experiments and regression analysis.

20.2 Wire Bonding Systems

Bonding systems for brass wire should develop satisfactory initial wire adhesion as well as high levels of retained adhesion despite exposure to heat, humidity, dynamic flexing and other environmental aging conditions. To achieve this degree of adhesion performance, rubber compound ingredients must be properly selected for any given brass plating composition and vulcanization condition.

Natural-rubber ply compounds are generally formulated with higher than normal levels of insoluble sulfur and reduced levels of accelerator to ensure adequate formation of copper sulfide(Cu_xS) at the brass wire surface. Because wire-bonding formulations are sulfur-vulcanization dependent, higher cure temperatures can adversely affect wire sulfidation and adhesion development.

To achieve satisfactory levels of adhesion with different plating compositions and cure conditions, adhesion promoters or bonding agents such as organocobalt salts and resorcinol-based resins are incorporated into the compound formulation. Adhesion can be further improved by the partial replacement of carbon black with precipitated silica.

20.2.1   Cobalt

Brass-plated steel cords typically have copper/zinc plating compositions in the range of 60 to 70% Cu and 0.1 to 0.5 microns thickness. With increasing copper content and plating thickness, copper sulfide formation at the wire-rubber interface can be excessive and weaken interfacial adhesion.

The addition of organocobalt salts such as cobalt naphthenate, cobalt neodecanoate, cobalt-boron complex, and cobalt stearate to the rubber compound moderates copper sulfide formation and improves both initial and aged wire adhesion. These cobalt salts provide comparable adhesion when added to the compound on an equal cobalt-metal basis.

Because increased levels of cobalt salts can reduce humidity-aged adhesion and accelerate sulfur cure, ply compounds are formulated with higher sulfur/accelerator ratios and optimal sulfur/cobalt ratios for balanced cure and adhesion properties.

20.2.2   RF Resin-Cobalt

In addition to organocobalt salts, an array of bonding systems based on resorcinol, resorcinolic resin, melamine resin, and various nitro and amino derivatives may be considered for brass wire adhesion. Resorcinol and resorcinol-formaldehyde(RF) resin are crosslinked *in situ* with methylene donors such as hexamethoxymethylmelamine (HMMM) and hexamethylenetetramine (HMT) during vulcanization. Chemical structures of these bonding agents are shown in Fig. 20.1.

The resulting performance of the NR ply compound depends on the bonding-agent type, resin/donor ratio, and loading level [7]. Although the cobalt-resin bonding systems in Table 20.1 yield comparable cure times, hardnesses, and initial adhesion, Resorcinol/

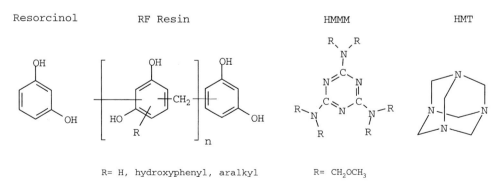

Figure 20.1   Chemical structures of bonding agents.

Table 20.1   Comparison of Resorcinol and RF Resin-Cobalt Bonding Systems*

Wire bonding system	R/HMMM	RF/HMMM	R/HMT	RF/HMT
Bonding agent, phr	2.0/2.0	3.0/2.0	2.6/1.4	3.8/1.2
Cobalt naphthenate (12% Co)	0.83	0.83	0.83	0.83
Process fuming @ 125 °C	high	low	high	low
Properties				
T90, min	16.6	15.8	15.0	15.6
Shore A hardness	83.0	84.0	84.0	84.0
Unaged coverage, %	87.0	86.0	90.0	95.0
Humidity coverage, %	88.0	86.0	10.0	10.0

* Formulation, phr: NR-100; N326 black-55; zinc oxide-8; stearic acid-1.2; antidegradant-3; inhibitor-0.2; sulfur-3.75; TBBS-0.6; bonding system-variable.

HMMM and RF Resin/HMMM in combination with organocobalt salt are preferred for their high levels of adhesion retention after humidity aging.

Because the addition of bonding agents to the compound can affect both cure and mechanical properties, adjustments in curatives and fillers are required to achieve target performance.

20.3   The Adhesion Mechanism

Combined cobalt and RF Resin/HMMM bonding systems promote brass wire-to-rubber bonding by the following proposed adhesion mechanism [1,2,4,5,6].

1. Cobalt stimulates copper sulfide growth at the wire surface, which helps anchor the sulfurated rubber to the wire.
2. Cobalt moderates the growth of copper sulfide through the formation of interfacial ZnO, ZnS, and CoS, which has a beneficial effect on both initial and aged adhesion.
3. Partial replacement of carbon black with precipitated silica also enhances ZnO formation and aged adhesion.
4. RF Resin/HMMM in combination with cobalt improves initial adhesion and stabilizes aged adhesion by protecting the wire interface from attack by moisture during aging.

The primary function of organocobalt salt and RF Resin/HMMM bonding agents is to promote satisfactory initial wire adhesion during vulcanization and to maintain this adhesion throughout the service life of the ply compound.

20.4   Compound Ingredient Effects

When formulating rubber compounds for specific applications, such as brass wire-to-rubber adhesion, fundamental knowledge is required about the effect of elastomers, fillers,

Table 20.2   Model NR Ply Compound Variables

Formulation variables	phr
Natural rubber CV60	100
Carbon black N326	40 to 80
Precipitated silica	0 to 16
Zinc oxide	7.5 to 9.5
Stearic acid	0.5 to 2.5
Antidegradant 6PPD	2.0
Inhibitor CTP	0.2
Aromatic oil	3.0
Insoluble sulfur/DCBS (80/20)	3.0 to 7.0
Bonding agents	
Cobalt naphthenate (12% Co)	0 to 2.5
RF resin/HMMM (60/40)	0 to 6.0

antidegradants, processing aids, curatives, and bonding agents on compound performance. The model NR ply compound formulation variables in Table 20.2 were evaluated by experimental design and regression analysis for their effect on optimum cure time, T90, Shore A hardness, unaged adhesion, and retention of adhesion after humidity aging.

20.4.1   Mixing

Rubber compound mixing was performed by the three-stage procedure outlined in Table 20.3. The masterbatch containing NR CV60, N326 carbon black, precipitated silica, zinc oxide, stearic acid, antidegradant 6PPD, inhibitor CTP, and aromatic process oil was mixed to 170 °C in the first internal mixing stage. In the second stage, cobalt naphthenate and RF Resin were mixed into the masterbatch on the roll mill at 125 °C. Insoluble sulfur, accelerator DCBS, and HMMM were mixed in the final stage on the roll mill at 95 °C.

20.4.2   Testing

ASTM testing methods for optimum cure time, T90, Shore A hardness, and brass wire adhesion are listed in Table 20.4. Optimum cure time, T90, was measured with an oscillating disk rheometer at 1.67 Hz and 1.0 degree arc at 150 °C. Test compounds for the measurement of mechanical and adhesion properties were cured to T90 at 150 °C.

The hardness levels of 4 mm thick rubber sheets were measured with a Shore A durometer. Wire pullout adhesion tests were conducted with $3 \times 0.20 + 6 \times 0.35$ mm

Table 20.3   Compound Mixing Procedure

Mixing stage		Ingredients	Temperature, °C
1st stage	Internal	NR, fillers, powders, oil	170
2nd stage	Roll mill	RF resin, cobalt naphthenate	125
Final stage	Roll mill	Sulfur, DCBS, HMMM	95

Table 20.4   Compound Testing Methods

ASTM D2084	Vulcanization using oscillating disk cure meter
ASTM D2240	Durometer hardness
ASTM D2229	Adhesion between steel tire cords and rubber

brass-plated truck cords with 64% Cu plating and nominal 0.2 microns plating thickness, which were embedded 19.5 mm in 12.5 mm thick rubber blocks. Wires pulled from the unaged and humidity-aged (14 days at 85 °C/95% RH) rubber blocks were observed for rubber coverage with an optical microscope.

20.4.3   Regression Plots

Figures 20.2 to 20.13 show the effects of carbon black, zinc oxide/stearic acid, sulfur/DCBS, cobalt, RF Resin/HMMM, and carbon black/silica on cure time, hardness, and brass wire adhesion.

20.4.3.1 Carbon Black

Figures 20.2 and 20.3 show the effect of carbon black level at 50 to 80 phr. The ingredients held constant are Silica-0, ZnO/StA-9.5/0.5, S/DCBS-5.6/1.4, Cobalt-0, and RF Resin/HMMM-0.

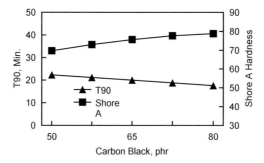

Figure 20.2   Effect of carbon black on cure time and hardness.

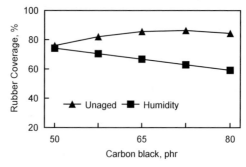

Figure 20.3   Effect of carbon black on unaged and humidity adhesion.

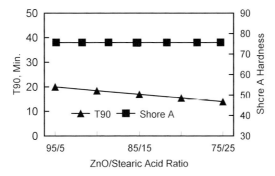

Figure 20.4   Effect of ZnO/stearic acid ratio on cure time and hardness.

Increased carbon black levels result in increased hardness and unaged adhesion and reduced humidity-aged adhesion.

20.4.3.2 Zinc Oxide/Stearic Acid

Figures 20.4 and 20.5 show the effect of the zinc oxide to stearic acid ratio at a constant zinc oxide/stearic acid level of 10 phr. Ingredients are CB-65, Silica-0, S/DCBS-5.6/1.4, Cobalt-0, and RF Resin/HMMM-0.

Higher stearic acid fractions reduce T90 and unaged and humidity-aged adhesion.

20.4.3.3 Sulfur/DCBS

Figures 20.6 and 20.7 show the effect of sulfur/DCBS level at a constant sulfur/DCBS ratio of 80/20. Ingredients are CB-65, Silica-0, ZnO/StA-9.5/0.5, Cobalt-0, and RF Resin/HMMM-0.

Increased sulfur/DCBS levels reduce T90 and increase hardness and unaged and humidity-aged adhesion.

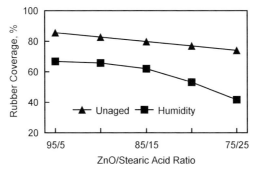

Figure 20.5   Effect of ZnO/stearic acid ratio on brass adhesion.

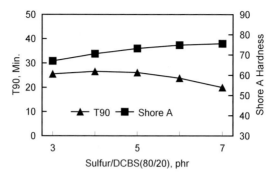

Figure 20.6   Effect of sulfur/DCBS on cure time and hardness.

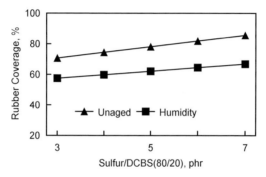

Figure 20.7   Effect of sulfur/DCBS on brass wire adhesion.

20.4.3.4   Cobalt

Figures 20.8 and 20.9 show the effect of cobalt naphthenate level at 0.0 to 0.3 phr cobalt metal. Ingredients are CB-65, Silica-0, ZnO/StA-9.5/0.5, S/DCBS-5.6/1.4, and RF Resin/HMMM-0.

Increased cobalt levels at constant sulfur increase hardness and unaged adhesion and reduce T90 and humidity-aged adhesion.

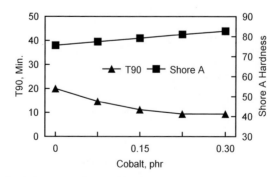

Figure 20.8   Effect of cobalt on cure time and hardness.

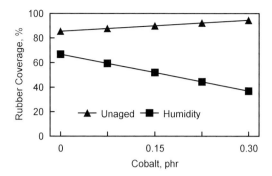

Figure 20.9   Effect of cobalt on brass wire adhesion.

20.4.3.5   RF Resin/HMMM

Figures 20.10 and 20.11 show the effect of RF Resin/HMMM level at a constant RF Resin/
HMMM ratio of 60/40. Ingredients are CB-55, Silica-0, ZnO/StA-9.5/0.5, S/DCBS-5.6/
1.4, and Cobalt-0.10.

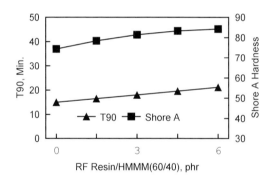

Figure 20.10   Effect of RF resin/HMMM on cure time and hardness.

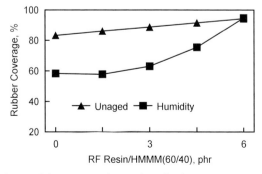

Figure 20.11   Effect of RF resin/HMMM on brass wire adhesion.

Increased RF Resin/HMMM levels increase T90, hardness, and unaged and humidity-aged adhesion.

20.4.3.6   Carbon Black/Silica

Figures 20.12 and 20.13 show the effect of carbon black to silica ratio at a constant carbon black/silica level of 55 phr. Ingredients are ZnO/StA-9.5/0.5, S/DCBS-5.6/1.4, Cobalt-0.10, and RF Resin/HMMM-3/2.

Higher silica fractions reduce hardness and increase T90 and unaged and humidity-aged adhesion.

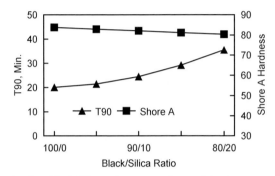

Figure 20.12   Effect of carbon black/silica ratio on cure time and hardness.

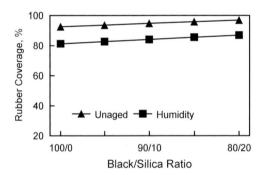

Figure 20.13   Effect of carbon black/silica ratio on brass wire adhesion.

20.4.3.7   Summary of Test Results

The regression plots in Figs. 20.2 to 20.13 show that fillers, curatives, and bonding agents have varied effects on cure time, hardness, and wire adhesion. Cure time is reduced with higher levels of cobalt, sulfur/accelerator, and stearic acid and lower levels of RF Resin/HMMM and silica. Hardness is increased with higher levels of carbon black, sulfur/DCBS, cobalt, and RF Resin/HMMM. Brass wire adhesion increases with higher levels of sulfur/DCBS, Resin/HMMM, and silica and lower levels of cobalt and stearic acid.

Table 20.5   Model NR Ply Compound Formulations and Properties

Compound	Black control	Black cobalt	Black resin/cobalt	Black/silica resin/cobalt
Formulation, phr				
Natural rubber CV60	100.0	100.00	100.00	100.00
Carbon black N326	80.0	70.00	50.00	50.00
Precipitated silica	0.0	0.00	0.00	5.00
Zinc oxide	9.5	9.50	9.50	9.00
Stearic acid	0.5	0.50	0.50	1.00
Antidegradant 6PPD	2.0	2.00	2.00	2.00
Inhibitor CTP	0.2	0.20	0.20	0.20
Aromatic oil	3.0	3.00	3.00	3.00
Insoluble sulfur	5.2	6.50	5.60	5.60
DCBS	1.3	0.50	1.40	1.40
Bonding system				
Cobalt naphthenate (12% Co)	0.0	1.66	0.83	0.83
RF resin	0.0	0.00	3.00	3.00
HMMM	0.0	0.00	2.00	2.00
Properties				
T90, min	20.1	21.50	20.90	21.90
Shore A hardness	79.0	80.00	82.00	82.00
Unaged coverage, %	80.0	90.00	90.00	95.00
Humidity coverage, %	60.0	70.00	80.00	85.00

20.5   Model NR Ply Compounds

The model NR ply compounds in Table 20.5 were formulated to give maximum brass wire-to-rubber adhesion while maintaining a cure time, T90, of 20 minutes and Shore A hardness of 80.

20.5.1   Black Control Compound

The black control compound, which did not include any bonding agents, has relatively low levels of unaged and humidity-aged adhesion. High carbon black levels are required for target hardness.

20.5.2   Black/Cobalt Compound

With the addition of cobalt naphthenate to the black control compound, adjustments in sulfur/DCBS ratio and carbon black are required to achieve target properties. Unaged and humidity-aged wire adhesion improved with optimal sulfur level, sulfur/cobalt ratio, and sulfur/DCBS ratio.

20.5.3   Black/Cobalt/RF Resin

The addition of RF Resin/HMMM to the black/cobalt compound results in increased cure time and hardness, which are adjusted with reduced carbon black, cobalt, and sulfur/DCBS ratio. The combined RF Resin-cobalt bonding system has improved unaged and humidity-aged adhesion.

20.5.4   Black/Silica/Cobalt/RF Resin

The partial replacement of carbon black with silica increased the cure time of the RF Resin-cobalt bonding system, and is adjusted with increased stearic acid level. Unaged and humidity-aged adhesion increased with the addition of silica.

The model NR ply compound formulations and performance properties shown in Table 20.5 indicate that highest levels of unaged and humidity-aged adhesion are achieved with balanced levels of zinc oxide/stearic acid, sulfur/DCBS, cobalt, RF Resin/HMMM, and silica in the compound.

20.6   Summary

The successful bonding of brass-plated steel cord to NR-based ply compounds by sulfur vulcanization is highly specific to the physical and chemical characteristics of the brass wire and rubber compound formulation. Optimal levels of adhesion development are compromised to some extent because other performance criteria, such as processability, cure, and mechanical properties also depend on the formulation.

Ply compounds may be formulated with organocobalt salt, resorcinol-based resin bonding systems, and silica to achieve target performance properties while maintaining a satisfactory level of brass wire adhesion.

Acknowledgments

The author would like to thank Mr. Michael O. Enright for his dedicated assistance with the development and evaluation of bonding systems for steel cord reinforcements.

References

1.   Van Ooij, W. J., *Rubber Chem. Technol.*, (1979), 52, 605
2.   Haemers, G., *Rubber World*, (1980), 182, 40

3.   Ishikawa, Y., *Rubber Chem. Technol.*, (1984), 57, 855
4.   Hamed, G. R., Huang, J., *Rubber Chem. Technol.*, (1991), 64, 285
5.   Okel, T. A., Waddell, W. H., Evans, L. R., *Tire Technol. Int. '96*, (1996), 104
6.   Seo, G., *J. Adhesion Sci. Technol.*, (1997), 11, 1433
7.   Peterson, A., *Tire Technol. Int. '98*, (1998), 63

21 Chemical Blowing Agents

Ralph A. Annicelli

21.1 Introduction

Chemical blowing agents for rubber and plastics are available in a wide range of formulations depending on the polymer involved, processing temperature, and end use. For simplicity, this discussion is limited to those commercially available blowing agents commonly employed for producing cellular rubber products. This chapter summarizes their characteristics, uses, and processing.

There are two distinct classes of chemical blowing agents: inorganic (endothermic) and organic (exothermic). The inorganic types are mainly bicarbonate and carbonate salts and generally release carbon dioxide gas upon decomposition. The most important among these is sodium bicarbonate, which has for many years been used to manufacture open-cell sponge based on natural and synthetic rubbers. The organic types are nitrogen compounds that decompose to release predominately nitrogen gas. These organic materials are capable of producing fine, closed-cell structures in rubber and plastics.

The type of cell structure formed by a blowing agent is a chief factor in determining its properties and use. Open-cell materials consist of a network of interconnecting cells formed by the expansion of gases that ruptures the wall between cells. Closed cell foam differs from open cell sponge in that the cells are individual and non-interconnecting, formed by the entrapment of gases as discrete bubbles in a polymer matrix.

21.2 Terminology

The term cellular rubber is used to apply broadly to both natural and synthetic rubber materials containing cells or small hollow bubbles or pores. Cellular rubber is a generic term covering latex foam, urethane foam, sponge rubber, and expanded rubber containing many cells (either open, closed, or both) dispersed throughout the mass [1].

Latex foam rubber is made from rubber latices or liquid rubbers, either natural or synthetic. The structure of foam rubber consists of a network of open or interconnecting cells. Sponge rubber refers to those products in which the cells are interconnected (open cell) and expanded rubber refers to materials in which the cell structure consists of a myriad of individual, non-connecting gas-tight cells (closed cell). Only open cell sponge rubber and closed cell expanded rubber are discussed in this chapter.

21.2.1   Open Cell Structure

In open cell sponge rubber, the walls between the cells are ruptured, allowing air to pass freely between them. The resulting open cell structure consists of a self-supporting network of bridges and arches, resulting in a material with excellent load-bearing characteristics. Thus, open cell sponge has main applications in cushioning and sealing. This type of cell structure is made by incorporating into the polymeric compound a gas-producing chemical, such as sodium bicarbonate, which decomposes, releasing CO_2 during vulcanization, thus providing the structure from which the name is derived. Open cell sponge usually has low resilience and compression set properties, has poor mechanical strength and insulation properties, and high water absorption.

Open cell products are manufactured in sheets, molded strips, and special shapes. Sheets and parts cut from sheets often have a surface impression because they are usually molded against a fabric surface that allows air to be vented during the expansion of the sponge.

These products are commonly used for vibration damping or shock absorption because they can dissipate a relatively large fraction of the energy generated during a compressive deformation, i.e., they are characterized by a relatively low rebound. The energy is dissipated not only by the inherent hysteresis of the matrix material, but also by the physical process of forcing air to flow in and out of the cell. Applications consists of load-bearing items such as rug underlay, seals, and pads.

21.2.2   Closed Cell Structure

A closed cell structure consists of a network of non-interconnecting cells formed by the entrapment of gas as discrete bubbles in a matrix of rubber or plastic. This cell structure is obtained by the proper balance of blowing and curing systems. These systems are regulated so that sufficient cure is developed to encapsulate the gas formed as the decomposition of the blowing agent proceeds. If the cure progresses too far before the release of the gases, the expansion is restricted and a high density product is formed. Premature decomposition of the blowing agent generally results in the rupture of cell walls. The effect of this condition may vary from producing a high density material with a blistered surface to a low density sponge with aggregates of open cell structure.

Because the cells are not interconnected, closed cell materials have low water absorption, high resiliency, and excellent thermal properties, making them good for flotation, shock absorption, and insulating materials.

Closed cell, expanded rubber is manufactured in sheets, strips, and special shapes by molding or extruding. Closed cell molded sheets are split to the prescribed thickness and then die-cut for gaskets used in air conditioners, heaters, closures, seals, etc. Die cut parts have exposed cells on all cut edges. Extruded shapes have a skin on all surfaces except cut ends. The choice of polymer is a function of the desired end use requirements, such as cost, oil and ozone resistance, and other factors.

21.3   Inorganic Blowing Agents

Chemical blowing agents provide gas for the expansion of polymers by undergoing a chemical reaction which results in the thermal decomposition of the molecule, yielding one or more gases. These agents are now more widely used than ever before in the manufacture of cellular rubber and plastic products for use in various automotive, industrial, and consumer applications. For the production of open cell and closed cell products, both inorganic (endothermic) and organic (exothermic) chemical blowing agents are used.

Sodium bicarbonate (soda) is by far the most well known of this type. Other inorganic blowing agents occasionally used are sodium carbonate, ammonium carbonate, and ammonium bicarbonate. Ammonium carbonate and bicarbonate tend to produce a coarser and more uneven blow than sodium bicarbonate, are unstable, and give off ammonia readily when exposed to air.

Sodium bicarbonate, also called ''soda'', is the most widely used blowing agent for open cell sponge. Soda begins to decompose over a broad temperature range (130 to 180 °C) to yield carbon dioxide and a small amount of water. In the thermal decomposition of soda, only half of the available carbon dioxide in the molecule is evolved [2]. Total decomposition of the bicarbonate occurs when acidic materials are used. For open cell sponge, soda is activated with fatty acids, such as stearic acid or oleic acid, to obtain a higher gas yield.

Ordinary soda is difficult to disperse in rubber and the strongly alkaline decomposition residue may be objectionable in some applications where staining of wood surfaces and water resistance are important. Because of its hydroscopic nature, soda tends to absorb moisture and cake badly. Current products coated with oil or wax can be dispersed more easily and decompose more rapidly, meaning that less blowing agent is needed for an equivalent degree of blow and less unreacted blowing agent remains in the sponge. In elastomeric compounds (NR, polychloroprene, SBR, EPDM, nitrile, and butyl), soda is generally used in amounts ranging from 1.0 to 15 phr.

21.4   Organic Blowing Agents

The organic types are nitrogen releasing compounds, which are stable at normal storage and mixing temperatures, but undergo controllable gas evolution at reasonably well defined decomposition temperatures. Decomposition temperature refers to the point at which the material begins to produce gas at a reasonably rapid rate. The efficiency of the blowing agent is expressed as gas evolution in cubic centimeters per gram measured at standard temperature and pressure.

Nitrogen, which is nonflammable, non-toxic, and inert, has a slow rate of diffusion through polymers and is a very efficient gas for expanding most materials. Other gases, such as carbon monoxide, carbon dioxide, and ammonia, may also be given off in lesser amounts and have a faster diffusion rate. The organic blowing agents have more narrow

Table 21.1   Chemical Blowing Agents

Chemical name	Common trade name	Manufacturer	Decomposition temperature (°C)	Gas yield (cc/m)
Unmodified azodicarbonamide (1)	Celogen AZ	Uniroyal Chemical	205–212	220
	Porofor ADC	Bayer		
	Unicell D	Dong Jin		
	Vinyfor AC	Eiwa		
	Cellcom AC	Kum Yang		
	Azofoam	Otsuka		
Preactivated azodicarbonamide (2)	Celogen 754A	Uniroyal Chemical	165–180	200
	Porofor ADC-K	Bayer		
	Unicell DX	Dong Jin		
	Cellcom CF	Kum Yang		
	Celogen 765A	Uniroyal Chemical	152–160	180
	Celogen 780	Uniroyal Chemical	140–150	190
p,p′-oxybis(benzene sulfonylhydrazide)	Celogen OT	Uniroyal Chemical	158–160	125
	Unicell OH	Dong Jin		
	Cellcom OBSH	Kum Yang		
Dinitrosopentamethylene tetramine	Opex-80	Uniroyal Chemical	190–195	190
	Cellcom A-80	Kum Yang		

(1) Available in particle size from 2.0–15.0 microns.
(2) Preactivated form available to as low as 140 °C.

decomposition temperature ranges compared to the inorganic types, and generally produce more gas per unit weight. They include a wide range of chemical compositions, physical properties, and decomposition characteristics. Table 21.1 lists some of the more commonly used commercial blowing agents.

The organic class of blowing agents is used in closed cell products. However, with certain types of accelerator systems and curing conditions, they yield open cell sponge. In comparison to sodium bicarbonate, they disperse better, have better processing safety, and give a more uniform cell structure.

While a wide range of organic chemical blowing agents are patented, few are of commercial significance. For simplicity, this discussion is limited to a review of blowing agents currently used to produce expanded rubber. Organic blowing agents are generally classified according to their chemical composition. The most important commercial blowing agents include pure and modified azodicarbonamides, sulfonyl hydrazides, and dinitrosopentamethylenetetramine.

21.4.1   Azodicarbonamide (ADC)

$$NH_2-CO-N = N-CO-NH_2$$

The first use of ADC as a chemical blowing agent was in Germany during World War II to expand PVC [3]. Its decomposition temperature was initially considered too high for use at typical curing temperatures for expanding rubber. A research and development team at Naugatuck Chemical headed by Dr. Byron Hunter discovered that glycols and glycerol effectively lowered the decomposition temperature. The patent application by Dr. Hunter and Wesley Curtis was filed in 1953 and approved in 1957 [4,5]. This work was critical to the use of ADC in the rubber and plastics industry. The first patent (U.S. Patent 2,692,281) for the manufacture of Azodicarbonamide in the U.S. was to Naugatuck Chemical, now Uniroyal Chemical Company Inc. The product, Celogen AZ, was first produced in Naugatuck, CT, in 1954 [5,6]. ADC and its modified forms continues to be the predominant chemical blowing agent for rubber and plastics.

21.4.1.1   Properties

ADC is a yellow-orange powder with a decomposition temperature range of 205 to 215 °C. It is one of the most efficient blowing agents, generating 220 cc of gas per gram of product. The gas is mainly nitrogen with lesser amounts of carbon monoxide, carbon dioxide, and ammonia.

Upon complete decomposition, its residues are white to off-white in color. The reactions involved in the thermal decomposition of ADC are complex. More detailed information on the chemistry and decomposition mechanisms of ADC and other blowing agents discussed in this chapter can be found in Reference [3]. ADC produces odorless, non-staining, low density expanded products of extremely fine and uniform cell structure.

21.4.1.2   Activation

The decomposition temperature of ADC is clearly outside the processing window for rubber. Fortunately, this temperature can be lowered by compounds known as blow activators. Many chemicals usually employed as compounding ingredients for non-cellular rubber compounds also function as activators for ADC. Zinc oxide and salts of zinc are especially effective. Effective activation can lower the decomposition temperature to 140 °C. As a result a number of preactivated ADC blends have been developed for applications where the process temperature falls between 140 and 180 °C [18].

21.4.1.3   Factors Affecting Performance

The rate of decomposition of ADC is a function of temperature, activator type and amount used, and the particle size of the grade of ADC decomposed. A number of activators can be used with ADC to lower the decomposition temperature, including zinc oxide, zinc stearate, zinc octoate, organic metal complexes, glycols, and surface treated urea, as well as vinyl stabilizers, such as those containing zinc or barium. Adding such activators to rubber formulations is a practical method of altering the decomposition characteristics of ADC to meet special requirements.

Glycols and surface activated urea have a pronounced activating effect on the curing rate of natural and synthetic rubber. Therefore, the amount of acceleration must be reduced to compensate for their activity and to maintain the proper cure-blow balance. The effect

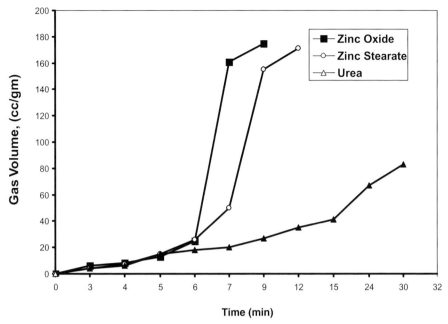

Figure 21.1   Gas evolution at 170 °C.

of activators on the rate of decomposition of ADC may be shown by gas evolution data obtained at various concentrations and temperatures. Because it is the compound additives, and generally not the polymer itself, which alter the decomposition mechanism, tests can be run in the absence of a polymer matrix. The rate of decomposition of a blowing agent system can be monitored by simply capturing and measuring the gas evolved. The gas evolution data is measured by heating a sample in an inert dispersant (e.g., dioctylphthalate or mineral oil) in a flask and collecting the non-condensable gas. The data is then corrected to standard conditions. The gas evolution curves in Figs. 21.1 and 21.2 demonstrate the effectiveness of zinc activation on ADC at 170 °C and 185 °C [7].

An activator classification system (Table 21.2) has been suggested in previous work by D. G. Rowland [8]. Based on their activating effect, both zinc oxide and zinc stearate are classified as ''very strong'' activators, while urea is classified as a ''strong'' activator. Zinc compounds are not unique in their ability to modify the decomposition kinetics of ADC. Table 21.3 characterizes the relative effectiveness of various materials as activators of ADC.

21.4.1.4   Effect of Particle Size

Besides activation, particle size is the most significant factor determining the rate of ADC decomposition. Surface area increases very rapidly as the particle diameter decreases. Heat transfer to the particle is more effective. Therefore, the finer the particle size, the faster the rate of decomposition. The efficiency of the activator is also increased with finer particle size. Because activation involves a chemical reaction between the activator and the particle

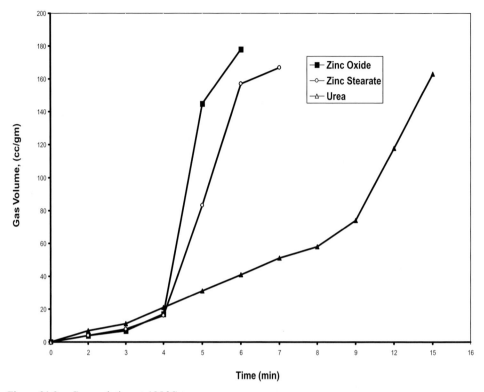

Figure 21.2   Gas evolution at 185 °C.

surface of the ADC, the rate of decomposition depends on the ADC particle size range. Therefore, the decomposition temperature range is controlled by the selection of activator and rate of decomposition, which is largely a function of the ADC grade. The rate of decomposition also has some effect on the apparent decomposition temperature. Therefore, strict control of the particle size distribution of ADC is very important to maintain uniform expansion under production conditions.

Particle size is selected to provide the proper balance between the development of cure and the decomposition of the blowing agent. The average particle size of commercially available ADC typically ranges from as small as 2 microns to as large as 15 microns.

Table 21.2   Activation Classification System (10 Parts of Activator per 100 Parts of ADC)

Relative effectiveness	Gas evolution (cc/gm)	Temperature (°C)
Very Strong	>150	in 15 minutes @ 170
Strong	>150	in 15 minutes @ 185
Moderate	>50	in 15 minutes @ 185
Weak	<50	in 15 minutes @ 185
	>100	in 30 minutes @ 185

Table 21.3   Decomposition Activators

Activators	ADC	OBSH
Zinc oxide, zinc stearate	Very strong	Weak
Urea	Strong	Very strong
Triethanol amine	Moderate	Very strong
Diphenyl guanadine	Weak	Strong
Calcium oxide	Weak	Weak
Benzoic/Citric/Salicylic acids	Weak	Weak
Barium/Calcium stearates	Moderate	Moderate
Stearic acid	Weak	Moderate
Polyethylene glycol	Moderate	Weak

21.4.1.5   Effect of Temperature

The decomposition rate of ADC varies directly with temperature. Lower temperatures require longer times for complete decomposition.

21.4.1.6   ADC Activation and Cell Size

Cell structure is the result of the interaction between the rate of development of the cure modulus with the evolution of gas. Thus, for a given cure system, a specific particle size is optimum. A faster decomposition rate requires a somewhat faster cure rate to capture the gas as it is released. Therefore, a change from a 3 micron to a smaller 2 micron particle size requires a somewhat faster cure rate. Conversely, a change to a larger 5 micron particle size necessitates a substantial reduction in the cure rate. In the same manner, given a set particle size, ADC activated with urea requires a faster cure system than unactivated ADC. The activated forms of ADC require a faster cure rate to encapsulate the gas and to prevent blow-through or cell blowout. Faster cures tend to give finer cell structures. When the cure rate is balanced with the blow rate, a very fine cell structure is produced.

Table 21.4   Particle Size Grades of ADC (microns)

2	Finest particle size grade, fastest rate of decomposition with activation, used for low temperature precures (132 to 143 °C) and expansion at 157 to 166 °C.
3	Choice for most closed cell sheets because it provides rapid, controlled gas development.
5	Used when the cure rate of the polymer system is too slow for the decomposition rate of the 3 micron size. Useful in extruded EPDM closed cell profiles.
8	Used in press molded and continuous sheet applications where slow cure rates are desired.
10–15	Used in automotive, extruded, closed cell profiles based on EPDM, EPDM – Neoprene, or SBR blends where high temperature precure (172 to 182 °C) and expansion temperatures (204 to 220 °C) are desired for very fast production rates.

21.4.2   Sulfonyl Hydrazides

The sulfonyl hydrazide class of blowing agents was developed independently in Germany by Farbenfabriken Bayer A. G. and in the U.S. by Naugatuck Chemical Company [2,3]. Of the sulfonyl hydrazides, p,p'-oxybis(benzenesulfonylhydrazide) (OBSH) is the most significant and by far the most popular. OBSH was discovered by Dr. Loren Schoene of Naugatuck Chemical Company (now Uniroyal Chemical Company, Inc.) in 1951. U.S. Patent 2,552,065 was awarded to Dr. Schone at that time [9].

21.4.2.1   Properties

Physically, OBSH is a white powder with a decomposition temperature of 160 °C, yielding 125 cc/gm of nitrogen gas and a small amount of water vapor. It is an efficient, non-discoloring, non-staining, and odorless nitrogen type blowing agent [10]. OBSH disperses easily in all elastomers on an open mill. An internal mixer can be used, provided the batch temperature does not exceed 110 °C. The effect of particle size on the decomposition of OBSH is not as critical as with ADC systems because of its lower decomposition temperature and much faster rate of decomposition.

21.4.2.2   Activation

OBSH decomposition is a function of temperature and time and may start as low as 125 °C with appropriate activation. The temperature range for the most rapid and efficient gas evolution is from 143 to 160 °C. This range is sufficiently low to allow the material to be milled directly into the rubber. OBSH is generally used in rubber and plastic in the temperature range of 135 to 170 °C. The decomposition point may be lowered by the addition of activators such as urea and triethanolamine (TEA). Both materials are cure activators as well as blow activators for lowering the decomposition temperature for OBSH, and should be recognized for their bifunctional nature. The effect of activators on OBSH is shown in Table 21.3 [7].

21.4.2.3   Applications

OBSH is used in both molded and continuous vulcanization of expanded elastomers such as EPDM, CR, SBR, NBR/PVC, and thermoplastics, notably PVC and LDPE. The most important applications for OBSH include pipe insulation, athletic padding, molded gaskets, and flotation products. Typical OBSH levels range from 1 to 15 phr, depending on the application. Several representative compounding formulations utilizing OBSH as a blowing agent can be found in Reference [7].

21.4.3   Dinitrosopentamethylenetetramine (DNPT)

DNPT was first used in 1947 with the introduction of Vulcacel BN by ICI [2], and was for a long time the most economical blowing agent available. DNPT is a nitrosoamine capable of splitting off the nitroso group (-NO) and creating a toxic nitrosamine as well as giving a ''fishy'' amine type odor. For this reason, DNPT is slowly disappearing from the marketplace in industrialized countries and being replaced with activated ADC, such as Celogen 754A and Celogen 765A (registered trademarks of Uniroyal Chemical, Inc.). DNPT is a flammable solid easily ignited by an open flame and sensitive to impact and friction. Because of the high energy content of DNPT, this blowing agent can only be handled without danger in a diluted form. It is commercially available as an 80% active dispersion to help conform to international shipping regulations. The dilution of DNPT facilitates handling and reduces the fire hazard associated with the active material. Typical applications include microcellular shoe soles and other molded and extruded products where its characteristic odor is not objectionable. DNPT has been used as a blowing agent in the rubber industry but is of limited use in plastics because of its high decomposition exotherm.

21.4.3.1 Properties

DNPT is a fine light yellow powder with a decomposition temperature range of 190 to 200 °C. Gas yield of the 80% active material is 190 cc/gm. The gas is essentially nitrogen with small amounts of other gases such as CO_2 and NH_3 [3]. DNPT is nondiscoloring, nonstaining, and produces a fine cell structure in both open and closed cell products. At curing temperatures of 130 °C and higher, DNPT demonstrates cure activation in natural and synthetic rubber. Therefore, some adjustments in curative concentrations may be required in compounds containing DNPT.

21.4.3.2 Activation

The decomposition of pure DNPT occurs at a temperature somewhat higher than the normal curing temperature for rubber. However, many rubber compounding ingredients, such as acid retarders, basic secondary accelerators, cure activators, as well as fatty and rosin acids present in natural and synthetic rubber, can lower the decomposition temperature and increase the rate of gas evolution.

Acid materials, such as phthalic anhydride, provide excellent activation for DNPT at vulcanization temperatures. However, the by-products of the decomposition of DNPT produce certain amines in addition to nitrogen, which have an unpleasant fish-like odor undesirable in some products. Using a basic material to activate DNPT can considerably reduce the odor [11].

Some basic activator systems, such as urea, are effective. Crystalline urea is the most economical of the various forms available. However, in this form, urea is very difficult to disperse because of its high melting point. For compounds that must be mixed at lower temperatures, fine ground treated forms of urea, such as BIK-OT (Uniroyal Chemical), are most effective. Urea does however, develop a slight ammonia odor that may be objectionable. If the compound is mill mixed, urea should be added prior to fillers and softeners for better dispersion. In an internal mixer, urea may be added with other dry ingredients. In products where odor must be reduced to an absolute minimum, activated ADC and OBSH may be used.

21.5   Methods of Expansion

Cellular rubber products are manufactured in simple and complex shapes by molding or extrusion. Manufacturing methods in commercial use include low pressure molding, high pressure molding, and continuous vulcanization.

21.5.1   Low Pressure Molding Process

In the low pressure process, mold cavities are partially filled with uncured compound containing sufficient blowing agent to allow for the expansion to fill the cavity and produce the final shape. It is necessary to vent the mold cavity of air displaced by the expanding rubber. A fabric sheet and/or dusting agent such as talc help remove trapped air. A low viscosity compound is essential for good mold flow and ease of expansion. A product containing open or closed cells can be produced by this process, depending on the type and amount of blowing agent and cure rate of the compound.

21.5.2   High Pressure Molding Process

The high pressure molding process is different in that it employs a two-stage cure: the precure stage and the final cure stage.

21.5.2.1 Precure Stage

At this point, an excess amount of compound is loaded into the mold cavity as in conventional solid rubber molding. As the press is closed, the compound completely fills the mold, expelling air and sealing the cavity. A typical compound is relatively low in viscosity, flows easily in the mold, coalesces, and eliminates trapped air. During the cure, the blowing agent decomposes, releasing nitrogen gas which form the cells as the pressure in the mold increases.

The press is opened before the rubber is completely cured. The high internal gas pressure in the partially cured rubber causes rapid expansion of the molded part when the press is opened, producing a fine closed cell structure in the molded part or sheet. The extent of expansion is controlled through compound design and regulation of the cure time and temperature. The molded part or sheet is expanded to its final size and the cure is completed in a second stage.

Care must be taken to ensure that molds open fully and quickly with no obstructions. Mold cavities should have beveled edges (15°) to assist ejection and the press platen size should be large enough to allow full sideways expansion without obstruction from the tie bars or other parts of the press structure when the press is opened. If expansion is restricted, the part usually distorts itself irreversibly or splits.

21.5.2.2 Final Cure Stage

A second curing operation known as a postcure is normally used to complete the cure and expansion and to stabilize the dimensions of the precured part or sheet. A typical precure time for a sheet from a 2.5 cm thick mold is 25 to 35 minutes at 135 to 145 °C. The sheet is typically expanded to its full size in an oven or larger mold at a temperature 15 to 25 °C above the precure temperature. This final cure step generally takes 60 to 90 minutes to complete, depending on thickness. Molded, closed cell sheet materials prepared by this method are usually skived into sheets of prescribed thickness, which are then die cut for gaskets used in air conditioners, heaters, closures, seals, etc. Cellular shoe soling is made in a similar manner, except that the sheets are cured and expanded to full size in the first stage. After molding, the sheets are normalized for 10 to 24 hours at 100 to 121 °C to stabilize dimensions. If stabilization is not done, excessive shrinkage in service results. A typical curing cycle is generally 15 to 20 minutes at 155 to 165 °C.

21.5.3   Continuous Vulcanization (CV)

The manufacture of cellular rubber by CV systems is one of the most economical and efficient processes in the rubber industry. Extruded open and closed cell profiles and calendered sheets are continuously vulcanized in any one of several heat transfer media such as hot air ovens, liquid curing medium (LCM), microwave ovens, or fluidized beds. All of these processes require curing at atmospheric pressure and high temperatures. The faster decomposition rate of the blowing agent at the higher temperatures associated with CV curing systems requires a faster cure rate to adequately contain the gas as it is released. This requires selection of a fast curing polymer base and a faster than normal vulcanization system to ensure containment of the gas during expansion.

Typical applications for closed cell profiles include automotive weather strips, architectural gaskets, and pipe insulation, among others. Elastomers most commonly used in these applications include EPDM, CR, SBR, and their blends. NBR/PVC blends are used extensively in pipe insulation and sheet because of flame resistance requirements.

References

1. *Rubber Handbook for Cellular Rubber Products*, Rubber Manufacturers Association, Washington, D.C., Fifth Edition, 1992, pp. 41–45.
2. Lasman, H. R. *Encyclopedia of Polymer Science and Technology*, Vol. 2, (1965) John Wiley & Sons, New York, pp. 532–565.
3. *Chemical Blowing Agents Chemistry and Decomposition Mechanisms, Bulletin ASP-4455*, Uniroyal Chemical Company, Inc., Middlebury, CT, Sept. 1977.
4. Curtis, N. B. and Hunter, B. A., U.S. Patent 2 806 073, September 10, 1957, (Uniroyal Chemical Company).
5. Rowland D. G., *Expanded Rubber Through the Years*, Paper No. 37, Presented at the Spring, 1992 Meeting of the Rubber Div. ACS.

6. Newby, T. H. and Allen, J. M., US Patent 2 692 281, Oct. 19, 1954 (Uniroyal Chemical Company).
7. *Celogen Chemical Blowing Agents Methods of Expanding Rubber, Bulletin VO0590*, Uniroyal Chemical Company, Inc., Middlebury, CT, Oct. 1993.
8. Rowland, D. G., *Activators of Chemical Blowing Agents,* Paper No. 12, presented at the Spring, 1987 Meeting of the Rubber Div. ACS.
9. Dr. Loren Schoene, U.S. Patent 2 552 065, May 8, 1951, (Uniroyal Chemical Company).
10. *Celogen OT, Bulletin ASP4335B*, Uniroyal Chemical Company, Inc., Middlebury, CT, Oct. 1991.
11. *Unicel ND and NDX, Blowing Agents for Elastomeric Sponge Products, Chemical Bulletin No.6,* E.I. DuPont DE Nemours and Co. Inc., Wilmington, DE, Mar. 1962.
12. Hofmann, W., *Rubber Technology Handbook*, (1989) Hanser Publishers, Munich, pp. 308–311.
13. *Unicel S, A Blowing Agent for Open Cell Sponge, Chemical Bulletin No. 33,* DuPont DE Nemours and Co. Inc., Wilmington, DE, Mar. 1962.
14. Barnhart, R. R., *Rubber Compounding (reprint from) Kirk-Othmer: Encyclopedia of Chemical Technology*, Vol. 20, (1982) John Wiley & Sons, New York, pp. 415–418.
15. *The R. T. Vanderbilt Rubber Handbook*, R. T. Vanderbilt Company, Inc., Norwalk CT, Thirteenth Edition, 1990, pp. 678–680.
16. *Technology of Celogen Blowing Agents, Bulletin ASR 3917R*, Uniroyal Chemical Company, Inc., Middlebury, CT, Aug. 1992.
17. *Celogen AZ, A Nitrogen Blowing Agent for Sponge Rubber, Bulletin ASP4909R*, Uniroyal Chemical Company, Inc., Middlebury, CT, Mar. 1996.
18. *Celogen 754A, A Low Temperature Nitrogen Blowing Agent for Rubber and Plastics, Bulletin ASP7092*, Uniroyal Chemical Company, Inc., Middlebury, CT, Feb. 1997.

22   Flame Retardants

Kelvin K. Shen and David R. Schultz

22.1   Introduction

Hydrocarbon elastomers, like all other organic substances, are highly flammable. As a result, in some applications, special attention must be paid to fire resistance. This chapter covers the practical aspects of preparing flame retardant elastomeric compounds. Major applications of flame retardant elastomers, common fire tests, the flame retardants used, and most importantly, the effects of flame retardant additives on elastomer processing and physical properties are reviewed.

The use of chlorine or bromine halogen sources to impart flame retardancy is well known in the elastomer industry [1–4]. In the additive approach, a halogen source, such as chlorinated paraffin or decabromodiphenyl oxide, is added at the compounding stage. In the reactive approach, halogen is introduced to a polymer backbone, such as in polychloroprene(CR), chlorinated polyethylene(CPE), or chlorosulfonated polyethylene(CSM). To enhance flame retardancy, antimony oxide is frequently used as a synergistic additive. This halogen/antimony oxide combination functions as a flame retardant by inhibiting chain-branching free radical reactions in the gas phase effectively. However, it suffers from the disadvantage of generating large amounts of smoke during combustion.

In recent years, considerable effort has been expended to find a partial or complete substitute for antimony oxide. This effort has been spurred by the desire to reduce smoke and improve cost performance, as well as concern about the toxicity of antimony oxide. The major substitute used is zinc borate [5–8].

More recently, using halogen-containing polymers under confined conditions, such as in wire and cables in subways, has been found to be inadequate, because the HCl or HBr generated from polymer combustion is highly corrosive and can damage expensive electronic equipment. Thus, halogen-free flame retardant polymers are now also in demand [9].

It should be emphasized that there are no perfectly flame retardant elastomeric compounds. Elastomers treated with flame retardants can still burn if they are exposed to enough heat and oxygen for a long enough time.

22.2   Fire Standards, Testing, and Applications

Small scale fire tests measure different stages of fire behavior, and involve:

Table 22.1   Common Flammability Standards and Test Methods

Test method	Purpose	Description
Oxygen index (ASTM D 2863)	General measurement of flammability	Measures the minimum level (%) of oxygen necessary to sustain combustion in a nitrogen-oxygen controlled atmosphere
UL 94 (vertical)	Test method for flammability classification of small appliances and electronic devices	A vertically mounted sample is subjected to two successive 10-sec. ignitions from a 3/4-inch Bunsen gas burner flame. The material classification (V-O, V-1, V-2) is based on afterflame time, afterglow time, and if any burning drips ignite the cotton
ASTM E 662 NBS smoke chamber test	National Institute of Standards & Technology test for optical smoke density	The optical density of smoke produced is measured from a vertical 7.6×7.6 cm sample exposed to a radiant furnace emitting 2.5 watts/cm^2 of heat flux

- Ease of ignition
- Flame spread
- Rate of heat release/flashover
- Smoke development
- Toxicity of gas
- Afterglow
- Corrosivity

Some of the most commonly used, small-scale, fire test methods are listed in Table 22.1. It should be noted that new fire test methods are continuously being developed and promulgated. In recent years, the bench scale Cone Calorimeter (ASTM E1354) for studying the rate of heat release from a polymer has become a powerful research tool. The rate of heat release is an important parameter for characterizing the contribution of a polymer to fire growth. This chapter however, covers mainly the test methods mentioned in Table 22.1.

Because all fire tests are conducted under controlled conditions, they are only valid as measures of the performances of the test materials under those specific conditions. Accordingly, they cannot be considered as accurate indicators of the hazard of the tested materials under all actual fire conditions. All flame retardant formulations mentioned in this chapter are

Table 22.2   Typical Flame Retardant Elastomeric Compounds

Applications	Polymer base
Wire & Cable	EPDM, polychloroprene, chlorinated polyethylene, ethylene-propylene rubber, ethylene-vinyl acetate, chlorosulfonated polyethylene, silicone, urethane
Conveyor belting	SBR, polychloroprene
Flooring	SBR
Roofing membranes	EPDM, chlorosulfonated polyethylene
Foam insulation	EPDM
Tubing	NBR
Coated fabrics	Polychloroprene, urethane

presented as starting-point guidelines. They must be tested and adjusted to achieve specific end use requirements and meet fire standards.

Major applications of flame retardant elastomeric compounds are listed in Table 22.2.

22.3   Commonly Used Flame Retardants in Elastomers

22.3.1   Aliphatic and Alicyclic Halogen Sources

Chlorinated paraffin, an aliphatic material, is available with chlorine contents from 40 to 70% and with viscosities from 2 to 1500 poise at 25 °C. Liquid chlorinated paraffin is also used to replace or reduce flammable processing oils. It is considered a low cost flame retardant that can also sometimes function as a plasticizer. Solid chlorinated paraffin (70% Cl) is also commercially available. It not only provides flame retardancy, but also can improve the tensile and tear properties of polychloroprene, styrene-butadiene rubber and nitrile rubber. Usually it is used in conjunction with antimony oxide, zinc borate, and alumina trihydrate.

Aliphatic material is less thermally stable than aromatic material, but it is adequate for most elastomer processing. The chlorinated alicyclic material, Dechlorane Plus*, however, is stable up to about 290 °C.

22.3.2   Aromatic Halogen Sources

Bromine sources are generally more effective than chlorine sources as flame retardants. Decabromodiphenyl oxide (DBDPO), which contains 83% bromine by weight, has been the most commonly used aromatic halogen source in elastomers. It is a white powder and melts at around 300 °C. Because it is very thermally stable, it can be used for applications in which resistance to high temperatures is required. Normally one part of antimony oxide (by weight) is used in conjunction with every three parts of brominated flame retardant to achieve the best results. The use of decabromdiphenylethane to replace DBDPO has become increasingly popular, mainly because of concerns about toxic polybrominated dibenzofuran and dioxin formation derived from DBDPO during polymer combustion. When good UV stability is needed, ethylene bis-tetrabromophthalimide (Saytex BT-93†) containing 67% bromine should be considered. This material also has less tendency to bloom than DBDPO.

22.3.3   Synergists of Halogen Sources

22.3.3.1   Antimony Oxide

Antimony oxide is the most common synergist with halogen sources. It functions predominately as a gas phase flame retardant. The disadvantage of using antimony

---

* Dechlorane Plus is a registered trademark of Oxychem.
† Saytex is a registered trademark of Albemarle.

oxide is that it tends to promote smoke formation during polymer combustion. In recent years, the toxicity of antimony oxide has also been a subject of some concern, particularly in Europe.

22.3.3.2 Zinc Borate

Zinc borate is commonly referred to as a class of compounds with different ratios of $ZnO:B_2O_3:H_2O$. Among the different zinc borates, the one with the molecular formula of $2ZnO·3B_2O_3·3.5H_2O$ is the most commonly used synergist in elastomers. This zinc borate is a multifunctional flame retardant. It is stable up to 290 °C and can function as a flame retardant, smoke suppressant, and afterglow suppressant. In contrast to antimony oxide, zinc borate works as a flame retardant predominantly in the condensed (solid) phase (i.e., promoting char formation and stabilizing the char). Depending on the halogen source and fire standard, zinc borate can either partially or completely replace antimony oxide. In some systems, zinc borate displays a synergistic effect with antimony oxide in fire test performances.

22.3.3.3 Phosphorus Compounds

Organophosphate esters, such as triaryl phosphates, provide not only flame retardancy, but also plasticizing effect to a compound. They have been used to replace flammable phthalate esters and paraffinic oil to improve flame retardancy mostly as one-for-one partial replacements. Phosphate esters can function as flame retardants in both the gas phase and condensed phase [10]. Inorganic ammonium polyphosphate has also been used in halogen-free elastomers.

22.3.4   Flame Retardant Fillers

22.3.4.1 Alumina Trihydrate (ATH)

ATH functions as a flame retardant and smoke suppressant by releasing 34.6% by weight of water starting at around 210–220 °C. It is generally not used in polymers with processing temperatures exceeding 180 °C. In halogen-containing elastomers, it is used in conjunction with antimony oxide and/or zinc borate. In halogen-free elastomers, ATH loading must range between 80 and 250 phr to be effective.

22.3.4.2 Magnesium Hydroxide

This material, like ATH, can function both as a flame retardant and smoke suppressant. It can release 31.0% by weight of water starting at around 310 °C. It is generally more expensive than ATH; however, it is also generally more effective than alumina trihydrate on a per weight basis in fire test performances. In halogen-free systems, high loadings of magnesium hydroxide (100 to 250 phr) again are generally required to be effective.

22.3.4.3 Calcium Carbonate

This material is a low-cost filler, which may exhibit slight flame retardant effects at high loadings. However, it can also reduce the effectiveness of halogenated flame retardants by scavenging hydrogen halide. Thus, using calcium carbonate may necessitate the adjustment of the loadings of other flame retardants in the formulation. Depending on the fire standard, the use of calcium carbonate should be avoided in halogen-containing elastomers.

22.3.4.4 Clay, Talc, and Silica

These mineral fillers are commonly incorporated into elastomers for cost reduction and/or reinforcement. They generally display low levels of flame retardancy by displacing combustible elastomers.

22.3.4.5 Carbon Black

Carbon black is commonly incorporated into elastomers for reinforcement. Carbon blacks with high surface areas however, tend to prolong afterglow combustion. This can normally be offset by adding zinc borate or phosphate.

22.4   Compounding and Dispersion Considerations

Depending on the elastomers used and fire standards to be met, flame retardant loadings generally range between 10 and 250 phr. Most of the flame retardant additives are powders that do not melt during elastomer processing. To develop optimal physical properties and flame retardancy in the base polymer, it is essential to have a homogeneous dispersion of the additives in the polymer matrix. The key factors which determine dispersion efficiency are the quality of additive (particle size, morphology, and surface treatment), mixing procedures, processing equipment, and composition of the formulation [11]. The elastomer should be of sufficient viscosity at processing temperatures to ensure adequate shear input to achieve good deagglomeration and uniform dispersion. Thus, during the preparation of a batch in an internal mixer, the flame retardant should be charged early in the procedure, ahead of such ingredients as plasticizers and processing aids that can soften the compound.

It should be noted that plasticizers can affect how much flame retardant is needed. For example, the flame retardants required would be higher with paraffinic oil than with phosphate esters or chlorinated paraffin as plasticizers.

In halogen-free flame retardant elastomers, high loadings of alumina trihydrate or magnesium hydroxide are normally treated with a stearate, silane, or titanate coupling agent to improve the dispersion. The fillers can be either pretreated with coupling agents or treated *in situ*.

The following paragraphs give a brief description of each of the common flame retardant elastomeric systems. The discussion starts with elastomers containing halogens in the polymer backbones.

22.4.1   Polychloroprene (CR)

Polychloroprene (neoprene) inherently contains about 40% chlorine in its backbone. As a result, it does not burn as readily as hydrocarbon elastomers. It has been the polymer of choice for applications demanding high levels of flame retardancy such as in conveyor belting, arm rests in mass transit vehicles, and wire and cables. In these applications, a combination of antimony oxide (5 to 15 phr), zinc borate (5 to 15 phr), and ATH (15 to 40 phr) is normally recommended. A typical "flame retardant" polychloroprene compound is described in Table 22.3 [12].

Antimony oxide is an effective synergist, but it tends to promote smoke formation. Zinc borate can also function as a synergist and promote char formation. In addition, zinc borate may be able to replace the zinc oxide commonly used for regulating the scorch and cure rate. Low molecular weight polychloroprene can exhibit a softening action, similar to that of petroleum plasticizer, and therefore, is one of the preferred flame retardant plasticizers. Other good plasticizers include phosphate esters and chlorinated paraffins.

In the mining belt application (Table 22.4), one of the major functions of zinc borate is to inhibit afterglow combustion [13]. To achieve low smoke, antimony oxide should be minimized and zinc borate and ATH maximized.

Inert fillers, such as clay, can raise the Oxygen Index by diluting the combustible moiety of polychloroprene. Conversely, calcium carbonate decreases the Oxygen Index by absorbing the released hydrogen chloride. It has been noted that barium sulfate, which has an inherently high specific gravity, should be avoided because of its effect on the cure rate.

Table 22.3   Typical "Flame Retardant" Polychloroprene Compound

Additive	Parts by weight
Polychloroprene	75.0
Polychloroprene (low ml.wt.)	25.0
Sb_2O_3	15.0
Zinc borate	10.0
Alumina trihydrate	30.0
Hard clay	20.0
Magnesium oxide	4.0
Carbon black	5.0
Ethylene thiourea	0.5
N,N'-diphenyl-p-phenylenediamine	2.0

Table 22.4   Typical Polychloroprene Mining Belt Compound

Components	Cover compound (parts by weight)	Skim compound (parts by weight)
Polychloroprene	100.0	100.0
Butadiene rubber	8.0	10.0
Chlorinated paraffin	15.0	30.0
Phosphate plasticizer	–	10.0
Sb_2O_3	5.0	10.0
Zinc borate	15.0	–
Alumina trihydrate	25.0	30.0
Carbon black	40.0	20.0
Soft clay	–	80.0
Silica	–	20.0
Antioxidant/curatives/processing aids	19.5	15.2
ML 1 + 4 (100 °C)	66.0	50.0
Vulcanizate properties		
Hardness (shore A)	68.0	66.0
Tensile strength (Mpa)	20.0	15.0
Elongation (%)	480.0	610.0
300% Modulus (Mpa)	10.0	6.0

22.4.2   Chlorinated Polyethylene (CM)

Chlorinated polyethylenes (CM/CPE) are based on high density polyethylene (HDPE) with chlorine contents ranging from 25 to 42%. In addition to different chlorine contents, CPEs can also have different crystallinities and molecular weights. Because of its chlorine content, CPE usually shows good results in small-scale combustibility tests in the laboratory. Other attractive properties of CPE include its chemical resistance, good low temperature properties, and its ability to accept high filler loadings. CPE elastomers typically are compounded with plasticizers and fillers, and crosslinked by thiadiazole, peroxide, or electron beam. The thiadiazole-cured system should not contain chlorinated plasticizers, as they tend to interfere with this type of cure system.

To increase the flame retardancy of CPE, one or a combination of the following approaches are possible [14]:

1. Use a CPE with higher chlorine content.
2. Add chlorinated or brominated flame retardant additives.
3. Use chlorinated paraffin or phosphate esters to replace a more flammable plasticizer.
4. Add a synergist, such as antimony oxide/ATH.
5. Add fillers, such as clays and talc, to improve ignition resistance, although higher loadings are required.

A combination of antimony oxide (up to 5 parts by weight), phosphate derivatives (up to 5 parts by weight), and ATH (20 to 50 parts by weight) is generally a good starting point for maximum flame test performance.

Table 22.5   Chlorosulfonated Polyethylene Formulation

Components	Parts by weight
Chlorosulfonated polyethylene[a]	70.0
EPDM	30.0
Alicyclic chlorine compound[b]	6.0
Antimony oxide	6.0
Dibasic lead phthalate	30.0
Calcined clay	90.0
Triallyl cyanurate	1.0
Antioxidant	3.5
Vinyl silane	1.0
Peroxide curing agent	2.5
Performances	
Oxygen index (%)	30.0
UL-44 (VW-1 flame test)	Pass
Tensile strength (Mpa)	16.9
Elongation (%)	245.0

[a] Hypalon 45 (Hypalon is a registered trademark of DuPont Dow).
[b] Dechlorane Plus (Dechlorane is a registered trademark of Oxychem).

22.4.3   Chlorosulfonated Polyethylene (CSM)

This elastomer is produced by treating polyethylene with a mixture of sulfur dioxide and chlorine. A typical elastomer contains 25 to 42% chlorine and 1 to 3% sulfonyl chloride groups. It is usually cured with either sulfur or peroxide. To improve flame retardancy, antimony oxide and ATH (or clay) can be used. When necessary, a halogen-containing compound can be added. Flame retardant grades of rubber compounds are normally used for roofing membranes and wire and cable applications. Table 22.5 suggests a combination of antimony oxide, an alicyclic chlorine source, and clay in a CSM/EPDM blend to possibly meet the VW-1 requirement in wire and cable applications [15].

22.4.4   Ethylene-Propylene-Diene-Monomer (EPDM)

Flame retardant grade EPDM compounds are used extensively in the building industry, public transportation, and electrical and electronic applications. In wire and cable applications, EPDM cables are more flexible than crosslinked polyethylene (XLPE) and show better dimensional stability over a broad temperature range. Because EPDM does not contain halogen in the polymer backbone, it is very combustible.

In the halogenated approach, a halogen source, such as chlorinated paraffin, a chloro-alicyclic compound, decabromodiphenyl oxide or ethylene bis(tetrabromophthalimide), and synergists, such as antimony oxide and zinc borate are commonly used. For example, decabromodiphenyl oxide, antimony oxide, and zinc borate in the presence of talc/clay filler reportedly can meet a UL roofing test [16]. Zinc borate is known to suppress the formation of

Table 22.6   Typical EPDM Formulation

Components[a]	Parts by weight
EPDM	100.0
Chlorinated paraffin (71.5% Cl)	50.0
Parrafinic oil	60.0
Sb_2O_3	8.0
Zinc borate	8.0
Talc	60.0
Carbon black	125.0
Zinc oxide	5.0
Stearic acid	1.0
Sulfur	0.5
Performances	
UL-94 (3.0 mm)	V-O
Tensile strength (Mpa)	959.0
Elongation (%)	540.0
Hardness, Shore A	72.0

[a] Plus 4,4′-dithiodimorpholine (0.7), tetramethylthiuram disulfide (1), methylzimate (0.3).

flying embers during roofing membrane combustion. As a starting point, Table 22.6 suggests chlorinated paraffin, antimony oxide, zinc borate, and talc in EPDM to possibly achieve V-O in samples 3.0 mm thick [17]. Table 22.7 suggests an alicyclic chlorine compound, DBDPO, and ethylene bis(tetrabromophthalimide) as halogen sources in EPDM. All three formulations reportedly have significant afterglow combustions in the UL 94 test (>60 s) [10]. Zinc borate and phosphate ester reduce the afterglow combustion.

Table 22.7   Typical EPDM Formulation

Components[a]	Parts by weight		
	1	2	3
EPDM	100.0	100.0	100.0
Process oil	15.0	15.0	15.0
Alicyclic chlorine compd.[b]	30.0	–	–
Decabromodiphenyl oxide	–	30.0	–
Ethylene bis-tetrabromophthalimide	–	–	30.0
Sb_2O_3	15.0	15.0	15.0
Carbon black	55.0	55.0	55.0
Performance			
Oxygen index (%)	26.9	27.6	27.3
UL-94 (burn times, s)[c]	0–0	0–0	0–0
Oven aged (70 h, at 121 °C) Tensile strength(N/mm^2)	12.5	14.7	16.5
Elongation (%)	310.0	340.0	400.0

[a] Plus peroxide (6), zinc oxide (5) and antioxidant (1.5).
[b] Dechlorane Plus.
[c] Burn times after 1st and 2nd ignition; all samples glowed longer than 60 s.

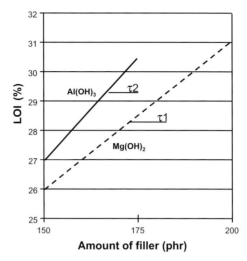

Figure 22.1   Effects of alumina trihydrate and magnesium hydroxide on the oxygen index.

One of the attributes of EPDM is that it can tolerate high filler loadings. In the halogen-free approach, high loadings (100 to 250 phr) of hydrated fillers, such as alumina trihydrate or magnesium hydroxide, impart some degree of flame retardancy. Figure 22.1 illustrates the effects of alumina trihydrate and magnesium hydroxide on the Oxygen Index (O.I.) of an amorphous EPDM. The O.I. increased with increasing filler loadings in a linear relationship between 150 to 200 phr filler. The relationship can be expressed in Eq. (22.1):

$$\text{O.I.} (\%) = \text{O.I. (initial filler loading)} + \tau[\text{increase in filler loading}] \qquad (22.1)$$

This study showed that the higher the value of the factor τ, the greater the effect of the flame retardant filler in the Oxygen Index test (Fig. 22.2) [18]. ATH is usually more effective than magnesium hydroxide in a vertical downward test, such as O.I.; magnesium

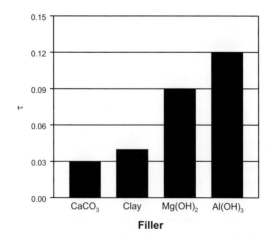

Figure 22.2   Comparison of factor values for cited experiment.

hydroxide is more effective than ATH in a vertical upward (higher enthalpy) fire test, such as UL 94. The combination of magnesium hydroxide (100 to 150 phr) and carbon black (30 to 40 phr) in halogen-free EPDM roofing membrane was reported to meet European standards [18].

22.4.5   Styrene-Butadiene (SBR)

SBR is typically characterized by its styrene content, which ranges from a low of 9.5% to a high of 40.0%, with a standard of approximately 23.5%. SBR has been used extensively in flooring, conveyor belting, and adhesives, as well as in wire and cables, because of its good strength, flexibility, and ability to tolerate high filler loadings. A combination of chlorinated paraffin, zinc borate, clay, ATH, and calcium carbonate can usually provide some degree of flame retardancy in flooring. Table 22.8 lists additives for a "flame retardant" grade conveyor belting [8]. In this system, a combination of chlorinated paraffin (15 to 20 phr) and DBDPO (15 to 20 phr) can also be used as the halogen source. Synergistic effects between the chlorine and bromine sources have been reported in other polymer systems [19].

With SBR latex, the stability of the additives under different pH conditions must be monitored. For example, zinc borate is not stable at a pH of 9.5 or higher.

22.4.6   Nitrile-Butadiene Rubber (NBR) and Hydrogenated-Nitrile-Butadiene Rubber (HNBR)

Because of its acrylonitrile content, NBR tends to be more resistant to a wide variety of liquids, solvents, petroleum oil, and hydrocarbon fuels. Acrylonitrile contents can range from 20 to 45%.

Table 22.8   SBR Conveyor Belting

Components	Parts by weight
SBR/Natural Rubber	100
Chlorinated paraffin (70% Chlorine)	20–25
Sb_2O_3	2–4
Zinc borate	6–10
Alumina trihydrate	35–60
Clay	35–50
Carbon black	40–50
Phosphate plasticizer	3–10
Other plasticizer	10–15
Antiozonant	2–5
Phenolic resin	2–4
Zinc oxide	3–5
Sulfur	0.5–2.5
Accelerator	1

Table 22.9   Low Density NBR/PVC Foam

Components	Parts by weight
NBR/PVC (50:50)	100
Diisodecycl phthalate	25
Chlorinated paraffin	20
2-Ethylhexyl diphenyl-phosphate	30
Sb_2O_3	7
Zinc borate	7
Alumina trihydrate	30
Super clay	30
Azodicarbonamide	20
Zinc oxide	5
Magnesium oxide	5
Other curatives	–
Properties	
Final density	0.08 g/ml

Blends of nitrile elastomer and PVC are known to combine the oil, chemical, and ozone resistance of PVC with the solvent resistance and elastic properties of nitrile elastomer. In addition, PVC can provide some degrees of flame retardancy to NBR. Nitrile/PVC blends with different flame retardants are used in applications such as fire hose, roll cover, electrical connector plugs, automotive components, and foam insulation. Table 22.9 includes antimony oxide, zinc borate, phosphate plasticizer, chlorinated paraffin, and ATH in a "flame retardant" NBR/PVC foam [20].

For improved resistance to liquids, gases, and heat, specialized nitrile elastomers can be selectively hydrogenated to form hydrogenated nitrile butadiene rubber (HNBR). The higher mechanical strength of HNBR versus that of NBR allows high loadings of flame retardant filler to be used without significant reductions in physical properties. Table 22.10 describes a halogen-free, highly filled, HNBR cable jacketing, which is claimed to meet the UK Naval Engineering Standard [21].

22.4.7   Silicone Elastomer

Silicone elastomers exhibit excellent oxidative stability, high degree of chemical inertness, resistance to weathering, good electrical properties, and low surface tension. They do ignite, but their combustion product is siliaceous, not carbonaceous, with no burning drips. Improvement of thermal stability and flame retardancy can be made by:

1. Replacing the methyl group in the polymer with phenyl or vinyl groups.
2. Using platinum or platinum-based compounds to promote crosslinking.
3. Adding zinc borate and/or alumina trihydrate.

In wire and cable application, zinc borate is known to promote the formation of highly insulative char in silicone elastomers [22].

Table 22.10   Halogen-Free Hydrogenated Nitrile Butadiene Rubber for Cable Jackets

Components[a]	Party by weight		NES[b]
	1	2	
HNBR (30% acrylonitrile)	100.0	–	
HNBR (40% acrylonitrile)	–	100.0	
ATH	190.0	190.0	
Zinc borate	10.0	10.0	
Peroxide	6.0	6.0	
Properties			
Oxygen index (%)	42.0	45.0	>29.0
Temperature index NES 715 (°C)	290.0	310.0	>250.0
Toxicity index NES 713	3.2	3.7	<5.0
Smoke density (D max., flaming)	170.0	120.0	
Corrosivity of smoke (pH)	8.3	8.5	

[a] Plus zinc stearate (1 part), antioxidant (1.9), dioctylsebacate (6), TRIM (0.7).
[b] U.K. Naval Engineering Standard specifications.

22.4.8   Ethylene-Vinyl Acetate (EVM)

Ethylene-vinyl acetate elastomers (EVM) are copolymers of ethylene and vinyl acetate with the proportion of the latter in the range of 40 to 80%. The saturated backbone of the polymer gives EVM high ozone and weather resistance. EVM also exhibits excellent thermal stability and good oil resistance. It can be treated with high levels of alumina trihydrate or magnesium hydroxide to meet most fire standards. Table 22.11 suggests the use of alumina trihydrate, zinc borate, and magnesium carbonate in a possible halogen-free wire and cable application [23]. Zinc borate and alumina trihydrate at high temperatures can form a highly insulative and foamy ceramic residue [9].

Table 22.11   Low Smoke Cable Material for Offshore Applications

Components	Parts by weight
Ethylene-vinyl acetate[a]	100.0
Alumina trihydrate	170–240
Zinc borate	10.0
Mg carbonate	20.0
Vinyl silane	2.0
Mineral oil	8.0
Zinc stearate	1.0
Curatives	10.5

[a] Vinyl acetate content 40 to 50%.

22.4.9   Ethylene-Propylene Elastomer (EPR)

EPR copolymer is normally cured with peroxide. It is used mostly for roofing membrane and wire and cable applications. In wire and cable, EPR provides excellent electrical properties. For example, an EPR insulation material consisting of EPR (100 parts), decabromodiphenylethane (20 phr), Sb_2O_3 (7 phr), and clay (25 to 35 phr) could provide an O.I. of 27% and V-O at 1.6 mm.

In the roofing membrane application, a combination of a bromine source (BT-93) and antimony oxide is suggested. BT-93 provides much better UV stability than other aromatic bromine sources.

22.4.10   Thermoplastic Elastomers (TPE)

The attributes of TPEs are discussed in Chapter 10. In wire and cable applications, TPE is usually used as a jacketing material and is normally flame retarded. The most common method of imparting flame retardancy is to use a combination of antimony oxide, a halogen source, and ATH (or clays).

Acknowledgments

Special thanks to Stephen Hughes of DSM Elastomers Americas for allowing the use of Figs. 22.1 and 22.2. Thanks Dave Edenburn of Albemarle, John Kardos of Dover Chemical Corporation, and Tom Hofer of Zeon Chemicals for providing valuable information. Last but not the least, thanks to Mr. John Dick of Alpha Technologies for his helpful comments.

References

1. Lyons, J. W., *The Chemistry and Uses of Fire Retardants* (1970) Wiley-Interscience, New York.
2. Fabris, H. J., Sommer, J. G., In Fire Retardancy of Polymeric Materials, Kuryla & Papa (Ed.) (1973) Marcel Dekker, Monticello, New York.
3. Lawson, D., In *Flammability of Elastomeric Materials- Handbook of Polymer Science & Technology*, Cheremisinoff, N. P. (Ed.)(1989) Marcel Dekker, Monticello, New York.
4. Troitzsch, J. H., *International Plastics Flammability Handbook* (1990) Hanser, Munich.
5. Shen, K. K., Griffin T. S., In *Fire and Polymers*. Nelson, G. (Ed.) (1990) ACS Symposium Series 425, pp. 1575.
6. Shen, K. K., O'Connor R., In *Plastics Additives*, Pritchard, G. (Ed.) (1998), Chapman & Hall, London.
7. Schultz, D. R., *Rubber World*, August 1992.

8. Shen, K. K., *Educational Symposium on Fire Retardants, Rubber Division of American Chemical Society Fall Technical Meeting*, Nashville, TN (1992).

9. Shen, K. K., *Plastics Compounding*, Nov./Dec. (1988).

10. Green, J., *Rubber and Plastics News*, Jan. (1993).

11. Dean, P. R., *Dispersions/Masterbatches for the Rubber & Plastics, Tire Technology International* (1993).

12. Murray, R. M., Thomson, D. C., *The Neoprenes*, DuPont Book, 1963.

13. Tsou, D., Rhode, E., *143rd Rubber Division of American Chemical Society Meeting*, Denver, Colorado (1993) paper #9.

14. DuPont-Dow Elastomers, *Tyrin Resins for Versatility, Function, and Performance-Service Bulletin* (1980).

15. Markezich, R., Rubber Division of American Chemical Society Meeting, Nashville, Tennessee (1992).

16. Wasitis, W. A., Hoff, J. L., Eur. Pat. Appl. EP 301,176 (1988).

17. Dover Chemical – *Chlorinated Paraffin Service Bulletin* (1998).

18. Krans, J. and Eichler H. J., *Flame Retardancy of EPDM Compounds*, DSM Technical Information Bulletin, February (1997).

19. Markezich, R., *Proceedings of Flame Retardance 98'*, p. 93, February (1998).

20. Schwarz, H. F., *Rubber & Plastics News*, Dec. (1985).

21. Meisenheimer, H., *Rubber World*, p. 19, June (1991).

22. Sharp, D. W., GB 2,140,325A (1984).

23. Haag, F., *Rubber Offshore Proc. Conference*, 244 (1984).

23   Rubber Mixing

W.J. Hacker

23.1   Introduction

The mixing operation is one of the most important stages through which raw materials must pass in manufacturing elastomers. The processing stages subsequent to the mixing depend on the dispersion and the homogeneity of the mixed compound. Also, the economical manufacturing of quality elastomer products is directly affected by mixing.

The primary objectives of mixing are to:

- Attain a uniform blend of all the constituents in the mix; each portion of the mixed compound is of uniform composition
- Attain an adequate dispersion of fillers
- Produce consecutive batches which are uniform both in degree of dispersion and viscosity

Selection of the materials to be used in a particular product depends on the technical requirements of the product, the properties attainable by formulating, and economic factors. Most of the examples shown in this chapter are based on the use of general purpose elastomers.

23.2   History

Early compound mixing took place on two-roll mills or in single-rotor machines, such as Hancock's Pickle, which reportedly was used in the early 1820s.

The first reference to double-rotor internal mixers appeared around 1865: the Quartz mill of Nathaniel Goodwin, but there appears to be some doubt whether this machine was suitable for mixing rubber because of its apparent lack of strength. The first machine which appeared suitable for mixing rubber was a twin-rotor design patented by Paul Pfleidener in the late 1870s, which resembles a Z-Blade mixer. Refinements of this mixer resulted in the Gummi-Kneter (GK) mixing machine patented in 1913.

Fernley H. Banbury, while working for the Werner and Pfleiderer Company in Saginaw, Michigan, refined an idea for a mixing rotor for rubbers because of disappointment in the performance of early GK mixers. Because the Saginaw plant did not support Banbury's development, he took out personal patents on his machine in 1915 and resigned. The Birmingham Iron Foundry, which later merged with the Farrel Company, produced the Banbury mixer still made by this same company today.

23.3   Equipment

23.3.1   Mills

Prior to the internal mixer, the mill was extensively used for compounding. But in the modern tire plant or custom mixer, the mill has been relegated in many cases to sheeting and cooling compounds or warming compounds for further manufacturing stages (although the former functions are being replaced by roller die extruders and the latter by pin extruders). As mixing equipment, mills are primarily used to prepare colored, sticky, or very hard compounds in rubber industrial products. The rubber fabricator has a variety of mills available, varying in size from the small laboratory mill to the large 305 cm mill, with the 107 cm, 152 cm, and 213 cm mills the most popular production sizes. The selection of mills for a particular batch is affected by production volume, batch size, and the care with which the batch is to be handled. Very sensitive compounds must be mixed in much smaller sizes and on a mill where the mill operator has more control.

Conventionally, mills have a friction ratio from 1:1.05 to 1:1.2 between the rolls to keep the compound on one roll. Early natural rubber compounds invariably stayed on the slower roll, which is tradionally regarded as the front roll. Unfortunately, some problems with modern compounds of polychloroprene or EPDM are regularly encountered with materials going onto the back roll, which can result in severe delays in production. In practice, it may be better to utilize rolls running at the same speed, with temperature control systems on each roll, to drive material onto the front roll. It is still better to have individual roll drives, with drilled rolls and temperature control systems, such that both temperature and friction ratio can be set to direct the rubber onto the correct roll.

23.3.2   Internal Mixers

Trying to emulate the mixing action of a mill gave rise to both the Banbury, Tangential rotor type, and the Intermix, Intermeshing rotor type, mixing action (Fig. 23.1). These two mixing actions are significantly different as a result of the different emphases each inventor applied to the actions taking place on the mill while mixing a batch of rubber. Milling procedures typically utilize the shearing action at the nip, the rolling of the batch into the roll nip, and the cutting and folding of the sheet on the roll by the operator. The action within the nip and the approach to the nip are simulated in the Tangential rotor by the rotor shape acting on the compound being squeezed and sheared against the mixer side. The cutting and folding are simulated by the transfer of compound from one rotor to the other, and the movement of compound along each rotor results from the shape of the rotor. In an Intermeshing process, emphasis is placed on the cutting and folding enforced by the impingement of the radius of rotation of one rotor with the other and the Archimedean screw shape of each rotor driving material in opposite directions. The nip of the mill and the friction ratio of the rolls are simulated by the nip between the rotors being offset from the centerline of the machine, causing a difference in the surface speeds of the top of the nogs and root of the rotor and hence, the simulation of the friction ratio.

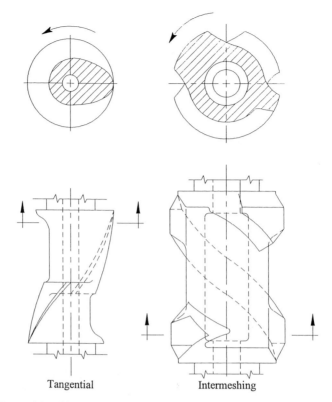

Tangential   Intermeshing

Figure 23.1   Tangential and intermeshing rotors (photo: Pomini).

23.3.2.1 Tangential Rotor Type

The Banbury mixer was initially developed by Fernley H. Banbury as the first truly successful batch mixer. Early machines were not produced with a ram and were expected to mix within the mixing chamber alone. It soon became apparent that the addition of a ram dramatically increased the efficiency of this mixer. The mixing principle relies on a tapering nip between the rotor and the sidewall of the mixer to give mix dispersion. It also relies on the transfer of material around the mixing chamber and from one rotor to the other to give mix distribution. Over the years, the Banbury was refined further by the addition of a hinged hopper door, the sight rod attached to the weight at the bottom of the ram to monitor batch weight, a means to measure the temperature, improved cooling of the machine, and the return to the drop door as a replacement of the sliding gate.

23.3.2.2 Intermeshing Rotor Type

The concept for the intermix was developed in the United Kingdom during the early 1930s. Construction and design of the machine were contracted to Francis Shaw &

Company of Manchester, U. K., who eventually acquired and patented the design. This approached the problem of mixing rubber differently. The emphasis was given to the transfer of material around the chamber by utilizing scrolls on what are basically mill rolls. Each rotor transfers material along its length and in the opposite direction from the other rotor. Transfer from rotor to rotor occurs because of the interlocking nature of the rotors. Mixing takes place initially in the nip between the two rotors.

23.3.2.3 Variable Internal Clearance Mixer

This machine was developed recently (1987) in Italy to address what some regard as problems in the designs of the intermeshing rotor mixer. On this machine, a device allows axial movement of the rotors to increase or decrease the gap between them during mixing. This is said to allow larger batches of rubber at the start of the cycle to feed more easily, and also adjustment of the rotors to an ideal mixing position for each type of rubber compound. This is analogous to the requirement for different nip settings during mill mixing (see Fig. 23.2).

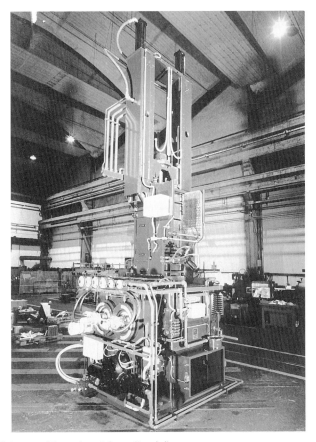

Figure 23.2   VIC intermeshing mixer (photo: Pomini).

23.3.2.4 Continuous Mixers

The continuous mixing of rubber compounds is still very much in its infancy. The EVK (Extruding, Venting, Kneading) machine made by Werner & Pfliederer was primarily used in the EPDM extrusion compound area using powdered polymer. The MVX (Mixing, Venting, Extruding) machine made by Farrel Bridge was used in cable compounding and in the production of tire compounds using granulated polymer. Successful tests were run with the MVX on compounds of the common rubbers, such as NR, SBR, IR, BR, CR, and EPDM, and also on specialty types, such as NBR and HNBR.

The latest work on continuous mixing of rubbers has centered on modified twin screw compounders. These machines have been used for many years in the plastics compounding industry. The problem to be overcome with all continuous compounding systems is to achieve good dispersion with adequate distribution of compound ingredients. All continuous mixers, by their very nature, mix only a small quantity of compound at any one time. Therefore, all ingredients used in a rubber compound must be present sometimes in very small amounts in this system. By contrast, in batch mixing, where batches are both extensively (distribution of materials over the entire volume) and intensively (subdivision of the materials, reduction of the size of the particles) mixed, the ingredients present in small amounts are adequately distributed in the mixer or on the mills which follow. Continuous mixers achieve only intensive mixing; extensive mixing must be done outside of the continuous mixer, either by blending all materials or by using a multiplicity of gravimetric feeders.

Most rubber compounds include between 7 and 20 ingredients. Some degree of premixing is essential to avoid the costs associated with using separate feeders for each material.

23.3.2.5 Extruders

The basic principle of a screw extruder is to continuously convert the rubber compound feed to a finished form or profile. The screw pulls the compound into a barrel and forces the rubber forward by its continuous rotating movement. As the compound moves forward, it is softened by the frictional heat developed through the shearing action of the screw. At the screw outlet, the compound has been transformed into a viscous state that can be forced through an orifice or die and formed into the desired profile. Developments which have taken place in rubber extruders over the years have moved the industry away from the original hot feed extruders that require the use of mills to blend and precondition or soften the compound prior to feeding the extruder to cold feed extruders.

By projecting pins into the barrel along the path of the screw, the flow of the compound is continuously divided and split, effectively mixing and homogenizing without using high shear rates. The pin barrel extruder is regarded as a great step forward when compared to the conventional method of hot feed extruders. The pin barrel extruder allows the compounder to run hard-to-process, high Mooney compounds. While a hot feed extruder receives the compound at near minimum viscosity, a cold feed extruder is generally fed the compound at room temperature. The screw must then transmit sufficient energy to soften the rubber to a minimum viscosity and provide a forward motion power to

Figure 23.3   MCTD cold-feed-extruder multi-cut transfer mix generation D (photo: A-Z Formen- und Maschinenbau GmbH).

overcome the back pressure caused by the orifice or die. Because the compound at ingestion on a pin barrel extruder is harder, a longer length barrel is needed to provide sufficient softening. Before the introduction of the Multi-Cut Transfer (MCT) extruder (the forerunner of the MCTD), other cold feed extruders had relied on friction for ''working'' the compound. The MCTD system introduced the concept of mixing by flow division and rearrangement of the rubber compound in the screw. The unique design of the MCTD extrusion system involves a double helix which causes the mixing of the compound to occur by a mechanical shear in the MCTD section of the screw and barrel. This results in minimum viscous heating which allows faster screw speeds (see Figs. 23.3 and 23.4).

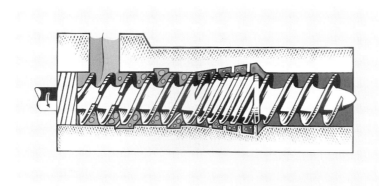

Figure 23.4   MCTD cold-feed-extruder screw (photo: A-Z Formen- und Maschinenbau GmbH).

23.4   Mixing

23.4.1   Mill Mixing

While dry fillers are worked into the batch, a continuous sheet must be maintained on the mill roll. For this reason, practically all the work is performed on the slow roll. Ingredients of the same softness blend more readily than do hard ones with soft ones. Thus, in preparing the "rubber portion" of the batch, the sequence of ingredient addition is important for satisfactory mixing. Generally, the object is to maintain the stiffness (not too much breakdown) of the elastomer until dispersion is achieved. Sometimes, this means the plasticizers and softeners are added later in the mixing cycle, after the fillers and powders. Fillers tend to work into the surface of the sheet on a mill. And, in highly loaded batches, it is necessary to stop adding fillers until the sheet is blended homogeneously. In batches containing high loadings of plasticizer, it is desirable to add the plasticizer concurrently with the fillers to prevent the stock from breaking up and falling off the mill roll. In batches where dispersion of high loadings of hard (small particle size) carbon black or other reinforcing agents is required, it is better to work in the dry fillers before adding the liquid plasticizers.

Typical mixing procedures for an open mill mix of a natural rubber compound are described as follow:

1. The mill is set to a 6.4 mm nip opening between the rolls with cooling water circulating through the rolls during mixing. The natural rubber is added to the mill, and after several passes, it sheets out sufficiently to cling to the front roll and to be fed back into the nip continuously. The rubber is then cut back and forth several times to allow the rubber in the rolling bank to pass through the nip and to facilitate blending the entire batch. Zinc oxide, antioxidants/antiozants, and stearic acid are then added. The rubber is then cut back and forth several times to distribute these materials throughout the batch. At this point, about eight minutes have elapsed.

2. The mill nip is opened slightly and carbon black is added slowly to the batch to prevent its excessive loading at the center. Strips of rubber are cut from the ends of the rolls several times during the time the black is added to the batch and are passed into the bank. When about 90% of the carbon black has been added, oil and remaining black are added slowly and alternately to the batch. When all the carbon black is in and no free carbon black is visible, the batch is cut back and forth several times. The carbon black and oil incorporation takes about ten minutes.

3. At this stage, the total batch volume has been appreciably increased by the carbon black and oil additions. As a result, the bank is larger now than at the start of the mixing. Again, the mill nip opening is increased until a rolling bank is obtained. The sulfur is added and worked into the batch, which takes about three minutes. To ensure thorough blending, the batch is cut back and forth each way six times. The finished batch is cut by the mill operator into sheets, which are then dipped in water containing an antitack agent and hung to dry and cool. This operation requires about eight minutes. The total time for mixing the batch is about 30 minutes.

23.4.2   Internal Mixer

23.4.2.1 Batch Size

When determining the batch size for a given compound, consideration must be given to the following:

- Type of mixer: tangential or intermeshing
- Rotor configuration: two-wing or four-wing
- Energy input: low energy or high energy
- Cooling system: flow or tempered water system
- Type of ram used: pneumatic or hydraulic
- Mixer drive system: fixed or variable
- Condition of the mixer rotors and chamber: new or worn
- Type of compound: masterbatch, remill, or final mix
- Compound properties: viscosity of the raw materials and the compound
- Density of the compound
- Mixing temperature

To establish the proper batch size for an internal mixer, a great deal of experimental work is usually required. Keeping in mind all of the above factors which can affect the volume of the batch that can be mixed, time can be saved by initially estimating the size of the batch (see Table 23.1).

Before the final decision of a optimal batch weight and resulting fill factor is made, trial runs establish optimum conditions and ensure that the stock is of a proper consistency and quality for processing in the subsequent operations.

The efficiency of an internal mixer is assumed to be optimum when the mixing time is the shortest for a given stock. It can be assumed the optimum batch size is obtained when an increase or decrease in that size increases the mixing time. The optimum batch weight can only be determined by trial and observation. When the weight ram remains at the bottom of its stroke during the mixing operation, indications are the batch is of insufficient volume to fill the mixing chamber. Depending on the mixer size, typically, the batch size is correct when the weight ram is rising and falling about 8 to 10 cm from the bottom of its stroke at the beginning of the mixing cycle and approximately 2 cm near the end of the cycle. However, ram positions can vary greatly depending on the type and size of internal mixer. Too large a batch causes some stock to remain in the mixer hopper neck, which results in poor and prolonged mixing. If the batch is the proper size, the ram pushes all the stock down into the

Table 23.1   Estimating the Size of a Batch

$$\text{NCV} \times \text{fill factor} \times \text{density of compound} = \text{batch weight (kg)}$$

$$\text{Example}: \quad 270 \times 0.75 \times 1.0 = 203\,\text{kg}$$

$$650 \times 0.75 \times 1.0 = 488\,\text{kg}$$

where

NCV is the net chamber volume, in liters, obtained from tables of mixer capacities; 0.75 is 75% of net chamber volume (fill factor)

Table 23.2   Calculations to Determine Optimum Throughput

Cycle time	Batches/hr	Batch Wt./kg	Kg/hr	Mix quality
3.0	20.0	216	4,320	Good
2.5	24.0	227	5,448	Good
2.0	30.0	239	7,170	Poor

mixing chamber within the specified time. As the chamber of the mixer wears over continuous operation, the batch weight must be increased to maintain adequate mixing. Subsequently, the batch weight must be reduced after a mixer is rebuilt.

23.4.2.1.1 Procedure to Determine Batch Size

The following is a procedure to determine an internal mixer volume and batch size. First take a compounded masterbatch with a density known to be accurate (verified). This batch should not consist of natural rubber only. Load the mixer with a batch estimated to be approximately the correct volume. The formula given in Table 23.1 can be used to make this estimate. Check the time that it takes the ram to seat after it is lowered. The batch is probably the correct size if this time is between 15 and 20 s for an F270 Banbury operating at 40 rpm. If the ram seats in less time, the batch may be too small. If more time is required, the batch may be too large. If the batch weight is not correct as determined in Table 23.1, try other batches by adjusting the weight up or down in 5 to 7 kg increments until the correct loading is obtained. When the correct weight is obtained, check several consecutive batches of the same weight to verify the size and cycle time. Mix ten batches, record the cycle times, and calculate the average cycle time. Divide 60 min by this value to obtain the number of batches that can be mixed in one hour. The number of batches per hour times the batch weight gives the kilograms of stock that can be mixed in one hour for that compound. This calculation is shown in Table 23.2. All batches must be tested for quality.

According to examples in Table 23.2, a batch weight of 227 kg maximizes the output of a quality product. The undersized batch (216 kg) produces a quality product, but requires a longer mix cycle. The 239 kg batch, while resulting in the maximum kg/hr throughput, does not constantly produce a quality product.

23.4.2.2   Batch Conversion Factor

This is the number or factor by which each individual ingredient weight in the 100 parts by weight of elastomer base is multiplied to obtain the weight in kilograms of each ingredient for a given mixer batch weight. The batch conversion factor is determined by dividing the batch weight by the total parts per hundred rubber (tphr) (see Table 23.3).

23.4.2.3   Density and Cost Calculations

Density is the weight per unit volume at a specified temperature. Details concerning this calculation are given in Chapter 4.

Detailed cost calculations and examples are given in Chapter 4.

Table 23.3   Calculation of Batch Conversion Factor and Weights of Individual Ingredients

$$\text{Batch conversion factor} = \frac{\text{Batch weight}}{\text{Total 100 parts by weight base}} = \frac{227.50}{175.00} = 1.30$$

Weight of individual ingredients = 100 parts by weight base × batch conversion factor

$$\text{Batch weight} = \sum \text{Weight of individual ingredients}$$

Ingredient	100 parts by weight base	Batch conversion factor		kg of each ingredient	% of each ingredient
Natural rubber	10.00	×1.30	=	13.00	5.7
SBR-1502	90.00	×1.30	=	117.00	51.4
N-234 black	50.00	×1.30	=	65.00	28.6
Hi Aro oil	20.00	×1.30	=	26.00	11.4
Zinc oxide	3.00	×1.30	=	3.90	1.7
6 PPD	2.00	×1.30	=	2.60	1.4

23.4.2.4   Mixing Procedures

Stock mixing is the first of the many operations required to make an elastomer product. It is therefore, very important that not only the specified ingredients and the correct weights be used, but that the specified mixing procedures be followed. The proportions of the batch must be correct when they enter the mixer or they will not be correct in the finished product. It may be possible, depending on the downstream process, to introduce some cross blending to "level out" minor inconsistencies. But the object must be from the start to avoid ever introducing such inconsistencies in the weighing and feed processes. Formulations, and the tolerances for variation in the target weight of the raw material ingredients in these formulations, must be established. It is important that these tolerances be realistic and tight enough to assure acceptable end-product quality. The order of adding ingredients is very important if satisfactory mixing is to be attained. Generally, the object is to maintain the "stiffness," not allowing too much breakdown of the elastomer until dispersion is attained. Normally, this means the plasticizers and softeners should be added later in the mixing cycle, after the fillers and powders have been incorporated. A normal sequence is as follows:

- Load elastomers and part of the filler
- Add remaining filler
- Add plasticizers and softeners

In some cases however, the batch may soften too quickly, before dispersion is complete. In such a case, it may be necessary to use the following sequence:

- Load all fillers
- Load rubber

This upside down mixing also prevents excessive filler from being extracted by the dust collector exhaust (see Section 23.5.9). Mixing procedures vary with the nature of the stock being handled. Each stock must be handled in accordance with its general composition and the desired result.

If a stock is too hard or stiff, one or more of the following steps can be taken:

- If the mixer has two speeds, use the higher speed
- Break the elastomer down more prior to using it or use a peptizer
- Vary the batch weight (with the appropriate ingredient weight changes)
- Vary the mixing time
- Vary the sequence and/or time of addition of the various ingredients while maintaining good dispersion of powders
- Add more softener or otherwise modify the formulation recipe while maintaining the desired physical properties

23.4.2.5 Mixing Temperatures

Whenever possible and contingent on quality requirements, it is desirable to establish a mixing cycle that allows the proper rate of temperature increase during mixing the masterbatch. During the masterbatch stage, the temperature rise must be controlled to allow sufficient time for the incorporation of all ingredients before the discharge temperature is reached. High structure carbon blacks, while improving dispersion also cause higher mixing and processing temperatures. These higher mixing temperatures can result in a reduction of the mix cycle because most batches are discharged when specific temperatures are reached. In severe cases, after the mixing of 10 to 20 batches, this rise in temperature can result in a discharged batch that is so undermixed it cannot be handled by the downstream processing equipment, resulting in the shutdown of the mixer.

In adjusting and maintaining stock mixing temperatures, the compounder has four variables to control. Actions that put more or less mechanical work into a mix increase or decrease the stock temperature, including:

- Increased or decreased batch size
- Increased or decreased ram pressure
- Increased or decreased rotor speed
- Increased or decreased amount or temperature of cooling water

Control of the batch temperature of final mixed compounds is very critical, and the time in an internal mixer should be minimized to reduce the ''heat history'' of the compound for increased scorch safety, control of bloom, and decreased scrap.

23.5   Mixing Methods

23.5.1   Natural Rubber Mastication

NR must be broken down one way or another. Prior to the use of technical grades, large bales of NR were conditioned in a hot room, then cut and masticated with a peptizer in a plasticator or in an internal mixer. With many grades of NR, this is still done. To limit heat generation during mastication as much as possible, the internal mixer is operated with full cooling. In addition, it is underfilled by about 10%. The operating conditions (speed, time) are chosen so that a rubber temperature of 150 to 160 °C (or even lower) is not exceeded, depending on the compound. The entire process lasts approximately 3 to 5 min., depending on the desired degree of mastication, which depends in turn on the amount of peptizing chemicals, the energy uptake (depending on the filling level and the rotor speed), and the rubber temperature. If NR has been masticated only to a small degree, it can accept higher filler loadings, and the vulcanizates also give better mechanical dynamic properties than those from NR that has been severely masticated. For these as well as economic reasons, it is desirable to keep mastication and mixing cycles as short as possible. The masticated rubber is then added to the mix in the masterbatch stage.

Now, however, with modern internal mixers and the technical grades of NR on the market, mastication of the NR is usually accomplished during the masterbatch mixing step itself. However, the reader should be cautioned that portions of the NR may remain unfilled, especially in cold weather. Remilling or final mixing a masterbatch containing lumps of unfilled NR result in their breaking into smaller unfilled NR particles in the compound. This condition may result in poor physical properties and/or premature failure sites in the finished product.

Also when processing NR or compounds composed of 50 parts or more of NR, it is very important that the material cool and stabilize at ambient temperatures before measuring its viscosity. Viscosity testing prior to this point does not always truly reflect the viscosity of the material in sequent processing steps.

23.5.2   Masterbatch Mixing

During the masterbatch mixing stage, fillers, plasticizers, and chemicals are incorporated into the rubber to enhance final performance and/or to improve handling and processability of the compound in the stages which follow. The two major objectives of the masterbatch mixing stage are maximum carbon black dispersion and proper final viscosity. Both of these objectives must be accomplished without exceeding the specified temperature limits. The complete incorporation of the fillers into the rubber and a good dispersion can be detected by ''sucking noises'' just prior to the discharge of the batch as the ram is cycling approximately 2 cm from the bottom of its cycle. The batch may be discharged on a mill or a mixing extruder beneath the internal mixer.

23.5.3   Phase Mixing

Elastomer blends are microheterogeneous and the continuous phase is either the polymer in highest concentration or the polymer of lowest viscosity. The size of the zones of the disperse phase vary according to the polymer characteristics, viscosity differential, and manner of blending.

A very effective mixing procedure to ensure that all the NR in a batch is properly filled with carbon black is to use a NR carbon black masterbatch (see Table 23.4).

Table 23.4   Phase Mixing

NR CBMB recipe				NR CBMB required for 80 part MB recipe
Ingredient	100 parts by weight base			= 100 parts by weight base
Natural rubber	100.00	×0.80	=	80.00
Carbon black	30.00	×0.80	=	24.00
Stearic acid	2.50	×0.80	=	2.00
Zinc oxide	6.25	×0.80	=	5.00
Total	138.75	×0.80	=	111.00

	Second stage MB		
Ingredient	First stage NR CBMB +	balance of ingredients =	final MB
NR	80.00	0.00	80.00
Cis Pb	–	20.00	20.00
Carbon black	24.00	31.00	55.00
Oil	–	10.00	10.00
Stearic acid	2.00	0.00	2.00
Zinc oxide	5.00	0.00	5.00
A.O.	–	1.50	1.50
Resin	–	3.00	3.00
Total	111.00	65.50	176.50

According to this table, a masterbatch of NR, carbon black, zinc oxide, and stearic acid is mixed. The ratio of the zinc oxide and stearic acid can be adjusted in the NR carbon black masterbatch to carry over the required amounts in the full masterbatch. Chapter 4 provides details to carry over these amounts.

This mixing procedure has the added advantage of providing a means of incorporating carbon black into lumps of NR in the NR carbon black masterbatch which were not filled during the mixing stage because of improper conditioning of the NR prior to use and/or an inadequate mixing cycle.

This procedure provides a method of remilling only the NR portion of the formulation, which results in a savings in mixing capacity because the total weight of the masterbatch is not passed through the mixer.

Phase inversion (the reversal of the continuous and disperse phases) occurs via an interpenetrating network of polymers, often at 50/50 blend ratio, but all such transitions depend very much on polymer viscosity and type.

Where elastomers of different cure rates are to be blended, a masterbatch process may be desirable. The component polymers are precompounded with vulcanizing agents and additives, and then the individual compounds are blended in the desired proportions to produce the finished compound. A carbon black masterbatch obtained from a polymer producer (from a wet mixing process such as SBR 1606, etc.) can be used very effectively in phase mixing to introduce a different polymer and carbon black at a fixed polymer/ black ratio without dry mixing a separate masterbatch.

Powdered rubber can be blended with a solid rubber by conventional mixing techniques. The blending of two rubbers, both in powdered form, is also possible in a continuous mixer. Each rubber can have a distinct and different composition of polymer, carbon black type, and loading.

23.5.4   Single-Stage Mix

In this method, the batch of compound is processed through the internal mixer only one time. The accelerators and sulfur are charged along with the synthetic rubber manufactured masterbatch, and other ingredients required to produce a final mixed compound. The compound, after discharge from the mixer, is blended on a mill, slabbed, cooled, and piled.

This type of mix is rarely used and is only suitable for compounds which do not generate high temperatures because of the temperature limitation of compounds containing curatives. An alternate procedure is to add sulfur directly to the hot batch on the discharge mill or a second mill. The compound is not cooled until the sulfur has been added to the compound. The compound is then blended on a mill, slabbed, cooled, and piled.

23.5.5   Single-Cycle Mix

As in single-stage mixing, the batch of compound is only processed through the internal mixer one time. However, single-cycle mixing differs slightly in that the batch contains all ingredients (including accelerators) except the sulfur.

The batch is sent from the discharge mill through a cooling process and is then returned to a mill where the sulfur is added to the batch without interrupting the process. The "sulfured" batch is then blended, slabbed off the mill, cooled, and piled. All operations are performed during one mixing cycle of the internal mixer. Generally, SBR tread type compounds can be mixed by the single-cycle method. Single-cycle mixing is not considered satisfactory for compounds of NR because it requires more mastication than a single-cycle process provides (see Phase Mixing, Section 23.5.3).

The beneficial effect obtained by subsequent internal mixing or mill mixing of cooled masterbatch in reducing stock viscosity and improving carbon black dispersion cannot be obtained by using a lower viscosity rubber blend in a single-cycle or single-stage mix.

23.5.6   Two-Stage Mix

Two internal mixing operations or cycles are involved. The first is the carbon black masterbatch stage, after which the compound is milled, slabbed, cooled, stacked, and aged. The masterbatch is then returned to the internal mixer and the curatives added. The final mix is then milled, slabbed, cooled, and stacked prior to subsequent use in calendering or extruding operations.

23.5.7   Tandem Mixing

This is a variation of a two-stage mix in which the first-stage masterbatch compound is discharged directly into a second ramless interlocking rotor internal mixer. The second mixer is larger in capacity than the masterbatch mixer and relies on the cooling ability of the interlocking rotor mixer to lower the temperature of the masterbatch to 105 to 110 °C in about two minutes. The curatives are then added, mixed, and the batch is discharged onto a mill, blended, slabbed, cooled, and piled.

Tandem mixing can also consist of two mixers operating as a single mixing line. The first of the two mixers produces masterbatch, which in turn feeds a second mixer which produces the final mix. The output from the masterbatch mixer is pelletized, cooled, and coated with an antitack agent. The pellets are then delivered to several storage bins for cross blending. The pellets are then discharged directly onto a charging conveyor, weighed, and delivered to the second mixer, along with the curatives and other chemicals. The batch is discharged onto a mill, blended, slabbed, cooled, and piled.

The major drawback in operating a tandem mixing operation is the loss of production from both mixers when one of them is shut down. For this reason, many mixing operations do not run mixers in tandem.

23.5.8   Three-Stage Mix

Because of the nature of the materials and/or the composition of the compounds to be mixed, it may not be possible to adequately disperse all the various ingredients in the masterbatch stage to produce a compound meeting the required viscosity and level of dispersion. Usually, the maximum temperature of the compound is reached before the required level of dispersion. Therefore, an intermediate mixing stage or remill of the compound is introduced prior to the final mixing stage. The cooled carbon black masterbatch compound is returned to an internal mixer and mixed again (generally referred to as a remill of the compound) to the maximum allowable temperature, discharged, slabbed, cooled, piled, and aged. If there are no further ingredients added during this stage, it is called a remill. But if additional black and/or other ingredients are added during this mixing stage, it is called an intermediate mix. A masterbatch is remilled to obtain the required level of viscosity and/or to further improve the dispersion of the fillers.

23.5.9   Upside Down Mix

In the case where a batch softens too quickly, before complete dispersion under a conventional mixing procedure, the compounder can use an upside down mixing method where the fillers and oils are added to the mixer first, followed immediately by the elastomers. This order prevents excessive filler from being extracted by the dust collector exhaust. This mixing procedure can be very hard on the dust stops, and the mixer is more prone to leakage than when the elastomers are added to the internal mixer prior to the fillers.

23.5.10   Variable Speed Mixing

A tool to optimize the mixing cycle is a variable speed drive. By using extremely high speeds at the beginning of the mixing cycle, high shear stress can be obtained, which is very effective for masticating the elastomer and dispersing the fillers. The rotor speed is then reduced and the batch is completed at normal mixing speeds. At high rotor speeds, the temperature increases and cannot be stopped (especially when mixing tread compounds using high structure carbon blacks) and the mixing cycle must be terminated before the proper level of dispersion is obtained when a critical temperature level has been reached. To achieve the level of dispersion required for a quality product, additional mixing stages must occur. Using a mixer equipped with a variable speed drive, the rotor speed can be reduced, which can decrease the rate of the upward slope of the compound temperature, allowing the batch to remain in the mixer until an adequate level of dispersion is obtained without recycling the compound.

The new tread compounds used by the tire industry featuring silica (with a surface activated by organosilanes which establish chemical bonds between the filler and polymer during the vulcanization of the tire) are very difficult to process in standard internal mixing equipment. Two chemical reactions have to take place during processing these new silica compounds. The reaction to bind organosilanes to the surface of the silica particles should be done within the internal mixer. The second reaction to bind the organosilanes to the polymer should not take place before vulcanization of the tire.

The first reaction within the internal mixer is relatively slow at moderate temperatures (120 °C). However, higher batch temperatures in the first step of mixing have to be established and must be maintained for up to 15 minutes to achieve a short reaction time. At elevated temperatures however, the second reaction may result in processing problems.

The typical mixers used by the tire industry are of the Tangential design. The temperature of the material in the mixer has a steady upwards slope as the mix cycle proceeds; it is not possible to run the mixer at a speed where the mix temperature stays constant. Because of this steady rise in temperature, the mix cycle is brought to a certain temperature and then terminated. Reaction times at the elevated temperatures are in the range of only 2 to 3 minutes per mixing stage. Thus, the material must pass through the internal mixer 4 to 5 times to achieve the 10 or 15 minutes of total reaction time. By mixing a silica compound first at a high rotor speed to disperse the silica and incorporate the fillers, oil, and other materials into the compound, and then reducing the rotor speed to a fixed level where the dissipation and the heat flow to the chamber sidewall and the rotor surface are at equilibrium, mixing can continue at the elevated temperature. Thus, additional organosilane bonding can be achieved in this mixing stage, resulting in eliminating several stages of mixing. The cooling ability of the larger rotor inherent in intermeshing mixers is well suited for mixing these silica compounds.

23.5.11   Final Mix

It is in this stage that the crosslinking agents, curatives, and other chemicals that can accelerate or retard the vulcanization process are added to the masterbatch. Final mixing can occur on mills or in an internal mixer. In the modern mixing facility, final mixing is assigned to specific mixers that are operated at slower speeds. Maximum distribution of the final mix ingredients must be accomplished before the temperature limits are exceeded to avoid premature crosslinking or scorch. Mixing times are short and batch temperatures usually do not exceed 120 °C. Compounds exposed to high mixing temperatures may exhibit surface bloom, resulting in poor tack.

The incorporation and dispersion of curatives into the masterbatch is not a difficult task in an internal mixer. The challenge is to distribute the curatives uniformly within the batch and then to consistently maintain this batch-to-batch uniformity along with adequate scorch properties.

The final mixed compound is usually discharged from the internal mixer onto a mill where it is cooled, blended, and transferred to a second mill where final blending takes place. The mixed compound is then slabbed, cooled, and piled prior to subsequent use in calendering or extruding operations.

23.5.12   Continuous Mixing

Continuous mixers operate on small volumes of material at a time and therefore, require that the feed stream be quite precisely metered into the feed throat of the mixer. This dictates that all of the feed stocks be finely divided particles.

Continuous mixing will play an ever-increasing role in rubber compounding in the future. This method of mixing rubber compounds is still very much in its infancy. The reality is that the vast majority of today's elastomers remain available only in bale form, or in chips too large for small volume continuous compounding. Rubbers in powdered form are excellent candidates for use in a continuous mixer. With the advent of advanced, gas phase, polymerization technology which produces polymers in a unique, free-flowing, grandular form, a new era in mixing compounds appears to be beginning.

23.5.13   E-SBR Carbon Black Masterbatch

E-SBR carbon black masterbatch is produced by dispersing carbon black in a slurry, which may also contain an emulsion of oil, which is then added to the latex. The resulting blend is coagulated, dried, and baled. This carbon black masterbatch can be readily mixed either on a mill or in an internal mixer with a few additional ingredients to make a finished compound.

The use of carbon black masterbatch can help the compounder meet increased demands placed on the quality of the compound and its effects on the performance of the end product. Because the carbon black is already partially dispersed in the rubber, the initial shear in the internal mixer is higher and ultimate dispersion can be reached in a shorter mix time. Carbon black masterbatches are very useful for adding carbon black to a compound where free black would create a housekeeping or contamination problem. Carbon black masterbatches are an economical alternative to increase mixing capacity by reducing the number of mixing stages to obtain a finished compound (see Table 23.5).

Adding carbon black masterbatch to the dry mixed masterbatch during the final mixing step can also increase mixing capacity (see Table 23.5). The major advantages resulting from the use of carbon black masterbatch include:

- Reduced mixing time
- Improved carbon black dispersion
- Improved housekeeping

Table 23.5   Use of Carbon Black Masterbatch to Reduce Mixing Stages

Mixing stage	3-Stage dry mix			Mixing stage	2-Stage CBMB mix		
	First	Second	Final		Masterbatch		Final
Ingredients				Ingredients	CBMB	Dry	
SBR 1712H	(137.50)			SBR 9296	(225.00)		
Polymer	100.00		(100.00)	Polymer	100.00		(100.00)
Hi Aro oil	37.50	12.50	(50.00)	Hi Aro oil	50.00		(50.00)
N-339 black	60.00	30.00	(90.00)	N-339 black	75.00	15.00	(90.00)
Zinc oxide	3.00	–	(3.00)	Zinc oxide	–	3.00	(3.00)
Stearic acid	–	1.50	(1.50)	Stearic acid	–	1.50	(1.50)
MB first stage	–	200.50	–	MB first stage	–	–	244.50
MB second stage	–		244.50				
Total MB	**200.50**	**244.50**	**244.50**	**Total MB**	**225.00**	**19.50**	**244.50**
Santocure NS	–	–	0.65				0.65
Sulfur	–	–	2.00				2.00
Total			**247.15**				**247.15**

- Increased mixing capacity
- Improved control of black and oil loadings

The three key ingredients, SBR, carbon black, and oil, affect compounding options, processing, and end product performance. The use of carbon black masterbatches can add flexibility in meeting existing market needs as well as allowing the compounder to design solutions for future tire and industrial rubber product applications. This can be accomplished by utilizing the technical advantages carbon black masterbatch has to offer.

23.5.14   Energy Mixing

The production of consecutive batches that are uniform both in degree of dispersion and viscosity are critical to the rubber manufacturing process. Mixing a uniform compound is very difficult to achieve from batch to batch day after day. The compounder must deal with a large number of variables (see Section 23.4.2.1) which are very difficult to control during mixing. Most mixers are poorly equipped to control the process. At present, only motor speed, mixing time, power integration, and/or compound temperature are the standard methods used to control the mixer. Measurement of mix properties online is another area which is developing fast. PC and PLC (programmable logic controller) hardware components are available for controlling the mixing cycle. A standard control program normally consists of the following:

- Opening/closing of loading door, loading belt
- Temperature control of bearings and wear rings
- Temperature control of mixer cooling (walls, rotors, gate)
- Weighing of pigments, carbon black, and oils
- Cycle, loading of rubber and pigments, ram action, energy, temperature, and discharge

Control is carried out by dividing the mix cycle into operations. A macro instruction corresponds to each operation and contains all micro instructions to let the requested operation take place. An operation can start by setting time, energy, or temperature. These three parameters can be individually used or linked together in an "and" or "or" function. The function "and" means that the passage to the next operation takes place when both parameters are satisfied. The "or" function means that the next operation takes place only when one of the two conditions is satisfied. A third possibility exists: "if"; in this case, to the temperature parameter, a second parameter is linked which "must" be reached before the temperature to let this operation end and to pass to the next one.

Example: 122 °C (temperature) if 360 kWh (energy)

According to the above conditions, if the energy set point is reached first, the system waits for the temperature to be reached before moving to the next operation. The mixing cycle can proceed to the next operation only if the previous one has been carried out according to the program. Energy mixing by providing improved control of the mixing cycle from batch to batch and day to day allows a more uniform mixed compound to be produced.

References

1. Melotto, M. A., "The History and State of Art in the Internal Mixer". Paper at the 120th ACS-Meeting, Rubber-Div., October 13–16, 1981, Cleveland, OH.
2. Sheehan, E. and Pomini, L., "Mixers With Variable Intermeshing Clearance". Rubber World 215, No. 6, March 1997.
3. Gooch, L. R., Musco, M. V. and Pascuzzo, A., "Setting up a Modern Mixing Facility". Rubber World, 1996.
4. Wiedmann, W. M. and Schmid, H. R., "Optimization of Rubber Mixing in Internal Mixers". Rubber Chemical Technology, Vol. 55, pp. 363–381.
5. Pohl, J. W., "Are There Alternatives to Internal Mixers? " Rubber World 211, No. 5, February 1995, pp. 33–34.
6. White, J. L., "Development of Internal Mixer Technology for the Rubber Industry". Rubber Chemical Technology, 65, No. 3, July/August 1992, pp. 527–579.
7. Dongian, G. S., Canedo, E. L. and Valsamis, L. N., "Optimizing Mixing in the Banbury Mixer With Synchronous Technology (ST) Rotors". Paper at the 140th ACS-Meeting, Rubber-Div., October 8–11, 1991, Detroit, MI.
8. Nekola, K. and Asada, M., "Advanced Technology Internal Mixers". Rubber World 205, No. 5, February 1992, pp. 34–37.
9. Elwood, H., "Continuous Compounding of Rubber". Paper at the 136th ACS-Meeting, Rubber Div., October 17–20, 1989, Detroit, MI.
10. Melotto, M. A., "Internal Batch Mixer: Recent Developments in Designs and Controls". Rubber World 199, No. 5, February 1989, pp. 34–38.
11. Meyer, P., Frenkel C-D Central, Co. Ltd. US Patent No. 5-421-650, Mixing Machinery of the Transfermix Type, June 6, 1995.

Index

Editor's Biography

John Dick has over thirty years of experience in the rubber industry. He was with BF Goodrich and later Uniroyal Goodrich Tire Co. as a Section Manager and Development Scientist in R & D until 1991 when he joined Monsanto's Rubber Instruments Group (now Alpha Technologies) as a Senior Marketing Technical Service specialist. Mr. Dick has authored over 40 journal and magazine publications and two books on polymer technology. He received the Monsanto Master Technical Service Award in 1994, the ACS Rubber Division "Best Paper Award" in 1995 and a University of Akron Appreciation Award in 1998 for Teaching Polymer Compounding Courses in their Continuing Education Program. He is a Fellow in the American Society for Testing and Materials (ASTM) receiving the Award of Merit in 1990. Also he has represented the United States as a delegate to the International Standards Organization (ISO) for the last 18 years. He was appointed in 1992 to be Leader of the U.S.A. Delegation to ISO TC-45 on Rubber. He is a member of the American Chemical Society, Society of Rheology, and ASQ with a CQE. He is also a representative to the RMA. Mr. Dick received his B.S. degree from Virginia Polytechnic Institute in 1970 and an M.A. from the University of Akron in 1979. He is married with two children and his hobbies include photography and amateur radio.